Fundamentals in Nuclear Physics

T0211186

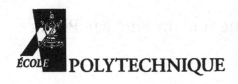

ÉCOLE POLYTECHNIQUE

The Ecole Polytechnique, one of France's top academic institutions, has a long-standing tradition of producing exceptional scientific textbooks for its students. The original lecture notes, the *Cours de l'Ecole Polytechnique*, which were written by Cauchy and Jordan in the nineteenth century, are considered to be landmarks in the development of mathematics.

The present series of textbooks is remarkable in that the texts incorporate the most recent scientific advances in courses designed to provide undergraduate students with the foundations of a scientific discipline. An outstanding level of quality is achieved in each of the seven scientific fields taught at the *Ecole*: pure and applied mathematics, mechanics, physics, chemistry, biology, and economics. The uniform level of excellence is the result of the unique selection of academic staff there which includes, in addition to the best researchers in its own renowned laboratories, a large number of world-famous scientists, appointed as part-time professors or associate professors, who work in the most advanced research centers France has in each field.

Another distinctive characteristics of these courses is their overall consistency; each course makes appropriate use of relevant concepts introduced in the other textbooks. This is because each student at the Ecole Polytechnique has to acquire basic knowledge in the seven scientific fields taught there, so a substantial link between departments is necessary. The distribution of these courses used to be restricted to the 900 students at the Ecole. Some years ago we were very successful in making these courses available to a larger French-reading audience. We now build on this success by making these textbooks also available in English.

Jean-Louis Basdevant
James Rich
Michel Spiro

Fundamentals In Nuclear Physics

From Nuclear Structure to Cosmology

With 184 Figures

 Springer

Prof. Jean-Louis Basdevant
Ecole Polytechnique
Département de Physique
Laboratoire Leprince-Ringuet
91128 Palaiseau
France
jean-louis.basdevant@polytechnique.edu

Dr. James Rich
Dapnia-SPP
CEA-Saclay
91191 Gif-sur-Yvette
France
rich@hep.saclay.cea.fr

Dr. Michel Spiro
IN2P3-CNRS
3 Rue Michel-Ange
75794 Paris cedex 16
France
mspiro@admin.in2p3.fr

Cover illustration: Background image—Photograph of Supernova 1987A Rings. Photo credit Christopher Burrows (ESA/STScI) and NASA, Hubble Space Telescope, 1994. Smaller images, from top to bottom—Photograph of Supernova Blast. Photo credit Chun Shing Jason Pun (NASA/GSFC), Robert P. Kirshner (Harvard-Smithsonian Center for Astrophysics), and NASA, 1997. Interior of the JET torus. Copyright 1994 EFDA-JET. See figure 7.6 for further description. The combustion chamber at the Nova laser fusion facility (Lawrence Livermore National Laboratory, USA). Inside the combustion chamber at the Nova laser fusion facility (Lawrence Livermore National Laboratory, USA) The Euratom Joint Research Centres and Associated Centre.

Library of Congress Cataloging-in-Publication Data
Basdevant, J.-L. (Jean-Louis)
    Fundamentals in nuclear physics / J.-L. Basdevant, J. Rich, M. Spiro.
      p. cm.
    Includes bibliographical references and index.

    1. Nuclear physics.   I. Rich, James, 1952–   II. Spiro, M. (Michel)   III. Title.
    QC173.B277   2004
    539.7—dc22   2004056544

ISBN 978-1-4419-1849-9   e-ISBN 978-0-387-25095-3   Printed on acid-free paper.

Printed in the United States of America.   (EB)

9 8 7 6 5 4 3 2 1

springeronline.com

Preface

Nuclear physics began one century ago during the "miraculous decade" between 1895 and 1905 when the foundations of practically all modern physics were established. The period started with two unexpected spinoffs of the Crooke's vacuum tube: Roentgen's X-rays (1895) and Thomson's electron (1897), the first elementary particle to be discovered. Lorentz and Zeemann developed the the theory of the electron and the influence of magnetism on radiation. Quantum phenomenology began in December, 1900 with the appearance of Planck's constant followed by Einstein's 1905 proposal of what is now called the photon. In 1905, Einstein also published the theories of relativity and of Brownian motion, the ultimate triumph of Boltzman's statistical theory, a year before his tragic death. For nuclear physics, the critical discovery was that of radioactivity by Becquerel in 1896.

By analyzing the history of science, one can be convinced that there is some rationale in the fact that all of these discoveries came nearly simultaneously, after the scientifically triumphant 19th century. The exception is radioactivity, an unexpected baby whose discovery could have happened several decades earlier.

Talented scientists, the Curies, Rutherford, and many others, took the observation of radioactivity and constructed the ideas that are the subject of this book. Of course, the discovery of radioactivity and nuclear physics is of much broader importance. It lead directly to quantum mechanics via Rutherford's planetary atomic model and Bohr's interpretation of the hydrogen spectrum. This in turn led to atomic physics, solid state physics, and material science. Nuclear physics had the important by-product of elementary particle physics and the discovery of quarks, leptons, and their interactions. These two fields are actually impossible to dissociate, both in their conceptual and in their experimental aspects.

The same "magic decade" occurred in other sectors of human activity. The second industrial revolution is one aspect, with the development of radio and telecommunications. The automobile industry developed at the same period, with Daimler, Benz, Panhard and Peugeot. The Wright brothers achieved a dream of mankind and opened the path of a revolution in transportation. Medicine and biology made incredible progress with Louis Pasteur and many others. In art, we mention the first demonstration of the "cinématographe" by Auguste and Louis Lumière on december 28 1895, at the Grand Café, on Boulevard des Capucines in Paris and the impressionnist exhibition in Paris in 1896.

Nowadays, is is unthinkable that a scientific curriculum bypass nuclear physics. It remains an active field of fundamental research, as heavy ion accelerators of Berkeley, Caen, Darmstadt and Dubna continue to produce new nuclei whose characteristics challenge models of nuclear structure. It has major technological applications, most notably in medicine and in energy production where a knowledge of some nuclear physics is essential for participation in decisions that concern society's future.

Nuclear physics has transformed astronomy from the study of planetary trajectories into the astrophysical study of stellar interiors. No doubt the most important result of nuclear physics has been an understanding how the observed mixture of elements, mostly hydrogen and helium in stars and carbon and oxygen in planets, was produced by nuclear reactions in the primordial universe and in stars.

This book emerged from a series of topical courses we delivered since the late 1980's in the Ecole Polytechnique. Among the subjects studied were the physics of the Sun, which uses practically all fields of physics, cosmology for which the same comment applies, and the study of energy and the environment. This latter subject was suggested to us by many of our students who felt a need for deeper understanding, given the world in which they were going to live. In other words, the aim was to write down the fundamentals of nuclear physics in order to explain a number of applications for which we felt a great demand from our students.

Such topics do not require the knowledge of modern nuclear theory that is beautifully described in many books, such as *The Nuclear Many Body Problem* by P. Ring and P Schuck. Intentionally, we have not gone into such developments. In fact, even if nuclear physics had stopped, say, in 1950 or 1960, practically all of its applications would exist nowadays. These applications result from phenomena which were known at that time, and need only qualitative explanations. Much nuclear phenomenology can be understood from simple arguments based on things like the Pauli principle and the Coulomb barrier. That is basically what we will be concerned with in this book. On the other hand, the enormous amount of experimental data now easily accesible on the web has greatly facilitated the illustration of nuclear systematics and we have made ample use of these resources.

This book is an introduction to a large variety of scientific and technological fields. It is a first step to pursue further in the study of such or such an aspect. We have taught it at the senior undergraduate level at the Ecole Polytechnique. We believe that it may be useful for graduate students, or more generally scientists, in a variety of fields.

In the first three chapters, we present the "scene" , i.e. we give the basic notions which are necessary to develop the rest. Chapter 1 deals with the basic concepts in nuclear physics. In chapter 2, we describe the simple nuclear models, and discuss nuclear stability. Chapter 3 is devoted to nuclear reactions.

Chapter 4 goes a step further. It deals with nuclear decays and the fundamental electro-weak interactions. We shall see that it is possible to give a comparatively simple, but sound, description of the major progress particle physics and fundamental interactions made since the late 1960's.

In chapter 5, we turn to the first important practical application, i.e. radioactivity. We shall see examples of how radioactivity is used be it in medicine, in food industry or in art.

Chapters 6 and 7 concern nuclear energy. Chapter 6 deals with fission and the present aspects of that source of energy production. Chapter 7 deals with fusion which has undergone quite remarkable progress, both technologically and politically in recent years with the international ITER project.

Fusion brings us naturally, in chapter 8 to the subject of nuclear astrophysics and stellar structure and evolution. Finally, we present an introduction to present ideas about cosmology in chapter 9. A more advanced description can be found in *Fundamentals of Cosmology*, written by one of us (J. R.).

We want to pay a tribute to the memory of Dominique Vautherin, who constantly provided us with ideas before his tragic death in December 2000. We are grateful to Martin Lemoine, Robert Mochkovitch, Hubert Flocard, Vincent Gillet, Jean Audouze and Alfred Vidal-Madjar for their invaluable help and advice throughout the years. We also thank Michel Cassé, Bertrand Cordier, Michel Cribier, David Elbaz, Richard Hahn, Till Kirsten, Sylvaine Turck-Chièze, and Daniel Vignaud for illuminating discussions on various aspects of nuclear physics.

Palaiseau, France   *Jean-Louis Basdevant, James Rich, Michel Spiro*
April, 2005

Contents

Introduction

Nuclear physics started by accident in 1896 with the discovery of radioactivity by Henri Becquerel who noticed that photographic plates were blackened when placed next to uranium-sulfide crystals. He, like Poincaré and many others, found the phenomenon of "Becquerel rays" fascinating, but he nevertheless lost interest in the subject within the following six months. We can forgive him for failing to anticipate the enormous amount of fundamental and applied physics that would follow from his discovery.

In 1903, the third Nobel prize for Physics was awarded to Becquerel, and to Pierre and Marie Curie. While Becquerel discovered radioactivity, it was the Curies who elucidated many of its characteristics by chemically isolating the different radioactive elements produced in the decay of uranium. Ernest Rutherford, became interested in 1899 and performed a series of brilliant experiments leading up to his discovery in 1911 of the nucleus itself. Arguably the founder of nuclear physics, he was, ironically, awarded the Nobel prize in Chemistry in 1908.

It can be argued, however, that the first scientists to observe and study radioactive phenomena were Tycho Brahe and his student Johannes Kepler. They had the luck in 1572 (Brahe) and in 1603 (Kepler) to observe bright *stellae novae*, i.e. new stars. Such *supernovae* are now believed to be explosions of *old* stars at the end of their normal lives.[1] The post-explosion energy source of supernovae is the decay of radioactive nickel (^{56}Ni, half-life 6.077 days) and then cobalt (^{56}Co, half-life 77.27 days). Brahe and Kepler observed that the luminosity of their supernovae, shown in Fig. 0.1, decreased with time at a rate that we now know is determined by the nuclear lifetimes.

Like Becquerel, Brahe and Kepler did not realize the importance of what they had seen. In fact, the importance of supernovae dwarfs that of radioactivity because they are the culminating events of the process of nucleosynthesis. This process starts in the cosmological "big bang" where protons and neutron present in the primordial soup condense to form hydrogen and helium. Later, when stars are formed the hydrogen and helium are processed through

---

[1] Such events are extremely rare. In the last millennium, only five of them have been seen in our galaxy, the Milky Way. The last supernova visible to the naked eye was seen on February 23, 1987, in the Milky Way's neighbor, the Large Magellanic Cloud. The neutrinos and γ-rays emitted by this supernova were observed on Earth, starting the subject of extra-solar nuclear astronomy.

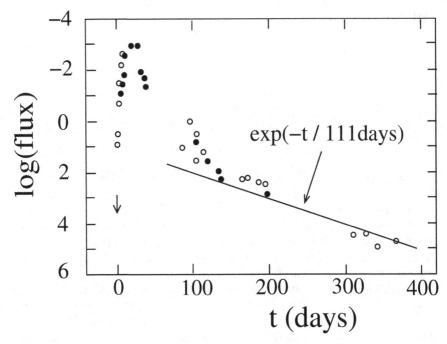

Fig. 0.1. The luminosity of Kepler's supernova as a function of time, as reconstructed in [4]. Open circles are European measurements and filled circles are Korean measurements. Astronomers at the time measured the evolution of the luminosity of the supernovae by comparing it to known stars and planets. It has been possible to determine the positions of planets at the time when they were observed, and, with the notebooks, to reconstruct the luminosity curves. The superimposed curve shows the rate of ^{56}Co decay using the laboratory-measured half life. The vertical scale gives the visual magnitude V of the star, proportional to the logarithm of the photon flux. $V = 0$ corresponds to a bright star, while $V = 5$ is the dimmest star that can be observed with the naked eye.

nuclear reactions into heavier elements. These elements are ejected into the interstellar medium by supernovae. Later, some of this matter condenses to form new stellar systems, now sometimes containing habitable planets made of the products of stellar nucleosynthesis.

Nuclear physics has allowed us to understand in considerable quantitative detail the process by which elements are formed and what determines their relative abundances. The distribution of nuclear abundances in the Solar System is shown in Fig. 8.9. Most ordinary matter[2] is in the form of hydrogen ($\sim 75\%$ by mass) and helium ($\sim 25\%$). About 2% of the solar system material is in heavy elements, especially carbon, oxygen and iron. To the extent that nuclear physics explains this distribution, it allows us to understand why we

---

[2] We leave the question of the nature of the unknown cosmological "dark matter" for Chap. 9.

live near a hydrogen burning star and are made primarily of elements like hydrogen, carbon and oxygen.

A particularly fascinating result of the theory of nucleosynthesis is that the observed mix of elements is due to a number of delicate inequalities of nuclear and particle physics. Among these are

- The neutron is slightly heavier than the proton;
- The neutron–proton system has only one bound state while the neutron–neutron and proton–proton systems have none;
- The ^{8}Be nucleus is slightly heavier than two ^{4}He nuclei and the second excited state of ^{12}C is slightly heavier than three ^{4}He nuclei.

We will see in Chaps. 8 and 9 that modifying any of these conditions would result in a radically different distribution of elements. For instance, making the proton heavier than the neutron would make ordinary hydrogen unstable and none would survive the primordial epoch of the Universe.

The extreme sensitivity of nucleosynthesis to nuclear masses has generated a considerable amount of controversy about its interpretation. It hinges upon whether nuclear masses are fixed by the fundamental laws of physics or are accidental, perhaps taking on different values in inaccessible regions of the Universe. Nuclear masses depend on the strengths of the forces between neutrons and protons, and we do not now know whether the strengths are uniquely determined by fundamental physics. If they are not, we must consider the possibility that the masses in "our part of the Universe" are as observed because other masses give mixes of elements that are less likely to provide environments leading to intelligent observers. Whether or not such "weak-anthropic selection" had a role in determining the observed nuclear and particle physics is a question that is appealing to some, infuriating to others. Resolving the question will require better understanding of the origin of observed physical laws.

Some history

The history of nuclear physics can be divided into three periods. The first begins with the discovery radioactivity of the nucleus and ends in 1939 with the discovery of fission. During this period, the basic components (protons and neutrons) of the nucleus were discovered as well as the quantum law governing their behavior. The second period from 1947 to 1969 saw the development of nuclear spectroscopy and of nuclear models. Finally, the emergence of a microscopic unifying theory starting in the 1960s allowed one to understand the structure and behavior of protons and neutrons in terms of the fundamental interactions of their constituent particles, quarks and gluons. This period also saw the identification of subtle non-classical mechanisms in nuclear structure.

Since the 1940s, nuclear physics has seen important developments, but most practical applications and their simple theoretical explanations were in place by the mid 1950s. This book is mostly concerned with the simple models from the early period of nuclear physics and to their application in energy production, astrophysics and cosmology.

The main stages of this first period of nuclear physics are the following [5, 6].

- **1868** Mendeleev's periodic classification of the elements.
- **1895** Discovery of X-rays by Roentgen.
- **1896** Discovery of radioactivity by Becquerel.
- **1897** Identification of the electron by J.J. Thomson.
- **1898** Separation of the elements polonium and radium by Pierre and Marie Curie.
- **1908** Measurement of the charge +2 of the α particle by Geiger and Rutherford.
- **1911** Discovery of the nucleus by Rutherford; "planetary" model of the atom.
- **1913** Theory of atomic spectra by Niels Bohr.
- **1914** Measurement of the mass of the α particle by Robinson and Rutherford.
- **1924–1928** Quantum theory (de Broglie, Schrödinger, Heisenberg, Born, Dirac).
- **1928** Theory of barrier penetration by quantum tunneling, application to α radioactivity, by Gamow, Gurney and Condon.
- **1929–1932** First nuclear reactions with the electrostatic accelerator of Cockcroft and Walton and the cyclotron of Lawrence.
- **1930–1933** Neutrino proposed by Pauli and named by Fermi in his theory of beta decay.
- **1932** Identification of the neutron by Chadwick.
- **1934** Discovery of artificial radioactivity by F. and I. Joliot-Curie.
- **1934** Discovery of neutron capture by Fermi.
- **1935** Liquid-drop model and compound-nucleus model of N. Bohr.
- **1935** Semi-empirical mass formula of Bethe and Weizsäcker.
- **1938** Discovery of fission by Hahn and Strassman.
- **1939** Theoretical interpretation of fission by Meitner, Bohr and Wheeler.

To these fundamental discoveries we should add the practical applications of nuclear physics. Apart from nuclear energy production beginning with Fermi's construction of the first fission reactor in 1942, the most important are astrophysical and cosmological. Among them are

- **1938** Bethe and Weizsäcker propose that stellar energy comes from thermonuclear fusion reactions.
- **1946** Gamow develops the theory of cosmological nucleosynthesis.
- **1953** Salpeter discovers the fundamental solar fusion reaction of two protons into deuteron.

- **1957** Theory of stellar nucleosynthesis by Burbidge, Burbidge, Fowler and Hoyle.
- **1960–** Detection of solar neutrinos
- **1987** Detection of neutrinos and γ-rays from the supernova SN1987a.

The scope of nuclear physics

In one century, nuclear physics has found an incredible number of applications and connections with other fields. In the most narrow sense, it is only concerned with bound systems of protons and neutrons. From the beginning however, progress in the study of such systems was possible only because of progress in the understanding of other particles: electrons, positrons, neutrinos and, eventually quarks and gluons. In fact, we now have a more complete theory for the physics of these "elementary particles" than for nuclei as such.[3]

A nuclear species is characterized by its number of protons Z and number of neutrons N. There are thousands of combinations of N and Z that lead to nuclei that are sufficiently long-lived to be studied in the laboratory. They are tabulated in Appendix G. The large number of possible combinations of neutrons and protons is to be compared with the only 100 or so elements characterized simply by Z.[4]

A "map" of the world of nuclei is shown in Fig. 0.2. Most nuclei are unstable, i.e. radioactive. Generally, for each $A = N + Z$ there is only one or two combinations of (N, Z) sufficiently long-lived to be naturally present on Earth in significant quantities. These nuclei are the black squares in Fig. 0.2 and define the bottom of the *valley of stability* in the figure.

One important line of nuclear research is to create new nuclei, both high up on the sides of the valley and, especially, super-heavy nuclei beyond the heaviest now known with $A = 292$ and $Z = 116$. Phenomenological arguments suggest that there exists an "island of stability" near $Z = 114$ and 126 with nuclei that may be sufficiently long-lived to have practical applications.

The physics of nuclei as such has been a very active domain of research in the last twenty years owing to the construction of new machines, the heavy ion accelerators of Berkeley, Caen (GANIL), Darmstadt and Dubna. The physics of atomic nuclei is in itself a domain of *fundamental research*. It constitutes a true many-body problem, where the number of constituents is too large for exact computer calculations, but too small for applying the methods of statistical physics. In heavy ion collisions, one discovers subtle effects such as local superfluidity in the head-on collision of two heavy ions.

---

[3] This is of course a false paradox; the structure of DNA derives, in principle, completely from the Schrödinger equation and Quantum Electrodynamics. However it is not studied it that spirit.

[4] Different isotopes of a same element have essentially the same chemical properties.

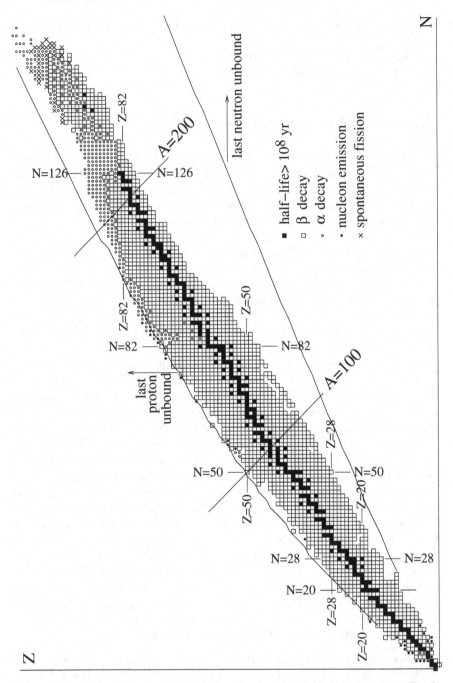

Fig. 0.2. The nuclei. The black squares are long-lived nuclei present on Earth. Unbound combinations of (N, Z) lie outside the lines marked "last proton/neutron unbound." Most other nuclei β-decay or α-decay to long-lived nuclei.

Nuclear physics has had an important by-product in elementary particle physics and the discovery of the elementary constituents of matter, quarks and leptons, and their interactions. Nuclear physics is essential to the understanding of the structure and the origin of the world in which we live. The birth of nuclear astrophysics is a decisive step forward in astronomy and in cosmology. In addition, nuclear technologies play an important role in modern society. We will see several examples. This book is intended to be a first introduction to a large variety of scientific and technological fields. It can be a first step in the study of the vast field of nuclear physics.

Bibliography

On the history of nuclear and particle physics:

1. Abraham Pais Inward Bound, Oxford University Press, Oxford, 1986.
2. Emilio Segré, From X rays to Quarks, Freeman, San Francisco, 1980.

Introductory textbooks on nuclear physics

1. B. Povh, K. Rith, C. Scholz and F. Zetsche, Particles and Nuclei, Springer, Berlin, 2000.
2. W.N. Cottingham and D.A.Greenwood, Nuclear Physics, Cambridge University Press, Cambridge, 2002.
3. P.E. Hodgson, E. Gadioli and E. Gadioli Erba, Introductory Nuclear Physics, Clarendon Press, Oxford, 1997.
4. Harald Enge, Introduction to Nuclear Physics, Addison-Wesley, Reading, 1966.
5. J. S. Lilley, Nuclear Physics, Wiley, Chichester, 2001.

Advanced textbooks on nuclear physics

1. Nuclear Structure A. Bohr and B. Mottelson, Benjamin, New York, 1969.
2. M.A. Preston and R.K. Bhaduri, Structure of the Nucleus, Addison-Wesley, Reading, 1975.
3. S.M. Wong, Nuclear Physics, John Wiley, New York, 1998.
4. J.D. Walecka,Theoretical Nuclear and Subnuclear Physics, Oxford University Press, Oxford, 1995.
5. A. de Shalit and H. Feshbach,Theoretical Nuclear Physics, Wiley, New York, 1974.
6. D.M. Brink, Nuclear Forces, Pergamon Press, Oxford, 1965.
7. J.M. Blatt and V.F. Weisskopf, Theoretical Nuclear Physics, John Wiley and Sons, New-York, 1963.

1. Basic concepts in nuclear physics

In this chapter, we will discuss the basic ingredients of nuclear physics. Section 1.1 introduces the elementary particles that form nuclei and participate in nuclear reactions. Sections 1.2 shows how two of these particles, protons and neutrons, combine to form nuclei. The essential results will be that nuclei have volumes roughly proportional to the number of nucleons, $\sim 7\,\mathrm{fm}^3$ per nucleon and that they have binding energies that are of order $8\,\mathrm{MeV}$ per nucleon. In Sect. 1.3 we show how nuclei are described as *quantum* states. The forces responsible for binding nucleons are described in Sect. 1.4. Section 1.5 discusses how nuclei can be transformed through nuclear reactions while Sect. 1.6 discusses the important conservation laws that constrain these reactions and how these laws arise in quantum mechanics. Section 1.7 describes the isospin symmetry of these forces. Finally, Sect. 1.8 discusses nuclear shapes.

1.1 Nucleons and leptons

Atomic nuclei are quantum bound states of particles called *nucleons* of which there are two types, the positively charged proton and the uncharged neutron. The two nucleons have similar masses:

$$m_\mathrm{n}c^2 = 939.56\,\mathrm{MeV} \qquad m_\mathrm{p}c^2 = 938.27\,\mathrm{MeV} \quad , \tag{1.1}$$

i.e. a mass difference of order one part per thousand

$$(m_\mathrm{n} - m_\mathrm{p})c^2 = 1.29\,\mathrm{MeV} . \tag{1.2}$$

For nuclear physics, the mass difference is much more important than the masses themselves which in many applications are considered to be "infinite." Also of great phenomenological importance is the fact that this mass difference is of the same order as the electron mass

$$m_\mathrm{e}c^2 = 0.511\,\mathrm{MeV} . \tag{1.3}$$

Nucleons and electrons are *spin 1/2 fermions* meaning that their intrinsic angular momentum projected on an arbitrary direction can take on only the values of $\pm\hbar/2$. Having spin 1/2, they must satisfy the *Pauli exclusion principle* that prevents two identical particles (protons, neutrons or electrons) from having the same spatial wavefunction unless their spins are oppositely aligned.

Nucleons and electrons generate magnetic fields and interact with magnetic fields with their *magnetic moment*. Like their spins, their magnetic moments projected in any direction can only take on the values $\pm\mu_p$ or $\pm\mu_n$:

$$\mu_p = 2.792\,847\,386\,(63)\,\mu_N \qquad \mu_n = -1.913\,042\,75\,(45)\,\mu_N\,, \qquad (1.4)$$

where the *nuclear magneton* is

$$\mu_N = \frac{e\hbar}{2m_p} = 3.152\,451\,66\,(28) \times 10^{-14}\text{MeV}\,\text{T}^{-1}\,. \qquad (1.5)$$

For the electron, only the mass and the numerical factor changes

$$\mu_e = 1.001\,159\,652\,193\,(40)\,\mu_B\,, \qquad (1.6)$$

where the *Bohr magneton* is

$$\mu_B = \frac{q\hbar}{2m_e} = 5.788\,382\,63\,(52) \times 10^{-11}\text{MeV}\,\text{T}^{-1}\,. \qquad (1.7)$$

Nucleons are bound in nuclei by *nuclear forces*, which are of short range but are sufficiently strong and attractive to overcome the long-range Coulomb repulsion between protons. Because of their strength compared to electromagnetic interactions, nuclear forces are said to be due to the *strong interaction* (also called the *nuclear interaction*).

While protons and neutrons have different charges and therefore different electromagnetic interactions, we will see that their strong interactions are quite similar. This fact, together their nearly equal masses, justifies the common name of "nucleon" for these two particles.

Some spin 1/2 particles are not subject to the strong interaction and therefore do not bind to form nuclei. Such particles are called *leptons* to distinguish them from nucleons. Examples are the electron e^- and its antiparticle, the positron e^+. Another lepton that is important in nuclear physics is the *electron–neutrino* ν_e and *electron-antineutrino* $\bar{\nu}_e$. This particle plays a fundamental role in nuclear *weak interactions*. These interactions, as their name implies, are not strong enough to participate in the binding of nucleons. They are, however, responsible for the most common form of radioactivity, β-decay.

It is believed that the ν_e is, in fact, a quantum-mechanical mixture of three neutrinos of differing mass. While this has some interesting consequences that we will discuss in Chap. 4, the masses are sufficiently small ($m_\nu c^2 < 3\,\text{eV}$) that for most practical purposes we can ignore the neutrino masses:

$$m_{\nu i} \sim 0 \quad i = 1, 2, 3\,. \qquad (1.8)$$

As far as we know, leptons are *elementary particles* that cannot be considered as bound states of constituent particles. Nucleons, on the other hand, are believed to be bound states of three spin 1/2 fermions called *quarks*. Two species of quarks, the *up-quark* u (charge 2/3) and the *down quark* d (charge -1/3) are needed to construct the nucleons:

$$\text{proton} = \text{uud}, \qquad \text{neutron} = \text{udd}.$$

The constituent nature of the nucleons can, to a large extent, be ignored in nuclear physics.

Besides protons and neutrons, there exist many other particles that are bound states of quarks and antiquarks. Such particles are called *hadrons*. For nuclear physics, the most important are the three *pions*: (π^+, π^0, π^+). We will see in Sect. 1.4 that strong interactions between nucleons result from the *exchange* of pions and other hadrons just as electromagnetic interactions result from the exchange of photons.

1.2 General properties of nuclei

Nuclei, the bound states of nucleons, can be contrasted with atoms, the bound states of nuclei and electrons. The differences are seen in the units used by atomic and nuclear physicists:

$$\text{length}: \quad 10^{-10}\text{m (atoms)} \quad \rightarrow \quad 10^{-15}\text{m} = \text{fm (nuclei)}$$

$$\text{energy}: \quad \text{eV (atoms)} \quad \rightarrow \quad \text{MeV (nuclei)}$$

The typical nuclear sizes are 5 orders of magnitude smaller than atomic sizes and typical nuclear binding energies are 6 orders of magnitude greater than atomic energies. We will see in this chapter that these differences are due to the relative strengths and ranges of the forces that bind atoms and nuclei.

We note that nuclear binding energies are still "small" in the sense that they are only about 1% of the nucleon rest energies mc^2 (1.1). Since nucleon binding energies are of the order of their kinetic energies $mv^2/2$, nucleons within the nucleus move at non-relativistic velocities $v^2/c^2 \sim 10^{-2}$.

A nuclear species, or *nuclide*, is defined by N, the number of neutrons, and by Z, the number of protons. The mass number A is the total number of nucleons, i.e. $A = N + Z$. A nucleus can alternatively be denoted as

$$(A, Z) \leftrightarrow {}^A X \leftrightarrow {}^A_Z X \leftrightarrow {}^A_Z X_N,$$

where X is the chemical symbol associated with Z (which is also the number of electrons of the corresponding neutral atom). For instance, ^{4}He is the helium-4 nucleus, i.e. $N = 2$ and $Z = 2$. For historical reasons, ^{4}He is also called the α particle. The three nuclides with $Z = 1$ also have special names

$${}^1\text{H} = \text{p} = \text{proton} \qquad {}^2\text{H} = \text{d} = \text{deuteron} \qquad {}^3\text{H} = \text{t} = \text{triton}$$

While the numbers (A, Z) or (N, Z) define a nuclear species, they do not determine uniquely the nuclear *quantum state*. With few exceptions, a nucleus (A, Z) possesses a rich spectrum of excited states which can decay to the ground state of (A, Z) by emitting photons. The emitted photons are often called γ-*rays*. The excitation energies are generally in the MeV range and their lifetimes are generally in the range of 10^{-9}–10^{-15} s. Because of their high energies and short lifetimes, the excited states are very rarely seen on Earth and, when there is no ambiguity, we denote by (A, Z) the *ground state* of the corresponding nucleus.

Some particular sequences of nuclei have special names:

- *Isotopes* : have same charge Z, but different N, for instance $^{238}_{92}$U and $^{235}_{92}$U. The corresponding atoms have practically identical chemical properties, since these arise from the Z electrons. Isotopes have very different nuclear properties, as is well-known for ^{238}U and ^{235}U.
- *Isobars* : have the same mass number A, such as ^{3}He and ^{3}H. Because of the similarity of the nuclear interactions of protons and neutrons, different isobars frequently have similar nuclear properties.

Less frequently used is the term *isotone* for nuclei of the same N, but different Z's, for instance ^{14}C$_6$ and ^{16}O$_8$.

Nuclei in a given quantum state are characterized, most importantly, by their size and binding energy. In the following two subsections, we will discuss these two quantities for nuclear ground states.

1.2.1   Nuclear radii

Quantum effects inside nuclei are fundamental. It is therefore surprising that the volume $\mathcal{V}$ of a nucleus is, to good approximation, proportional to the number of nucleons A with each nucleon occupying a volume of the order of $\mathcal{V}_0 = 7.2$ fm^3. In first approximation, stable nuclei are spherical, so a volume $\mathcal{V} \simeq A\mathcal{V}_0$ implies a radius

$$R = r_0 A^{1/3} \quad \text{with} \quad r_0 = 1.2 \text{ fm} \quad . \tag{1.9}$$

We shall see that r_0 in (1.9) is the order of magnitude of the range of nuclear forces.

In Chap. 3 we will show how one can determine the spatial distribution of nucleons inside a nucleus by scattering electrons off the nucleus. Electrons can penetrate inside the nucleus so their trajectories are sensitive to the charge distribution. This allows one to reconstruct the proton density, or equivalently the proton probability distribution $\rho_p(r)$. Figure 1.1 shows the charge densities inside various nuclei as functions of the distance to the nuclear center.

We see on this figure that for $A > 40$ the charge density, therefore the proton density, is roughly constant inside these nuclei. It is independent of the nucleus under consideration and it is roughly 0.075 protons per fm^3. Assuming the neutron and proton densities are the same, we find a nucleon density inside nuclei of

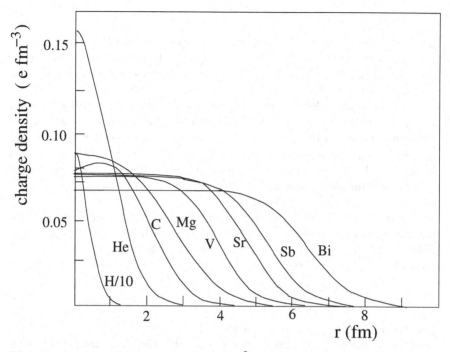

Fig. 1.1. Experimental charge density $(e\,fm^{-3})$ as a function of $r(fm)$ as determined in elastic electron–nucleus scattering [8]. Light nuclei have charge distributions that are peaked at $r = 0$ while heavy nuclei have flat distributions that fall to zero over a distance of $\sim 2\,fm$.

Table 1.1. Radii of selected nuclei as determined by electron–nucleus scattering [8]. The size of a nucleus is characterized by r_{rms} (1.11) or by the radius R of the uniform sphere that would give the same r_{rms}. For heavy nuclei, the latter is given approximately by (1.9) as indicated in the fourth column. Note the abnormally large radius of ^{2}H.

nucleus	r_{rms} (fm)	R (fm)	$R/A^{1/3}$ (fm)	nucleus	r_{rms} (fm)	R (fm)	$R/A^{1/3}$ (fm)
^{1}H	0.77	1.0	1.0	^{16}O	2.64	3.41	1.35
^{2}H	2.11	2.73	2.16	^{24}Mg	2.98	3.84	1.33
^{4}He	1.61	2.08	1.31	^{40}Ca	3.52	4.54	1.32
^{6}Li	2.20	2.8	1.56	^{122}Sb	4.63	5.97	1.20
^{7}Li	2.20	2.8	1.49	^{181}Ta	5.50	7.10	1.25
^{9}Be	2.2	2.84	1.37	^{209}Bi	5.52	7.13	1.20
^{12}C	2.37	3.04	1.33				

$$\rho_0 \simeq 0.15 \text{ nucleons fm}^{-3} \quad . \tag{1.10}$$

If the nucleon density were exactly constant up to a radius R and zero beyond, the radius R would be given by (1.9). Figure 1.1 indicates that the density drops from the above value to zero over a region of thickness ~ 2 fm about the nominal radius R.

In contrast to nuclei, the size of an atom does *not* increase with Z implying that the electron density does increase with Z. This is due to the long-range Coulomb attraction of the nucleus for the electrons. The fact that nuclear densities do not increase with increasing A implies that a nucleon does not interact with all the others inside the nucleus, but only with its nearest neighbors. This phenomenon is the first aspect of a very important property called the *saturation* of nuclear forces.

We see in Fig. 1.1 that nuclei with $A < 20$ have charge densities that are not flat but rather peaked near the center. For such light nuclei, there is no well-defined radius and (1.9) does not apply. It is better to characterize such nuclei by their rms radius

$$(r_{\text{rms}})^2 = \frac{\int d^3 r \, r^2 \rho(\boldsymbol{r})}{\int d^3 r \rho(\boldsymbol{r})} . \tag{1.11}$$

Selected values of r_{rms} as listed in Table 1.1.

Certain nuclei have abnormally large radii, the most important being the loosely bound deuteron, ^{2}H. Other such nuclei consist of one or two loosely bound nucleons orbiting a normal nucleus. Such nuclei are called *halo nuclei* [7]. An example is ^{11}Be consisting of a single neutron around a ^{10}Be core. The extra neutron has wavefunction with a rms radius of ~ 6 fm compared to the core radius of ~ 2.5 fm. Another example is ^{6}He consisting of two neutrons outside a ^{4}He core. This is an example of a *Borromean nucleus* consisting of three objects that are bound, while the three possible pairs are unbound. In this case, ^{6}He is bound while n-n and n-^{4}He are unbound.

1.2.2 Binding energies

The saturation phenomenon observed in nuclear radii also appears in nuclear binding energies. The binding energy B of a nucleus is defined as the negative of the difference between the nuclear mass and the sum of the masses of the constituents:

$$B(A, Z) = N m_n c^2 + Z m_p c^2 - m(A, Z) c^2 \tag{1.12}$$

Note that B is defined as a positive number: $B(A, Z) = -E_B(A, Z)$ where E_B is the usual (negative) binding energy.

The binding energy per nucleon B/A as a function of A is shown in Fig. 1.2. We observe that B/A increases with A in light nuclei, and reaches a broad maximum around $A \simeq 55 - 60$ in the iron-nickel region. Beyond, it decreases slowly as a function of A. This immediately tells us that energy

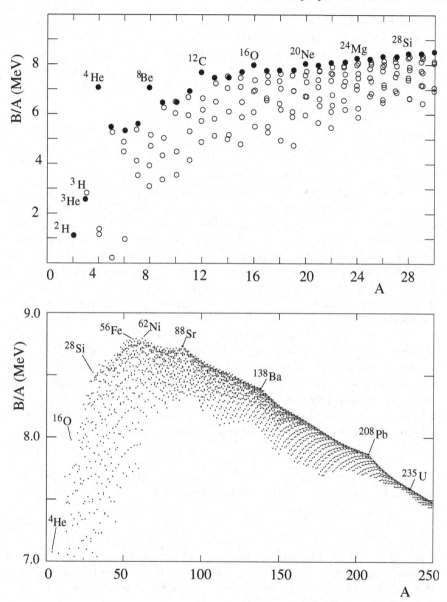

Fig. 1.2. Binding energy per nucleon, $B(A, Z)/A$, as a function of A. The upper panel is a zoom of the low-A region. The filled circles correspond to nuclei that are not β-radioactive (generally the lightest nuclei for a given A). The unfilled circles are unstable (radioactive) nuclei that generally β-decay to the lightest nuclei for a given A.

can be released by the "fusion" of light nuclei into heavier ones, or by the "fission" of heavy nuclei into lighter ones.

As for nuclear volumes, it is observed that for stable nuclei which are not too small, say for $A > 12$, the binding energy B is in first approximation *additive*, i.e. proportional to the number of nucleons :

$$B(A, Z) \simeq A \times 8 \text{ MeV} ,$$

or more precisely

$$7.7 \text{ MeV} < B(A, Z)/A < 8.8 \text{ MeV} \qquad 12 < A < 225 .$$

The numerical value of ~ 8 MeV per nucleon is worth remembering!

The additivity of binding energies is quite different from what happens in atomic physics where the binding energy of an atom with Z electrons increases as $Z^{7/3}$, i.e. $Z^{4/3}$ per electron. The nuclear additivity is again a manifestation of the saturation of nuclear forces mentioned above. It is surprising from the quantum mechanical point of view. In fact, since the binding energy arises from the pairwise nucleon–nucleon interactions, one might expect that $B(A, Z)/A$ should increase with the number of nucleon pairs $A(A-1)/2$.[1] The additivity confirms that nucleons only interact strongly with their nearest neighbors.

The additivity of binding energies and of volumes are related via the uncertainty principle. If we place A nucleons in a sphere of radius R, we can say that each nucleon occupies a volume $4\pi R^3 A/3$, i.e. it is confined to a linear dimension of order $\Delta x \sim A^{-1/3} R$. The uncertainty principle [2] then implies an uncertainty $\Delta p_i \sim \hbar A^{1/3}/R$ for each momentum component. For a bound nucleon, the expectation value of p_i must vanish, $\langle p_i \rangle = 0$, implying a relation between the momentum squared and the momentum uncertainty

$$(\Delta p_i)^2 = \langle p_i^2 \rangle - \langle p_i \rangle^2 = \langle p_i^2 \rangle . \tag{1.13}$$

Apart from numerical factors, the uncertainty principle then relates the mean nucleon kinetic energy with its position uncertainty

$$\left\langle \frac{p^2}{2m} \right\rangle \sim \frac{\hbar^2}{2m} \frac{A^{2/3}}{R^2} . \tag{1.14}$$

Since $R \simeq r_0 A^{1/3}$ this implies that the average kinetic energy per nucleon should be approximately the same for all nuclei. It is then not surprising that the same is true for the binding energy per nucleon.[3] We will see in Chap. 2 how this comes about.

---

[1] In the case of atoms with Z electrons, it increases as $Z^{4/3}$. In the case of pairwise harmonic interactions between A fermions, the energy per particle varies as $A^{5/6}$.

[2] See for instance J.-L. Basdevant and J. Dalibard, Quantum mechanics , chapter 16, Springer-Verlag, 2002.

[3] The virial theorem only guarantees that for power-law potentials these two energies are of the same order. Since the nuclear potential is not a power law, exceptions occur. For example, many nuclei decay by dissociation, e.g. $^8\text{Be} \to {}^4\text{He} \, {}^4\text{He}$. Considered as a "bound" state of two ^4He nuclei, the binding energy is, in fact,

As we can see from Fig. 1.2, some nuclei are exceptionally strongly bound compared to nuclei of similar A. This is the case for ^{4}He, ^{12}C, ^{16}O. As we shall see, this comes from a filled shell phenomenon, similar to the case of noble gases in atomic physics.

1.2.3 Mass units and measurements

The binding energies of the previous section were defined (1.12) in terms of nuclear and nucleon masses. Most masses are now measured with a precision of $\sim 10^{-8}$ so binding energies can be determined with a precision of $\sim 10^{-6}$. This is sufficiently precise to test the most sophisticated nuclear models that can predict binding energies at the level of 10^{-4} at best.

Three units are commonly used to described nuclear masses: the atomic mass unit (u), the kilogram (kg), and the electron-volt (eV) for rest energies, mc^2. In this book we generally use the energy unit eV since energy is a more general concept than mass and is hence more practical in calculations involving nuclear reactions.

It is worth taking some time to explain clearly the differences between the three systems. The atomic mass unit is a purely microscopic unit in that the mass of a ^{12}C *atom* is defined to be 12 u:

$$m(^{12}\text{C atom}) \equiv 12 \text{ u} . \qquad (1.15)$$

The masses of other atoms, nuclei or particles are found by measuring ratios of masses. On the other hand, the kilogram is a macroscopic unit, being defined as the mass of a certain platinum-iridium bar housed in Sèvres, a suburb of Paris. Atomic masses on the kilogram scale can be found by assembling a known (macroscopic) number of atoms and comparing the mass of the assembly with that of the bar. Finally, the eV is a hybrid microscopic-macroscopic unit, being defined as the kinetic energy of an electron after being accelerated from rest through a potential difference of 1 V.

Some important and very accurately known masses are listed in Table 1.2.

Mass spectrometers and ion traps. Because of its purely microscopic character, it is not surprising that masses of atoms, nuclei and particles are most accurately determined on the atomic mass scale. Traditionally, this has been done with mass spectrometers where ions are accelerated by an electrostatic potential difference and then deviated in a magnetic field. As illustrated in Fig. 1.3, mass spectrometers also provide the data used to determine the isotopic abundances that are discussed in Chap. 8.

The radius of curvature R of the trajectory of an ion in a magnetic field B after having being accelerated from rest through a potential difference V is

---

negative and ^{8}Be exists for a short time ($\sim 10^{-16}$s) only because there is an energy barrier through which the ^{4}He must tunnel.

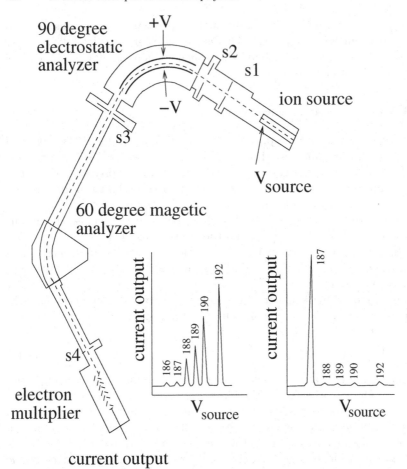

current output

Fig. 1.3. A schematic of a "double-focusing" mass spectrometer [9]. Ions are accelerated from the source at potential V_{source} through the beam defining slit S_2 at ground potential. The ions are then electrostatically deviated through 90 deg and then magnetically deviated through 60 deg before impinging on the detector at slit S_4. This combination of fields is "double focusing" in the sense that ions of a given mass are focused at S_4 independent of their energy and direction at the ion source. Mass ratios of two ions are equal to the voltage ratios leading to the same trajectories. The inset shows two mass spectra [10] obtained with sources of OsO_2 with the spectrometer adjusted to focus singly ionized molecules OsO_2^+. The spectra show the output current as a function of accelerating potential and show peaks corresponding to the masses of the long-lived osmium isotopes, $^{186}Os - ^{192}Os$. The spectrum on the left is for a sample of terrestrial osmium and the heights of the peaks correspond to the natural abundances listed in Appendix G. The spectrum on the right is for a sample of osmium extracted from a mineral containing rhenium but little natural osmium. In this case the spectrum is dominated by ^{187}Os from the β-decay $^{187}Re \rightarrow {}^{187}Os e^- \bar{\nu}_e$ with $t_{1/2} = 4.15 \times 10^{10}$yr (see Exercise 1.15).

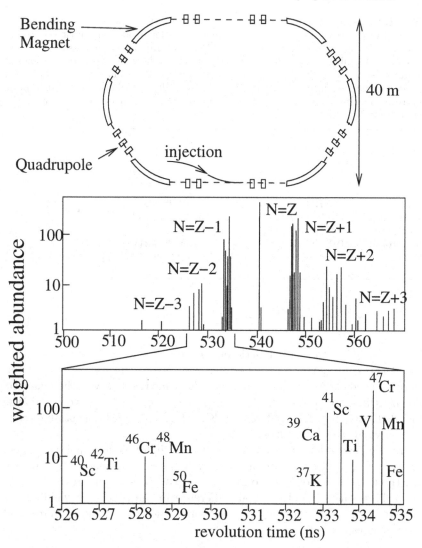

Fig. 1.4. Measurement of nuclear masses with isochronous mass spectroscopy [11]. Nuclei produced by fragmentation of 460 MeV/u ^{84}Kr on a beryllium target at GSI laboratory are momentum selected [12] and then injected into a storage ring [13]. About 10 fully ionized ions are injected into the ring where they are stored for several hundred revolutions before they are ejected and a new group of ions injected. A thin carbon foil (17 µg cm^{-2}) placed in the ring emits electrons each time it is traversed by an ion. The detection of these electrons measures the ion's time of passage with a precision of $\sim$ 100 ps. The periodicity of the signals determines the revolution period for each ion. The figure shows the spectrum of periods for many injections. The storage ring is run in a mode such that the non-relativistic relation for the period, $T \propto q/m$ is respected in spite of the fact that the ions are relativistic. The positions of the peaks for different q/m determine nuclide masses with a precision of $\sim$ 200 keV (Exercise 1.16).

Table 1.2. Masses and rest energies for some important particles and nuclei. As explained in the text, mass ratios of charged particles or ions are most accurately determined by using mass spectrometers or Penning trap measurements of cyclotron frequencies. Combinations of ratios of various ions allows one to find the ratio of any mass to that of the ^{12}C atom which is defined as $12\,\mathrm{u}$. Masses can be converted to rest energies accurately by using the theoretically calculable hydrogen atomic spectrum. The neutron mass is derived accurately from a determination of the deuteron binding energy.

particle	mass m (u)	mc^2 (MeV)
electron e	$5.485\,799\,03\,(13) \times 10^{-4}$	$0.510\,998\,902\,(21)$
proton p	$1.007\,276\,470\,(12)$	$938.271\,998\,(38)$
neutron n	$1.008\,664\,916\,(82)$	$939.565\,33\,(4)$
deuteron d	$2.013\,553\,210\,(80)$	$1875.612\,762\,(75)$
^{12}C atom	12 (exact)	$12 \times 931.494\,013\,(37)$

$$ R = \frac{\sqrt{2Em}}{qB} = \frac{\sqrt{2V}}{B}\sqrt{\frac{m}{q}} \, , \tag{1.16} $$

where $E = qV$ is the ion's kinetic energy and q and m are its charge and mass. To measure the mass ratio between two ions, one measures the potential difference needed for each ion that yields the same trajectory in the magnetic field, i.e. the same R. The ratio of the values of q/m of the two ions is the ratio of the two potential differences. Knowledge of the charge state of each ion then yields the mass ratio.

Precisions of order 10^{-8} can be obtained with double-focusing mass spectrometers if one takes pairs of ions with similar charge-to-mass ratios. In this case, the trajectories of the two ions are nearly the same in an electromagnetic field so there is only a small difference in the potentials yielding the same trajectory. For example, we can express the ratio of the deuteron and proton masses as

$$ \frac{m_{\mathrm{d}}}{m_{\mathrm{p}}} = 2\,\frac{m_{\mathrm{d}}}{2m_{\mathrm{p}} + m_{\mathrm{e}} - m_{\mathrm{e}}} $$

$$ = 2\left[\frac{m_{\mathrm{d}}}{2m_{\mathrm{p}} + m_{\mathrm{e}}}\right]\left[1 - \frac{m_{\mathrm{e}}/m_{\mathrm{p}}}{2(1 + m_{\mathrm{e}}/m_{\mathrm{p}})}\right]^{-1} . \tag{1.17} $$

The first factor in brackets, $m_{\mathrm{d}}/(2m_{\mathrm{p}}+m_{\mathrm{e}})$, is the mass ratio between a deuterium ion and singly ionized hydrogen molecule.[4] The charge-to-mass ratio of these two objects is nearly the same and can therefore be very accurately measured with a mass spectrometer. The second bracketed term contains a

---

[4] We ignore the small ($\sim$ eV) electron binding energy.

small correction depending on the ratio of the electron and proton masses. As explained below, this ratio can be accurately measured by comparing the electron and proton cyclotron frequencies. Equation (1.17) then yields m_d/m_p.

Similarly, the ratio between m_d and the mass of the ^{12}C atom (= 12 u) can be accurately determined by comparing the mass of the doubly ionized carbon atom with that of the singly ionized ^{2}H$_3$ molecule (a molecule containing 3 deuterons). These two objects have, again, similar values of q/m so their mass ratio can be determined accurately with a mass spectrometer. The details of this comparison are the subject of Exercise 1.7. The comparison gives the mass of the deuteron in atomic-mass units since, by definition, this is the deuteron-^{12}C atom ratio. Once m_d is known, m_p is then determined by (1.17).

Armed with m_e, m_p, m_d and $m(^{12}$C atom$) \equiv 12$ u it is simple to find the masses of other atoms and molecules by considering other pairs of ions and measuring their mass ratios in a mass spectrometer.

The traditional mass-spectrometer techniques for measuring mass ratios are difficult to apply to very short-lived nuclides produced at accelerators. While the radius of curvature in a magnetic field of ions can be measured, the relation (1.16) cannot be applied unless the kinetic energy is known. For non-relativistic ions orbiting in a magnetic field, this problem can be avoided by measuring the orbital period $T = m/qB$. Ratios of orbital periods for different ions then yield ratios of charge-to-mass ratios. An example of this technique applied to short-lived nuclides is illustrated in Fig. 1.4.

The most precise mass measurements for both stable and unstable species are now made through the measurement of ionic cyclotron frequencies,

$$\omega_c = \frac{qB}{m} \ . \tag{1.18}$$

For the proton, this turns out to be $9.578 \times 10^7 \mathrm{rad\,s^{-1}\,T^{-1}}$. It is possible to measure ω_c of individual particles bound in a Penning trap. The basic configuration of such a trap in shown in Fig. 1.5. The electrodes and the external magnetic field of a Penning trap are such that a charged particle oscillates about the trap center. The eigenfrequencies correspond to oscillations in the z direction, cyclotron-like motion in the plane perpendicular to the z direction, and a slower radial oscillation. It turns out that the cyclotron frequency is sum of the two latter frequencies.

The eigenfrequencies can be determined by driving the corresponding motions with oscillating dipole fields and then detecting the change in motional amplitudes with external pickup devices or by releasing the ions and measuring their velocities. The frequencies yielding the greatest energy absorptions are the eigenfrequencies.

If two species of ions are placed in the trap, the system will exhibit the eigenfrequencies of the two ions and the two cyclotron frequencies determined.

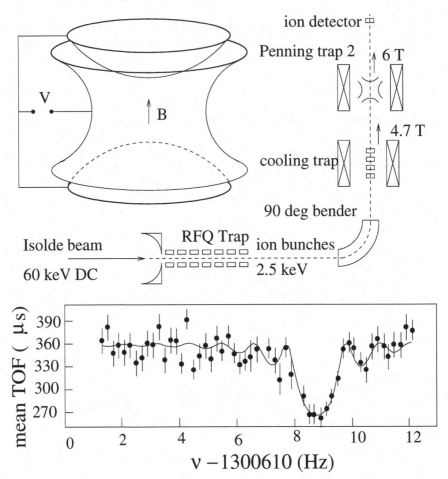

Fig. 1.5. The Isoltrap facility at CERN for the measurement of ion masses. The basic configuration of a Penning trap is shown in the upper left. It consists of two end-cap electrodes and one ring electrode at a potential difference. The whole trap is immersed in an external magnetic field. A charged particle oscillates about about the center of the trap. The cyclotron frequency, qB/m can be derived from the eigenfrequencies of this oscillation and knowledge of the magnetic field allows one to derive the charge-to-mass ratio. In Isoltrap, the 60 keV beam of radioactive ions is decelerated to 3 keV and then cooled and isotope selected (e.g. by selective ionization by laser spectroscopy) in a first trap. The selected ions are then released into the second trap where they are subjected to an RF field. After a time of order 1 s, the ions are released and detected. If the field is tuned to one of the eigenfrequencies, the ions gain energy in the trap and the flight time from trap to detector is reduced. The scan in frequency on the bottom panel, for singly-ionized ^{70}Cu, $t_{1/2} = 95.5$ s [15], demonstrates that frequency precisions of order 10^{-8} can be obtained.

The ratio of the frequencies gives the ratio of the masses. Precisions in mass ratios of 10^{-9} have been obtained [14].

The neutron mass. The one essential mass that cannot be determined with these techniques is that of the neutron. Its mass can be most simply derived from the proton and deuteron masses and the deuteron binding energy, $B(2,1)$

$$m_d = m_p + m_n - B(2,1)/c^2 . \tag{1.19}$$

The deuteron binding energy can be deduced from the energy of the photon emitted in the capture of neutrons by protons,

$$np \rightarrow {}^2H\gamma . \tag{1.20}$$

For slow (thermal) neutrons captured by stationary protons, the initial kinetic energies are negligible (compared to the nucleon rest energies) so to very good approximation, the ^2H binding energy is just the energy of the final state photon (Exercise 1.8):

$$B(2,1) = E_\gamma \left(1 + \frac{E_\gamma}{2m_d c^2}\right) . \tag{1.21}$$

The correction in parenthesis comes from the fact that the ^2H recoils from the photon and therefore carries some energy. Neglecting this correction, we have

$$m_n = m_d - m_p + E_\gamma/c^2 . \tag{1.22}$$

Thus, to measure the neutron mass we need the energy of the photon emitted in neutron capture by protons.

The photon energy can be deduced from its wavelength

$$E_\gamma = \frac{2\pi\hbar c}{\lambda_\gamma} , \tag{1.23}$$

so we need an accurate value of $\hbar c$. This can be found most simply by considering photons from transitions of atomic hydrogen whose energies can be calculated theoretically. Neglecting calculable fine-structure corrections, the energy of photons in a transition between states of principal quantum numbers n and m is

$$E_{nm} = (1/2)\alpha^2 m_e c^2 \left(n^{-2} - m^{-2}\right) , \tag{1.24}$$

corresponding to a wavelength

$$\lambda_{nm} = \frac{2\pi\hbar c}{(1/2)\alpha^2 m_e c^2 \left(n^{-2} - m^{-2}\right)} = \frac{1}{R_\infty \left(n^{-2} - m^{-2}\right)} , \tag{1.25}$$

where $R_\infty = \alpha^2 m_e c^2/4\pi\hbar c$ is the Rydberg constant and $\alpha = e^2/4\pi\epsilon_0\hbar c \sim 1/137$ is the fine-structure constant. This gives

$$2\pi\hbar c = \lambda_{mn} \left(n^{-2} - m^{-2}\right)(1/2)\alpha^2 m_e c^2 = R_\infty^{-1}(1/2)\alpha^2 m_e c^2 . \tag{1.26}$$

The value of R_∞ can be found from any of the hydrogen lines by using (1.25) [16]. The currently recommended value is [17]

$$R_\infty = 10\,973\,731.568\,549(83)\mathrm{m}^{-1} \, . \tag{1.27}$$

Substituting (1.26) into (1.23) we get a formula relating photon energies and wavelengths

$$\frac{E_\gamma}{m_e c^2} = \frac{R_\infty^{-1}}{\lambda_\gamma}(1/2)\alpha^2 \, . \tag{1.28}$$

The fine-structure constant can be determined by a variety of methods, for example by comparing the electron cyclotron frequency with its spin-precession frequency.

The wavelength of the photon emitted in (1.20) was determined [18] by measuring the photon's diffraction angle (to a precision of 10^{-8} deg) on a silicon crystal whose interatomic spacing is known to a precision of 10^{-9} yielding

$$\lambda_\gamma = 5.576\,712\,99(99) \times 10^{-13}\mathrm{m} \, . \tag{1.29}$$

Substituting this into (1.28), using the value of m_e (Table 1.2) and then using (1.21) we get

$$B(2,1)/c^2 = 2.388\,170\,07(42) \times 10^{-3}\,\mathrm{u} \, . \tag{1.30}$$

Substituting this into (1.19) and using the deuteron and proton masses gives the neutron mass.

The eV scale. To relate the atomic-mass-unit scale to the electron-volt energy scale we can once again use the hydrogen spectrum

$$m_e c^2 = \frac{4\pi\hbar c R_\infty}{\alpha^2} \, . \tag{1.31}$$

The electron-volt is by definition the potential energy of a particle of charge e when placed a distance $r = 1\,\mathrm{m}$ from a charge of $q = 4\pi\epsilon_0 r$, i.e.

$$1\,\mathrm{eV} = \frac{eq}{4\pi\epsilon_0 r} \qquad r = 1\,\mathrm{m}, \ \ q = 1.112 \times 10^{-10}\,\mathrm{C} \, . \tag{1.32}$$

Dividing (1.31) by (1.32) we get

$$\frac{m_e c^2}{1\,\mathrm{eV}} = \frac{4\pi}{\alpha^3}\frac{e}{1.112 \times 10^{-10}\mathrm{C}}\frac{R_\infty}{1\,\mathrm{m}^{-1}} \, . \tag{1.33}$$

We see that in order to give the electron rest-energy on the eV scale we need to measure the atomic hydrogen spectrum in meters, e in units of Coulombs, and the unit-independent value of the fine-structure constant. The currently accepted value is given by (1.3). This allows us to relate the atomic-mass-unit scale to the electron-volt scale by simply calculating the rest energy of the ^{12}C atom:

$$mc^2\,(^{12}\mathrm{C\ atom}) = m_e c^2 \frac{12\,\mathrm{u}}{m_e} = 12 \times 931.494\,013\,(37)\,\mathrm{MeV} \, , \tag{1.34}$$

or equivalently $1\,\mathrm{u} = 931.494013\,\mathrm{MeV}/c^2$.

The kg scale. Finally, we want to relate the kg scale to the atomic mass scale. Conceptually, the simplest way is to compare the mass of a known number of particles (of known mass on the atomic-mass scale) with the mass of the platinum-iridium bar (or one of its copies). One method [19] uses a crystal of ^{28}Si with the number of atoms in the crystal being determined from the ratio of the linear dimension of the crystal and the interatomic spacing. The interatomic spacing can determined through laser interferometry. The method is currently limited to a precision of about 10^{-5} because of uncertainties in the isotopic purity of the ^{28}Si crystal and in uncertainties associated with crystal imperfections. It is anticipated that once these errors are reduced, it will be possible to define the kilogram as the mass of a certain number of ^{28}Si atoms. This would be equivalent to fixing the value of the Avogadro constant, N_A, which is defined to be the number of atoms in $12\,$g of ^{12}C.

1.3 Quantum states of nuclei

While (A, Z) is sufficient to denote a nuclear *species*, a given (A, Z) will generally have a large number of quantum states corresponding to different wavefunctions of the constituent nucleons. This is, of course, entirely analogous to the situation in atomic physics where an atom of atomic number Z will have a lowest energy state (ground state) and a spectrum of excited states. Some typical nuclear spectra are shown in Fig. 1.6.

In both atomic and nuclear physics, transitions from the higher energy states to the ground state occurs rapidly. The details of this process will be discussed in Sect. 4.2. For an isolated nucleus the transition occurs with the emission of photons to conserve energy. The photons emitted during the decay of excited nuclear states are called γ-*rays*. A excited nucleus surrounded by atomic electrons can also transfer its energy to an electron which is subsequently ejected. This process is called *internal conversion* and the ejected electrons are called *conversion electrons*. The energy spectrum of γ-rays and conversion electrons can be used to derive the spectrum of nuclear excited states.

Lifetimes of nuclear excited states are typically in the range $10^{-15} - 14^{-10}$s. Because of the short lifetimes, with few exceptions only nuclei in the ground state are present on Earth. The rare excited states with lifetimes greater than, say, $1\,$s are called *isomers*. An extreme example is the first exited state of ^{180}Ta which has a lifetime of $10^{15}\,$yr whereas the ground state β-decays with a lifetime of $8\,$hr. All ^{180}Ta present on Earth is therefore in the excited state.

Isomeric states are generally specified by placing a m after A, i.e.

$$^{180\text{m}}\text{Ta} \,. \tag{1.35}$$

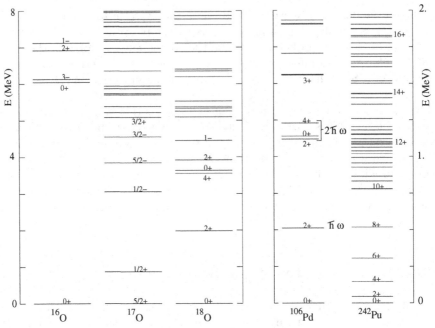

Fig. 1.6. Spectra of states of ^{16}O, ^{17}O ^{18}O (scale on the left) and of ^{106}Pd, and ^{242}Pu (scale on the right). The spin-parities of the lowest levels are indicated. ^{17}O has the simplest spectrum with the lowest states corresponding to excitations a a single neutron outside a stable ^{16}O core. The spectrum of ^{106}Pd exhibits collective vibrational states of energy $\hbar\omega$ and $2\hbar\omega$. The spectrum of ^{242}Pu has a series of rotational states of $J^P = 0^+$, 2^+.....16^+ of energies given by (1.40).

While exited states are rarely found in nature, they can be produced in collisions with energetic particles produced at accelerators. An example is the spectrum of states of ^{64}Ni shown in Fig. 1.7 and produced in collisions with 11 MeV protons.

A quantum state of a nucleus is defined by its energy (or equivalently its mass via $E = mc^2$) and by its *spin* J and *parity* P, written conventionally as

$$J^P \equiv \text{spin}^{\text{parity}} . \tag{1.36}$$

The spin is the total angular momentum of the constituent nucleons (including their spins). The parity is the sign by which the total constituent wavefunction changes when the spatial coordinates of all nucleons change sign. For nuclei, with many nucleons, this sounds like a very complicated situation. Fortunately, identical nucleons tends to pair with another nucleon of the opposite angular momentum so that in the *ground state*, the quantum numbers are determined by unpaired protons or neutrons. For N-even, Z-even nuclei there are none implying

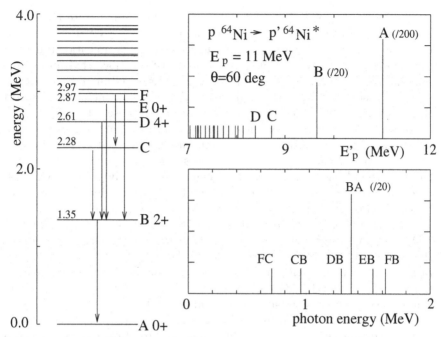

Fig. 1.7. Spectra of states of ^{64}Ni as shown in the left. The excited states can be produced by bombarding ^{64}Ni with protons since the target nucleus can be placed in an excited state if the proton transfers energy to it. The spectrum on the top right is a schematic of the energy spectrum of final-state protons (at fixed scattering angle of 60 deg) for an initial proton energy of 11 MeV (adapted from [20]). Each proton energy corresponds to an excited state of ^{64}Ni: $E'_p \sim 11\,\text{MeV} - \Delta E$ where ΔE is the energy of the excited state relative to the ground state (Exercise 1.17). Once produced, the excited states decay by emission of photons or conversion electrons as indicated by the arrows on the left. The transitions that are favored are determined by the spins and parities of each state. The photon spectrum for the decays of the 6 lowest energy states is shown schematically on the lower right.

$$J^P = 0^+ \quad \text{even} - \text{even nuclei (ground state)} . \tag{1.37}$$

For even–odd nuclei the quantum numbers are determined by the unpaired nucleon

$$J = l \pm 1/2 \quad P = -1^l \quad \text{even} - \text{odd nuclei (ground state)} , \tag{1.38}$$

where l is the angular momentum quantum number of the unpaired nucleon. The $\pm$ is due to the fact that the unpaired spin can be either aligned or anti-aligned with the orbital angular momentum. We will go into more detail in Sect. 2.4 when we discuss the nuclear shell model.

Spins and parities have important phenomenological consequences. They are important in the determination the rates of β-decays (Sect. 4.3.2) and γ-decays (Sect. 4.2) because of *selection rules* that favor certain angular momentum and parity changes. This is illustrated in Fig. 1.7 where one sees that excited states do not usually decay directly to the ground state but rather proceed through a cascade passing through intermediate excited states. Since the selection rules for γ-decays are known, the analysis of transition rates and the angular distributions of photons emitted in transitions that are important in determining the spins and parities of states. Ground state nuclear spins are also manifest in the *hyperfine splitting* of atomic atomic spectra (Exercise 1.12) and *nuclear magnetic resonance* (Exercise 1.13).

In general, the spectra of nuclear excited states are much more complicated than atomic spectra. Atomic spectra are mostly due to the excitations of one or two external electrons. In nuclear spectroscopy, one really faces the fact that the physics of a nucleus is a genuine many-body problem. One discovers a variety of subtle collective effects, together with individual one- or two-nucleon or one α-particle effects similar to atomic effects.

The spectra of five representative nuclei are shown in Fig. 1.6. The first, ^{16}O, is a very highly bound nucleus as manifested by the large gap between the ground and first excited states. The first few excited states of ^{17}O have a rather simple one-particle excitation spectrum due to the unpaired neutron that "orbits" a stable ^{16}O core. Both the ^{16}O and ^{18}O spectra are more complicated than the one-particle spectrum of ^{17}O.

For heavier nuclei, *collective excitations* involving many nucleons become more important. Examples are vibrational and rotational excitations. An example of a nucleus with vibrational levels is ^{106}Pd in Fig. 1.6. For this nucleus, their are groups of excited states with energies

$$E_n = \hbar\omega(n + 3/2) \quad n = 0, 1, 2..... . \tag{1.39}$$

More striking are the rotational levels of ^{242}Pu in the same figure. The classical kinetic energy of a rigid rotor is $L^2/2I$ where L is the angular momentum and I is the moment of inertia about the rotation axis. For a quantum rotor like the ^{242}Pu nucleus, the quantization of angular momentum then implies a spectrum of states of energy

$$E_J = \frac{L^2}{2I} = \frac{\hbar^2 J(J+1)}{2I} \quad J = 0, 2, 4...... , \tag{1.40}$$

where J is the angular momentum quantum number. For A-even-Z-even nuclei, only even values of J are allowed because of the symmetry of the nucleus. Many heavy nuclei have a series of excited states that follow this pattern. These states form a *rotation band*. If a nucleus is produced in a high J in a band, it will generally cascade down the band emitting photons of energies

$$E_J - E_{J-1} = \frac{\hbar^2 J}{I} . \tag{1.41}$$

The spectrum of photons of such a cascade thus consists of a series of equally spaced energies. One such spectrum is illustrated in Fig. 1.9.

The energy spectrum allows one to deduce the nuclear moment of inertia through (1.41). The deduced values are shown in Fig. 1.8 for intermediate

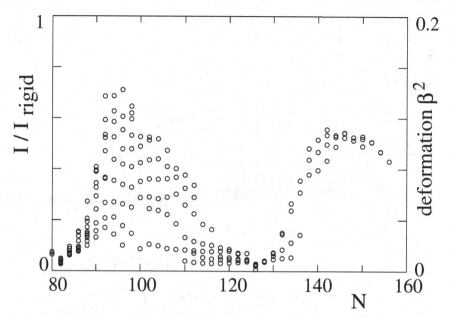

Fig. 1.8. The nuclear moments of inertia divided by the moment of a rigid sphere, $(2/5)mR^2 \sim (2/5)Am_\mathrm{p} \times (1.2\,\mathrm{fm}A^{1/3})^2$. The moments are deduced from the (1.40) and the energy of the first 2^+ state. Nuclei far from the magic number $N = 80$ and $N = 126$ have large moments of inertial, implying the rotation is due to collective motion. The scale on the right shows the nuclear deformation (square of the relative difference between major and minor axes) deduced from the lifetime of the 2^+ state [24]. As discussed in Sect. 4.2, large deformations lead to rapid transitions between rotation levels. We see that nuclei far from magic neutron numbers are deformed by of order 20%.

mass nuclei. We see that for nuclei with $N \sim 100$ and $N \sim 150$, the moment of inertia is near that for a rigid body where all nucleons rotate collectively, $I = (2/5)mR^2$. If the angular momentum where due to a single particle revolving about a non-rotating core, the moment of inertia would be a factor $\sim A$ smaller and the energy gap a factor $\sim A$ larger. We will see in Section 1.8 that these nuclei are also non-spherical so that the rotational levels are analogous to those of diatomic molecules.

Many nuclei also possess excited rotation bands due to a metastable deformed configuration that has rotational levels. An example of such a spectrum is shown in Fig. 1.9. This investigation of such bands is an important area of research in nuclear physics.

1.4 Nuclear forces and interactions

One of the aims of nuclear physics is to calculate the energies and quantum numbers of nuclear bound states. In atomic physics, one can do this starting

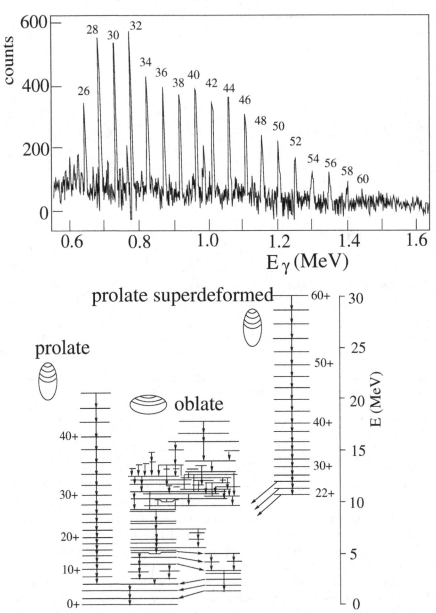

Fig. 1.9. Excited states of ^{152}Dy. The bottom panel [21] shows the spectrum of states, including (on the left) the ground state rotation band for a prolate nucleus and (on the right) a super-deformed band extending from $J^P \sim 22^+$ to 60^+. Nuclei in states of a rotation band generally cascade rapidly down the band, giving coincident photons that are evenly spaced in energy. The top panel [22] shows the "picket fence" spectrum of photons produced in a cascade of the super-deformed band.

from first principles. In fact Coulomb's law (or more generally the equations of electromagnetism) determines the interactions between electrons and nuclei. Spin corrections and relativistic effects can be calculated perturbatively to very good accuracy because of the smallness of the fine structure constant $\alpha = e^2/4\pi\epsilon_0\hbar c \sim 1/137$. Together with the Pauli principle which leads to the shell structure of electron orbitals, these facts imply that one can calculate numerically spectra of complex atoms despite the difficulties of the many-body problem.

Unfortunately, none of this holds in nuclear physics. Forces between nucleons are neither simple nor fully understood. One of the reasons for this is that the interactions between nucleons are "residuals" of the fundamental interactions between quarks inside the nucleons. In that sense, nuclear interactions are similar to Van der Waals forces between atoms or molecules, which are also residual or "screened" Coulomb interactions. For these reasons, forces between nucleons are described by semi-phenomenological forms, e.g. (1.56), which are only partly deduced from fundamental principles.

Reversing the order of inference, physicists could have derived the form of Coulomb's law from the spectrum of bound states of the hydrogen atom. This is not possible in nuclear physics because there is only one two-nucleon bound state, the deuteron. In the next subsection, we will find that there is much to be learned from this fact but it will not be sufficient to derive the nucleon–nucleon potential in all its detail. To do this, we will need to attack the more difficult problem of nucleon–nucleon scattering. This will be done in Chap. 3. When we do this, we will be confronted with the other major difficulty of nuclear forces: the coupling constants are large so perturbative treatments do not apply as systematically as in atomic physics.

1.4.1 The deuteron

There is only one $A = 2$ nucleus, the deuteron, and it has no excited states. Its binding energy, quantum numbers, and magnetic moments are

$$B(2,1) = 2.225\,\text{MeV} \qquad J^P = 1^+ \qquad \mu_d = 0.857\mu_N \,. \qquad (1.42)$$

Note that $B(2,1)$ is quite small compared to typical nuclear binding energies, 8 MeV per nucleon.

We also note that to good approximation $\mu_d = \mu_p + \mu_n$. This suggests the the magnetic moment comes only from the spins of the constituents, implying that the nucleons are in a state of vanishing orbital angular momentum, $l = 0$. In fact, this turns out only to be a good first approximation since the deuteron is slightly deformed, possessing a small quadrupole moment. This requires that the wavefunction have a small admixture of $l = 2$. Both $l = 0$ and $l = 2$ are consistent with the parity of the deuteron since for two-nucleon states, the parity is -1^l.

Since the deuteron has spin-1 and is (mostly) in an $l = 0$ state, the spins of the nucleons must be aligned, i.e. the total spin, $s_{\text{tot}} = s_n + s_p$, must take on the quantum number $s = 1$. The other possibility is $s = 0$. Since the deuteron is the only bound state, we conclude

$$\text{n p } (s = 1) \text{ bound} \qquad \text{n p } (s = 0) \text{ unbound} .$$

We conclude that the neutron–proton potential is spin-dependent.

What about the neutron–neutron and proton–proton potentials and the fact that there are no bound states for these two systems? If the strong interactions do not distinguish between neutrons and protons, the non-existence of a $s = 0$ neutron–proton state is consistent with the non-existence of the analogous pp and nn states:

$$\text{n n } (s = 0) \text{ unbound} \qquad \text{p p } (s = 0) \text{ unbound} ,$$

i.e. all nucleon–nucleon $s = 0$, $l = 0$ states are unbound. On the other hand, the non-existence of pp and nn $s = 1$, $l = 0$ states is explained by the Pauli principle. This principle requires that the total wavefunction of pairs of identical fermions be antisymmetric. Loosely speaking, this is equivalent to saying that when two identical fermions are at the same place ($l = 0$), their spins must be anti-aligned. Thus, the $l = 0$, $s = 1$ proton–proton and neutron–neutron states are forbidden.

We make the important conclusion that the existence of a n p bound state and the non-existence of n n and p p bound states is consistent with the strong force not distinguishing between neutrons and protons but only if the force is spin-dependent.

The existence of an $s = 1$ state and non-existence of $s = 0$ states would naively suggest that the nucleon–nucleon force is attractive for $s = 1$ and repulsive for $s = 0$. This is incorrect. In fact the nucleon–nucleon force is attractive in both cases. For $s = 1$ it is sufficiently attractive to produce a bound state while for $s = 0$ it is not quite attractive enough.

We can understand how this comes about by considering the 3-dimensional square well potential shown in Fig. 1.10:

$$V(r) = -V_0 \quad r < R \qquad V(r) = 0 \quad r > R . \tag{1.43}$$

While it is hardly a realistic representation of the nucleon–nucleon potential, its finite range, R, and its depth V_0 can be chosen to correspond more or less to the range and depth of the real potential. In fact, since for the moment we only want to reproduce the deuteron binding energy and radius, we have just enough parameters to do the job.

The bound states are found by solving the Schrödinger eigenvalue equation

$$\left(\frac{-\hbar^2}{2m} \nabla^2 + V(r) \right) \psi(r) = E\psi(r) , \tag{1.44}$$

where m is the reduced mass $m \sim m_{\rm p}/2 \sim m_{\rm n}/2$. The $l = 0$ solutions depend only on r, so we set $\psi(r) = u(r)/r$ and find a simpler equation

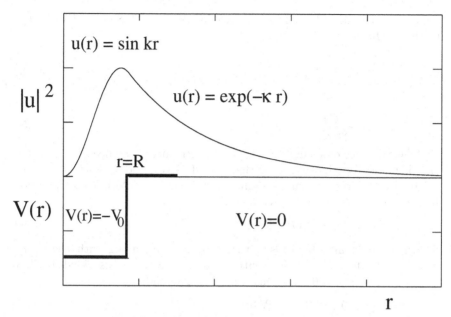

Fig. 1.10. A square-well potential and the square of the wavefunction $\psi(r) = u(r)/r$. The depth and width of the well are chosen to reproduce the binding energy and radius of the deuteron. Note that the wavefunction extends far beyond the effective range of the $s = 1$ nucleon–nucleon potential.

$$\left(\frac{-\hbar^2}{2m}\frac{\mathrm{d}^2}{\mathrm{d}r^2} + V(r)\right)u(r) \;=\; Eu(r)\,. \tag{1.45}$$

The solutions for $E < 0$ oscillate for $r < R$

$$u(r < R) \;\propto\; A\sin kr + B\cos(kr) \quad k(E) = \frac{\sqrt{2m(V_0 + E)}}{\hbar}\,, \tag{1.46}$$

and are exponentials for $r > R$

$$u(r > R) \;\propto\; C\exp(-\kappa r) + D\exp(\kappa r) \quad \kappa(E) = \frac{\sqrt{-2mE}}{\hbar}\,. \tag{1.47}$$

We set $B = 0$ to prevent ψ from diverging at the origin and $D = 0$ to make the wavefunction normalizable. Requiring continuity at $r = R$ of $u(r)$ and $u'(r)$ we find the condition that determines the allowed values of E

$$k(E)\cot k(E)R \;=\; -\kappa(E)\,. \tag{1.48}$$

We note that for $r \to 0$, the function on the left, $k\cot kr$, is positive and remains so until $kr = \pi/2$. Since the quantity on the left is negative the requirement for at least one bound state is that there exist an energy $E < 0$ such that

$$k(E)R \;>\; \pi/2\,, \tag{1.49}$$

i.e. that we can fit at least $1/4$ of a wave inside the well. Since $k(E) < k(E = 0)$ the condition is

$$\frac{\sqrt{2mV_0R^2}}{\hbar} > \pi/2 , \qquad (1.50)$$

i.e.

$$V_0R^2 > \frac{\pi^2\hbar^2c^2}{8mc^2} = 109\,\text{MeV}\,\text{fm}^2 . \qquad (1.51)$$

The existence of a single $s = 1$ state and the non-existence of $s = 0$ states can be understood by supposing that the effective values of V_0R^2 are respectively slightly greater than or slightly less than $109\,\text{MeV}\,\text{fm}^2$. The deuteron binding energy is correctly predicted if

$$V_0R^2\,(s = 1) = 139.6\,\text{MeV}\,\text{fm}^2 . \qquad (1.52)$$

This can be verified by substituting it into (1.48) along with the energy $E = -2.225\,\text{MeV}$. Data from neutron–proton scattering discussed in Chap. 3 shows that the $s = 0$ states just miss being bound

$$V_0R^2\,(s = 0) \sim 93.5\,\text{MeV}\,\text{fm}^2 . \qquad (1.53)$$

Including scattering data, Sect. 3.6, we can determine both V_0 and R:

$$s = 1 : \quad V_0 = 46.7\,\text{MeV} \quad R = 1.73\,\text{fm} . \qquad (1.54)$$

The wavefunction in shown in Fig. 1.10. The fact that $B(2,1)$ is small results causes the wavefunction to extend far beyond the effective range R of the potential and explains the anomalously large value of the deuteron radius (Table 1.1).

The scattering data discussed in Sect. 3.6 allow one to estimate the values of V_0 and R for $s = 0$ (Table 3.3). One finds for the proton–neutron system

$$s = 0 : \quad V_0 = 12.5\,\text{MeV} \quad R = 2.79\,\text{fm} . \qquad (1.55)$$

This potential is quite different from the $s = 1$ potential.

In summary, the strong interactions have a strength and range that places them precisely at the boundary between the the interactions that have no bound states (e.g. the weak interactions) and those that have many bound states (e.g. the electromagnetic interactions). The phenomenological implications are of a cosmological scale since the fabrication of heavy elements from nucleons must start with the production of the one 2-nucleon bound state. As we will see in Chaps. 8 and 9, if there were no 2-nucleon bound states the fabrication of multi-nucleon nuclei would be extremely difficult. On the other hand, the existence of many 2-nucleon states would make it considerably simpler. The actual situation is that heavy elements can be slowly formed, leaving, for the time being, a large reserve of protons to serve as fuel in stars.

1.4.2 The Yukawa potential and its generalizations

We want to consider more realistic potentials than the simple potential (1.43). This is necessary to completely describe the results of nucleon–nucleon scattering (Chap. 3) and to understand the saturation phenomenon on nuclear binding. The subject of nucleon–nucleon potentials is very complex and we will give only a qualitative discussion. Our basic guidelines are as follow:

- Protons and neutrons are spin 1/2 *fermions*, and therefore obey the Pauli principle. We have already seen in the previous subsection how this restricts the number of 2-nucleon states.
- Nuclear forces are *attractive* and *strong*, since binding energies are roughly a 10^6 times the corresponding atomic energies. They are however *short range forces* : a few fm. The combination of strength and short range makes 2-nucleon systems only marginally bound but creates a rich spectrum of many-nucleon states.
- They are *"charge independent."* Nuclear forces are blind to the electric charge of nucleons. If one were to "turn off" Coulomb interactions, the nuclear proton–proton potential would be the same as the neutron–neutron potential. A simple example is given by the binding energies of isobars such as tritium and helium 3 : $B(^3\mathrm{H}) = 8.492\,\mathrm{MeV} > B(^3\mathrm{He}) = 7.728\,\mathrm{MeV}$. If the difference $\Delta B = 0.764\,\mathrm{MeV}$ is attributed to the Coulomb interaction between the two protons in $^3\mathrm{He}$, $\Delta B = e^2 < 1/r_{12} > /4\pi\epsilon_0$, one obtains a very reasonable value for the mean radius of the system : $R \approx 2\,\mathrm{fm}$ (this can be calculated or measured by other means). We shall come back to this question in a more quantitative way when we discuss isospin.
- Nuclear forces *saturate*. As we have already mentioned, this results in the volumes and binding energies of nuclei being *additive* and, in first approximation, proportional to the mass number A. This is a remarkable fact since it is reminiscent of a *classical* property and not normally present in quantum systems. It appears as if each nucleon interacts with a given fixed number of neighbors, whatever the nucleus.

The theoretical explanation of the saturation of nuclear forces is subtle. The physical ingredients are the short range attractive potential ($r \sim 1\,\mathrm{fm}$), a hard core repulsive force at smaller distances $r < 0.5\,\mathrm{fm}$, and the Pauli principle. Being the result of these three distinct features, there is no simple explanation for saturation. It is simple to verify that the Pauli principle alone cannot suffice, and that any power law force does not lead to saturation (Exercise 1.9).

Many properties of nuclear forces can, be explained quantitatively by the potential proposed by Yukawa in 1939:

$$V(r) = g\frac{\hbar c}{r} \exp(-r/r_0) \quad . \tag{1.56}$$

The factor $\hbar c$ is present so that the coupling constant g is dimensionless. As we will see in the next section, forces between particles are due, in quantum field theory, to the quantum exchange of *virtual particles*. The range r_0 of a force is the Compton wavelength $\hbar/mc$ of the exchanged particle of mass m. Yukawa noticed that the range of nuclear forces $r_0 \simeq 1.4\,\mathrm{fm}$, corresponds to the exchange of a particle of mass $\simeq 140\,\mathrm{MeV}$. This is how he predicted the existence of the π meson. The discovery of that particle in cosmic rays was a decisive step forward in the understanding of nuclear forces.

When applied to a two-nucleon system, the potential (1.56) is reduced by a factor of $(m_\pi/2m_p)^2 \sim 10^{-2}$ because of spin-parity considerations. A dimensionless coupling constant of $g \simeq 14.5$ explains the contribution of the π meson to the nucleon–nucleon force.

It is necessary to add other exchanged particles to generate a realistic potential. In fact, the general form of nucleon–nucleon strong interactions can be written as a linear superposition of Yukawa potentials :

$$V(r) = \sum_i g_i \frac{\hbar c}{r} \exp(-\mu_i r) \tag{1.57}$$

where the sum is over a discrete or continuous set of exchanged particles with masses given by $\mu_i = m_i c/\hbar$.

However, we need some more elements to explain saturation and prevent a nuclear "pile-up" where all nucleons collapse to an object of the size of the order of the range of the strong interactions.

First, one must add a strong *repulsive* shorter range potential, called a "hard core" interaction. This potential is of the form (1.56) with a negative coupling constant g and a range $r_0 \simeq 0.3\,\mathrm{fm}$. The physical origin of the repulsive core is not entirely understood but it certainly includes the effect of the Pauli principle that discourages placing the constituent quarks of nucleons near each other.

Second, spin effects and relativistic effects must be taken into account. One writes the potential as the sum of central potentials with spin-dependent coefficients:

$$V = V_{\mathrm{C}}(r) + \Omega_{\mathrm{T}} V_{\mathrm{T}}(r) + \Omega_{\mathrm{SO}} V_{\mathrm{SO}}(r) + \Omega_{\mathrm{SO2}} V_{\mathrm{SO2}}(r) , \tag{1.58}$$

where V_{C} is a pure central potential and the other terms are spin dependent. The most important is the *tensor potential* V_{T} with

$$\Omega_{\mathrm{T}} = [3(\boldsymbol{\sigma_1} \cdot \boldsymbol{r}/r)(\boldsymbol{\sigma_2} \cdot \boldsymbol{r}/r) - \boldsymbol{\sigma_1} \cdot \boldsymbol{\sigma_2}] \tag{1.59}$$

where $\boldsymbol{\sigma_1}$ and $\boldsymbol{\sigma_2}$ are the Pauli spin matrices for the two nucleon spins. Figure 1.11 shows how this term has the important effect of inducing a correlation between the position and spin of the two nucleons. This results in a permanent quadrupole moment for the deuteron. It is also the dominant force in binding the deuteron. However, it averages to zero for multi-nucleon systems where it plays a minor role.

The last two terms in (1.58) are spin-orbit interactions:

$$\hbar \Omega_{\mathrm{SO}} = (\boldsymbol{\sigma_1} + \boldsymbol{\sigma_2}) \cdot \boldsymbol{L} \tag{1.60}$$

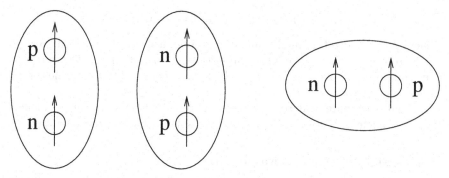

Fig. 1.11. The tensor potential for the $s = 1$ state (1.59) makes the configuration on the right ($\boldsymbol{\sigma} \cdot \boldsymbol{r} = 0$) have a different potential energy than the two configurations on the left ($\boldsymbol{\sigma} \cdot \boldsymbol{r} \neq 0$). This results in the permanent quadrupole moment of the deuteron.

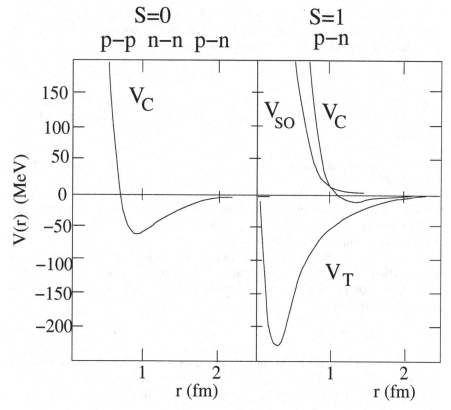

Fig. 1.12. The most important contributions to the nucleon–nucleon potentials in the $s = 0$ state (left) and the $s = 1$ state (right) (the so-called Paris potential). The two central potentials V_C depend only on the relative separations. The tensor potential V_T is of the form (1.59) is responsible for the deuteron binding and for its quadrupole moment. The spin-orbit potential is V_{SO}.

$$\hbar^2 \Omega_{SO2} = (\boldsymbol{\sigma_1} \cdot \boldsymbol{L})(\boldsymbol{\sigma_2} \cdot \boldsymbol{L}) + (\boldsymbol{\sigma_2} \cdot \boldsymbol{L})(\boldsymbol{\sigma_1} \cdot \boldsymbol{L}) \tag{1.61}$$

where $\boldsymbol{L}$ is the orbital angular momentum operator for the nucleon pair.

Figure 1.12 shows the most important contributions to the nucleon–nucleon potential [23]. For spin-anti-aligned nucleons ($s = 0$), only the central potential contributes.

Finally, we note that to correctly take into account the charge independence of nuclear forces the formalism of isospin (Sect. 1.7) must be used.

1.4.3 Origin of the Yukawa potential

The form (1.56), which is derived from quantum field theory, can be understood quite simply. Consider a de Broglie wave

$$\psi = \exp\left(-i(Et - \boldsymbol{p}.\boldsymbol{r})/\hbar\right) \quad . \tag{1.62}$$

The Schrödinger equation in vacuum is obtained by using $E = p^2/2m$ and then taking the Laplacian and the time derivative. Assume now that we use the relativistic relation between energy and momentum :

$$E^2 = p^2 c^2 + m^2 c^4$$

(this is how Louis de Broglie proceeded initially). By taking second-order derivatives of (1.62) both in time and in space variables we obtain the Klein—Gordon equation :

$$\frac{1}{c^2}\frac{\partial^2 \psi}{\partial t^2} - \nabla^2 \psi + \mu^2 \psi = 0 \tag{1.63}$$

where we have set $\mu = mc/\hbar$. Originally, this equation was found by Schrödinger. He abandoned it because it did not lead to the correct relativistic corrections for the levels of the hydrogen atom.[5] It was rediscovered later on by Klein and Gordon.

Forgetting about the exact meaning of ψ in this context, (1.63) is the propagation equation for a relativistic free particle of mass m. In the case $m = 0$, i.e., the photon, we recover the propagation equation for the electromagnetic potentials :

$$\frac{1}{c^2}\frac{\partial^2 \psi}{\partial t^2} - \nabla^2 \psi = 0 \quad . \tag{1.64}$$

The classical electrostatic potential produced by a point charge is obtained as a static, isotropic, time-independent solution to this equation with a source term added to represent a point-like charge at the origin, i.e.,

$$\nabla^2 V(r) = -\frac{q}{\epsilon_0}\delta(\boldsymbol{r}) \ , \tag{1.65}$$

---

[5] Schrödinger did not have in mind spin and magnetic spin-orbit corrections. These effects are accounted for by the Dirac equation, the relativistic wave equation for a spin 1/2 particle

whose solution is

$$V(r) = \frac{q}{4\pi\epsilon_0 \ r} \quad .$$

(1.66)

Similarly, by looking for static, isotropic solutions of the Klein–Gordon equation 1.63 with a point-like source at the origin :

$$\nabla^2 V(r) - \mu^2(r) = -\frac{g(\hbar c)}{4\pi}\delta(\boldsymbol{r}) \quad ,$$

(1.67)

one can readily check that the Yukawa potential (1.56)

$$V(r) = g(\hbar c/r)\exp(-\mu r) \, ,$$

(1.68)

satisfies this equation. The solution $V(r) = g(\hbar c/r)\exp(+\mu r)$ is discarded since it diverges at infinity.

1.4.4 From forces to interactions

We have emphasized that, in quantum field theory, forces between particles are described by the exchange of virtual particles. The interactions can be described by (Feynman) "diagrams" like those shown in Fig. 1.13. Each diagram corresponds to a scattering amplitude that can be calculated according to the rules of quantum field theory. As we will see in Chap. 3, the effective potential is Fourier transform of the amplitude.

In quantum electrodynamics, the exchange of massless photons leads to the Coulomb potential. The exchange of massive particles leads to Yukawa-like potentials. One example is the exchange of pions as shown in the first two diagrams in Fig. 1.13. These diagrams contribute to the nucleon–nucleon potential of Fig. 1.12 and lead to the binding of nucleons and to nucleon–nucleon scattering.

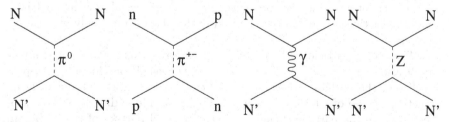

Fig. 1.13. Diagrams contributing to the nucleon–nucleon interaction due to the exchange of a pion, a photon and a Z^0. In these diagrams N and N' represent either a proton or a neutron. The two first diagrams with the exchange of a neutral or charged pion generates the long-range part of the nucleon–nucleon potential. The photon exchange diagram generates the Coulomb interaction between protons and the magnetic interaction between any nucleons. The diagram with Z exchange is negligible for the nucleon–nucleon interactions but dominates for neutrino-neutron scattering since the neutrino has no strong interactions.

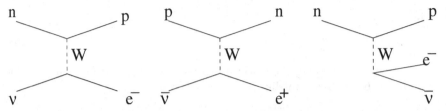

Fig. 1.14. Some diagrams with the exchange of a W. The two diagrams on the left contribute to inelastic neutrino scattering. The diagram on the right leads to neutron decay.

Other massive particles can be exchanged between particles, for example the Z^0 boson

$$m_Z\,c^2 \; = \; 91.188\,\text{GeV} \qquad \frac{\hbar c}{m_Z\,c^2} \; = \; 2.11 \times 10^{-3}\,\text{fm} \,. \qquad (1.69)$$

This particle is the mediator of the *neutral current* sector of the weak interactions as illustrated in the fourth diagram of Fig. 1.13. It thus leads to a Yukawa-like potential between particles of the form (1.56). Compared to pion exchange, the Yukawa potential has a range about 10^{-3} times smaller. The effective coupling is also much smaller being of order

$$g_Z \; \sim \; \alpha \,. \qquad (1.70)$$

To compare with the strong interactions, we estimate an effective value of $V_0 R^2$ by taking $R \sim 2 \times 10^{-3}$fm and $V_0 \sim \alpha \hbar c / R$:

$$V_0 R^2 \; \sim \; 3 \times 10^{-3}\,\text{MeV}\,\text{fm}^2 \,, \qquad (1.71)$$

i.e. about 5 orders of magnitude smaller than that of the strong interactions (1.52) and (1.53). We conclude that Z^0 can play no role in nucleon–nucleon binding or scattering.

On the other hand, Z^0 exchange plays the essential role in neutrino–nucleon elastic scattering. This is because the neutrino has only weak interactions.

The use of particle exchange to describe potentials obviously has direct applications in bound states and elastic scattering. If one includes the possibility of the exchange of charged particles, we can also describe *inelastic* reactions, some of which are shown in Fig. 1.14. These diagrams show processes where the particle exchanged is the charged partner of the Z^0 boson, the $W^{\pm}$

$$m_W c^2 \; = \; 80.42\,\text{GeV} \,. \qquad (1.72)$$

This particle mediates the so-called *charged-current* weak interactions. They play the essential cosmological role of changing neutrons into protons and vice versa. They thus play an essential role in the cosmological and astrophysical synthesis of heavy elements. On Earth, they are mostly seen in neutrino scattering and β-decay as illustrated in Fig. 1.14 and discussed further in the next section.

Table 1.3 summarizes the interactions important in nuclear physics and their exchanged particles. The weak and electromagnetic interactions are *fundamental interactions* that will be discussed in more detail in Chap. 4. The ordinary strong interactions that lead to binding of nucleons in nuclei are "effective" in the sense that, as previously emphasized, they are the screened residuals due to the fundamental chromodynamic interactions between quarks.

Table 1.3. The interactions involving elementary particles. The strong interactions are considered to be screened quantum chromodynamic interactions.

interaction	examples of reactions	exchanged particle
weak neutral current	$\nu n \rightarrow \nu n$	Z
weak charged current	$\nu n \rightarrow e^- p$	W
electromagnetic	$ep \rightarrow ep$	γ
strong	$np \rightarrow np$	π and other hadrons
chromodynamic	quark–quark scattering	gluons

1.5 Nuclear reactions and decays

Nuclear species can be transformed in a multitude of nuclear reactions. In nuclear reactions involving only strong and electromagnetic interactions, the number of protons and the number of neutrons are conserved separately. An important example is neutron absorption followed by photon emission, the so-called " (n, γ) " reaction:

$$n(A, Z) \rightarrow \gamma(A + 1, Z) \quad \text{i.e.} \quad (A, Z) (n, \gamma) (A + 1, Z) . \qquad (1.73)$$

The second form is a standard way of denoting the reaction. Other reactions are " (p, γ) " reactions

$$p(A, Z) \rightarrow \gamma(A + 1, Z + 1) \quad \text{i.e.} \quad (A, Z) (p, \gamma) (A + 1, Z + 1) , \qquad (1.74)$$

"(n, p)" reactions

$$n(A, Z) \rightarrow p(A, Z - 1) \quad \text{i.e.} \quad (A, Z) (n, p) (A, Z - 1) , \qquad (1.75)$$

and "(p, n)" reactions

$$p(A, Z) \rightarrow n(A, Z + 1) \quad \text{i.e.} \quad (A, Z) (p, n) (A, Z + 1) . \qquad (1.76)$$

In all these reactions, the final state nucleus may be produced in an excited state so additional photons are produced in de-excitation.

Many nuclei can decay to lighter nuclei by redistributing the nucleons among "daughter" nuclei. The most important are "α-decays" with the emission of a ^{4}He nucleus :

$$(A, Z) \rightarrow (A - 4, Z - 2) \, ^4\text{He} , \qquad (1.77)$$

for instance $^{238}\text{U} \rightarrow {}^{234}\text{Th} \, ^4\text{He}$. The condition for a nucleus to be stable against α-decay is

$$m(A, Z) < m(A - 2, Z - 2) + m(2, 2) . \qquad (1.78)$$

Alpha decay will be studied in more detail in Sect. 2.6.

α-decay is a special case of *spontaneous fission* into two lighter nuclei with, perhaps, the emission of k neutrons, such as:

$$(A, Z) \rightarrow (A', Z') \, (A'', Z'') + k\,\text{n} , \qquad (1.79)$$

with $A' + A'' + k = A$ and $Z' + Z'' = Z$. Such decays occur mostly for very heavy nuclei. They will be studied in more detail in Chap. 6.

While the strong and electromagnetic interactions responsible for the above reactions separately conserve the number of neutrons and the number of protons, neutrons and protons can be transformed into each other by the weak interactions mentioned in the previous section. These interactions are weak in the sense that the rates of reactions they induce are in general much slower than those due to the electromagnetic or strong interactions. Weak interactions will be studied in some detail in Chap. 4. Here, we note that weak nuclear interactions are usually signaled by the participation of an electron–neutrino, ν_e. Two examples are electron capture by a proton and neutrino capture by a neutron

$$\text{e}^- \, \text{p} \leftrightarrow \nu_\text{e} \, \text{n} . \qquad (1.80)$$

We see that when the proton is transformed to a neutron or vice-versa, an electron or positron is created or absorbed to conserve charge. Since all particles in these reactions have spin $1/2$, the neutrino is necessary to conserve angular momentum. All other possible weak interactions can be found by moving particles from left to right and changing them to their antiparticles, e.g.

$$\bar{\nu}_\text{e} \, \text{p} \leftrightarrow \text{e}^+ \, \text{n} . \qquad (1.81)$$

Another possibility is spontaneous neutron decay

$$\text{n} \rightarrow \text{p} \, \text{e}^- \, \bar{\nu}_\text{e} . \qquad (1.82)$$

This decay is allowed by energy conservation, $(m_\text{n} - m_\text{p} - m_\text{e})c^2 = 0.78\,\text{MeV}$, and the free neutron has a mean-lifetime of

$$\tau_\text{n} = 886.7 \pm 1.9\,\text{s} . \qquad (1.83)$$

Of course, the fact that $m_\text{p} < m_\text{n} + m_\text{e}$ means that the corresponding decay of *free* protons is forbidden

$$p \rightarrow n\,e^+\,\nu_e \qquad \text{energy not conserved.} \tag{1.84}$$

Since the weak interactions can turn protons into neutrons and vice versa while conserving the total number of nucleons, their most important role is the decay of radioactive isobars via so-called β-decays:

$$(A, Z) \rightarrow (A, Z+1)\,e^-\bar{\nu}_e \qquad (A, Z) \rightarrow (A, Z-1)\,e^+\nu_e\,. \tag{1.85}$$

These two reactions are the nuclear equivalents of the two fundamental decays (1.82) and (1.84). Another form of β-decay (or transmutation) is electron capture

$$e^-\,(A, Z) \rightarrow (A, Z-1)\,\nu_e\,, \tag{1.86}$$

i.e. the nuclear equivalent of (1.80) This latter reaction is possible if the nucleus is surrounded by atomic electrons but cannot happen in the case of an isolated nucleus.

A nucleus is stable against the β-decays (1.85) if it is sufficiently light compared to its two neighboring isobars:

$$m(A, Z) \;<\; m(A, Z\pm 1) + m_e\,, \tag{1.87}$$

where $m(A, Z)$ stands for the mass of the (A, Z) nucleus. The condition for it to be stable against both β-decay and electron capture is slightly different: the mass of the *atom* containing $A - Z$ neutrons, Z protons and Z electrons must be smaller than the masses of its two neighboring isobars :

$$m(A, Z) + Z m_e \;<\; m(A, Z\pm 1) + (Z\pm 1)m_e\,. \tag{1.88}$$

(In the above formula, we neglect the electronic binding energies which are comparatively much smaller.) In general, for each value of the mass number A, there exist only one or two stable isobars. The other nuclei are radioactive and decay to the stable isobars by a series of β-decays and/or electron captures.

While the weak interactions do not conserve separately the number of protons and the number of neutrons, they do (as do the strong and electromagnetic interactions) conserve electric charge and *baryon number*, i.e. the total number of nucleons. Baryon number is conserved in nuclear reactions but it is believed that very rare, as yet unobserved, reactions allow baryon number non-conservation. On the other hand, electric charge conservation is believed to be absolute.

1.6 Conservation laws

The investigation of the fundamental constituents of matter and their interactions comes from the experimental and theoretical analysis of reactions. These reactions can be scattering experiments with or without production of particles, and decays of the unstable particles produced in these reactions.

Various fundamental *conservation laws* govern nuclear reactions. The laws allow the identification of particles, i.e. the determination of their masses, spins, energies, momenta etc.

The most important laws are energy-momentum conservation, angular momentum conservation and electric charge conservation. In nuclear physics, other laws play an important role such as lepton number, baryon number and isospin conservation.

In this book, we shall mainly make use of simple "selection rules" implied by these conservation laws. In this section, we will first discuss the experimental and phenomenological consequences of the most important laws. We will then show how the conservation laws are related to *invariance* properties of transition operators between initial and final states, or, equivalently, invariance laws of Hamiltonians of the systems under consideration.

1.6.1 Energy-momentum conservation

By far the most important conservation law is that for Energy-momentum. For example, in nuclear β-decay

$$(A, Z) \rightarrow (A, Z+1) \, e^- \, \bar{\nu}_e \tag{1.89}$$

we require

$$E_{A,Z} = E_{A,Z+1} + E_e + E_{\bar{\nu}_e} , \tag{1.90}$$

and

$$\boldsymbol{p}_{A,Z} = \boldsymbol{p}_{A,Z+1} + \boldsymbol{p}_e + \boldsymbol{p}_{\bar{\nu}_e} . \tag{1.91}$$

These two laws are only constraints. As discussed in later chapters, the way that momentum and energy are distributed between the decay products depends on the details of the interaction responsible for the reaction.

When one applies energy-momentum conservation, it is of course necessary to take into account the masses of initial and final particles by using the relativistic expression for the energy

$$E = (p^2 c^2 + m^2 c^4)^{1/2} \tag{1.92}$$

for a free particle of mass m. The square root in this formula often makes calculations very difficult. However, in nuclear physics, nuclei and nucleons are usually non-relativistic, $v = pc^2/E \ll c$, and one can use the non-relativistic approximation :

$$E = \sqrt{p^2 c^2 + m^2 c^4} \simeq mc^2 + p^2/2m , \tag{1.93}$$

i.e. the energy is the sum of the rest energy mc^2 and the non-relativistic kinetic energy $p^2/2m$. On the other hand, photons and neutrinos are relativistic:

$$E = \sqrt{p^2 c^2 + m^2 c^4} \simeq pc + m^2 c^4/2pc , \tag{1.94}$$

where the mass term $m^2c^4/2pc$ can usually be neglected for neutrinos and always for the massless photon $E = pc$.

The presence of non-relativistic and relativistic particles in a given reaction results in the very useful fact that, viewed in the center-of-mass, the momentum is shared democratically between all final state particles whereas the kinetic energy is carried mostly by the relativistic particles. This is most easily seen in the decay of an excited nucleus:

$$(A, Z)^* \rightarrow (A, Z)\gamma. \tag{1.95}$$

Energy conservation in the initial rest frame implies

$$m^*c^2 = mc^2 + \frac{p^2}{2m} + pc, \tag{1.96}$$

where m^* and m are the masses of the excited and unexcited nuclei and p is the common momentum of the final nucleus and photon. (Momentum conservation requires that these two momenta be equal.) It is clear that the photon energy pc is much greater than the nuclear kinetic energy:

$$\frac{p^2}{2m} = pv/2 \ll pc \quad \text{for } v \ll c. \tag{1.97}$$

Neglecting $p^2/2m$ in (1.96), we see that the photon energy is then to good approximation proportional to the mass difference

$$pc \sim (m^* - m)c^2. \tag{1.98}$$

Using this value for the momentum, we find that the ratio between the nuclear kinetic energy and the photon energy is

$$\frac{p^2/2m}{pc} = (1/2)\frac{m^* - m}{m}. \tag{1.99}$$

This is at most of order 10^{-3} in transitions between nuclear states.

In more complicated reactions like three-body decays, one generally finds that the momentum is evenly distributed on average among the final-state particles. Once again, this implies that the kinetic energy is taken by the lightest particles.

In nuclear physics, one often mentions explicitly the energy balance in writing reactions

$$A + B \rightarrow a_1 + a_2 + a_n + Q \tag{1.100}$$

where

$$Q = (\Sigma m_i - \Sigma m_f)c^2. \tag{1.101}$$

If the reaction can take place when A and B are at rest, Q is the total kinetic energy of the particles in the final state. If Q is negative, the reaction is endothermic and it can only take place if the energy in the center-of-mass is above the energy threshold.

An important example in producing heavy elements is neutron capture accompanied by the production of k photons:

$$n + (A, Z) \rightarrow (A + 1, Z) + k\gamma + Q \qquad (1.102)$$

The fact that binding energies per nucleon are $\sim 8\,\text{MeV}$ means that Q is positive and of order $8\,\text{MeV}$ (near the bottom of the stability valley). Since the final state photons are the only relativistic particles, we can expect that they take all the energy, $\sum E_\gamma \sim Q$. A detailed calculation of the constraints of energy-momentum conservation confirms this (Exercise 1.10).

Of course, some reactions involve no relativistic particles, for example

$$d\,t \rightarrow n\,^4\text{He} + 17.58\,\text{MeV}. \qquad (1.103)$$

We leave it to Exercise 1.11 to show that in the limit of low-center-of-mass energy, the final state neutron takes the majority of the $17.58\,\text{MeV}$.

1.6.2 Angular momentum and parity (non)conservation

Angular momentum conservations plays a different role than that of energy-momentum conservation. The latter can by verified to a useful precision in *individual events* where the energies and momenta of final-state particles can be compared with those of the initial-state particles. This is because there is a relatively well-defined correspondence between momentum wavefunctions (plane waves) and the classical tracks of particles that are actually observed, i.e. a plane wave of wave vector $\boldsymbol{k}$ and angular frequency ω generates a detector response that appears to be due to a classical particle of momentum $\hbar\boldsymbol{k}$ and energy $E = \hbar\omega$.

On the other hand, the wavefunctions corresponding to a definite angular momentum, correspond to certain angular dependence of the function about the origin. This information is lost when an individual track going in a particular direction is measured. It can be recovered only by observing many events and reconstructing the angular distribution.

The same consideration applies to parity which gives the behavior of a wavefunction under reversal of all coordinates. Its conservation can only be verified in the distribution of tracks. As it turns out, parity is not in fact conserved in the weak interactions, as we will see in Chap. 4.

1.6.3 Additive quantum numbers

As we have already emphasized nuclear reactions may or may not respect certain additive conservations laws. The most important is electric charge conservation which is believed to be absolutely respected in all interactions.

The second most respectable conservation law is that of *baryon number*, i.e. the total number of nucleons (neutrons and protons), minus the total number of anti-nucleons (antiprotons and antineutrons). For instance, the reaction $pd \rightarrow pp\gamma$ conserves electric charge but not baryon number and is therefore not observed. Conversely, the photo-disintegration of the deuteron, $\gamma d \rightarrow pn$, is allowed and observed. The baryon number conservation forbids the decay of the proton, such as in the reaction $p \rightarrow e^+\gamma$. The present experimental limit on the proton lifetime, $\tau_p > 10^{32}$ yr shows that the baryon number is conserved to very good accuracy.

Finally, we mention lepton number. The electron and the neutrinos are called *leptons*. The *lepton number* (more precisely the "electron lepton number") L_e is defined as the difference between the total number of leptons (electrons and neutrinos) and the total number of anti-leptons (anti-electrons and antineutrinos)

$$L_e \equiv N(e^-) + N(\nu_e) - N(e^+) - N(\bar{\nu}_e) , \qquad (1.104)$$

where $N(e^-)$ is the number of electrons, $N(\nu_e)$ the number of neutrinos, etc. L_e is conserved in nuclear reactions involving electrons and neutrinos. In neutron decay, for instance, $n \rightarrow p + e^- + \bar{\nu}_e$, an *antineutrino* is produced together with the electron in order to conserve the lepton number. This antineutrino then can interact and produce a positron but *not* an electron:

$$\bar{\nu}_e + p \rightarrow e^+ + n \quad \text{but not} \quad \bar{\nu}_e + n \rightarrow e^- + p . \qquad (1.105)$$

There exist two other types of charged leptons, the $\mu^\pm$ ($m_\mu = 105.66$ MeV) and the $\tau^\pm$ ($m_\tau = 1777.03$ MeV). Each of these leptons is associated with its own neutrino ν_μ and ν_τ both of which have very small masses. The "muon" and "tauon" lepton numbers are defined in the same way as the electron lepton number:

$$L_\mu \equiv N(\mu^-) + N(\nu_\mu) - N(\mu^+) - N(\bar{\nu}_\mu) , \qquad (1.106)$$

$$L_\tau \equiv N(\tau^-) + N(\nu_\tau) - N(\tau^+) - N(\bar{\nu}_\tau) . \qquad (1.107)$$

They are both conserved separately. Since m_μ and m_τ are much larger than characteristic nuclear energy scales ($\sim$ 1 MeV), the μ and τ leptons have fewer applications in nuclear physics than in particle physics.

In the same way as the strong nuclear interactions show that the neutron is a neutral partner of the proton, weak interactions show that the ν_e acts as a neutral partner of the electron (and similarly for the ν_μ and μ, and for the ν_τ and τ). We have been careful to use the words "acts as" because it is now believed the separate conservation of electron, muon and tauon lepton numbers is only an effective conservation law. Recent experiments on "neutrino oscillations" that are discussed in Chap. 4 indicate that the only truly conserved number is the sum of the 3 lepton numbers:

$$L = L_e + L_\mu + L_\tau . \qquad (1.108)$$

1.6.4 Quantum theory of conservation laws

A quantum system described by a state vector $|\psi\rangle$ evolves with time in a way governed by the Schrödinger equation and the system's Hamiltonian operator H:

$$\frac{\mathrm{d}|\psi(t)\rangle}{\mathrm{d}t} = H|\psi(t)\rangle . \tag{1.109}$$

It is interesting to see how the dynamics defined by (1.109) conspires to conserve a quantity associated with a time-independent operator A. One answer is given by Ehrenfest's theorem relating the time development of an operator's expectation value to the operator's commutator with H:

$$\frac{\mathrm{d}}{\mathrm{d}t}\langle\psi(t)|A|\psi(t)\rangle = \frac{1}{i\hbar}\langle\psi(t)|[A,H]|\psi(t)\rangle , \tag{1.110}$$

which follows simply from (1.109) and its Hermitian conjugate. We see that the expectation value is time independent if the operator commutes with H.

In nuclear physics, we are generally interested in transitions between states and we would like to see how selection rules that constrain the transitions are generated. Consider the transition amplitude of a system from an initial state $|i\rangle$ to a final state $|f\rangle$, both eigenstates of a Hamiltonian H_0, due to a transition Hamiltonian H_T. The total Hamiltonian of the system is $H = H_0 + H_T$. In first-order perturbation theory, i.e. in Born approximation, for $f \neq i$, the amplitude $\gamma_{i\to f}$ is proportional to the matrix element of H_T between initial and final states

$$\gamma_{i\to f} \propto \langle f|H_T|i\rangle . \tag{1.111}$$

If A commutes with H_0 we can take the initial and final states to be also eigenstates of A with eigenvalues a_i and a_f, respectively,

$$A|i\rangle = a_i|i\rangle , \quad A|f\rangle = a_f|f\rangle . \tag{1.112}$$

Since we suppose that A commutes with H_T, we obtain

$$\langle f|[A,H_T]|i\rangle = (a_f - a_i)\langle f|H_T|i\rangle \propto (a_f - a_i)\gamma_{i\to f} = 0 . \tag{1.113}$$

Therefore, either $\gamma_{i\to f} = 0$, i.e. the transition is forbidden, or, if it is not forbidden, $a_f = a_i$. The quantity A is conserved in any transition if A commutes with the transition Hamiltonian. While we have shown this only in Born approximation, the property can be extended to all orders of perturbation theory.

The conservation of a quantity described by the operator A is equivalent to the *invariance* of the Hamiltonian with respect to unitary transformations defined by the operator

$$D(\alpha) = \exp(i\alpha A) , \tag{1.114}$$

where α is an arbitrary real number. The operator D is unitary, i.e. $D^+D = 1$, because A is Hermitian. The unitary transformation of states $|\psi\rangle$ and of operator B, associated with D is

$$|\psi'\rangle = D|\psi\rangle \qquad B' = DBD^+ . \tag{1.115}$$

From the commutation relation $[A, H] = 0$, we deduce

$$[D, H] = 0 \Rightarrow H = DHD^\dagger . \tag{1.116}$$

In other words, the Hamiltonian H is invariant under the unitary transformation D associated with the operator A.

We see that the *conservation* of A in the transitions induced by H is a result of the *invariance* of H in the unitary transformation D. To this *symmetry* property of the Hamiltonian there corresponds a *conservation law* of the quantity A. The operator A is called the infinitesimal generator of the transformation D.

We also notice that equation (1.114) defines a group of transformations $D(\alpha)D(\beta) = D(\alpha + \beta)$. In the case of (1.114), the group is Abelian, or commutative.

It is interesting to associate the various conservation laws of the previous subsections with invariances. The Schrödinger equation (1.109) follows from

$$|\psi(t)\rangle = \exp(iHt)|\psi(t = 0)\rangle , \tag{1.117}$$

if we assume that the Hamiltonian is time-independent. Equation (1.117) tells us that the Hamiltonian is the generator of translations in time. Energy is trivially conserved because the Hamiltonian commutes with itself so we can conclude that energy conservation occurs for any time-invariant system. In other words, it occurs in any isolated system.

Total momentum conservation comes from the invariance under translations in space. The infinitesimal generator of the group of space translations is the total momentum $\boldsymbol{P}$. Let $D(\boldsymbol{x}_0)$ be an element of this group

$$D(\boldsymbol{x}_0) = \exp(i\boldsymbol{x}_0 \cdot \boldsymbol{P}/\hbar) , \tag{1.118}$$

the translation invariance of the Hamiltonian can be written equivalently as $[H, D] = 0 \Leftrightarrow [H, \boldsymbol{P}] = 0$. If H is translation invariant, in a process leading from an initial state of total momentum $\boldsymbol{P}_i$ to a final state of total momentum $\boldsymbol{P}_f$, one has

$$[H, D] = 0 \Rightarrow \boldsymbol{P}_i = \boldsymbol{P}_f . \tag{1.119}$$

The conservation of the total angular momentum follows from the *rotation invariance* of the transition Hamiltonian. Consider a rotation of an angle α around an axis along the unit vector $\hat{u}$ and set $\boldsymbol{\alpha} = \alpha\hat{u}$. The corresponding unitary operator is

$$D_\alpha = \exp(i\boldsymbol{\alpha} \cdot \boldsymbol{J}/\hbar) . \tag{1.120}$$

Consider an initial state which is an eigenstate of J^2 and J_z with eigenvalues j_i, m_i, the transition to a final state $|f\rangle$ can happen only if it is also an eigenstate of J^2 and J_z with the same eigenvalues :

$$\langle f, j_f, m_f | H_T | i, j_i, m_i \rangle = \delta_{j_f j_i} \delta_{m_f m_i} \langle \tilde{h} \rangle \tag{1.121}$$

Since in nuclear processes one measures momenta of particles, it is through the angular distributions of particles that one measures the conservation of angular momentum.

We note that the infinitesimal generators J_x, J_y and J_z do not commute, i.e. the group of rotations is non-Abelian. A consequence of this is that the end result of two successive rotations (about different axes) depends on the order that the two rotations are performed. This can be confirmed by rotating macroscopic objects.

The invariances associated with the additive quantum numbers can be understood by considering transformations generated by one component of the angular momentum which we take to be the z component:

$$D(\varphi) = \exp(i\varphi J_z/\hbar) . \tag{1.122}$$

These transformations form an Abelian group.

The invariance of H under rotations around the z axis yields the conservation of the component J_z of the angular momentum along this axis. Consider a set of n subsystems whose state vector is factorized :

$$|\psi\rangle = |\alpha_1, m_1\rangle \otimes |\alpha_2, m_2\rangle \otimes \ldots \otimes |\alpha_n, m_n\rangle . \tag{1.123}$$

The z component of the angular momentum is

$$M = \sum m_i \Rightarrow J_z = \sum_i J_{iz} \tag{1.124}$$

and the rotation operator around the z axis is

$$D(\varphi)|\psi\rangle = e^{iM\varphi}|\psi\rangle \quad . \tag{1.125}$$

Therefore, M appears as an *additive quantum number*. The conservation of $M = \sum m_i$ can also be seen as the invariance of the matrix element $\langle \chi | H_T | \psi \rangle$ under the *phase* transformations :

$$|\psi\rangle \to e^{iM\varphi}|\psi\rangle, \quad |\chi\rangle \to e^{iM'\varphi}|\chi\rangle \quad . \tag{1.126}$$

The group of phase transformations defined in this way is called the unitary group $U(1)$. It is the rotation group in the plane : $x + iy \to e^{i\varphi}(x + iy)$.

This is how one can represent the conservation of any *additive quantum number*. Consider for instance the electric charge Q and a union of subsystems as in (1.123)

$$|\psi\rangle = |\alpha_1, q_1\rangle \otimes |\alpha_2, q_2\rangle \otimes \ldots \otimes |a_n a_q\rangle \tag{1.127}$$

whose total charge is $Q = \sum q_i$. Under the phase transformation $|\psi\rangle \to e^{i\lambda Q}|\psi\rangle$, $|\chi\rangle \to e^{i\lambda Q'}|\chi\rangle$ where λ is an arbitrary real number, the invariance of $\langle \chi | H | \psi \rangle \to e^{i\lambda(Q-Q')}\langle \chi | H | \psi \rangle$ implies the conservation of charge : $Q = Q'$.

This type of invariance is called *gauge invariance* of the first kind. In 1929, Hermann Weyl remarked that the choice $\lambda = \text{constant}$ is non-natural from the point of view of relativity. In fact, space-like separated points are not related, therefore it doesn't have much physical meaning to change the phase of states in these points in the same way. If λ depends on the point under consideration : $\lambda(\boldsymbol{x}, t)$, one deals with a gauge transformation of the second kind. In this latter formulation, gauge invariance has dynamical consequences which underly all present theories of fundamental interactions. At present, the conservation of electric charge appears to be an absolute conservation law. On the other hand, baryon number conservation is expected to be only approximately conserved.

We end this discussion by noting that in a more formal approach to quantum mechanics, one can *start* with invariance principles, next define observables by making use of the infinitesimal generators of various transformations, and finally *deduce* the commutation relations of the observables from the structure of the invariance groups under consideration (Galileo group, Poincaré group, etc.). The proof of the fundamental commutation relation $[x, p_x] = i\hbar$ starting form (1.118) is an example.

In many cases, the Hamiltonian is invariant under other transformation groups than translations or rotations. In all cases, the same property will appear, i.e. a symmetry induces a conservation law. Such properties are very important in the theories of fundamental interactions and, because of this, group theory plays a crucial role in elementary particle physics.

1.7 Charge independence and isospin

We argued in Sect. 1.4 that nuclear forces are rather insensitive to the electric charge. For instance, we have compared the binding energies of ^{3}H and ^{3}He, whose difference can be understood as originating solely from the Coulomb repulsion of the protons. Figure 1.15 shows that this is a systematic effect for mirror nuclei, i.e. pairs of nuclei for which N and Z are interchanged.

Many other observations confirm charge independence. Spectroscopy provides many spectacular examples. For example, the spectra of the mirror nuclei for $A = 11$, 12 and 13, shown on Fig. 1.16, are remarkably similar.

1.7.1 Isospin space

Charge independence is more subtle than just saying that protons and neutrons can be replaced by one another in nuclear forces. It is formalized by using the concept of "isotopic spin" or, better, *isospin T*, that was introduced by Heisenberg in 1932.[6]

A spin 1/2 particle is a two-state system, since a measurement of the projection of its spin along an axis can lead to one of the two results $\pm\hbar/2$. Similarly, the proton and the neutron can be considered as two different *states* $T_3 = \pm1/2$ of a *single* physical object of *isospin $T = 1/2$*, the nucleon. We

---

[6] The translation of Heisenberg's original paper can be found in D.M. Brink, *Nuclear Forces*.

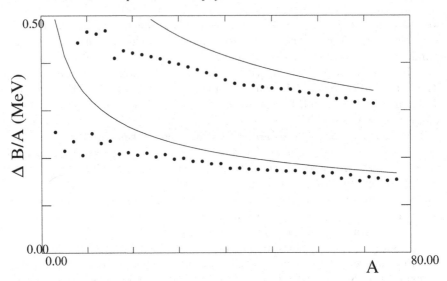

Fig. 1.15. The difference in binding energy per nucleon, B/A, for pairs of "mirror nuclei", i.e. pairs with Z and N interchanged. The lower points are pairs with $|Z - N| = 1$ and the upper points are pairs with $|Z - N| = 2$. The lines show the prediction of the semi-empirical mass formula (2.13) which supposes that neutrons and protons have identical strong interactions and that the difference in binding energy comes only from the Coulomb repulsion of the protons.

therefore introduce an abstract three-dimensional Euclidean-like space called "isospin space." We will choose operators that allow us to perform rotations in this space. The spin formalism in isospin space is the same as that for normal spin in the Euclidean space.

Let $\boldsymbol{T}$ be the vector isospin operator, i.e. a set of three operators $\{T_1, T_2, T_3\}$. These three operators have the commutation relations of a usual angular momentum, up to the $\hbar$ factor :

$$\boldsymbol{T} \wedge \boldsymbol{T} = i\boldsymbol{T} \quad . \tag{1.128}$$

The eigenvalues $T(T + 1)$ and T_3 of the commuting observables T^2 and T_3 are the same as those of usual angular momentum (up to $\hbar$ factors).

Heisenberg's idea is to assume that nuclear strong interactions are rotation invariant in isospin space. The families of "hadrons", i.e. particles which have strong nuclear interactions, can be classified in isospin multiplets ($T = 0, 1/2, 1, 3/2 \ldots$). Different members of each multiplet can be distinguished by their value of T_3 which is linearly related to their charge.

1.7.2 One-particle states

The nucleon has isospin $1/2$. In other words, each of the operators T_1, T_2 and T_3 which are associated with this particle have eigenvalues $\pm 1/2$. The

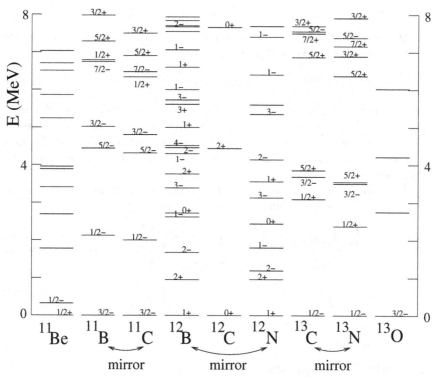

Fig. 1.16. Spectra of the low-lying levels for nuclei with $A = 11$, 12, and 13. The pairs of nuclei with N and Z interchanged (mirror nuclei) have remarkably similar spectra

operator $T^2 = T_1^2 + T_2^2 + T_3^2$ is proportional to the identity with eigenvalue 3/4.

The states $|p\rangle$ and $|n\rangle$, are, by definition, the eigenstates of the particular operator T_3

$$T_3|p\rangle = (1/2)|p\rangle , \quad T_3|n\rangle = (-1/2)|n\rangle \quad . \tag{1.129}$$

In actual physics, the operator T_3 plays a special role since electric charge is related to T_3 by

$$Q = T_3 + 1/2 \quad . \tag{1.130}$$

The action of T_1 and T_2 on these states, with $T_\pm = T_1 \pm T_2$, can be written as

$$T_+|p\rangle = 0 \qquad T_-|n\rangle = 0 \tag{1.131}$$
$$T_1|p\rangle = (1/2)|n\rangle \qquad T_1|n\rangle = (1/2)|p\rangle \tag{1.132}$$
$$T_2|p\rangle = (i/2)|n\rangle \qquad T_2|n\rangle = (-i/2)|p\rangle \quad . \tag{1.133}$$

An arbitrary nucleon state $|N\rangle$ is written

$$|N\rangle = \alpha\,|\mathrm{p}\rangle + \beta\,|\mathrm{n}\rangle \qquad |\alpha|^2 + |\beta|^2 = 1 \quad . \tag{1.134}$$

We remark that all of this is an abstraction applicable only to a world without electromagnetism. A state such as

$$\frac{1}{\sqrt{2}}\,(|T_3 = 1/2\rangle + |T_3 = -1/2\rangle)\,, \tag{1.135}$$

which is oriented along the direction T_2 cannot be observed physically. Since it is a superposition of a proton and a neutron, it is *both* of charge 0 and 1; *at the same time* it creates and doesn't create an electrostatic field. As such, it is a superposition of two macroscopically different states, an example of a "Schrödinger cat."

It is often convenient to use matrix representations for the states and operators. For the single-nucleon space we have

$$|\mathrm{p}\rangle = \begin{pmatrix} 1 \\ 0 \end{pmatrix} \quad,\quad |\mathrm{n}\rangle = \begin{pmatrix} 0 \\ 1 \end{pmatrix} \quad,\quad |\mathrm{N}\rangle = \begin{pmatrix} \alpha \\ \beta \end{pmatrix} \tag{1.136}$$

We use the Pauli matrices $\boldsymbol{\tau} \equiv \{\tau_1, \tau_2, \tau_3\}$,

$$\tau_1 = \begin{pmatrix} 0 & 1 \\ 1 & 0 \end{pmatrix} \qquad \tau_2 = \begin{pmatrix} 0 & -i \\ i & 0 \end{pmatrix} \qquad \tau_3 = \begin{pmatrix} 1 & 0 \\ 0 & -1 \end{pmatrix} \tag{1.137}$$

which satisfy the commutation relations

$$\boldsymbol{\tau} \wedge \boldsymbol{\tau} = 2i\,\boldsymbol{\tau} \quad . \tag{1.138}$$

The nucleon isospin operators are:

$$\boldsymbol{T} = (1/2)\boldsymbol{\tau} \quad . \tag{1.139}$$

Rotation invariance in isospin space amounts to saying that the nucleon–nucleon interaction is invariant under the transformations

$$\begin{aligned}
|\mathrm{p}'\rangle &= \mathrm{e}^{-i\phi/2}\cos(\theta/2)|\mathrm{p}\rangle + \mathrm{e}^{i\phi/2}\sin(\theta/2)|\mathrm{n}\rangle\,, \\
|\mathrm{n}'\rangle &= -\mathrm{e}^{-i\phi/2}\sin(\theta/2)|\mathrm{p}\rangle + \mathrm{e}^{i\phi/2}\cos(\theta/2)|\mathrm{n}\rangle\,.
\end{aligned} \tag{1.140}$$

Besides nucleons, there are hundreds of other strongly interacting particles. These particles, most of which are highly unstable, are called *hadrons*. Hadrons are characterized by their isospin, their electric charge, and other additive more exotic quantum numbers such as "strangeness," "charm," etc, that we will study in Chap. 4. For instance, the π mesons, which mediate some of the nuclear strong forces, have zero baryon number and form a triplet of charge states (π^+, π^0, π^-) whose masses are close. It is natural to consider them as the three states of an isospin triplet, the π meson, of isospin $T = 1$. Following (1.129) we write

$$T_3|\pi^\pm\rangle = \pm\,|\pi^\pm\rangle\,, \qquad T_3|\pi^0\rangle = 0 \quad . \tag{1.141}$$

The relation (1.130) between electric charge and T_3 can be generalized to

$$Q = T_3 + Y/2 \quad , \tag{1.142}$$

where this relation defines the "hypercharge" Y for a multiplet. Clearly we have $Y = 1$ for nucleons and $Y = 0$ for π mesons.

1.7.3 The generalized Pauli principle

The Pauli principle states that two identical fermions must be in an antisymmetric state. If the proton and the neutron were truly identical particles up to the projection of their isospin along the axis T_3, a state of several nucleons should be completely antisymmetric under the exchange of all variables, including isospin variables. If we forget about electromagnetic interactions, and assume exact invariance under rotations in isospin space, the Pauli principle is generalized by stating that an A-nucleon system is completely antisymmetric under the exchange of space, spin and isospin variables. This assumption does not rest on as firm a foundation as the normal Pauli principle and is only an approximation. However, we can expect that it is a good approximation, up to electromagnetic effects.

The generalized Pauli principle restricts the number of allowed quantum states for a system of nucleons. We shall see below how this determines the allowed states of the deuteron.

1.7.4 Two-nucleon system

The isospin states of a two-nucleon system are constructed in the same manner as the states of two spin $1/2$ particles.

The total isospin T of the system corresponds to:

$$T = 1 \quad \text{or} \quad T = 0$$

and the four corresponding eigenstates are :

$$|T = 1, T_3\rangle : \begin{cases} |T = 1, T_3 = 1\rangle = |\text{pp}\rangle \\ |T = 1, T_3 = 0\rangle = (|\text{pn}\rangle + |\text{np}\rangle)/\sqrt{2} \\ |T = 1, T_3 = -1\rangle = |\text{nn}\rangle \end{cases} \quad (1.143)$$

$$|T = 0, T_3 = 0\rangle : \quad |0,\ 0\rangle = (|\text{pn}\rangle - |\text{np}\rangle)/\sqrt{2} \quad . \quad (1.144)$$

We recall that, just as for spin, the three states $|T = 1, M\rangle$ are collectively called the isospin *triplet*. They are *symmetric* under the exchange of the components of the two particles along T_3. The state $|T = 0, 0\rangle$ is called the isospin *singlet* state. It is *antisymmetric* in that exchange. The triplet state transforms as a vector under rotations in isospin space. The singlet state is invariant under those rotations.

A series of simple but important consequences follow from these considerations.

- The Hamiltonian of two (or more) nucleons is invariant under rotations in isospin space so we can expect that the energies of all states in a given multiplet are equal (neglecting electromagnetic effects). In the two-nucleon system, rather than three independent Hamiltonians (i.e. one for p−p, one for p − n and one for n − n), there are only two, one Hamiltonian for the three $T = 1$ states and an independent Hamiltonian for the $T = 0$ state.

- The antisymmetric isospin $T = 0$ state is the state of the deuteron, the only nucleon–nucleon bound state. The deuteron has a symmetric spatial wavefunction. It is mainly an s–wave.[7] Owing to the generalized Pauli principle, it must have a symmetric spin state, i.e. $S = 1$, and a total angular momentum $J = 1$.
- There are no bound symmetric isospin states $(T = 1)$. The interaction is only slightly weaker than in the $T = 0$ state, there exists what is called technically a "virtual state," nearly bound.
- We see with this example that charge independence is more subtle than a simple invariance with respect to interchange of neutrons and protons. Otherwise, we would be sure to observe a neutron–neutron bound state (possibly unstable under β decay) in addition to the deuteron.

Isospin states of a system of A nucleons are constructed in the same way as total spin states of A spin-1/2 particles. If a nucleus has isospin T, we expect to observe $2T + 1$ isobars which have similar physical properties. This is the case for the isobars $^{11}B_5$ and $^{11}C_6$ whose spectra are shown on Fig. 1.16, and which form an isospin 1/2 doublet.

A nucleus (Z, N) has an isospin T at least equal to $|N - Z|/2$. We expect to observe at least $2T + 1 = |N - Z| + 1$ isobars of different charges, but with similar nuclear properties.

The electric charge of a system of A nucleons, of total isospin T is, according to (1.130),

$$Q = T_3 + A/2 .$$

A is (obviously) the baryon number of the system.

1.7.5 Origin of isospin symmetry; n-p mass difference

The near-equality of the proton and neutron masses is a necessary ingredient for isospin symmetry to appear. This symmetry can be understood quite naturally in the context of the quark model where nucleons are states of three quarks. The proton is a (uud) bound state of two u quarks of charge 2/3 and one d quark of charge -1/3. The neutron is a (udd) bound state, with two d quarks and one u quark.

Quarks interact according to the laws of "quantum chromodynamics" or QCD. In this theory, forces are universal in the sense that they make strictly no distinction between types, or *flavors*, of quarks involved. The only difference between the u and d quarks are their masses or charges and we can expect that the proton and neutron masses differ because of the differing quark masses and/or from electromagnetic effects.

It is tempting to suppose that isospin is an exact symmetry of strong nuclear interactions and that electromagnetism is a calculable, and comparatively small, correction. In that framework, it would be natural to assume

---

[7] A small *d*-wave component is necessary in order to explain that the deuteron is not spherical, as mentioned previously.

that, in the absence of electromagnetic forces, the proton and neutron masses should be equal and that their difference originates from calculable electromagnetic effects.

We know experimentally that the proton and neutron are extended objects; as we shall see in Chap. 3, the proton has a radius of the order $r \sim 1\,\mathrm{fm}$. To first approximation, the neutron does not have an electrostatic energy. The electrostatic energy of the proton is of the order of

$$E_{es} \simeq \frac{q_e^2}{4\pi\varepsilon_0\,r} \simeq 1.3\,\mathrm{MeV}\;,\tag{1.145}$$

which is indeed very close to the observed value neutron–proton mass difference, except for the wrong *sign*!

This argument shows that, unfortunately, within our assumptions, the proton should be *heavier* than the neutron. This is a very old problem, and nobody has ever been able to give an answer, except that there must be an additional contribution which reverses the sign of this number.

The only way out, at present, is to shift the problem down to quark masses. Since the proton is a (uud) state and the neutron a (udd) state, we simply assume that the d-quark mass is larger than the u-quark mass. This is completely arbitrary. The origin of mass, i.e. quark masses and more generally all particle masses, is one of the great issues of contemporary physics.

Of course, this *ad hoc* explanation is not at all satisfactory. We need an explanation because the *sign* of the neutron–proton mass difference has tremendous consequences. If the proton were heavier than the neutron, it would be unstable either by β-decay

$$\mathrm{p} \to \mathrm{n}\,\mathrm{e}^+\nu\quad,\tag{1.146}$$

or by electron capture:

$$\mathrm{e}^-\,\mathrm{p} \to \nu\,\mathrm{n}\;.\tag{1.147}$$

From the point of view of chemistry and biology, this could have serious consequences since all existing life relies on the existence of molecules that contain $^1\mathrm{H}$ which could not be stable if $m_\mathrm{p} > m_\mathrm{n}$.

The forms of life we know of would only exist if one could replace $^1\mathrm{H}$ by deuterium $^2\mathrm{H}$ which is chemically quite similar. As we shall see in Chap. 9, most of the $^1\mathrm{H}$ and $^2\mathrm{H}$ in the Universe is a remnant of the nuclear reactions that occurred in the first three minutes of the Universe. Very little deuterium was produced (in the actual situation with $m_\mathrm{p} < m_\mathrm{n}$) and it is difficult to see how much more would have been produced if the mass hierarchy were reversed. To summarize the argument, most of the nuclei produced were $^1\mathrm{H}$ and $^4\mathrm{He}$. Interchanging the proton and neutron masses would amount to replacing the $^1\mathrm{H} - {}^4\mathrm{He}$ mixture by a similar mixture of free stable neutrons and $^4\mathrm{He}$. The $^2\mathrm{H}$ necessary for life would have to be produced later (for instance during the formation of the solar system) via neutron fusion $nn \to$

^{2}H $e^-\bar{\nu}$. Compared to the enormous supply of ^{1}H kindly supplied by the big bang in the case of $m_n > m_p$, this source of hydrogen seems problematic.

If sufficient hydrogen could not be produced, it is not clear whether another form of chemistry could exist and be rich enough to generate life. At the very least, it is probable that the absence of hydrogen would considerably increase the time scale necessary in order to build living systems.

1.8 Deformed nuclei

In our discussion of nuclear radii, we implicitly assumed that nuclei have a spherical shape. This is a good approximation for nuclei that have "magic numbers" of neutrons or protons: 8, 20, 28, 50, 82, or 126. We will see in Sect. 2.4, that these numbers come from the shell structure of the nucleus that is analogous to the shell structure of atomic electrons. Nuclei with magic numbers of neutrons or protons have a "closed shell" that encourages a spherical shape.

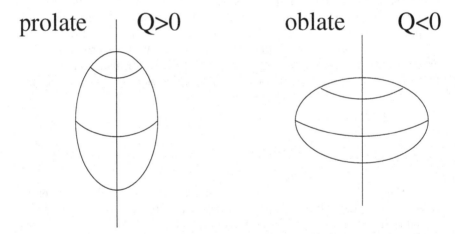

Fig. 1.17. Two charge distributions, one with a positive quadrupole moment $Q > 0$ (prolate) and one with a negative moment $Q < 0$ (oblate). The vertical lines shown the axis of cylindrical symmetry. Nuclear states with angular momentum quantum numbers (j, m) will have the symmetry axis oriented randomly according to the appropriate spherical harmonic, $Y_{lm}(\theta, \phi)$ with $l = j$. For example, a $j = 0$ deformed nucleus has a randomly oriented symmetry axis.

Nuclei with Z or N far from a magic number are generally deformed. The simplest deformations are so-called *quadrupole* deformations where the nucleus can take either an prolate shape (rugby ball) or oblate shape (cushion), as illustrated in Fig. 1.17. A quadrupole deformation retains one symmetry axis (z axis), and the electric quadrupole is defined by

$$Q = \frac{1}{V} \int d^3r (3z^2 - r^2) \rho(\boldsymbol{r}) , \tag{1.148}$$

where $\rho(\boldsymbol{r})$ is the charge density and V is the volume of the nucleus. Spherically symmetric distribution have $Q = 0$ while rugby balls have $Q > 0$ and cushions have $Q < 0$. For an ellipsoid of uniform density, the relative difference between the lengths of the major and minor axes, β, is simply related to the the quadrupole moment:

$$Q \sim \frac{6}{\sqrt{5\pi}} ZeR^2 \beta \quad \beta \ll 1 . \tag{1.149}$$

As shown in Fig. 1.8, deformations of order 10% are common in the ground states of nuclei far from magic numbers. Excited "super-deformed bands" with higher deformations are also observed. They are currently an important topic of research.

It is important to distinguish between the intrinsic deformation of a classical charge distribution from that of a nucleus in a state of definite angular momentum quantum numbers, (j, m). In quantum mechanics, the symmetry axis of a quadrupole cannot be taken to point in a fixed direction. Rather it has an amplitude to point in any direction that is determined by the appropriate spherical harmonic $Y_{lm}(\theta, \phi)$ with $j = l$. For example, an intrinsically deformed nucleus with $j = 0$ has its symmetry axis pointed in a random direction, since $Y_{00} = 1/\sqrt{4\pi}$. Thus, this quantum state has a spherically symmetric charge distribution in spite of the fact that the nucleus is deformed. This is analogous to the case of the ground state atomic hydrogen where the vector pointing from the proton to the electron is randomly oriented.

Nuclear deformation has several physical manifestations that are mostly related to the non-spherical distribution of electric charge.

- Rapid transitions between nuclear rotation levels of energy (1.40). As we will discuss more thoroughly in Section 4.2, this can be understood classically since a spinning spherically symmetric charge distribution does not create a classical radiation field whereas a spinning asymmetric distribution does. Thus, bands of rotation states in deformed nuclei like ^{242}Pu in Fig. 1.6 exhibit rapid cascades where an excited state decays via a series of $(J + 2) \to J$ decays down to the ground $(j = 0)$ state. This results in a "picket fence" spectrum of evenly spaced lines, as shown in Fig. 1.9 for ^{152}Dy. More complicated deformations, octopole, hexapole, and so on, result in more complicated spectra [25].
- The elimination of diffraction minima in electron–nucleus scattering, as discussed in Sect. 3.4.4. Diffraction minima that are pronounced in spherical nuclei are washed out for deformed nuclei because the orientation of the nucleus must be averaged over, leading to a "fuzzier" nucleus with no well-defined surface.
- The hyperfine splitting of the energy levels of atoms and molecules (Exercise 1.12). As illustrated in Fig. 1.17, this comes about if the nucleus

is surrounded by a non-spherically symmetric electron cloud. This effect adds to the normal hyperfine splitting due to the interaction of the nuclear magnetic moment with the magnetic field created by atomic electrons. It should be noted that the quadrupole effect allows one to determine the sign of the quadrupole moment. It turns out that the most deformed nuclei have $Q > 0$ (prolate deformation).

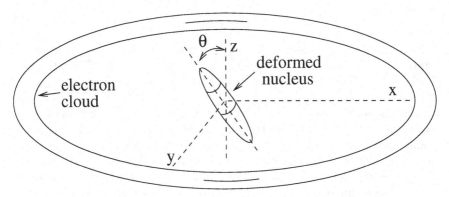

Fig. 1.18. The energy of a non-spherical nucleus surrounded by a non-spherical distribution of electrons depends on the orientation of the nuclear symmetry axis with respect to the electron cloud. The figure shows a prolate nucleus at the center of a donut-shaped electron cloud in the xy-plane. The nuclear symmetry axis makes an angle θ with respect to the z axis. Orientations with $\theta \sim 0$ have a larger electrostatic energy than those with $\theta \sim \pi/2$ since the mean nucleon-electron distance is greatest for $\theta = 0$. Quantum-mechanically, the symmetry axis cannot be taken to point in a fixed direction. For a rigid object of definite J^2 and J_z, the amplitude for the symmetry axis to make and angle θ with the z axis is given by the appropriate spherical harmonic, $Y_{lm}(\theta, \phi)$ where $l = J$ and $m = J_z$. The spherical harmonics with $m = l$ are maximized at $\theta = \pi/2$ while those with $m = 0$ are maximized at $\theta = 0$. We can therefore expect that, in general, the electrostatic energy of a prolate nucleus is lowest for the largest $|J_z|$ states. The magnitude of the effect is estimated in Exercise 1.12. Note that there can be no quadrupole hyperfine splitting for nuclei with $J = 0$ or $J = 1/2$ since in these two cases there is only one possible value of $|J_z|$.

Muonic atoms

The small hyperfine effects due to nuclear deformation become dominant effects in *muonic atoms*. The muon μ, which was discovered in 1937 and whose existence is still somewhat of a mystery, is basically a heavy electron. It is elementary, or point-like in the same way as the electron. It has the same electric charge and the same spin, but it is 200 times heavier, $m_\mu = 206.8 m_e$. It is unstable, decaying into an electron and two neutrinos:

$$\mu^- \to e^- \bar{\nu}_e \nu_\mu \qquad\qquad (1.150)$$

with a lifetime of $\tau = 2 \times 10^{-6}$s. The muon-neutrino ν_μ is associated with the muon in the same way that the electron neutrino ν_e is associated with the electron.

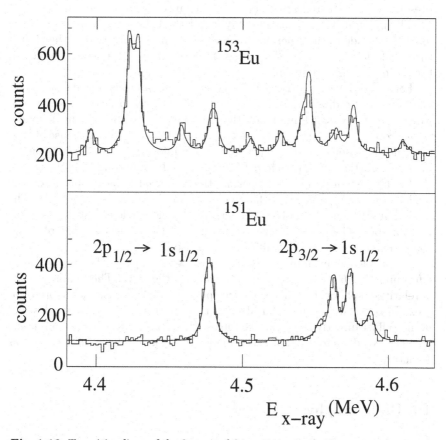

Fig. 1.19. Transition lines of the $2p_{1/2}$ and $2p_{3/2}$ states to the $2s_{1/2}$ state in muonic europium ($Z = 63$) [26]. As detailed in Exercise 1.14, the energy of all transition photons are near the expected value $\sim (3/8)(Z\alpha)^2 m_\mu c^2$. In the ^{151}Eu spectrum, one sees the expected fine-structure splitting between the $2p_{1/2}$ and $2p_{3/2}$ states of $\sim (Z\alpha)^4 m_\mu c^2$ and the hyperfine splitting of the $2p_{3/2}$ level due mostly to the slight deformation of the ^{151}Eu nucleus. On the other hand, the ^{153}Eu nucleus is strongly deformed, resulting in a very complicated spectrum. The complication is due to the fact that the eigenstates of the muonic atom are are mixtures of the ground state of the ^{153}Eu nucleus and its excited states.

In particle accelerators, one produces muons, one can slow then down in matter and have them captured by atoms, inside which they form hydrogen-like atoms.

In a multi-electron atom, the muon is not constrained by the Pauli principle. The μ thus cascades down to the lowest energy orbitals where it is in the direct vicinity of the nucleus at a distance $a_\mu \sim \hbar^2/Z m_\mu e^2$, 200 times smaller than the corresponding Bohr radius of internal electrons. It therefore forms a hydrogen-like atom of charge Z around the nucleus, oblivious to the presence of the other electrons at larger distances from the nucleus.

The muon lifetime is considerably larger than the total time $\approx 10^{-14}$ sec for it to cascade to the inner orbitals and even larger compared to the atomic time scale $\hbar^3/m_\mu e^4 \approx 10^{-19}$ s (the orbital period of the muon). The μ can therefore be considered stable.

For heavy nuclei, the Bohr radius of a muonic atom is of the same order as the nuclear radius. In lead, for instance, $Z = 82$, of radius $R \approx 8.5\,\text{fm}$, the Bohr radius is $a_\mu \approx 3.1\,\text{fm}$. The μ therefore penetrates the nucleus, having a 90 % probability to be inside the nucleus in the ground state. Because of this, the study of muonic atom spectra gives useful information on the structure of nuclei, in particular on the charge (i.e. proton) distribution inside the nuclei.

In the case of a spherical nucleus, the potential is harmonic inside the nucleus (assuming a uniform charge density) and Coulomb-like outside. The deviation of the position of levels compared to the Coulomb case gives information about the radial charge distribution. If the nucleus is non-spherical, the states of the same l but different m are split.

A spectacular example is given by the two isotopes ^{151}Eu and ^{153}Eu of europium $Z = 63$, whose spectra are given on Fig. 1.19. The lighter isotope is relatively spherical. Conversely, the spectrum of ^{153}Eu is much more complex. In other words, if ^{151}Eu absorbs two neutrons (neutral particles and do not affect directly the Coulomb forces) the proton distribution completely changes. Needless to say that this provides very useful information on nuclear structure.

1.9 Bibliography

1. *The Nuclear Many-Body Problem*, P. Ring and P. Schuck, Springer Verlag (1980).
2. *Nuclear Forces* D.M. Brink, Pergamon Press, Oxford (1965).

Exercises

1.1 Compare the mass of 1 mm^3 of nuclear matter and the mass of the Earth ($\sim 6 \times 10^{24}$ kg).

1.2 A commonly used quantity is the *mass excess* defined as

$$\Delta \equiv m(A, Z) - A \times 1\,\mathrm{u}\,. \tag{1.151}$$

Derive an expression for Δ in terms of the nuclear binding energy $B(A, Z)$.

1.3 Figure 1.2 shows that ^{3}H has more tightly bound than ^{3}He. Why is it, then, that ^{3}H β-decays to ^{3}He?

1.4 Calculate the recoil energy of ^{208}Pb in the c.m. decay ^{212}Po $\rightarrow$ ^{208}Pb $+ \alpha$. (The masses of the nuclei can be calculated from the data in Appendix G.)

1.5 Consider the reaction $\gamma + {}^{12}\mathrm{C} \rightarrow 3\,{}^4\mathrm{He}$. What is the threshold energy of the reaction? If two α-particles have the same momentum in the c.m. system, what fraction of the energy is carried by the third particle ?

1.6 The $A = 40$ isotopes of calcium ($Z = 20$), potassium ($Z = 19$) and argon ($Z = 18$) have respective binding energies $-332.65\,\mathrm{MeV}$, $-332.11\,\mathrm{MeV}$ and $-335.44\,\mathrm{MeV}$. What β decays are allowed between these nuclei? Specify the available energy Q in the final state. What peculiarity appears?

1.7 In Sect. 1.2.3 we showed how the proton–deuteron mass ratio can be determined accurately from the ratio of the masses the ionized deuterium atom and a singly ionized hydrogen molecule. To place these two nuclei on the atomic-mass-unit scale, we need the mass ratio of the deuteron and the non-ionized ^{12}C atom ($\equiv 12\,\mathrm{u}$). This ratio can be accurately determined from the ratio of the masses of the singly-ionized ^{2}H$_3$ molecule and the doubly-ionized ^{12}C atom. To do this, show that

$$\frac{3m_{\mathrm{d}}}{12\,\mathrm{u}} = \left[\frac{3m_{\mathrm{d}} + 2m_{\mathrm{e}}}{m(12, 6) + 4m_{\mathrm{e}}}\right]\left[\frac{1 - m_{\mathrm{e}}/6u}{1 + 2m_{\mathrm{e}}/(3m_{\mathrm{d}})}\right], \tag{1.152}$$

where the first bracketed term on the right is the aforementioned ratio. The second bracketed factor differs from unity only by small factors depending on the ratio of electron and nuclear masses and are therefore second-order corrections that need not be known as accurately as the first term.

1.8 Verify (1.21).

1.9 Consider a quantum system of A pairwise interacting fermions with two-body attractive interactions of the form $V(r) = -g^2/r$. Using the uncertainty relation [8] $\langle p^2 \rangle \geq A^{2/3}\hbar^2\langle 1/r \rangle^2$ show that $< E > /A \simeq -A^{4/3} \cdot g^4\,\mathrm{m}/8\hbar^2$, and

---

[8] See J.-L. Basdevant and J. Dalibard, <u>Quantum mechanics</u>, chapter 16, Springer-Verlag, 2002.

$r \simeq 2\hbar^2 A^{-1/3}/mg^2$. This is a usual cumulative effect for attractive forces: energies per particle increase and radii decrease as the number of particles increases. The powers $4/3$ and $-1/3$ are specific to the Coulomb-type interactions but one can verify that, for a harmonic potential, one obtains $|E/A| \sim A^{5/6}$ and for a spherical well, $|E/A| \sim A^{2/3}$. This shows that the Pauli principle alone cannot lead to saturation of nuclear forces.

1.10 Write the momentum- and energy-conservation equations for reaction (1.102). Show that if the initial neutron and nucleus are at rest, the final-state photons takes nearly all the released energy, Q.

1.11 Write the momentum- and energy-conservation equations for reaction (1.103). Show that in the limit where the initial particles are at rest, the neutron takes most of the released energy, $Q = 17.58\,\text{MeV}$.

1.12 The energies associated with the magnetic-dipole and electric-quadrupole moments of nuclei contribute to the *hyperfine structure* of atoms and molecules. The purpose of this exercise is to estimate the size of these energies compared to atomic binding energies $\sim \alpha^2 m_e c^2$.

Nuclei interact with a magnetic field via the Hamiltonian

$$H = -\boldsymbol{B} \cdot \boldsymbol{\mu} \tag{1.153}$$

where $\boldsymbol{\mu}$ is the nuclear magnetic moment

$$\boldsymbol{\mu} = g_A \frac{Ze\,\boldsymbol{J}}{2m_A}, \tag{1.154}$$

where $\boldsymbol{J}$ is the nuclear spin and and g_A is of order unity. The energies of the $(2j+1)$ states are

$$E_m = \frac{g_A\,Ze|\boldsymbol{B}|}{2m_A}\,m\hbar \quad m = -j, -(j-1).....(j-1), j\,. \tag{1.155}$$

Consider a nucleus surrounded by a non-spherically symmetric electron cloud as in Fig. 1.18. Atomic electrons move at velocities $\sim \alpha c$ with respect to the nucleus. Argue that the magnetic field seen by the nucleus is of order

$$|\boldsymbol{B}| \sim \frac{\mu_0\,\alpha c e}{a_0^2} \sim \frac{\alpha^2 \hbar}{e a_0^2}, \tag{1.156}$$

where $a_0 = \hbar/(\alpha m_e c)$ is the Bohr radius giving the typical nuclear–electron distances. Show that this leads to a splitting of nuclear levels of order

$$\Delta E \sim g_A \alpha^4 \frac{m_e}{m_p} m_e c^2\,. \tag{1.157}$$

This splitting is a factor $\alpha^2(m_e/m_p)$ smaller than the binding energy and a factor m_e/m_p smaller than the fine structure (spin orbit) splittings.

The energy associated with the nuclear electric-quadrupole moment can be estimated by calculating the electrostatic energy of the configuration in Fig. 1.18 as a function of the angle θ. ($\theta \sim 0$ corresponds to $|J_z| \sim 0$ and $\theta \sim \pi/2$ corresponds to $|J_z| \sim J$.) Argue that energy difference between $\theta = 0$ and $\theta = \pi/2$ for a highly deformed nucleus is of order

$$\Delta E \sim Z^2 \alpha^2 m_e c^2 \frac{R^2}{a_0^2} \tag{1.158}$$

where R is the length of the longest nuclear dimension. Compare ΔE with that due the magnetic moment (1.157). What energy splitting would you expect for a slightly deformed nucleus? How would the splitting be different for an oblate nucleus?

1.13   Consider a proton-rich material that is placed in a magnetic field of $|\boldsymbol{B}| = 1\,\mathrm{T}$. What is the energy difference between a proton with its spin aligned with $\boldsymbol{B}$ and one with its spin anti-aligned? Supposing thermal equilibrium at $kT = 300\,\mathrm{K}$, what is the relative number of proton spins aligned and anti-aligned with the magnetic field. What frequency oscillating field can induce transitions from the more populated state to the less populated one?

The absorption of energy by such a field tuned to the correct frequency is called *nuclear magnetic resonance* (NMR). It is the basis of *magnetic resonance imaging* (MRI) where the number of protons in a sample is deduced from the absorbed power. In medical applications, why is this technique mostly useful for deducing the amount of hydrogen and not other elements?

Consider two protons in a molecule separated by 10^{-10}m. Compare the external magnetic field of 1 T with the magnetic field seen by a proton due to the spin of its neighbor. The extra magnetic field shifts the resonant frequency making it sensitive to the molecular structure of the sample.

1.14 Estimate the transition energies of the ^{151}Eu atom shown in Fig. 1.19 by supposing that the muon forms a hydrogen-like atom.

1.15 The osmium mass spectrum on the right of Fig. 1.3 is due to osmium extracted from a mineral containing 0.32% rhenium and 0.00161% osmium [10]. Using the half-life and isotopic abundance of ^{187}Re from Table 5.2 derive the age of the mineral. Discuss what is meant by this "age".

1.16 The data of Fig. 1.4 gave the first precise measurement of the atomic mass of ^{48}Mn. Taking the binding energies of ^{46}Cr and ^{50}Fe as known (from Appendix G), use the data to estimate the binding energy of ^{48}Mn. What precision can be obtained with this method?

1.17 The purpose of this exercise is to estimate the energy spectrum of the protons recoiling from collisions with ^{64}Ni nuclei (Fig. 1.7). Since the proton is much lighter than ^{64}Ni, we can anticipate that most of the kinetic energy will be taken by the proton. Under this assumption, use energy conservation to show that the energy of the final state proton is $E'_\mathrm{p} = E_\mathrm{p} - \Delta E$ where ΔE is the energy of the produced excited state of ^{64}Ni relative to the ground state. Use momentum conservation to estimate the momentum and kinetic energy of the recoiling ^{64}Ni for protons recoiling at $60°$ from the beam direction. Finally, re-estimate the proton energy by imposing energy conservation taking into account the recoil of the ^{64}Ni.

2. Nuclear models and stability

The aim of this chapter is to understand how certain combinations of N neutrons and Z protons form bound states and to understand the masses, spins and parities of those states. The known (N, Z) combinations are shown in Fig. 2.1. The great majority of nuclear species contain excess neutrons or protons and are therefore β-unstable. Many heavy nuclei decay by α-particle emission or by other forms of spontaneous fission into lighter elements. Another aim of this chapter is to understand why certain nuclei are stable against these decays and what determines the dominant decay modes of unstable nuclei. Finally, forbidden combinations of (N, Z) are those outside the lines in Fig. 2.1 marked "last proton/neutron unbound." Such nuclei rapidly (within $\sim 10^{-20}$s) shed neutrons or protons until they reach a bound configuration.

The problem of calculating the energies, spins and parities of nuclei is one of the most difficult problems of theoretical physics. To the extent that nuclei can be considered as bound states of nucleons (rather than of quarks and gluons), one can start with empirically established two-nucleon potentials (Fig. 1.12) and then, in principle, calculate the eigenstates and energies of many nucleon systems. In practice, the problem is intractable because the number of nucleons in a nucleus with $A > 3$ is much too large to perform a direct calculation but is too small to use the techniques of statistical mechanics. We also note that it is sometimes suggested that intrinsic *three-body* forces are necessary to explain the details of nuclear binding.

However, if we put together all the empirical information we have learned, it is possible to construct efficient *phenomenological models* for nuclear structure. This chapter provides an introduction to the characteristics and physical content to the simplest models. This will lead us to a fairly good explanation of nuclear binding energies and to a general view of the stability of nuclear structures.

Much can be understood about nuclei by supposing that, inside the nucleus, individual nucleons move in a potential well defined by the mean interaction with the other nucleons. We therefore start in Sect. 2.1 with a brief discussion of the *mean potential model* and derive some important conclusions about the relative binding energies of different isobars. To complement the mean potential model, in Sect. 2.2 we will introduce the liquid-drop model that treats the nucleus as a semi-classical liquid object. When combined with

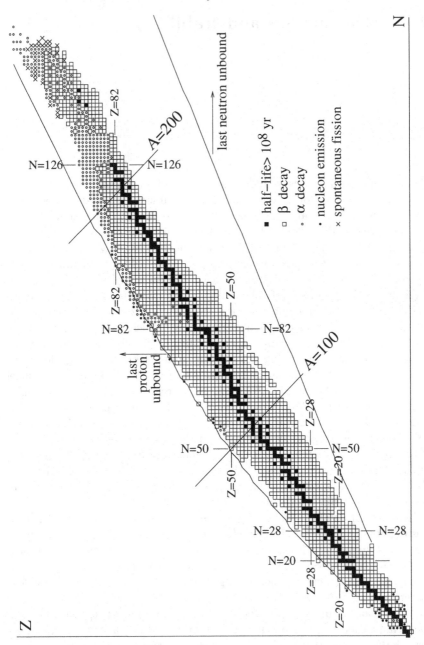

Fig. 2.1. The nuclei. The black squares are long-lived nuclei present on Earth. Combinations of (N, Z) that lie outside the lines marked "last proton/neutron unbound" are predicted to be unbound by the semi-empirical mass formula (2.13). Most other nuclei β-decay or α-decay to long-lived nuclei.

certain conclusions based on the mean potential model, this will allow us to derive Bethe and Weizsäcker's semi-empirical mass formula that gives the binding energy as a function of the neutron number N and proton number Z. In Sect. 2.3 we will come back to the mean potential model in the form of the Fermi-gas model. This model will allow us to calculate some of the parameters in the Bethe–Weizsäcker formula.

In Sect. 2.4 we will further modify the mean-field theory so as to explain the observed nuclear shell structure that lead to certain nuclei with "magic numbers" of neutrons or protons to have especially large binding energies.

Armed with our understanding of nuclear binding, in Sections 2.5 and 2.6 we will identify those nuclei that are observed to be radioactive either via β-decay or α-decay. Finally, in Sect. 2.8, we will discuss attempts to synthesize new metastable nuclei.

2.1 Mean potential model

The *mean potential model* relies on the observation that, to good approximation, individual nucleons behave inside the nucleus as *independent* particles placed in a mean potential (or mean field) due to the other nucleons.

In order to obtain a qualitative description of this mean potential $V(r)$, we write it as the sum of potentials $v(r - r')$ between a nucleon at r and a nucleon at r':

$$V(r) = \int v(r - r')\rho(r')dr' \quad .$$ (2.1)

In this equation, the nuclear density $\rho(r')$, is proportional to the probability per unit volume to find a nucleus in the vicinity of r'. It is precisely that function which we represented on Fig. 1.1 in the case of protons.

We now recall what we know about v and ρ. The strong nuclear interaction $v(r - r')$ is attractive and short range. It falls to zero rapidly at distances larger than $\sim 2\,\text{fm}$, while the typical diameter on a nucleus is "much" bigger, of the order of 6 fm for a light nucleus such as oxygen and of 14 fm for lead. In order to simplify the expression, let us approximate the potential v by a delta function (i.e. a point-like interaction)

$$v(r - r') \sim -v_0\delta(r - r') \quad .$$ (2.2)

The constant v_0 can be taken as a free parameter but we would expect that the integral of this potential be the same as that of the original two-nucleon potential (Table 3.3):

$$v_0 = \int \mathrm{d}^3r v(r) \sim 200\,\text{MeV fm}^3 \; ,$$ (2.3)

where we have used the values from Table 3.3. The mean potential is then simply

$$V(r) = -v_0\rho(r) \, . \tag{2.4}$$

Using $\rho \sim 0.15\,\mathrm{fm}^{-3}$ we expect to find a potential depth of roughly

$$V(r < R) \sim -30\,\mathrm{MeV} \, , \tag{2.5}$$

where R is the radius of the nucleus.

The shapes of charge densities in Fig. 1.1 suggest that in first approximation the mean potential has the shape shown in Fig. 2.2a. A much-used analytic expression is the Saxon–Woods potential

$$V(r) = -\frac{V_0}{1 + \exp(r - R)/R} \tag{2.6}$$

where V_0 is a potential depth of the order of 30 to 60 MeV and R is the radius of the nucleus $R \sim 1.2A^{1/3}$ fm. An even simpler potential which leads to qualitatively similar results is the harmonic oscillator potential drawn on Fig. 2.2b:

$$V(r) = -V_0[1 - \left(\frac{r}{R}\right)^2] = -V_0 + \frac{1}{2}M\omega^2 r^2 \qquad r < R \tag{2.7}$$

with $V_0 = \frac{1}{2}M\omega^2 R^2$, and $V(r > R) = 0$. Contrary to what one could believe from Fig. 2.2, the low-lying wave functions of the two potential wells (a) and (b) are very similar. Quantitatively, their scalar products are of the order of 0.9999 for the ground state and 0.9995 for the first few excited states for an appropriate choice of the parameter ω in b. The first few energy levels of the potentials a and b hardly differ.

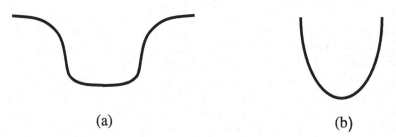

(a)   (b)

Fig. 2.2. The mean potential and its approximation by a harmonic potential.

In this model, where the nucleons can move independently from one another, and where the protons and the neutrons separately obey the Pauli principle, the energy levels and configurations are obtained in an analogous way to that for complex atoms in the Hartree approximation. As for the electrons in such atoms, the proton and neutron orbitals are independent fermion levels.

It is instructive, for instance, to consider, within the mean potential notion, the stability of various $A = 7$ nuclei, schematically drawn on Fig. 2.3. The figure reminds us that, because of the Pauli principle, nuclei with a large

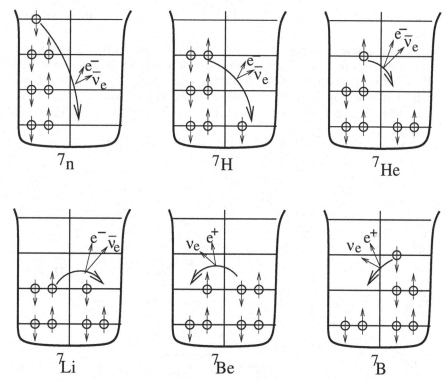

Fig. 2.3. Occupation of the lowest lying levels in the mean potential for various isobars $A = 7$. The level spacings are schematic and do not have realistic positions. The proton orbitals are shown at the same level as the neutron orbitals whereas in reality the electrostatic repulsion raises the protons with respect to the neutrons. The curved arrows show possible neutron–proton and proton–neutron transitions. If energetically possible, a neutron can transform to a proton by emitting a $e^-\bar{\nu}_e$ pair. If energetically possible, a proton can transform to a neutron by emitting a $e^+\nu_e$ pair or by absorbing an atomic e^- and emitting a ν_e. As explained in the text, which of these decays is actually energetically possible depends on the relative alignment of the neutron and proton orbitals.

excess of neutrons over protons or vice versa require placing the nucleons in high-energy levels. This suggests that the lowest energy configuration will be the ones with nearly equal numbers of protons and neutrons, ^{7}Li or ^{7}Be. We expect that the other configurations can β-decay to one of these two nuclei by transforming neutrons to protons or vice versa. The observed masses of the $A = 7$ nuclei, shown in Fig. 2.4, confirm this basic picture:

- The nucleus ^{7}Li is the most bound of all. It is stable, and more strongly bound than its mirror nucleus ^{7}Be which suffers from the larger Coulomb repulsion between the 4 protons. In this nucleus, the actual energy levels of the protons are increased by the Coulomb interaction. The physical properties of these two nuclei which form an isospin doublet, are very similar.

- The mirror nuclei ^{7}B and ^{7}He can β-decay, respectively, to ^{7}Be and ^{7}Li. In fact, the excess protons or neutrons are placed in levels that are so high that neutron emission is possible for ^{7}He and 3-proton emission for ^{7}B and these are the dominant decay modes. When nucleon emission is possible, the lifetime is generally very short, $\tau \sim 10^{-22}$s for ^{7}B and $\sim 10^{-21}$s for ^{7}He.
- No bound states of ^{7}n, ^{7}H, ^{7}C or ^{7}N have been observed.

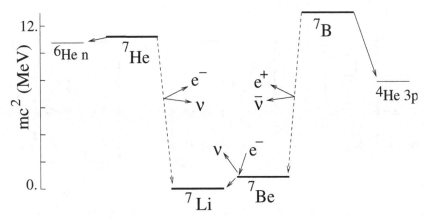

Fig. 2.4. Energies of the $A = 7$ isobars. Also shown are two unbound $A = 7$ states, ^{6}He n and ^{4}He 3p.

This picture of a nucleus formed with independent nuclei in a mean potential allows us to understand several aspects of nuclear phenomenology.

- For a given A, the minimum energy will be attained for optimum numbers of protons and neutrons. If protons were not charged, their levels would be the same as those of neutrons and the optimum would correspond to $N = Z$ (or $Z \pm 1$ for odd A). This is the case for light nuclei, but as A increases, the proton levels are increased compared to the neutron levels owing to Coulomb repulsion, and the optimum combination has $N > Z$. For mirror nuclei, those related by exchanging N and Z, the Coulomb repulsion makes the nucleus $N > Z$ more strongly bound than the nucleus $Z > N$.
- The binding energies are stronger when nucleons can be grouped into pairs of neutrons and pairs of protons with opposite spin. Since the nucleon–nucleon force is attractive, the energy is lowered if nucleons are placed near each other but, according to the Pauli principle, this is possible only if they have opposite spins. There are several manifestations of this pairing

effect. Among the 160 even-A, β-stable nuclei, only the four light nuclei, ^{2}H, ^{6}Li, ^{10}B, ^{14}N, are "odd-odd", the others being all "even-even." [1]

- The Pauli principle explains why *neutrons can be stable in nuclei while free neutrons are unstable*. Possible β-decays of neutrons in ^{7}n, ^{7}H, ^{7}He and ^{7}Li are indicated by the arrows in Fig. 2.3. In order for a neutron to transform into a proton by β-decay, the final proton must find an energy level such that the process n $\rightarrow$ pe$^-\bar{v}_e$ is energetically possible. If all lower levels are occupied, that may be impossible. This is the case for ^{7}Li because the Coulomb interaction raises the proton levels by slightly more than $(m_n - m_p - m_e)c^2 = 0.78\,\text{MeV}$. Neutrons can therefore be "stabilized " by the Pauli principle.

- Conversely, in a nucleus a proton can be "destabilized" if the reaction p $\rightarrow$ n+e^+v_e can occur. This is possible if the proton orbitals are raised, via the Coulomb interaction, by more than $(m_n + m_e - m_p)c^2 = 1.80\,\text{MeV}$ with respect to the neutron orbitals. In the case of ^{7}Li and ^{7}Be shown in Fig. 2.4, the proton levels are raised by an amount between $(m_n + m_e - m_p)c^2$ and $(m_n - m_e - m_p)c^2$ so that neither nucleus can β-decay. (The *atom* ^{7}Be is unstable because of the electron-capture reaction of an internal electron of the atomic cloud ^{7}Be e$^-$ $\rightarrow^7$ Li v_e.)

We now come back to (2.7) to determine what value should be assigned to the parameter ω so as to reproduce the observed characteristics of nuclei. Equating the two forms in this equation we find

$$\omega(A) = \left(\frac{2V_0}{M}\right)^{1/2} R^{-1}. \tag{2.8}$$

Equation (2.5) suggests that V_0 is independent of A while empirically we know that R is proportional to $A^{1/3}$. Equation (2.8) then tells us that ω is proportional to $A^{-1/3}$. To get the phenomenologically correct value, we take $V_0 = 20\,\text{MeV}$ and $R = 1.12A^{1/3}$ which yields

$$\hbar\omega = \left(\frac{2V_0}{m_p c^2}\right)^{1/2} \frac{\hbar c}{R} \sim 35\,\text{MeV} \times A^{-1/3}. \tag{2.9}$$

We can now calculate the binding energy $B(A = 2N = 2Z)$ in this model. The levels of the three-dimensional harmonic oscillator are $E_n = (n+3/2)\hbar\omega$ with a degeneracy $g_n = (n+1)(n+2)/2$. The levels are filled up to $n = n_{\text{max}}$ such that

$$A = 4 \sum_{n=0}^{n_{\text{max}}} g_n \sim 2n_{\text{max}}^3/3 \tag{2.10}$$

i.e. $n_{\text{max}} \sim (3A/2)^{1/3}$. (This holds for A large; one can work out a simple but clumsy interpolating expression valid for all A's.)

The corresponding energy is

---

[1] A fifth, ^{180m}Ta has a half-life of 10^{15}yr and can be considered effectively stable.

$$E = -AV_0 + 4 \sum_{n=0}^{n_{max}} g_n(n + 3/2)\hbar\omega \sim -AV_0 + \frac{\hbar\omega n_{max}^4}{2} . \qquad (2.11)$$

Using the expressions for $\hbar\omega$ and n_{max} we find

$$\sim -8\,\text{MeV} \times A \qquad (2.12)$$

i.e. the canonical binding energy of 8 MeV per nucleon.

2.2 The Liquid-Drop Model

One of the first nuclear models, proposed in 1935 by Bohr, is based on the short range of nuclear forces, together with the additivity of volumes and of binding energies. It is called the *liquid-drop model.*

Nucleons interact strongly with their nearest neighbors, just as molecules do in a drop of water. Therefore, one can attempt to describe their properties by the corresponding quantities, i.e. the radius, the density, the surface tension and the volume energy.

2.2.1 The Bethe–Weizsäcker mass formula

An excellent parametrization of the binding energies of nuclei in their ground state was proposed in 1935 by Bethe and Weizsäcker. This formula relies on the liquid-drop analogy but also incorporates two quantum ingredients we mentioned in the previous section. One is an *asymmetry* energy which tends to favor equal numbers of protons and neutrons. The other is a *pairing* energy which favors configurations where two identical fermions are paired.

The mass formula of Bethe and Weizsäcker is

$$B(A, Z) = a_v A - a_s A^{2/3} - a_c \frac{Z^2}{A^{1/3}} - a_a \frac{(N - Z)^2}{A} + \delta(A) . \quad (2.13)$$

The coefficients a_i are chosen so as to give a good approximation to the observed binding energies. A good combination is the following:

$$a_v = 15.753 \text{ MeV}$$

$$a_s = 17.804 \text{ MeV}$$

$$a_c = 0.7103 \text{ MeV}$$

$$a_a = 23.69 \text{ MeV}$$

and

$$\delta(A) = \begin{cases} 33.6A^{-3/4} & \text{if } N \text{ and } Z \text{ are even} \\ -33.6A^{-3/4} & \text{if } N \text{and } Z \text{ are odd} \\ 0 & \text{si } A = N + Z \text{ is odd} \end{cases} .$$

The numerical values of the parameters must be determined empirically (other than a_c), but the A and Z dependence of each term reflects simple physical properties.

- The first term is a *volume* term which reflects the nearest-neighbor interactions, and which by itself would lead to a constant binding energy per nucleon $B/A \sim 16$ MeV.
- The term a_s, which lowers the binding energy, is a *surface* term. Internal nucleons feel isotropic interactions whereas nucleons near the surface of the nucleus feel forces coming only from the inside. Therefore this is a *surface tension* term, proportional to the area $4\pi R^2 \sim A^{2/3}$.
- The term a_c is the *Coulomb repulsion* term of protons, proportional to Q^2/R, i.e. $\sim Z^2/A^{1/3}$. This term is calculable. It is smaller than the nuclear terms for small values of Z. It favors a neutron excess over protons.
- Conversely, the *asymmetry* term a_a favors symmetry between protons and neutrons (isospin). In the absence of electric forces, $Z = N$ is energetically favorable.
- Finally, the term $\delta(A)$ is a quantum *pairing* term.

The existence of the Coulomb term and the asymmetry term means that for each A there is a nucleus of maximum binding energy found by setting $\partial B/\partial Z = 0$. As we will see below, the maximally bound nucleus has $Z = N = A/2$ for low A where the asymmetry term dominates but the Coulomb term favors $N > Z$ for large A.

The predicted binding energy for the maximally bound nucleus is shown in Fig. 2.5 as a function of A along with the observed binding energies. The figure only shows even–odd nuclei where the pairing term vanishes. The figure also shows the contributions of various terms in the mass formula. We can see that, as A increases, the surface term loses its importance in favor of the Coulomb term. The binding energy has a broad maximum in the neighborhood of $A \sim 56$ which corresponds to the even-Z isotopes of iron and nickel.

Light nuclei can undergo exothermic fusion reactions until they reach the most strongly bound nuclei in the vicinity of $A \sim 56$. These reactions correspond to the various stages of nuclear burning in stars. For large A's, the increasing comparative contribution of the Coulomb term lowers the binding energy. This explains why heavy nuclei can release energy in fission reactions or in α-decay. In practice, this is observed mainly for very heavy nuclei $A > 212$ because lifetimes are in general too large for smaller nuclei.

For the even–odd nuclei, the binding energy follows a parabola in Z for a given A. An example of this is given on Fig. 2.6 for $A = 111$. The minimum of the parabola, i.e. the number of neutrons and protons which corresponds to the maximum binding energy of the nucleus gives the value $Z(A)$ for the most bound isotope :

$$\frac{\partial B}{\partial Z} = 0 \Rightarrow Z(A) = \frac{A}{2 + a_c A^{2/3}/2a_a} \sim \frac{A/2}{1 + 0.0075\, A^{2/3}} \, . \tag{2.14}$$

This value of Z is close to, but not necessarily equal to the value of Z that gives the stable isobar for a given A. This is because one must also take

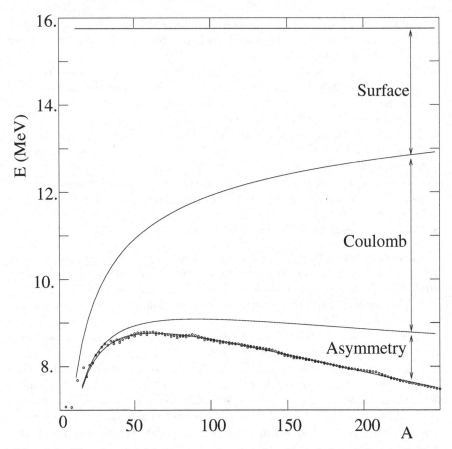

Fig. 2.5. The observed binding energies as a function of A and the predictions of the mass formula (2.13). For each value of A, the most bound value of Z is used corresponding to $Z = A/2$ for light nuclei but $Z < A/2$ for heavy nuclei. Only even–odd combinations of A and Z are considered where the pairing term of the mass formula vanishes. Contributions to the binding energy per nucleon of the various terms in the mass formula are shown.

into account the neutron–proton mass difference in order to make sure of the stability against β-decay. The only stable nuclei for odd A are obtained by minimizing the atomic mass $m(A, Z) + Zm_e$ (we neglect the binding energies of the atomic electrons). This leads to a slightly different value for the $Z(A)$ of the stable atom:

$$Z(A) = \frac{(A/2)(1 + \delta_{npe}/4a_a)}{1 + a_c A^{2/3}/4a_a} \sim 1.01 \frac{A/2}{1 + 0.0075A^{2/3}} \qquad (2.15)$$

where $\delta_{npe} = m_n - m_p - m_e = 0.75\,\mathrm{MeV}$. This formula shows that light nuclei have a slight preference for protons over neutrons because of their smaller mass while heavy nuclei have an excess of neutrons over protons because an extra amount of nuclear binding must compensate for the Coulomb repulsion.

For even A, the binding energies follow two parabolas, one for even–even nuclei, the other for odd–odd ones. An example is shown for $A = 112$ on Fig. 2.6. In the case of even–even nuclei, it can happen that an unstable odd-odd nucleus lies between two β-stable even-even isotopes. The more massive of the two β-stable nuclei can decay via 2β-decay to the less massive. The lifetime for this process is generally of order or greater than 10^{20} yr so for practical purposes there are often two stable isobars for even A.

The Bethe–Weizsäcker formula predicts the maximum number of protons for a given N and the maximum number of neutrons for a given Z. The limits are determined by requiring that the last added proton or last added neutron be bound, i.e.

$$B(Z+1, N) - B(Z, N) > 0, \quad B(Z, N+1) - B(Z, N) > 0, \quad (2.16)$$

or equivalently

$$\frac{\partial B(Z, N)}{\partial Z} > 0, \qquad \frac{\partial B(Z, N)}{\partial N} > 0. \qquad (2.17)$$

The locus of points (Z, N) where these inequalities become equalities establishes determines the region where bound states exist. The limits predicted by the mass formula are shown in Fig. 2.1. These lines are called the proton and neutron *drip-lines*. As expected, some nuclei just outside the drip-lines are observed to decay rapidly by nucleon emission. Combinations of (Z, N) far outside the drip-lines are not observed. However, we will see in Sect. 2.7 that nucleon emission is observed as a decay mode of many excited nuclear states.

2.3 The Fermi gas model

The Fermi gas model is a quantitative quantum-mechanical application of the mean potential model discussed qualitatively in Sect. 2.1. It allows one to account semi-quantitatively for various terms in the Bethe–Weizsäcker formula. In this model, nuclei are considered to be composed of two fermion gases, a neutron gas and a proton gas. The particles do not interact, but they are confined in a sphere which has the dimension of the nucleus. The interactions appear implicitly through the assumption that the nucleons are confined in the sphere.

The liquid-drop model is based on the saturation of nuclear forces and one relates the energy of the system to its geometric properties. The Fermi model is based on the quantum statistics effects on the energy of confined fermions. The Fermi model provides a means to calculate the constants a_v, a_s and a_a in the Bethe–Weizsäcker formula, directly from the density ρ of the nuclear matter. Its semi-quantitative success further justifies for this formula.

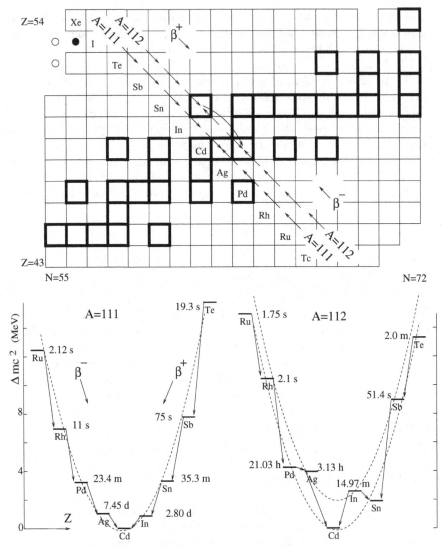

Fig. 2.6. The systematics of β-instability. The top panel shows a zoom of Fig. 2.1 with the β-stable nuclei shown with the heavy outlines. Nuclei with an excess of neutrons (below the β-stable nuclei) decay by β⁻ emission. Nuclei with an excess of protons (above the β-stable nuclei) decay by β⁺ emission or electron capture. The bottom panel shows the atomic masses as a function of Z for $A = 111$ and $A = 112$. The quantity plotted is the difference between $m(Z)$ and the mass of the lightest isobar. The dashed lines show the predictions of the mass formula (2.13) after being offset so as to pass through the lowest mass isobars. Note that for even-A, there can be two β-stable isobars, e.g. ¹¹²Sn and ¹¹²Cd. The former decays by 2β-decay to the latter. The intermediate nucleus ¹¹²In can decay to both.

The Fermi model is based on the fact that a spin $1/2$ particle confined to a volume V can only occupy a discrete number of states. In the momentum interval $\mathrm{d}^3\boldsymbol{p}$, the number of states is

$$\mathrm{d}\mathcal{N} = (2s+1)\frac{V\mathrm{d}^3\boldsymbol{p}}{(2\pi\hbar)^3} \quad , \tag{2.18}$$

with $s = 1/2$. This number will be derived below for a cubic container but it is, in fact, generally true. It corresponds to a density in *phase space* of 2 states per $2\pi\hbar^3$ of phase-space volume.

We now place $\mathcal{N}$ particles in the volume. In the ground state, the particles fill up the lowest single-particle levels, i.e. those up to a maximum momentum called the *Fermi momentum*, p_F, corresponding to a maximum energy $\varepsilon_\mathrm{F} = p_\mathrm{F}^2/2m$. The Fermi momentum is determined by

$$\mathcal{N} = \sum_{p<p_\mathrm{F}} \mathrm{d}\mathcal{N} = \frac{V\, p_\mathrm{F}^3}{3\pi^2\hbar^3} \quad . \tag{2.19}$$

This determines the Fermi energy

$$\varepsilon_\mathrm{F} = \frac{p_\mathrm{F}^2}{2m} = \frac{\hbar^2}{2m}(3\pi^2 n)^{2/3} \tag{2.20}$$

where n is the number density $n = \mathcal{N}/V$. The total (*kinetic*) energy $\mathcal{E}$ of the system is

$$\mathcal{E} = \sum_{p<p_\mathrm{F}} \frac{p^2}{2m} = \frac{3}{5}\mathcal{N}\varepsilon_\mathrm{F} \quad . \tag{2.21}$$

In a system of $A = Z+N$ nucleons, the densities of neutrons and protons are respectively $n_0(N/A)$ and $n_0(Z/A)$ where $n_0 \sim 0.15\,\mathrm{fm}^{-3}$ is the nucleon density. The total kinetic energy is then

$$\mathcal{E} = \mathcal{E}_Z + \mathcal{E}_N = 3/5\left[Z\frac{\hbar^2}{2m}(3\pi^2\frac{Zn_0}{A})^{2/3} + N\frac{\hbar^2}{2m}(3\pi^2\frac{Nn_0}{A})^{2/3}\right] . \tag{2.22}$$

In the approximation $Z \sim N \sim A/2$, this value of the nuclear density corresponds to a Fermi energy for protons and neutrons of

$$\varepsilon_\mathrm{F} = 35 \text{ MeV} \quad , \tag{2.23}$$

which corresponds to a momentum and a wave number

$$p_\mathrm{F} = 265\mathrm{MeV}/c \,, \qquad k_\mathrm{F} = p_\mathrm{F}/\hbar = 1.33\,\mathrm{fm}^{-1} \quad . \tag{2.24}$$

2.3.1 Volume and surface energies

In fact, the number of states (2.18) is slightly overestimated since it corresponds to the continuous limit $V \to \infty$ where the energy differences between levels vanishes. To convince ourselves, we examine the estimation of the number of levels in a cubic box of linear dimension a. The wavefunctions and energy levels are

$$\psi_{n_1,n_2,n_3}(x,y,z) = \sqrt{\frac{8}{a^3}} \, \sin(\frac{n_1\pi x}{a}) \, \sin(\frac{n_2\pi y}{a}) \, \sin(\frac{n_3\pi z}{a}) \qquad (2.25)$$

$$E = E_{n_1,n_2,n_3} = \frac{\hbar^2\pi^2}{2ma^2}(n_1^2 + n_2^2 + n_3^2) \quad , \qquad (2.26)$$

with $n_i > 0$, and one counts the number of states such that $E \leq E_0$, E_0 fixed, which corresponds to the volume of one eighth of a sphere in the space $\{n_1, n_2, n_3\}$. In this counting, one should not take into account the three planes $n_1 = 0$, $n_2 = 0$ and $n_3 = 0$ for which the wavefunction is identically zero, which does not correspond to a physical situation. When the number of states under consideration is very large, such as in statistical mechanics, this correction is negligible. However, it is not negligible here. The corresponding excess in (2.19) can be calculated in an analogous way to (2.18); one obtains

$$\Delta\mathcal{N} = \frac{p_F^2 \, S}{8\pi\hbar^2} = \frac{m\,\varepsilon_F \, S}{4\pi\hbar^2} \qquad (2.27)$$

where S is the external area of the volume V ($S = 6\,a^2$ for a cube, $S = 4\pi r_0^2$ for a sphere).[2]

The expression (2.19), after correction for this effect, becomes

$$\mathcal{N} = \frac{V\,p_F^3}{3\pi^2\hbar^3} - \frac{S\,p_F^2}{4\,\pi\hbar^2} \quad . \qquad (2.28)$$

The corresponding energy is

$$\mathcal{E} = \int_0^{p_F} \frac{p^2}{2m}\,\mathrm{d}\mathcal{N}(p) = \frac{V\,\varepsilon_F p_F^3}{5\pi^2\hbar^2} - \frac{S\,\varepsilon_F p_F^2}{8\pi\hbar^2} \quad . \qquad (2.29)$$

The first term is a volume energy, the second term is a surface correction, or a surface-tension term.

To first order in S/V the kinetic energy per particle is therefore

$$\frac{\mathcal{E}}{\mathcal{N}} = \frac{3}{5}\varepsilon_F\,(1 + \frac{\pi\,S\,\hbar}{8V\,p_F} + \ldots) \quad . \qquad (2.30)$$

In the approximation $Z \sim N \sim A/2$, the kinetic energy is of the form :

$$E_c = a_0 A + a_s A^{2/3} \qquad (2.31)$$

with

$$a_0 = \frac{3}{5}\varepsilon_F = 21\ \mathrm{MeV}\,, \qquad a_s = \frac{3}{5}\varepsilon_F\,\frac{3\pi\hbar}{8\,r_0\,p_F} = 16.1\,\mathrm{MeV}\,. \qquad (2.32)$$

---

[2] On can prove that these results, expressed in terms of V and S, are independent of the shape of the confining volume, see R. Balian and C. Bloch, Annals of Physics, **60**, p.40 (1970) and **63**, p.592 (1971).

The second term is the surface coefficient of (2.13), in good agreement with the experimental value. The mean energy per particle is the sum $a_v = a_0 - U$ of a_0 and of a potential energy U which can be determined experimentally by neutron scattering on nuclei (this is analogous to the Ramsauer effect). Experiment gives $U \sim -40$ MeV, i.e.

$$a_v \sim 19\,\text{MeV} , \tag{2.33}$$

in reasonable agreement with the empirical value of (2.13).

2.3.2 The asymmetry energy

Consider now the system of the two Fermi gases, with N neutrons and Z protons inside the same sphere of radius R. The total energy of the two gases (2.22) is

$$E = \frac{3}{5}\varepsilon_F \left(N(\frac{2N}{A})^{2/3} + Z(\frac{2Z}{A})^{2/3} \right) , \tag{2.34}$$

where we neglect the surface energy. Expanding this expression in the neutron excess $\Delta = N - Z$, we obtain, to first order in Δ/A,

$$E = \frac{3}{5}\varepsilon_F + \frac{\varepsilon_F}{3}\frac{(N-Z)^2}{A} + \dots . \tag{2.35}$$

This is precisely the form of the asymmetry energy in the Bethe–Weizsäcker formula. However, the numerical value of the coefficient $a_a \sim 12$ MeV is half of the empirical value. This defect comes from the fact that the Fermi model is too simple and does not contain enough details about the nuclear interaction.

2.4 The shell model and magic numbers

In atomic physics, the ionization energy E_I, i.e. the energy needed to extract an electron from a neutral atom with Z electrons, displays discontinuities around $Z = 2, 10, 18, 36, 54$ and 86, i.e. for noble gases. These discontinuities are associated with closed electron shells.

An analogous phenomenon occurs in nuclear physics. There exist many experimental indications showing that atomic nuclei possess a shell-structure and that they can be constructed, like atoms, by filling successive shells of an effective potential well. For example, the nuclear analogs of atomic ionization energies are the "separation energies" S_n and S_p which are necessary in order to extract a neutron or a proton from a nucleus

$$S_n = B(Z,N) - B(Z, N-1) \quad S_p = B(Z,N) - B(Z-1, N) . \tag{2.36}$$

These two quantities present discontinuities at special values of N or Z, which are called *magic numbers*. The most commonly mentioned are:

$$2 \quad 8 \quad 20 \quad 28 \quad 50 \quad 82 \quad 126 \, . \qquad\qquad (2.37)$$

As an example, Fig. 2.7 gives the neutron separation energy of lead isotopes ($Z = 82$) as a function of N. The discontinuity at the magic number $N = 126$ is clearly seen.

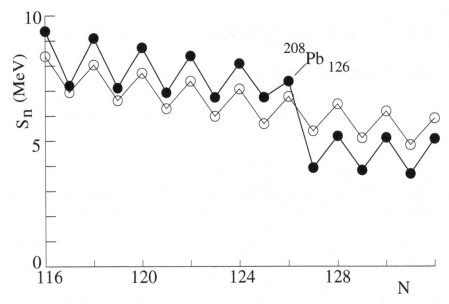

Fig. 2.7. The neutron separation energy in lead isotopes as a function of N. The filled dots show the measured values and the open dots show the predictions of the Bethe–Weizsäcker formula.

The discontinuity in the separation energies is due to the excess binding energy for magic nuclei as compared to that predicted by the semi-empirical Bethe–Weizsäcker mass formula. One can see this in Fig. 2.8 which plots the excess binding energy as a function of N and Z. Large positive values of B/A(experimental)-B/A(theory) are observed in the vicinity of the magic numbers for neutrons N as well as for protons Z. Figure 2.9 shows the difference as a function of N and Z in the vicinity of the magic numbers 28, 50, 82 and 126.

Just as the energy necessary to liberate a neutron is especially large at magic numbers, the difference in energy between the nuclear ground state and the first excited state is especially large for these nuclei. Table 2.1 gives this energy as a function of N (even) for Hg ($Z = 80$), Pb ($Z = 82$) and Po ($Z = 84$). Only even–even nuclei are considered since these all have similar nucleon structures with the ground state having $J^P = 0^+$ and a first excited state generally having $J^P = 2^+$. The table shows a strong peak at the doubly magic ^{208}Pb. As discussed in Sect. 1.3, the large energy difference between rotation

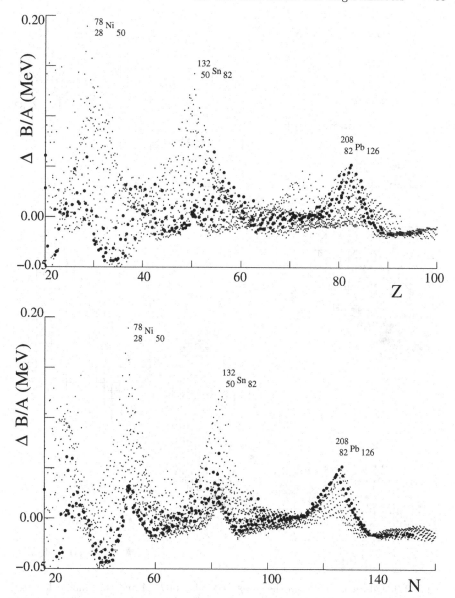

Fig. 2.8. Difference in MeV between the measured value of B/A and the value calculated with the empirical mass formula as a function of the number of protons Z (top) and of the number of neutrons N (bottom). The large dots are for β-stable nuclei. One can see maxima for the magic numbers $Z, N = 20, 28, 50, 82$, and 126. The largest excesses are for the doubly magic nuclides as indicated.

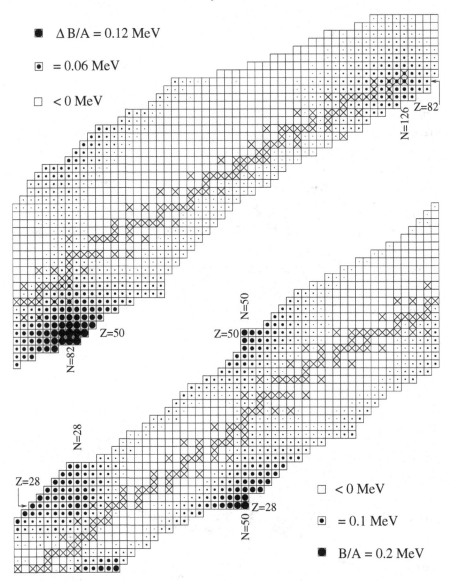

Fig. 2.9. Difference between the measured value of B/A and the value calculated with the mass formula as a function of N and Z. The size of the black dot increases with the difference. One can see the hills corresponding to the values of the magic numbers 28,50,82 and 126. Crosses mark β-stable nuclei.

states for ^{208}Pb is due to its sphericity. Sphericity is a general characteristic of magic nuclides, as illustrated in Fig. 1.8.

Table 2.1. The energy difference (keV) between the ground state (0^+) and the first excited state (2^+) for even–even isotopes of mercury $(Z = 80)$, lead $(Z = 82)$ and polonium $(Z = 84)$. The largest energy difference is for the the double-magic ^{208}Pb. As discussed in Sect. 1.3, this large difference between rotational states is due to the sphericity of ^{208}Pb.

N→	112	114	116	118	120	122	124	126	128	130	132
Hg	423	428	426	412	370	440	436	1068			
Pb	260	171	304	1027	960	900	803	2614	800	808	837
Po	463	605	665	677	684	700	687	1181	727	610	540

2.4.1 The shell model and the spin-orbit interaction

It is possible to understand the nuclear shell structure within the framework of a modified mean field model. If we assume that the mean potential energy is harmonic, the energy levels are

$$E_n = (n + 3/2)\hbar\omega \quad n = n_x + n_y + n_z = 0, 1, 2, 3 \ldots , \tag{2.38}$$

where $n_{x,y,z}$ are the quantum numbers for the three orthogonal directions and can take on positive semi-definite integers. If we fill up a harmonic well with nucleons, 2 can be placed in the one $n = 0$ orbital, i.e. the $(n_x, n_y, n_z) = (0, 0, 0)$. We can place 6 in the $n = 1$ level because there are 3 orbitals, $(1, 0, 0)$, $(0, 1, 0)$ and $(0, 0, 1)$. The number $N(n)$ are listed in the third row of Table 2.2.

We note that the harmonic potential, like the Coulomb potential, has the peculiarity that the energies depend only on the principal quantum number n and not on the angular momentum quantum number l. The angular momentum states, $|n, l, m\rangle$ can be constructed by taking linear combinations of the $|n_x, n_y, n_z\rangle$ states (Exercise 2.4). The allowed values of l for each n are shown in the second line of Table 2.2.

Table 2.2. The number N of nucleons per shell for a harmonic potential.

n	0	1	2	3	4	5	6
l	0	1	0,2	1,3	0,2,4	1,3,5	0,2,4,6
$N(n)$	2	6	12	20	30	42	56
$\sum N$	2	8	20	40	70	112	168

The magic numbers corresponding to all shells filled below the maximum n, as shown on the fourth line of Table 2.2, would then be 2, 8, 20, 40, 70, 112 and 168 in disagreement with observation (2.37). It might be expected that one could find another simple potential that would give the correct numbers. In general one would find that energies would depend on *two* quantum numbers: the angular momentum quantum number l and a second giving the number of nodes of the radial wavefunction. An example of such a *l-splitting* is shown in Fig. 2.10. Unfortunately, it turns out that there is no simple potential that gives the correct magic numbers.

The solution to this problem, found in 1949 by M. Göppert Mayer, and by D. Haxel J. Jensen and H. Suess, is to add a spin orbit interaction for each nucleon:

$$\hat{H} = V_{\text{s-o}}(r)\hat{\boldsymbol{\ell}} \cdot \hat{\boldsymbol{s}}/\hbar^2 . \tag{2.39}$$

Without the spin-orbit term, the energy does not depend on whether the nucleon spin is aligned or anti-aligned with the orbital angular momentum. The spin orbit term breaks the degeneracy so that the energy now depends on three quantum numbers, the principal number n, the orbital angular momentum quantum number l and the total angular momentum quantum number $j = l \pm 1/2$. We note that the expectation value of $\hat{\boldsymbol{\ell}} \cdot \hat{\boldsymbol{s}}$ is (Exercise 2.5) given by

$$\frac{\hat{\boldsymbol{\ell}} \cdot \hat{\boldsymbol{s}}}{\hbar^2} = \frac{j(j+1) - l(l+1) - s(s+1)}{2} \qquad s = 1/2 .$$

$$= l/2 \qquad \text{for } j = l + 1/2$$

$$= -(l+1)/2 \quad \text{for } j = l - 1/2 . \tag{2.40}$$

For a given value of n, the energy levels are then changed by an amount proportional to this function of j and l. For $V_{\text{s-o}} < 0$ the states with the spin aligned with the orbital angular momentum ($j = l + 1/2$) have their energies lowered while the states with the spin anti-aligned ($j = l - 1/2$) have their energies raised.

The orbitals with this interaction included (with an appropriately chosen $V_{\text{s-o}}$) are shown in Fig. 2.10. The predicted magic numbers correspond to orbitals with a large gap separating them from the next highest orbital. For the lowest levels, the spin-orbit splitting (2.40) is sufficiently small that the original magic numbers, 2, 8, and 20, are retained. For the higher levels, the splitting becomes important and the gaps now appear at the numbers 28, 50, 82 and 126. This agrees with the observed quantum numbers (2.37). We note that this model predicts that the number 184 should be magic.

Besides predicting the correct magic numbers, the shell model also correctly predicts the spins and parities of many nuclear states. The ground states of even–even nuclei are expected to be 0^+ because all nucleons are paired with a partner of opposite angular momentum. The ground states of

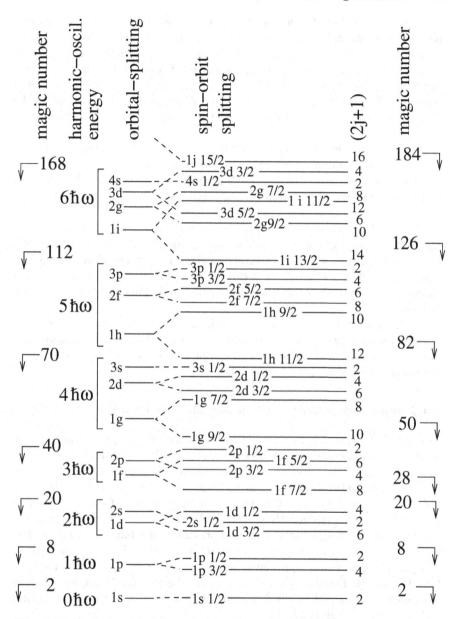

Fig. 2.10. Nucleon orbitals in a model with a spin-orbit interaction. The two left-most columns show the magic numbers and energies for a pure harmonic potential. The splitting of different values of the orbital angular momentum l can be arranged by modifying the central potential. Finally, the spin-orbit coupling splits the levels so that they depend on the relative orientation of the spin and orbital angular momentum. The number of nucleons per level $(2j + 1)$ and the resulting magic numbers are shown on the right.

odd–even nuclei should then take the quantum numbers of the one unpaired nucleon. For example, $^{17}_{9}F_8$ and $^{17}_{8}O_9$ have one unpaired nucleon outside a doubly magic $^{16}_{8}O_8$ core. Figure 2.10 tells us that the unpaired nucleon is in a $l = 2$, $j = 5/2$. The spin parity of the nucleus is predicted to by $5/2^+$ since the parity of the orbital is -1^l. This agrees with observation. The first excited states of $^{17}_{9}F_8$ and $^{17}_{8}O_9$, corresponding to raising the unpaired nucleon to the next higher orbital, are predicted to be $1/2^+$, once again in agreement with observation.

On the other hand, $^{15}_{7}N_7$ and $^{15}_{8}O_7$ have one "hole" in their ^{16}O core. The ground state quantum numbers should then be the quantum numbers of the hole which are $l = 1$ and $j = 1/2$ according to Fig. 2.10. The quantum numbers of the ground state are then predicted to be $1/2^-$, in agreement with observation.

The shell model also makes predictions for nuclear magnetic moments . As for the total angular momentum, the magnetic moments results from a combination of the spin and orbital angular momentum. However, in this case, the weighting is different because the gyromagnetic ratio of the spin differs from that of the orbital angular momentum. This problem is explored in Exercise 2.7.

Shell model calculations are important in many other aspects of nuclear physics, for example in the calculation of β-decay rates. The calculations are quite complicated and are beyond the scope of this book. Interested readers are referred to the advanced textbooks.

2.4.2 Some consequences of nuclear shell structure

Nuclear shell structure is reflected in many nuclear properties and in the relative natural abundances of nuclei. This is especially true for doubly magic nuclei like $^4He_2$, $^{16}O_8$ and $^{40}Ca_{20}$ all of which have especially large binding energies. The natural abundances of ^{40}Ca is 97% while that of $^{44}Ca_{24}$ is only 2% in spite of the fact that the semi-empirical mass formula predicts a greater binding energy for ^{44}Ca. The doubly magic $^{100}Sn_{50}$ is far from the stability line ($^{100}Ru_{56}$) but has an exceptionally long half-life of 0.94 s. The same can be said for , $^{48}Ni_{20}$, the mirror of $^{48}Ca_{28}$ which is also doubly magic. $^{56}Ni_{28}$ is the final nucleus produced in stars before decaying to ^{56}Co and then ^{56}Fe (Chap. 8). Finally, $^{208}_{82}Pb_{126}$ is the only heavy double-magic. It, along with its neighbors ^{206}Pb and ^{207}Pb, are the final states of the three natural radioactive chains shown in Fig. 5.2.

Nuclei with only one closed shell are called "semi-magic":

- isotopes of nickel, $Z = 28$;
- isotopes of tin, $Z = 50$;
- isotopes of lead, $Z = 82$;
- isotones $N = 28$ (^{50}Ti, ^{51}V, ^{52}Cr,^{54}Fe, etc.)
- isotones $N = 50$ (^{86}Kr, ^{87}Rb, ^{88}Sr, ^{89}Y, ^{90}Zr, etc.)

- isotones $N = 82$ (^{136}Xe, ^{138}Ba, ^{139}La, ^{140}Ce, ^{141}Pr, etc.)

These nuclei have

- a binding energy greater than that predicted by the semi-empirical mass formula,
- a large number of stable isotopes or isotones,
- a large natural abundances,
- a large energy separation from the first excited state,
- a small neutron capture cross-section (magic-N only).

The exceptionally large binding energy of doubly magic ^{4}He makes α decay the preferred mode of A non-conserving decays. Nuclei with $209 < A < 240$ all cascade via a series of β and α decays to stable isotopes of lead and thallium. Even the light nuclei ^{5}He, ^{5}Li and ^{8}Be decay by α emission with lifetimes of order 10^{-16} s.

While ^{5}He rapidly α decays, ^{6}He has a relatively long lifetime of 806 ms. This nucleus can be considered to be a three-body state consisting of 2 neutrons and an α particle. This system has the peculiarity that while being stable, none of the two-body subsystems (n-n or n-α) are stable. Such systems are called "Borromean" after three brothers from the Borromeo family of Milan. The three brothers were very close and their coat-of-arms showed three rings configured so that breaking any one ring would separate the other two.

Shell structure is a necessary ingredient in the explanation of nuclear deformation. We note that the Bethe–Weizsäcker mass formula predicts that nuclei should be spherical, since any deformation at constant volume increases the surface term. This can be quantified by a "deformation potential energy" as illustrated in Fig. 2.11. In the liquid-drop model a local minimum is found at vanishing deformation corresponding to spherical nuclei. If the nucleus is unstable to spontaneous fission, the absolute minimum is at large deformation corresponding to two separated fission fragments (Chap. 6).

Since the liquid-drop model predicts spherical nuclei, observed deformation must be due to nuclear shell structure. Deformations are then linked to how nucleons fill available orbitals. For instance, even–even nuclei have paired nucleons. As illustrated in Fig. 2.12, if the nucleons tend to populate the high-m orbitals of the outer shell of angular momentum l, then the nucleus will be oblate. If they tend to populate low-m orbitals, the nucleus will be prolate. Which of these cases occurs depends on the details of the complicated nuclear Hamiltonian. The most deformed nuclei are prolate.

Because of these quantum effects, the deformation energy in Fig. 2.11 will have a local minimum at non-vanishing deformation for non-magic nuclei. It is also possible that a local minimum occurs for super-deformed configurations. These metastable configurations are seen in rotation band spectra, e.g. Fig. 1.9.

We note that the shell model predicts and "island of stability" of super-heavy nuclei near the magic number $(A, N, Z) = (298, 184, 114)$ and $(310,$

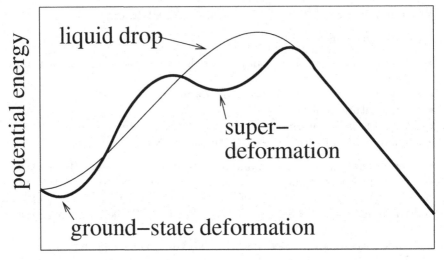

Fig. 2.11. Nuclear energies as a function of deformation. The liquid-drop model predicts that the energy has a local minimum for vanishing deformation because this minimizes the surface energy term. (As discussed in Chap. 6, in high-Z nuclei the energy eventually decreases for large deformations because of Coulomb repulsion, leading to spontaneous fission of the nucleus.) As explained in the text, the shell structure leads to a deformation of the ground state for nuclei with unfilled shells. Super-deformed local minima may also exist.

184, 126). The lifetimes are estimated to be as high as 10^6 yr making them of more than purely scientific interest. As discussed in Sect. 2.8, attempts to approach this island are actively pursued.

Finally, we mention that an active area or research concerns the study of magic numbers for neutron-rich nuclei far from the bottom of the stability valley. It is suspected that for such nuclides the shell structure is modified. This effect is important for the calculation of nucleosynthesis in the *r-process* (Sect. 8.3).

2.5 β-instability

As already emphasized, nuclei with a non-optimal neutron-to-proton ratio can decay in A-conserving β-decays. As illustrated in Fig. 2.6, nuclei with an excess of neutrons will β⁻ decay:

$$(A, Z) \rightarrow (A, Z + 1) \, e^- \, \bar{\nu}_e \tag{2.41}$$

which is the nuclear equivalent of the more fundamental particle reaction

$$n \rightarrow p + e^- + \bar{\nu}_e \, . \tag{2.42}$$

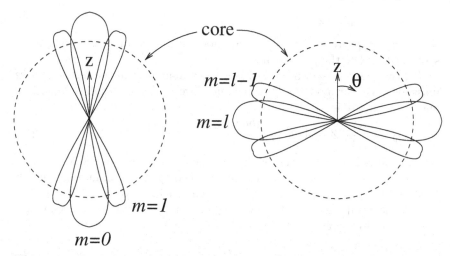

Fig. 2.12. The distribution of polar angle θ for high-l orbitals with respect to the symmetry (z) axis. In 0^+ nuclei, nucleons pair up in orbitals of opposite angular momentum. In partially-filled shells, only some of the m orbitals will be filled. If the pairs populate preferentially low $|m|$ orbitals, as on the left, the nucleus will take a deformed prolate shape. If the pairs populate preferentially high $|m|$ orbitals, as on the right, the nucleus will take an oblate shape. The spherical core of filled shells is also shown.

Nuclei with an excess of protons will either β^+ decay

$$(A, Z) \rightarrow (A, Z - 1)\, e^+\, \bar{\nu}_e \tag{2.43}$$

or, if surrounded by atomic electrons, decay by electron capture

$$e^-\, (A, Z) \rightarrow (A, Z - 1)\, \bar{\nu}_e \quad . \tag{2.44}$$

These two reactions are the nuclear equivalents of the particle reactions

$$e^-\, p \rightarrow n\, \nu_e \qquad p \rightarrow n\, e^+\, \nu_e \,. \tag{2.45}$$

In order to conserve energy-momentum, proton β^+-decay is only possible in nuclei.

The energy release in β^--decay is given by

$$\begin{aligned}
Q_{\beta^-} &= m(A, Z) - m(A, Z + 1) - m_e \\
&= (B(A, Z + 1) - B(A, Z)) + (m_n - m_p - m_e)
\end{aligned} \tag{2.46}$$

while that in β^+-decay is

$$\begin{aligned}
Q_{\beta^+} &= m(A, Z) - m(A, Z - 1) - m_e \\
&= (B(A, Z - 1) - B(A, Z)) - (m_n - m_p - m_e) \,.
\end{aligned} \tag{2.47}$$

The energy release in electron capture is larger than that in β^+-decay

$$Q_{ec} = Q_{\beta^+} + 2m_e \tag{2.48}$$

so electron capture is the only decay mode available for neighboring nuclei separated by less than m_e in mass.

The energy released in β-decay can be estimated from the semi-empirical mass formula. For moderately heavy nuclei we can ignore the Coulomb term and the estimate is

$$Q_\beta \sim \frac{8a_a}{A}|Z - A/2| \sim \frac{100\,\text{MeV}}{A} . \tag{2.49}$$

As with all reactions in nuclear physics, the Q values are in the MeV range.

β-decays and electron captures are governed by the weak interaction. The fundamental physics involved will be discussed in more detail in Chap. 4. One of the results will be that for $Q_\beta \gg m_e$, the decay rate is proportional to the fifth power of Q_β

$$\lambda_\beta \propto G_F^2 Q_\beta^5 \sim 10^{-4}\text{s}^{-1}\left(\frac{Q_\beta}{1\,\text{MeV}}\right)^5 \qquad Q_\beta \gg m_e c^2 \tag{2.50}$$

where the Fermi constant G_F, given by $G_F/(\hbar c)^3 = 1.166\,10^{-5}\,\text{GeV}^{-2}$, is the effective coupling constant for low-energy weak interactions. The constant of proportionality depends in the details of the initial and final state nuclear wavefunctions. In the most favorable situations, the constant is of order unity.

Figure 2.13 shows the lifetimes of β emitters as a function of Q_β. The line shows the maximum allowed rate, which, for $Q_\beta \gg m_e c^2$, is $\sim (G_F^2 Q^5)^{-1}$. For $Q_\beta < 1\,\text{MeV}$, lifetimes of β^+ emitters are shorter that those of β^- emitters because of the contribution of electron capture.

Electron captures are also governed by the weak interactions and, as such, capture rates are proportional to G_F^2. We will see in Chap. 4 that the decay rate is roughly

$$\lambda_{ec} \propto (\alpha Z m_e c^2)^3 G_F^2 Q_{ec}^2 , \tag{2.51}$$

where $\alpha \sim 1/137$ is the fine structure constant. The strong Z dependence comes from the fact that the decay rate is proportional to the probability that an electron is near the nucleus, i.e. the square of the wavefunction at the origin for the inner electrons. This probability is inversely proportional to the third power of the effective Bohr radius of the inner shell atomic electrons. This gives the factor in parentheses in the decay rate.

For nuclei that can decay by both electron capture and β^+-decay, the ratio between the two rates is given by

$$\frac{\lambda_{ec}}{\lambda_{\beta^+}} \sim (Z\alpha)^3 \frac{Q_{ec}^2(m_e c^2)^3}{(Q_{ec} - 2m_e c^2)^5} \qquad Q_{ec} > 2m_e c^2 . \tag{2.52}$$

We see that electron capture is favored for high Z and low Q_{ec}, while β^+ is favored for low Z and high Q_{ec}.

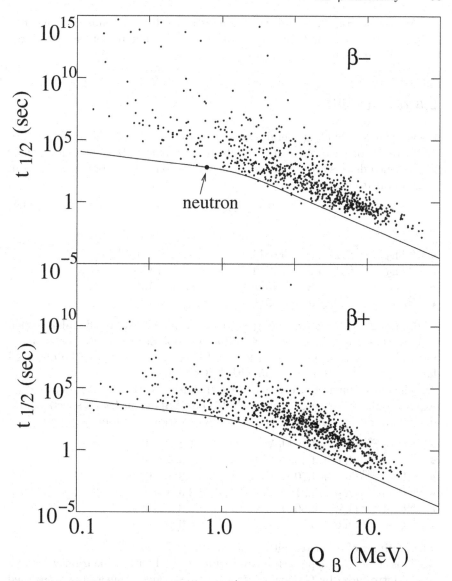

Fig. 2.13. The half-lives of β^- (top) and β^+ (bottom) emitters as a function of Q_β. The line corresponds to the maximum allowable β decay rate which, for $Q_\beta \gg m_e c^2$ is given by $t_{1/2}^{-1} \sim G_F^2 Q_\beta^2$. The complete Q_β dependence will be calculated in Chap. 4. For $Q_\beta < 1\,\text{MeV}$, the lifetimes of β^+ emitters are shorter than those for β^- emitters because of the contribution of electron-capture.

As seen in Fig. 2.13, β-decay lifetimes range from seconds to years. Examples are $\sim 10^{-5}$ s for ^{7}He and $\sim 10^{24}$ s for ^{50}V. The reasons for this large range will be discussed in Chap. 4.

2.6 α-instability

Because nuclear binding energies are maximized for $A \sim 60$, heavy nuclei that are β-stable (or unstable) can generally split into more strongly bound lighter nuclei. Such decays are called "spontaneous fission." The most common form of fission is α-decay:

$$(A, Z) \rightarrow (A - 4, Z - 2) + {}^4\text{He} \quad , \tag{2.53}$$

for example

- ^{232}Th$_{90} \rightarrow {}^{228}Ra_{88}$ α $+ 4.08$MeV   ; $t_{1/2} = 1.4 \ 10^{10}$ yr
- ^{224}Th$_{90} \rightarrow {}^{220}Ra_{88}$ α $+ 7.31$MeV   ; $t_{1/2} = 1.05$ s
- ^{142}Ce$_{58} \rightarrow {}^{138}Ba_{56}$ α $+ 1.45$MeV   ; $t_{1/2} \sim 5.10^{15}$ yr
- ^{212}Po$_{84} \rightarrow {}^{208}Pb_{82}$ α $+ 8.95$MeV   ; $t_{1/2} = 3.10^{-7}$s

Figure 2.14 shows the energy release, Q_α in α-decay for β-stable nuclei. We see that most nuclei with $A > 140$ are potential α-emitters. However, naturally occurring nuclides with α-half-lives short enough to be observed have either $A > 208$ or $A \sim 145$ with ^{142}Ce being lightest.

The most remarkable characteristic of α-decay is that the decay rate is an exponentially increasing function of Q_α. This important fact is spectacularly demonstrated by comparing the lifetimes of various uranium isotopes:

- ^{238}U $\rightarrow {}^{234}$Th α $+ 4.19$ MeV   ; $t_{1/2} = 1.4 \times 10^{17}$ s
- ^{236}U $\rightarrow {}^{232}$Th α $+ 4.45$ MeV   ; $t_{1/2} = 7.3 \times 10^{14}$ s
- ^{234}U $\rightarrow {}^{230}$Th α $+ 4.70$ MeV   ; $t_{1/2} = 7.8 \times 10^{12}$ s
- ^{232}U $\rightarrow {}^{228}$Th α $+ 5.21$ MeV   ; $t_{1/2} = 2.3 \times 10^{9}$ s
- ^{230}U $\rightarrow {}^{226}$Th α $+ 5.60$ MeV   ; $t_{1/2} = 1.8 \times 10^{6}$ s
- ^{228}U $\rightarrow {}^{224}$Th α $+ 6.59$ MeV   ; $t_{1/2} = 5.6 \times 10^{2}$ s

The lifetimes of other α-emitters are shown in Fig. 2.61.

This strong Q_α dependence can be understood within the framework of a model introduced by Gamow in 1928. In this model, a nucleus is considered to contain α-particles bound in the nuclear potential. If the electrostatic interaction between an α and the rest of the nucleus is "turned off," the α's potential is that of Fig. 2.16a. As usual, the potential has a range R and a depth V_0. Its binding energy is called E_α. In this situation, the nucleus is completely stable against α-decay.

If we now "turn on" the electrostatic potential between the α and the rest of the nucleus, E_α increases because of the repulsion. For highly charged heavy nuclei, the increase in E_α can be sufficient to make $E_\alpha > 0$, a situation shown in Fig. 2.16b. Such a nucleus, classically stable, can decay quantum

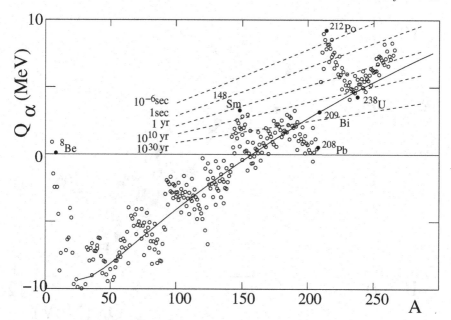

Fig. 2.14. Q_α vs. A for β-stable nuclei. The solid line shows the prediction of the semi-empirical mass formula. Because of the shell structure, nuclei just heavier than the doubly magic ^{208}Pb have large values of Q_α while nuclei just lighter have small values of Q_α. The dashed lines show half-lives calculated according to the Gamow formula (2.61). Most nuclei with $A > 140$ are potential α-emitters, though, because of the strong dependence of the lifetime on Q_α, the only nuclei with lifetimes short enough to be observed are those with $A > 209$ or $A \sim 148$, as well as the light nuclei ^{8}Be, ^{5}Li, and ^{5}He.

mechanically by the tunnel effect. The tunneling probability could be trivially calculated if the potential barrier where a constant energy V of width Δ:

$$P \propto \text{cte } e^{-2K\Delta} , \qquad K = \sqrt{\frac{2m(V - E_\alpha)}{\hbar^2}} . \qquad (2.54)$$

To calculate the tunneling probability for the potential of Fig. 2.16b, it is sufficient to replace the potential with a series of piece-wise constant potentials between $r = R$ and $r = b$ and then to sum:

$$P \propto e^{-2\gamma} \qquad \gamma = \int_R^b \sqrt{\frac{2(V(r) - E_\alpha)mc^2}{\hbar^2 c^2}} dr \qquad (2.55)$$

where $V(r)$ is the potential in Fig. 2.16b. The rigorous justification of this formula comes from the WKB approximation studied in Exercise 2.9.

The integral in (2.55) can be simplified by defining the dimensionless variable

$$u = \frac{E}{V(r)} = r \frac{E}{2(Z - 2)\alpha\hbar c} . \qquad (2.56)$$

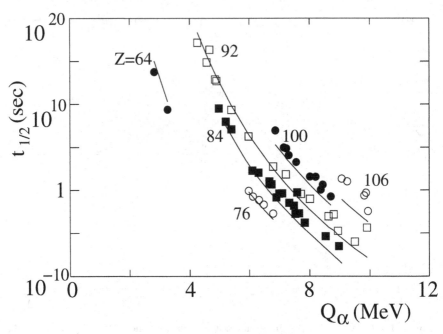

Fig. 2.15. The half-lives vs. Q_α for selected nuclei. The half-lives vary by 23 orders of magnitude while Q_α varies by only a factor of two. The lines shown the prediction of the Gamow formula (2.61).

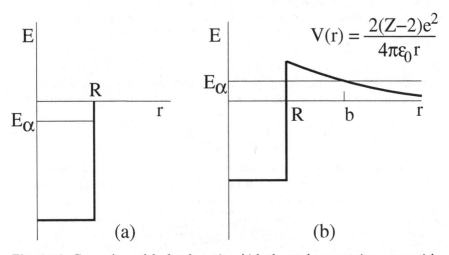

Fig. 2.16. Gamow's model of α-decay in which the nucleus contains a α-particle moving in a mean potential. If the electromagnetic interactions are "turned off", the α-particle is in the state shown on the left. When the electromagnetic interaction is turned on, the energy of the α-particle is raised to a position where it can tunnel out of the nucleus.

We then have

$$\gamma = \frac{2(Z-2)e^2}{4\pi\epsilon_0\hbar}\sqrt{\frac{2m_\alpha}{E}}\int_{u_{\min}}^1 \sqrt{u^{-1}-1}\,du\ . \tag{2.57}$$

For large Z, (2.56) suggests that it is a reasonably good approximation to take $u_{\min} = 0$ in which case the integral is $\pi/2$. This gives

$$\gamma = 2\pi(Z-2)\alpha\frac{c}{v} \tag{2.58}$$

where $v = \sqrt{2E/m_\alpha}$ is the velocity of the α-particle after leaving the nucleus. For ^{238}U we have $2\gamma \sim 172$ while for ^{228}U we have $2\gamma \sim 136$. We see how the small difference in energy leads to about 16 orders of magnitude difference in tunneling probability and, therefore, in lifetime.

To get a better estimate of the lifetime, we have to take into account the fact that $u_{\min} > 0$. This increases the tunneling probability since the barrier width is decreased. It is simple to show (Exercise 2.8) that to good approximation

$$\gamma = \frac{2Z}{\sqrt{(E_\alpha(MeV))}} - \frac{3}{2}\sqrt{ZR(fm)}\ \ . \tag{2.59}$$

The dependence of the lifetime of the nuclear radius provided one of the first methods to estimate nuclear radii.

The lifetime can be calculated by supposing that inside the nucleus the α bounces back and forth inside the potential. Each time it hits the barrier it has a probability P to penetrate. The mean lifetime is then just T/P where $T \sim R/v'$ is the oscillation frequency for the α of velocity $v' = \sqrt{2m_\alpha(E_\alpha + V_0)}$. This induces an additional Q_α dependence of the lifetime which is very weak compared to the exponential dependence on Q_α due to the tunneling probability. If we take the logarithm of the lifetime, we can safely ignore this dependence on Q_α, so, to good approximation, we have

$$\ln\tau(Q_\alpha, Z, A)\ =\ 2\gamma + \text{const}\ , \tag{2.60}$$

with γ given by (2.15). Numerically, one finds

$$\log(t_{1/2}/1\,\mathrm{s})\ \sim\ 2\gamma/\ln 10 + 25\ , \tag{2.61}$$

which is the formula used for the lifetime contours in Figs. 2.14 and 2.15.

One consequence of the strong rate dependence on Q_α is the fact that α-decays are preferentially to the ground state of the daughter nucleus, since decays to excited states necessarily have smaller values of Q_α. This is illustrated in Fig. 2.17 in the case of the decay ^{228}U $\rightarrow \alpha\,^{228}$Th. In β-decays, the Q_β dependence is weaker and many β-decays lead to excited states.

We note that the tunneling theory can also be applied to spontaneous fission decays where the nucleus splits into two nuclei of comparable mass and charge. In this case, the barrier is that of the deformation energy shown in Fig. 2.11. Note also that the decay

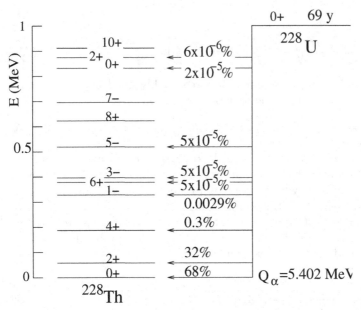

Fig. 2.17. The decay ^{228}U $\to$ α ^{228}Th showing the branching fractions to the various excited states of ^{228}Th. Because of the strong rate dependence on Q_α, the ground state his highly favored. There is also a slight favoring of spin-parities that are similar to that of the parent nucleus.

$$^{212}_{86}\text{Rn} \to {}^{198}_{80}\text{Hg}\,{}^{14}_{6}\text{C} \tag{2.62}$$

has also been observed, providing an example intermediate between α decay and spontaneous fission.

2.7 Nucleon emission

An extreme example of nuclear fission is the emission of single nucleons. This is energetically possible if the condition (2.16) is met. This is the case for the ground states of all nuclides outside the proton and neutron drip-lines shown in Fig. 2.1. Because there is no Coulomb barrier for neutron emission and a much smaller barrier for proton emission than for α emission, nuclei that can decay by nucleon emission generally have lifetimes shorter than $\sim 10^{-20}$s.

While few nuclides have been observed whose ground states decay by nucleon emission, states that are sufficiently excited can decay in this way. This is especially true for nuclides just inside the drip-lines. An example is the proton rich nuclide ^{43}V whose proton separation energy is only 0.194 MeV. All excited states above this energy decay by proton emission. The observation of these decay is illustrated in Fig. 2.18. Another example if that of the neutron-rich nuclide ^{87}Br (Fig. 6.13). We will see that such nuclei have an important role in the operation of nuclear reactors.

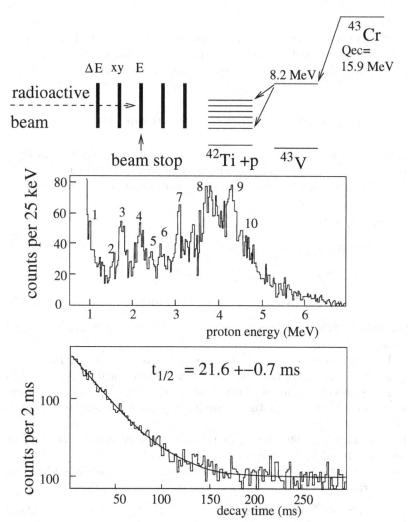

Fig. 2.18. The decay of the proton rich nuclide ^{43}Cr. Radioactive nuclei are produced in the fragmentation of 74.5 MeV/nucleon ^{58}Ni nuclei incident on a nickel target at GANIL [27]. After momentum selection, ions are implanted in a silicon diode (upper right). The (A, Z) of each implanted nucleus is determined from energy loss and position measurements as in Fig. 5.10. The implanted ^{43}Cr β-decays with via the scheme shown in the upper right, essentially to the 8.2 MeV excited state of ^{43}V. This state then decays by proton emission to ^{42}Ti. The proton deposits all its energy in the silicon diode containing the decaying ^{43}Cr and the spectrum of protons shows the position of excited states of ^{42}Ti. The bottom panel gives the distribution of time between the ion implantation and decay, indicating $t_{1/2}(^{43}$Cr$) = 21.6 \pm 0.7$ ms.

Highly excited states that can emit neutrons appear is resonances in the cross-section of *low-energy* neutrons. Examples are shown in Fig. 3.26 for states of ^{236}U and ^{239}U that decay by neutron emission to ^{235}U and ^{238}U. These states are also important in the operation of nuclear reactors.

2.8 The production of super-heavy elements

One of the most well-known results of research in nuclear physics has been the production of "trans-uranium" elements that were not previously present on Earth. The first trans-uranium elements, neptunium and plutonium, were produced by neutron capture

$$n \; ^{238}U \; \rightarrow \; ^{239}U \; \gamma \,, \tag{2.63}$$

followed by the β-decays

$$^{239}U \; \rightarrow \; ^{239}Np \; e^- \; \bar{\nu}_e \qquad t_{1/2} = 23.45 \, m \tag{2.64}$$

$$^{239}Np \; \rightarrow \; ^{239}Pu \; e^- \; \bar{\nu}_e \qquad t_{1/2} = 2.3565 \, day \,. \tag{2.65}$$

The half-life of ^{239}Pu is sufficiently long, 2.4×10^4 yr, that it can be studied as a chemical element.

Further neutron captures on ^{239}Pu produce heavier elements. This is the source of trans-uranium radioactive wastes in nuclear reactors. As shown in Fig. 6.12, this process cannot produce nuclei heavier than $^{258}_{100}$Fm which decays sufficiently rapidly ($t_{1/2} = 0.3\,$ms) that it does not have time to absorb a neutron.

Elements with $Z > 100$ can only be produced in heavy-ion collisions. Most have been produced by bombardment of a heavy element with a medium-A nucleus. Figure 2.19 shows how ^{260}Db (element 104) was positively identified via the reaction

$$^{15}N \; ^{249}Cf \; \rightarrow \; ^{260}Db \; + \; 4n \,. \tag{2.66}$$

Neutrons are generally present in the final state since the initially produced compound nucleus, in this case ^{264}Db, generally emits (evaporates) neutrons until reaching a bound nucleus. Such reactions are called *fusion-evaporation* reactions.

The heaviest element produced so far is the unnamed element 116, produced as in Fig. 2.20 via the reaction [29]

$$^{48}_{20}Ca \; ^{248}_{96}Cm \; \rightarrow \; ^{296}116 \,. \tag{2.67}$$

A beam of ^{48}Ca is used because its large neutron excess facilitates the production of neutron-rich heavy nuclei.[3] As shown in the figure, a beam of ^{48}Ca

---

[3] Planned radioactive beams (Fig. 5.5) using neutron-rich fission products will increase the number of possible reactions, though at lower beam intensity.

ions of kinetic energy 240 MeV impinges on a target of CmO_2. At this energy, the $^{296}116$ is produced in a very excited state that decays in $\sim 10^{-21}$ s by neutron emission

$$^{296}116 \rightarrow \, ^{292}116 \, 4n \, . \tag{2.68}$$

The target is sufficiently thin that the $^{292}116$ emerges from the target with most of its energy.

Only about 1 in 10^{12} collisions result in the production of element 116, most inelastic collisions resulting in the fission of the target and beam nuclei. It is therefore necessary to place beyond the target a series of magnetic and electrostatic filters so that only rare super-heavy elements reach a silicon-detector array downstream.

The $^{292}116$ ions then stop in the silicon-detector array where they eventually decay. The three sequences of events shown in Fig. 2.20 have been observed. The three sequences are believed to be due to the same nuclide because of the equality, within experimental errors, of the Q_α. The lifetimes for each step are also of the same order-of-magnitude so the half-lives can be estimated. The use of (2.61) then allows one to deduce the (A, Z) of the nuclei, confirming the identity of element 116.

Efforts to produce of super-heavy elements are being vigorously pursued. They are in part inspired by the prediction of some shell models to have an island of highly-bound nuclei around $N \sim 184$, $Z \sim 120$. There are speculations that elements in this region may be sufficiently long-lived to have practical applications.

2.9 Bibliography

1. *Nuclear Structure* A. Bohr and B. Mottelson, Benjamin, New York, 1969.
2. *Structure of the Nucleus* M.A. Preston and R.K. Bhaduri, Addison-Wesley, 1975.
3. *Nuclear Physics* S.M. Wong, John Wiley, New York, 1998.
4. *Theoretical Nuclear Physics* A. de Shalit and H. Feshbach, Wiley, New York, 1974.
5. *Introduction to Nuclear Physics* : Harald Enge, Addison-Wesley (1966).

Exercises

2.1 Use the semi-empirical mass formula to calculate the energy of α particles emitted by $^{235}U_{92}$. Compare this with the experimental value, 4.52 MeV. Note that while an α-particles are unbound in ^{235}U individual nucleons are bound. Calculate the energy required to remove a proton or a neutron from ^{235}U.

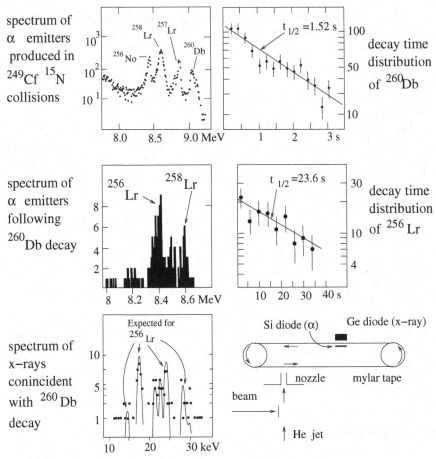

spectrum of α emitters produced in ^{249}Cf ^{15}N collisions

decay time distribution of ^{260}Db

spectrum of α emitters following ^{260}Db decay

decay time distribution of ^{256}Lr

spectrum of x−rays conincident with ^{260}Db decay

Fig. 2.19. The production and identification of element 105 ^{260}Db via the reaction ^{249}Cf(^{15}N, 4n)^{260}Db [28]. As shown schematically on the bottom right, a beam of 85 MeV ^{15}N nuclei (Oak Ridge Isochronous Cyclotron) is incident on a thin (635 µg cm^{-2}) target of ^{249}Cf (Cf$_2$O$_3$) deposited on a beryllium foil (2.35 mg cm^{-2}). Nuclei emerging from the target are swept by a helium jet through a nozzle where they are deposited on a mylar tape. After a deposition period of ∼ 1 s, the tape is moved so that the deposited nuclei are placed between two counters, a silicon diode to count α-particles and a germanium diode to count x-rays. The α-spectrum (upper left) shows three previously well-studied nuclides as well has that of ^{260}Db at ∼ 9.1 MeV. The top right panel shows the distribution of decay times indicating $t_{1/2} = 1.52 \pm 0.13$s. Confirmation of the identity of ^{260}Db comes from the α-spectrum for decays following the the 9.1 MeV α-decays. The energy spectrum indicates that the following decay is that of the previously well-studied ^{256}Lr. (The small amount of ^{258}Lr is due to accidental coincidences. Finally, the chemical identity of the Lr is confirmed by the spectrum of x-rays following the *atomic* de-excitation. (The Lr is generally left in an excited state after the decay of ^{260}Db.)

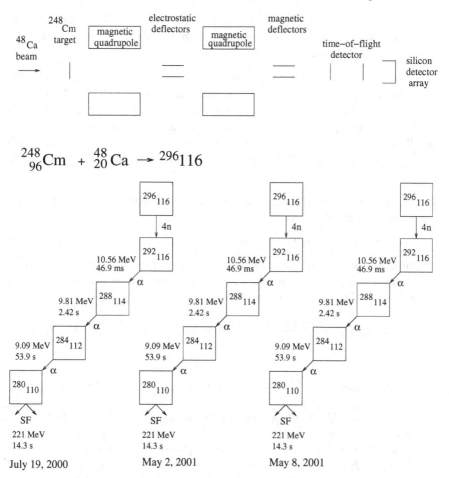

$$^{248}_{96}\text{Cm} + ^{48}_{20}\text{Ca} \rightarrow ^{296}116$$

Fig. 2.20. The production of element 116 in ^{48}Ca-^{248}Cm collisions. The top panel shows the apparatus used by [29]. Inelastic collisions generally produce light fission nuclei. Super-heavy nuclei are separated from these light nuclei by a system of magnetic and electrostatic deflectors and focusing elements. Nuclei that pass through this system stop in a silicon detector array (Chap. 5) that measures the time, position and deposited energy of the nucleus. Subsequent decays are then recorded and the decay energies measured by the energy deposited in the silicon. The bottom panel shows 3 observed event sequences that were ascribed in [29] to production and decay of $^{296}116$. The three sequences all have 3 α decays whose energies and decay times are consistent with the hypothesis that the three sequences are identical.

2.2 The radius of a nucleus is $R \simeq r_0 A^{1/3}$ with $r_0 = 1.2$ fm. Using Heisenberg's relations, estimate the mean kinetic momentum and energy of a nucleon inside a nucleus.

2.3 Consider the nucleon-nucleon interaction, and take the model $V(r) = (1/2)\mu\omega^2 r^2 - V_0$ for the potential. Estimate the values of the parameters ω and V_0 that reproduce the size and binding energy of the deuteron. We recall that the wave function of the ground state of the harmonic oscillator is $\psi(r, \theta, \phi) \propto \exp(-m\omega r^2/2\hbar)$. Is a $\ell = 1$ state bound predicted by this model?

2.4 Consider a three-dimensional harmonic oscillator, $V(r) = (1/2)m\omega^2 r^2$. the energy eigenvalues are

$$E_n = (n + 3/2)\hbar\omega \quad n = n_x + n_y + n_z , \tag{2.69}$$

where $n_{x,y,z}$ are the quantum numbers for the three orthogonal directions and can take on positive semi-definite integers. We denote the corresponding eigenstates as $|n_x, n_y, n_z\rangle$. They satisfy

$$x|n_x, n_y, n_z\rangle = \frac{\sqrt{n_x - 1}}{\alpha}|n_x - 1, n_y, n_z\rangle + \frac{\sqrt{n_x + 1}}{\alpha}|n_x + 1, n_y, n_z\rangle ,$$

and

$$ip|n_x, n_y, n_z\rangle = \frac{\sqrt{n_x - 1}}{\alpha}|n_x - 1, n_y, n_z\rangle - \frac{\sqrt{n_x + 1}}{\alpha}|n_x + 1, n_y, n_z\rangle ,$$

where $\alpha = \sqrt{2m\omega/\hbar}$. Corresponding relations hold for y and z. These states are not generally eigenstates of angular momentum but such states can be constructed from the $|n_x, n_y, n_z\rangle$. For example, verify explicitly that the $E = (1/2)\hbar\omega$ and $E = (3/2)\hbar\omega$ states can be combined to form $l = 0$ and $l = 1$ states:

$$|E = (1/2)\hbar\omega, l = 0, l_z = 0\rangle = |0, 0, 0\rangle ,$$

$$|E = (3/2)\hbar\omega, l = 1, l_z = 0\rangle = |1, 0, 0\rangle ,$$

$$|E = (3/2)\hbar\omega, l = 1, l_z = \pm 1\rangle = (1/\sqrt{2})(|0, 1, 0\rangle \pm i|0, 0, 1\rangle) .$$

2.5 Verify equations (2.40).

2.6 Use Fig. 2.10 to predict the spin and parity of ^{41}Ca.

2.7 The nuclear shell model makes predictions for the magnetic moment of a nucleus as a function of the orbital quantum numbers. Consider a nucleus consisting of a single unpaired nucleon in addition to a certain number of paired nucleons. The nuclear angular momentum is the sum of the spin and orbital angular momentum of the unpaired nucleon:

$$\boldsymbol{J} = \boldsymbol{S} + \boldsymbol{L} . \tag{2.70}$$

The total angular momentum (the nuclear spin) is $j = l \pm 1/2$ for the nucleon spin aligned or anti-aligned with the nucleon orbital angular momentum. The two types of angular momentum do not contribute in the same way to the magnetic moment:

$$\boldsymbol{\mu} = g_l \frac{e\boldsymbol{L}}{2m} + g_s \frac{e\boldsymbol{S}}{2m} \tag{2.71}$$

where $m \sim m_{\mathrm{p}} \sim m_{\mathrm{n}}$ is the nucleon mass, $g_l = 1$ ($g_l = 0$) for an unpaired proton (neutron) and $g_s = 2.792 \times 2$ ($g_s = -1.913 \times 2$) for a proton (neutron). The ratio between the magnetic moment and the spin of a nucleus is the *gyromagnetic ratio, g*. It can be defined as

$$g \equiv \frac{\boldsymbol{\mu} \cdot \boldsymbol{J}}{\boldsymbol{J} \cdot \boldsymbol{J}} \tag{2.72}$$

Use (2.40) to show that

$$g = (j - 1/2)g_l + (1/2)g_s \quad \text{for } l = j - 1/2 \tag{2.73}$$

$$g = \frac{j}{j+1}\left[(j + 3/2)g_l - (1/2)g_s\right] \quad \text{for } l = j + 1/2 . \tag{2.74}$$

Plot these two values as a function of j for nuclei with one unpaired neutron and for nuclei with one unpaired proton. Nuclei with one unpaired nucleon generally have magnetic moments that fall between these two values, known as the *Schmidt limits*.

Consider the two $(9/2)^+$ nuclides, $^{83}_{36}$Kr and $^{93}_{41}$Nb. Which would you expect to have the larger magnetic moment?

2.8 Verify (2.59) by using

$$\lim_{u_{\min} \to 0} \int_{u_{\min}}^{1} \sqrt{u^{-1} - 1}\,du \sim \int_{0}^{1} \sqrt{u^{-1} - 1}\,du - \int_{0}^{u_{\min}} \sqrt{u^{-1}}\,du .$$

2.9 To justify (2.55) write the wavefunction as

$$\psi(r) = C \exp(-\gamma(r)) + D \exp(+\gamma(r)) . \tag{2.75}$$

In the WKB approximation, we suppose that $\psi(r)$ varies sufficiently slowly that we can neglect $(\mathrm{d}^2/\mathrm{d}r^2)\gamma(r) \sim 0$. In this approximation, show that

$$\frac{\mathrm{d}^2\psi}{\mathrm{d}r^2} \sim \left(\frac{\mathrm{d}\gamma}{\mathrm{d}r}\right)^2 \psi \tag{2.76}$$

and that the Schrödinger becomes

$$\left(\frac{d\gamma}{dr}\right)^2 \psi + \frac{2M}{\hbar^2}\left(E - \frac{2Ze^2}{4\pi\epsilon_0 r}\right)\psi = 0 , \tag{2.77}$$

i.e.

$$\left(\frac{d\gamma}{dr}\right) = \sqrt{\frac{2M}{\hbar^2}\left(E - \frac{2Ze^2}{4\pi\epsilon_0 r}\right)} , \tag{2.78}$$

which is the desired result.

3. Nuclear reactions

In the last chapter we studied how nucleons could combine with each other to form bound states. In this chapter we consider how *free* particles and nuclei can interact with each other to scatter or initiate nuclear reactions. The concepts we will learn have great practical interest because they will allow us, in later chapters, to understand the generation of thermal energy in nuclear reactors and stars. However, in this chapter we will primarily be concerned with learning how to obtain information about nuclear interactions and nuclear structure from scattering experiments. In such experiments, a beam of free particles (electrons, nucleons, nuclei) traverses a target containing nuclei. A certain fraction of the beam particles will interact with the target nuclei, either scattering into a new direction or reacting in a way that particles are created or destroyed.

Classically, the character of a force field can be found by following the trajectory of test particles. The oldest example is the use of planets and comets to determine the gravitational field of the Sun. The nuclear force can not be studied directly with this technique because its short range makes it impossible to follow trajectories through the interesting region where, in any case, quantum mechanics limits the usefulness of the concept of trajectory. It is generally only possible to measure the probability of a certain type of reaction to occur. In such circumstances, it is natural to introduce the statistical concept of "cross-section" for a given reaction.

Cross-sections will be discussed in general terms in the Sect. 3.1. The sections that follow will present various ways of calculating cross-sections from knowledge of the interaction Hamiltonian.

Section 3.2 will tackle the simple problem of a particle moving into a fixed potential well. The cross-section will be calculated first supposing that the particle follows a classical trajectory and then by using quantum perturbation theory on plane waves. The use of perturbation theory will allow us to treat both elastic and simple "quasi-elastic" collisions due to electromagnetic or weak interactions. The section will end with a discussion of elastic scattering of wave packets that will allow us to better understand the angular distribution of scattered particles

The most important potential treated in Sect. 3.2 will be the Yukawa potential

$$V(r) = \frac{g\hbar c}{r} \exp(-r/r_0) \, . \tag{3.1}$$

As mentioned in Sect. 1.4, this potential describes the long-range component of the nucleon–nucleon potential. Unfortunately, the dimensionless coupling g is too large for the perturbative methods of Sect. 3.2 to be applicable. The interest of the Yukawa potential in this section will be rather the two limits $r_0 \to \infty$ and $r_0 \to 0$. In the first limit the Yukawa potential becomes the Coulomb potential

$$V(r) = \frac{g\hbar c}{r} \quad g = Z_1 Z_2 \alpha \, , \tag{3.2}$$

where Z_1 and Z_2 are the charges of the interacting particles and $\alpha = e^2/4\pi\epsilon_0\hbar c \sim 1/137$ is the fine structure constant. In the other limit $r_0 \to 0$, the potential is

$$V(r) = 4\pi G\delta^3(\boldsymbol{r}) \quad G = g\hbar c r_0^2 \, . \tag{3.3}$$

This potential is useful in discussing weak-interaction processes where G is of order the Fermi constant G_F. The techniques of Sect. 3.2 will therefore allow us to treat scattering and reactions due to the weak and electromagnetic interactions.

Section 3.4 will show how to take into account the fact that the scattering potential is not fixed, but due to a particle that will itself recoil from the collision. This will allow us to treat processes where the target is a complicated collection of particles that can be perturbed by the beam particle. With these techniques, we will learn how it is possible to determine the charge distribution in nuclei.

Section 3.5 will show how short-lived "resonances" can be produced during collisions.

Section 3.6 will introduce the more difficult problem of nucleon–nucleon and nucleon–nucleus scattering where the interaction potential can no longer be considered as weak. This problem will complete our treatment of the deuteron in Sect. 1.4.

Finally, in Sect. 3.7 we will learn how coherent forward scattering in a medium leads to a neutron refractive index. An application of this subject will be the production of ultra low-energy neutron beams.

3.1 Cross-sections

3.1.1 Generalities

To introduce cross-sections, it is conceptually simplest to consider a thin slice of matter of area L^2 containing N spheres of radius R, as shown in Fig. 3.1. A point-like particle impinging upon the slice at a random position will have

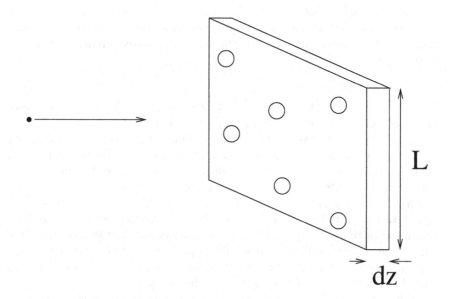

Fig. 3.1. A small particle incident on a slice of matter containing $N = 6$ target spheres of radius R. If the point of impact on the slice is random, the probability dP of it hitting a target particle is $dP = N\pi R^2/L^2 = \sigma n dz$ where the number density of scatterers is $n = N/(L^2 dz)$ and the cross section per sphere is $\sigma = \pi R^2$.

a probability dP of hitting one of the spheres that is equal to the fraction of the surface area covered by a sphere

$$dP = \frac{N\pi R^2}{L^2} = \sigma n dz \qquad \sigma = \pi R^2 . \qquad (3.4)$$

In the second form, we have multiplied and divided by the slice thickness dz and introduced the number density of spheres $n = N/(L^2 dz)$. The "cross-section" for touching a sphere, $\sigma = \pi R^2$, has dimensions of "area/sphere."

While the cross-section was introduced here as a classical area, it can be used to define a probability dP_r for any type of reaction, r, as long as the probability is proportional to the number density of target particles and to the target thickness:

$$dP_r = \sigma_r n dz . \qquad (3.5)$$

The constant of proportionality σ_r clearly has the dimension of area/particle and is called the cross-section for the reaction r.

If the material contains different types of objects i of number density and cross-section n_i and σ_i, then the probability to interact is just the sum of the probabilities on each type:

$$dP = \sum_i n_i \sigma_i \qquad (3.6)$$

Because nuclear radii are of order of a few femtometer we can anticipate that the cross-sections for nuclear reactions involving the strong interactions will often be of order $1 \, \text{fm}^2$. In fact, the units of cross-section most often used is the "barn,"

$$1 \, \text{b} = 100 \, \text{fm}^2 = 10^{-28} \, \text{m}^2 \, . \tag{3.7}$$

We will see in this chapter that nuclear weak interactions generally have cross-sections about 20 orders of magnitude smaller.

It should be emphasized that (3.5) supposes that the total probability for a type of reaction is found by summing the probabilities for reactions on each particle in the target. This assumption breaks down if interference is important, as in Bragg scattering on crystals or in elastic scattering at very small angles. In these cases, it is necessary to add *amplitudes* for scattering on target particles rather than *probabilities*. We emphasize that, in fact, adding amplitudes *always* gives the correct probability but in most cases the random phases for amplitudes from different target particles gives a $\mathrm{d}P$ that is proportional to the number of scatters rather than to its square. Equation (3.5) is therefore applicable except in special circumstances.

While we have introduced the cross-section in the context of particles incident upon a target, cross-sections are of more general applicability. For example, consider a pulse of classical electromagnetic radiation of a given energy density that impinges on a target. A cross-section can then be defined in terms of the fraction $\mathrm{d}F$ of *energy flux* that is scattered out of the original direction

$$\mathrm{d}F = \sigma n \mathrm{d}z \, . \tag{3.8}$$

We can take n to be the number density of atoms, so σ has dimensions of area/atom. This definition of the cross-section makes no reference to incident particles but only to incident energy.

The Thomson scattering cross-section of photons on electrons was originally derived in this manner by treating the interaction between a classical electromagnetic wave and free electrons. Consider a plane wave propagating in the z direction with the electric field oriented along the x direction:

$$E_x = E \cos(kz - \omega t) \, . \tag{3.9}$$

The (time averaged) electromagnetic energy energy flux (energy per unit area per unit time) is proportional to the square of the electric field:

$$\text{incident energy flux} = \frac{\epsilon_0 c E^2}{2} \, . \tag{3.10}$$

Suppose there is a free electron placed at the origin. It will be accelerated by the electric field and will oscillate in the direction of the electric field with its acceleration given by

$$\ddot{x}(t) = \frac{eE}{m_e} \cos(\omega t) \, . \tag{3.11}$$

The accelerated charge then radiates electromagnetic energy with a power given by the Larmor formula:

$$\text{radiated power} = \frac{e^2}{6\pi\epsilon_0 c^3}\langle\ddot{x}^2\rangle = c\frac{4\pi}{3}\epsilon_0 E^2 \left(\frac{e^2}{4\pi\epsilon_0 m_e c^2}\right)^2, \qquad (3.12)$$

where $\langle\rangle$ means time-average. The total cross-section defined by (3.8) is

$$\sigma = \frac{\text{power radiated}}{\text{incident energy flux}} = \frac{8\pi}{3}\left(\frac{e^2}{4\pi\epsilon_0 m_e c^2}\right)^2 = 0.665\,\text{b}. \qquad (3.13)$$

This is just the famous Thomson cross-section for the scattering of an electromagnetic wave on a free electron. Quantum mechanically, this can be interpreted as the scattering of photons on free electrons. Since the energy flux is proportional to the photon flux, the Thomson cross-section is the cross-section for the elastic scattering of photons on electrons. It turns out that the quantum-mechanical calculation gives the same result in the limit $\hbar\omega \ll m_e c^2$, i.e. that the photon energy be much less than the electron rest energy. The cross-section for higher energy photons and for photons scattering on bound electrons requires a quantum-mechanical calculation.

3.1.2 Differential cross-sections

The probability for elastic scattering is determined by the elastic scattering cross-section

$$dP_{\text{el}} = \sigma_{\text{el}}\, n\, dz. \qquad (3.14)$$

Going beyond this simple probability, we can ask what is the probability that the elastic scatter results in the particle passing through a detector of area dx^2 at a distance r from the target and angle θ with respect to the initial direction. The geometry in shown in Fig. 3.2 where the detector is oriented so that it is perpendicular to the vector between it and the target. The probability is proportional to the product of the probability of a scatter and the probability that the scattered particle goes through the detector. If the scattering angle is completely random, the second is just the ratio of dx^2 and the area of the sphere surrounding the target

$$dP_{\text{el},\theta} = \sigma_{\text{el}}\, n\, dz\frac{dx^2}{4\pi r^2} \qquad \text{isotropic scattering}. \qquad (3.15)$$

The solid angle covered by the detector is $d\Omega = dx^2/r^2$ so

$$dP_{\text{el},\theta} = \frac{d\sigma}{d\Omega}\, n\, dz\, d\Omega, \qquad (3.16)$$

where the differential scattering cross section is $d\sigma/d\Omega = \sigma_{\text{el}}/4\pi$ for isotropic scattering. In general, the scattering is not isotropic so $d\sigma/d\Omega$ is a function of θ. If the target or beam particles are polarized, it can be a function of the azimuthal angle ϕ.

The total elastic cross-section determines the total probability for elastic scattering so

$$\sigma_{\text{el}} = \int d\Omega \frac{d\sigma}{d\Omega} = \int_0^{2\pi} d\phi \int_0^\pi \sin\theta d\theta \frac{d\sigma}{d\Omega}(\theta, \phi) . \qquad (3.17)$$

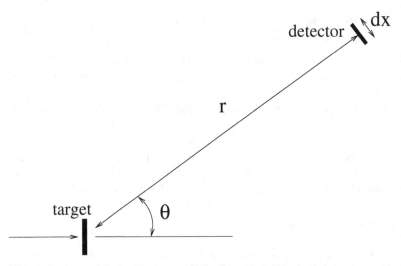

Fig. 3.2. A particle incident on a thin slice of matter containing n scatterers per unit volume of cross-section σ. A detector of area dx^2 is placed a distance r from the target and oriented perpendicular to r. If an elastic scatter results in a random scattering angle, the probability to detect the particle is $dP = ndz\sigma(dx^2/4\pi r^2) = ndz(\sigma/4\pi)d\Omega$, where $d\Omega = x^2/r^2$ is the solid angle covered by the detector.

3.1.3 Inelastic and total cross-sections

In general for a reaction creating N particles

$$a\, b \rightarrow x_1\, x_2 \ldots x_N \qquad (3.18)$$

the probability to create the particles x_i in the momentum ranges d^3p_i centered on the momenta p_i is given by

$$dP = \frac{d\sigma}{d^3p_1 \ldots d^3p_N}\, n_b\, dz\, d^3p_1 \ldots d^3p_N . \qquad (3.19)$$

The differential cross-section $d\sigma/d^3p_1 \ldots d^3p_N$ will be a singular function because only energy–momentum conserving combinations have non-vanishing probabilities.

The total probability for the reaction $a\, b \rightarrow x_1 \ldots x_N$ is

$$dP_{ab \rightarrow x_1 \ldots x_N} = \sigma_{ab \rightarrow x_1 \ldots x_N}\, n_b dz \qquad (3.20)$$

where the reaction cross-section is

$$\sigma_{ab \to x_1 \ldots x_N} = \int d^3 \boldsymbol{p}_1 \ldots \int d^3 \boldsymbol{p}_N \frac{d\sigma}{d^3 \boldsymbol{p}_1 \ldots d^3 \boldsymbol{p}_N} d^3 \boldsymbol{p}_1 \ldots d^3 \boldsymbol{p}_N . \quad (3.21)$$

The total probability that "anything" happens to the incident particle as it traverses the target of thickness dz is just the sum of the probabilities of the individual reactions

$$dP = \sigma_{tot} n_b dz \quad (3.22)$$

where the total cross-section is

$$\sigma_{tot} = \sum_i \sigma_i . \quad (3.23)$$

3.1.4 The uses of cross-sections

Cross-sections enter into an enormous number of calculations in physics. Consider a thin target (Fig. 3.1) containing a density n of target particles that is subjected to a flux of beam particles F (particles per unit area per unit time). If particles that interact in the target are considered to be removed from the beam (scattered out of the beam or changed into other types of particles), then the probability for interaction $dP = \sigma n_b dz$ implies that the F is reduced by

$$dF = -F\sigma n dz , \quad (3.24)$$

equivalent to the differential equation

$$\frac{dF}{dz} = -\frac{F}{l} , \quad (3.25)$$

where the "mean free path" l is

$$l = \frac{1}{n\sigma} . \quad (3.26)$$

For a thick target, (3.25) implies that the flux declines exponentially

$$F(z) = F(0)e^{-z/l} . \quad (3.27)$$

If the material contains different types of objects i of number density and cross-section n_i and σ_i, then (3.6) implies that the mean free path is given by

$$l^{-1} = \sum_i n_i \sigma_i . \quad (3.28)$$

The mean lifetime of a particle in the beam is the mean free path divided by the beam velocity v

$$\tau = \frac{l}{v} = \frac{1}{n_T \sigma_{tot} v} . \quad (3.29)$$

The inverse of the mean lifetime is the "reaction rate"

$$\lambda = n\sigma_{\text{tot}}v \, . \tag{3.30}$$

We will see that quantum-mechanical calculations most naturally yield the reaction rate from which one can derive the cross-section by dividing by nv.

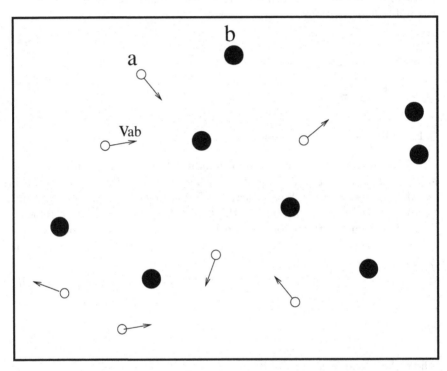

Fig. 3.3. A box containing two types of particles, a and b. The a particles move in random directions with velocity v_{ab} and can interact with the b particles (at rest) to form particles c and d with cross-section $\sigma_{ab \to cd}$. The time rate of change of the number density of particles a is determined by the Boltzmann equation (3.31).

The reaction rate enters directly into the "Boltzmann equation" governing the number density n_a of particles of type a confined to a region of space that contains particles of type b (Fig. 3.3). If the a particles are destroyed by the reaction $a\,b \to c\,d$, we have

$$\frac{\mathrm{d}n_a}{\mathrm{d}t} = -\frac{n_a}{\tau} = -n_a n_b \, \sigma_{ab \to cd} v_{ab} \, , \tag{3.31}$$

where v_{ab} is the relative velocity. (Of course it will be necessary to average the cross-section times velocity over the spectrum of particles.) The solution is just $n_a(t) = n_a(0)\exp(-t/\tau)$ as expected from (3.29).

If the region also contains particles of types c and d, particles of type a can also be created by the inverse reaction so the full Boltzmann equation is

$$\frac{dn_a}{dt} = -n_a n_b \sigma_{ab \to cd} v_{ab} + n_c n_d \sigma_{cd \to ab} v_{cd} .$$ (3.32)

3.1.5 General characteristics of cross-sections

The magnitude of a reaction cross-section depends on the energetics of the reaction (elastic, inelastic–endothermic, inelastic–exothermic) and the interaction responsible for the reaction (strong, electromagnetic, or weak). Additionally, at low energy, inelastic reactions between positively charged ions are strongly suppressed by the Coulomb barrier. In this section we review how these effects are manifested in the energy (Fig. 3.4) and angular dependences (Fig. 3.6) of cross-sections.

Elastic scattering The elastic cross-section depends on whether or not the scattering is due to long-range Coulomb interactions or to short-range strong interactions. As we will see in Sect. 3.2, the differential cross-section between two isolated charged particles diverges at small angles like $d\sigma/d\Omega \propto \theta^{-4}$. The total elastic cross-section is therefore infinite. For practical purposes, this divergence is eliminated because the Coulomb potential is "screened" at large distances by oppositely charged particles in the target. Nevertheless, the concept of total elastic cross-section for charged particles is not very useful.

Elastic neutron scattering is due to the short-range strong interaction so the differential cross-section does not diverge at small angles and the total elastic cross-section (calculated quantum mechanically) is finite. The elastic cross-sections are shown in Fig. 3.4 for neutron scattering on ^{1}H, ^{2}H and ^{6}Li. The ^{1}H cross-section is flat at low energy before decreasing slowly for $E > 1\,\text{MeV}$. The low energy value, $\sigma_{el} \sim 20\,\text{b}$, is surprisingly large compared to that expected from the range of the strong interaction, $\pi(2\,\text{fm})^2 \sim 0.1\,\text{b}$. We will see in Sect. 3.6 that this is due to the fact that the proton–neutron system is slightly unbound if the two spins are anti-aligned (and slightly bound if they are aligned). For neutron momenta greater than the inverse range r of the strong interactions, $p > \hbar/r$ $[p^2/2m_n > \hbar^2/(r^2 m_n) > 1\,\text{MeV}]$, the cross-section drops down to a value more in line with the value expected from the range of the strong interactions.

The elastic cross-section for ^{6}Li shows a resonance at $E_n \sim 200\,\text{keV}$ which results from the production of an excited state of ^{7}Li that decays back to n ^{6}Li. The level diagram of ^{7}Li is shown in Fig. 3.5. For heavy nuclei, there are many excited states leading to a very complicated energy dependence of the cross-section, as illustrated for uranium in Fig. 3.26. The process of resonant production will be discussed in Sect. 3.5.

The angular distribution for elastic neutron–nucleus scattering is isotropic as long as $p < \hbar/R$ (R=nuclear radius) as illustrated in Fig. 3.6 and explained in Sect. 3.6. For $p > \hbar/R$ the angular distribution approaches that expected for diffraction from a semi-opaque object of radius R.

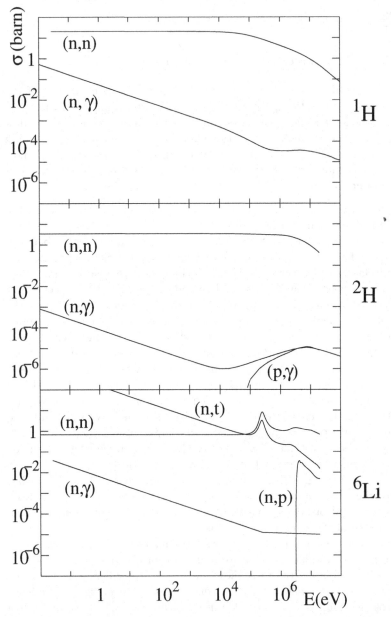

Fig. 3.4. Examples of reaction cross-sections on ^{1}H, ^{2}H, and ^{6}Li [30]. Neutron elastic scattering, (n,n), has a relatively gentle energy dependence while the exothermic reactions, (n,γ) and ^{6}Li(n,t)^{4}He (t=tritium=^{3}H), have a $1/v$ dependence at low energy. The exothermic (p,γ) reaction is suppressed at low energy because of the Coulomb barrier. The reaction ^{6}Li(n,p)^{6}Be has an energy threshold. The fourth excited state of ^{7}Li (Fig. 3.5) appears as a prominent resonance in n^{6}Li elastic scattering and in ^{6}Li(n,t)^{4}He.

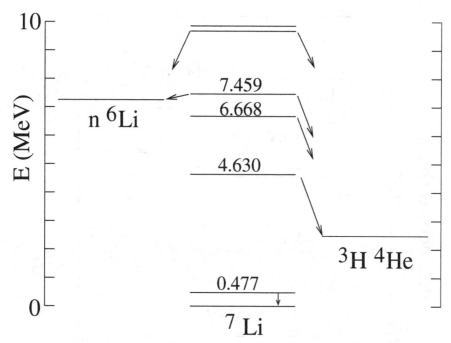

Fig. 3.5. The energy levels of ^{7}Li and two dissociated states n $-^6$ Li and ^{3}H $-^4$He. The first excited state of ^{6}Li decays to the ground state via photon emission while the higher excited states decay to ^{3}H$-^4$He. The fourth and higher excited states can also decay to n $-^6$ Li. The fourth excited state (7.459 MeV) appears prominently as a resonance in n ^{6}Li elastic scattering and in the exothermic (n,t) reaction n ^{6}Li $\to$ ^{3}H^{4}He. The resonance is seen at $E_n \sim 200\,\text{keV}$ in Fig. 3.4.

Inelastic scattering Inelastic reactions with no Coulomb barrier have cross-section dependences at low energy that depend on whether the reaction is exothermic or endothermic. Exothermic reactions generally have cross-section proportional to the inverse of the relative velocity, $\sigma \propto 1/v$. This leads to a velocity-independent reaction rate $\lambda \propto \sigma v$. Examples in the figures are neutron radiative capture (n, γ) reactions. The nucleus ^{7}Li also has an exothermic strong reaction n^{7}Li $\to$ ^{3}H ^{4}He. The resonance observed in elastic scattering is also observed in the inelastic channel since the resonant state (Fig. 3.5) can decay to ^{3}H ^{4}He.

Endothermic reactions have an energy threshold as illustrated in Fig. 3.4 by the (n,p) reaction n^{6}Li $\to$ p^{6}Be.

Coulomb barriers The low-energy cross-section for inelastic reactions are strongly affected by Coulomb barriers through which a particle must tunnel for the reaction to take place. Cross-sections for two exothermic reactions on ^{2}H are shown in Fig. 3.4. The barrier-free (n, γ) reaction n ^{2}H $\to$ γ ^{3}H has the characteristic $1/v$ behavior at low energy. On the other hand, the (p, γ) reaction between charged particles p ^{2}H $\to$ γ ^{3}He is strongly suppressed at

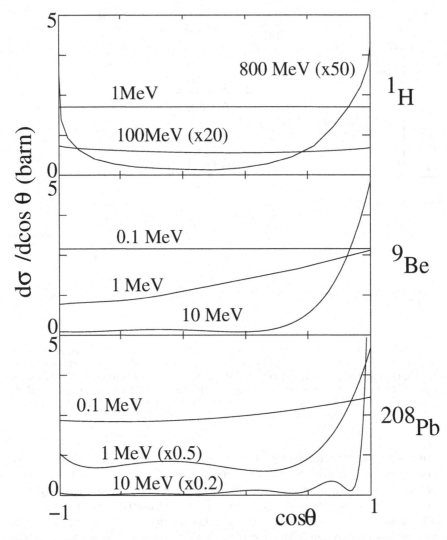

Fig. 3.6. The differential cross-section, $\mathrm{d}\sigma/\mathrm{d}\cos\theta = 2\pi\mathrm{d}\sigma/\mathrm{d}\Omega$, for elastic scattering of neutrons on ^{1}H, ^{9}Be and ^{208}Pb at incident neutron energies as indicated [30]. At low incident momenta, $p < \hbar/R_{\mathrm{nucleus}}$, the scattering is isotropic whereas for high momenta, the angular distribution resembles that of diffraction from a disk of radius R. Neutron scattering on ^{1}H at high-energy also has a peak in the backward directions coming from the exchange of charged pions (Fig. 1.13).

low energy. Its cross section becomes comparable to the (n, γ) reaction only for proton energies greater than the potential energy at the surface of the ^{6}Li nucleus $\sim 3\alpha\hbar c/2.4\,\text{fm} \sim 1.8\,\text{MeV}$.

High-energy inelastic collisions The Coulomb barrier becomes ineffective at sufficiently high energy, $E_{\text{cm}} > Z_1 Z_2 \alpha\hbar c/R$ where R is the sum of the radii of the nuclei of charges Z_1 and Z_2. In this case, the total inelastic cross-section becomes of order of the geometrical cross-section πR^2. At energies $< 1\,\text{GeV}$, most inelastic collisions involve a simple break up of one or both of the nuclei, leading to the production of the unstable nuclei present in Fig. 0.2. These are called *fragmentation* reactions for medium-A nuclei while the breakup of a heavy nucleus is called *collision-induced fission*. Fragmentation of a target by protons or neutrons is called *spallation*.

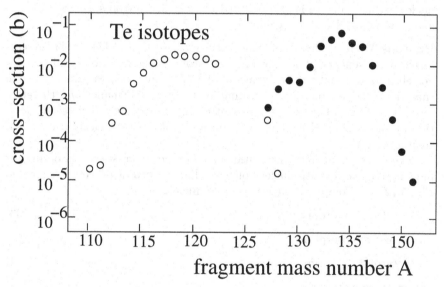

Fig. 3.7. The production of tellurium isotopes in the fragmentation of ^{129}Xe (790 MeV/nucleon) on a ^{27}Al target (open circles) and the collision-induced fission of ^{238}U (750 MeV/nucleon) on a Pb target (filled circles) [31]. Fragmentation leads to proton-rich isotopes while fission leads to neutron-rich isotopes.

Figure 3.7 gives the distribution of tellurium isotopes produced in the fragmentation of ^{129}Xe on a ^{27}Al target and the collision-induced fission of ^{238}U on a Pb target (filled circles) [31]. Fragmentation of Xe leads to proton-rich isotopes since mostly neutrons are ejected during the collision. Fission of uranium gives neutron-rich isotopes because of its large neutron-to-proton ratio. Reactions like these are the primary source of radioactive nuclides now used in the production of radioactive beams.

Occasionally, the target and projectile nuclei may fuse to form a much heavier nucleus. The produced nucleus is generally sufficiently excited to emit neutrons until a bound nucleus is produced. Such reactions are called *fusion–evaporation* reactions. This is the mechanism used to produce trans-uranium nuclei, as discussed in Sect. 2.8. The cross-section for the production of the heaviest elements is tiny, of order 10 pb for element 110.

For center-of-mass energies $> 1\,\mathrm{GeV/nucleon}$, inelastic nuclear collisions generally result in the production of pions and other hadrons. Collisions of cosmic-ray protons with nuclei in the upper atmosphere produce pions whose decays give rise to the muons that are the primary component of cosmic rays at the Earth's surface (Fig. 5.4).

Finally, we mention that for center-of-mass energies $> 100\,\mathrm{GeV/nucleon}$, certain collisions between heavy ions are believed to produce a state of matter called a *quark–gluon plasma* where the constituents of nucleons and hadrons are essentially free for a short time before recombining to form hadrons and nucleons. Such a state is also believed to exist in neutron stars (Sect. 8.1.2) and in the early Universe (Chap. 9).

Photons We have already calculated the cross-section (3.13) for elastic scattering of low-energy photons on free electrons. Since the cross-section is inversely proportional to the square of the electron mass, we can anticipate that the cross-section on free protons is 2000^2 times smaller and therefore negligible. This is because photon scattering is analogous to classical radiation of an accelerated charge, and a heavy proton is less easily accelerated than an electron.

The most important contributions to the photon cross-section on matter have nothing to do with nuclear physics. The important processes, shown in Fig. 5.12, are Compton scattering on atomic electrons

$$\gamma \,\mathrm{atom} \;\to\; \mathrm{atom}^+ \, e^- \, \gamma \,, \tag{3.33}$$

photoelectric absorption

$$\gamma \,\mathrm{atom} \;\to\; \mathrm{atom}^+ \, e^- \,, \tag{3.34}$$

and pair production

$$\gamma\,(A,Z) \;\to\; e^+ e^- \,(A,Z)\,. \tag{3.35}$$

Pair production dominates at energies above the threshold $E_\gamma = 2m_e c^2$.

Just as photons can breakup atoms through photoelectric absorption, they can excite or break up nuclei through photo-nuclear absorption. The cross-sections for this process on ^{2}H and ^{208}Pb are shown in Fig. 3.8. The cross-section for dissociation of ^{2}H exhibits a threshold at $E_\gamma = 2.22\,\mathrm{MeV}$, the binding energy of ^{2}H. The cross-section for dissociation of ^{208}Pb exhibits a broad *giant resonance* structure typical of heavy nuclei. Such resonances can be viewed semi-classically as the excitation of a *collective* oscillation of the proton in the nucleus with respect to the neutrons.

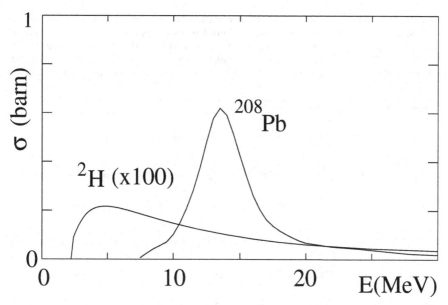

Fig. 3.8. The cross-sections for photo-dissociation of ^{2}H and of ^{208}Pb [30]. The cross-section of Pb exhibits a *giant resonance* typical of heavy nuclei.

Neutrinos Methods for calculating neutrino cross-sections will be presented in Sects. 3.2 and 3.4. Since neutrinos are subject to only weak interactions, their cross sections are considerably smaller than those of other particles. For neutrino energies much less than the masses of the intermediate vector bosons, $m_W c^2 = 80.4\,\mathrm{GeV}$ and $m_Z c^2 = 91.2\,\mathrm{GeV}$ the cross-sections are proportional to the square of the Fermi constant

$$\frac{G_F^2}{(\hbar c)^4} = 5.297 \times 10^{-48}\,\mathrm{m^2 MeV^{-2}} \ . \tag{3.36}$$

By dimensional analysis, this quantity must be multiplied by the square of an energy to make a cross-section. The cross-sections for several neutrino induced reactions are given in Table 3.1. For nuclear physics, neutrinos of energy $E_\nu \sim 1\,\mathrm{MeV}$ are typical so, multiplying (3.36) by $1\,\mathrm{MeV^2}$, gives cross-sections of order $10^{-48}\mathrm{m^2}$.

3.2 Classical scattering on a fixed potential

In this section, we consider the scattering of a particle in a fixed force field described by a potential $V(r)$. This corresponds to situations where the fact that the target particle recoils has little effect on the movement of the beam particle because the kinetic energy of the recoiling target particle can be neglected. For a beam particle of mass m_b and momentum p_b incident on a

Table 3.1. The cross-sections for selected neutrino-induced reactions important for nuclear physics. The energy range where the formulas are valid are given. Apart from G_F^2 given by (3.36), the cross-sections depend on the "weak mixing angle" $\sin^2 \theta_w = .2312 \pm 0.0002$, the "Cabibo angle" $\cos \theta_c = 0.975 \pm 0.001$, and the "axial-vector coupling" $g_A = 1.267 \pm 0.003$. The meaning of these quantities is discussed in Chap. 4.

reaction	cross-section	
$\nu_e \, e^- \rightarrow \nu_e \, e^-$	$\frac{G_F^2 E_{cm}^2}{4\pi(\hbar c)^4} \left[(2\sin^2 \theta_w + 1)^2 + \frac{4}{3}\sin^4 \theta_w \right]$	$E_{cm} \gg m_e c^2$
$\bar{\nu}_e \, e^- \rightarrow \bar{\nu}_e \, e^-$	$\frac{G_F^2 E_{cm}^2}{4\pi(\hbar c)^4} \left[\frac{1}{3}(2\sin^2 \theta_w + 1)^2 + 4\sin^4 \theta_w \right]$	$E_{cm} \gg m_e c^2$
$\nu_\mu \, e^- \rightarrow \nu_\mu \, e^-$	$\frac{G_F^2 E_{cm}^2}{4\pi(\hbar c)^4} \left[(2\sin^2 \theta_w - 1)^2 + \frac{4}{3}\sin^4 \theta_w \right]$	$E_{cm} \gg m_e c^2$
$\bar{\nu}_\mu \, e^- \rightarrow \bar{\nu}_\mu \, e^-$	$\frac{G_F^2 E_{cm}^2}{4\pi(\hbar c)^4} \left[\frac{1}{3}(2\sin^2 \theta_w - 1)^2 + 4\sin^4 \theta_w \right]$	$E_{cm} \gg m_e c^2$
$\nu_e \, n \rightarrow e^- \, p$	$\frac{G_F^2 E_\nu^2}{\pi(\hbar c)^4} \cos^2 \theta_c \left[1 + 3g_A^2 \right]$	$m_e c^2 \ll E_\nu \ll m_p c^2$
$\bar{\nu}_e \, p \rightarrow e^+ \, n$	$\frac{G_F^2 E_\nu^2}{\pi(\hbar c)^4} \cos^2 \theta_c \left[1 + 3g_A^2 \right]$	$m_e c^2 \ll E_\nu \ll m_p c^2$

target of mass m_t, it can be shown (Exercise 3.5) that the target recoil has negligible effect if the target rest-energy $m_t c^2$ is much greater than the beam energy

$$m_t c^2 \gg E_b = \sqrt{m_b^2 c^4 + p_b^2 c^2} \,. \tag{3.37}$$

Since nucleons and nuclei are so much heavier than electrons and neutrinos, these conditions will be satisfied in physically interesting situations. This fact, plus the mathematical simplification coming from ignoring target recoil, justifies spending some time on potential scattering. We will therefore first treat the problem classically by following the trajectories of particles through the force field. This will be followed by two quantum-mechanical treatments, the first using perturbation theory and plane waves, and the second using wave packets.

3.2.1 Classical cross-sections

Classically, cross-sections are calculated from the trajectories of particles in force fields. Consider a particle in Fig. 3.9 that passes through a spherically symmetric force field centered on the origin. The particle's original trajectory is parametrized by the "impact parameter" b which would give the particle's distance of closest approach to the force center if there were no scattering.

The scattering angle $\theta(b)$ depends on the impact parameter, as in the figure. The relation $\theta(b)$ or $b(\theta)$ can be calculated by integrating the equations

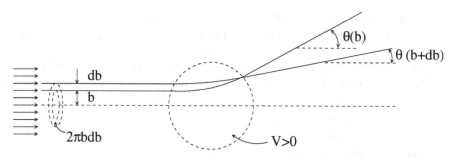

Fig. 3.9. The scattering of a particle of momentum p by a repulsive force. The trajectories for impact parameters b and $b + db$ are shown. The probability that a particle is scattered by an angle between $\theta(b)$ and $\theta(b + db)$ is proportional to the surface area $2\pi b db$.

of motion with the initial conditions $p_z = p$, $p_x = p_y = 0$. The probability that a particle is scattered into an interval $d\theta$ about θ is proportional to the area of the annular region between $b(\theta)$ and $b(\theta + d\theta) = b + db$, i.e. $d\sigma = 2\pi b db$. The solid angle corresponding to $d\theta$ is $d\Omega = 2\pi \sin \theta d\theta$. The differential elastic scattering cross-section is therefore

$$\frac{d\sigma}{d\Omega}(\theta) = \frac{2\pi b db}{2\pi \sin \theta d\theta} = \left| \frac{b(\theta)}{\sin \theta} \cdot \frac{db}{d\theta} \right| . \tag{3.38}$$

A measurement of $d\sigma/d\Omega$ determines the relation $b(\theta)$ which in turn gives information about the potential V.

3.2.2 Examples

We can apply (3.38) to several simple cases:

- Scattering of a point particle on a hard immovable sphere. The angle-impact parameter relation is

$$b = R \cos \theta/2 , \tag{3.39}$$

where R is the radius of the sphere. The cross section is then

$$\frac{d\sigma}{d\Omega} = R^2/4 \quad \Rightarrow \quad \sigma = \pi R^2 \tag{3.40}$$

so the total cross-section is just the geometrical cross section of the sphere. In the case of scattering of two spheres of the same radius, the total scattering cross-section is $\sigma = 4\pi R^2$.
- Scattering of a charged particle in a Coulomb potential

$$V(r) = \frac{Z_1 Z_2 e^2}{4\pi \epsilon_0 r} , \tag{3.41}$$

where Z_1 is the charge of the scattered particle, and Z_2 is the charge of the immobile target particle. This historically important reaction is called

"Rutherford scattering" after E. Rutherford who demonstrated the existence of a compact nucleus by studying α-particle scattering on gold nuclei. The unbound orbits in the Coulomb potential are hyperbolas so the scattering angle is well-defined in spite of the infinite range of the force. For an incident kinetic energy $E_k = mv^2/2$, the angle-impact parameter relation is

$$b = \frac{Z_1 Z_2 e^2}{8\pi\epsilon_0 E_k} \cot(\theta/2) \,. \tag{3.42}$$

The cross-section is then

$$\frac{d\sigma}{d\Omega} = \left(\frac{Z_1 Z_2 e^2}{16\pi\epsilon_0 E_k}\right)^2 \frac{1}{\sin^4 \theta/2} \,. \tag{3.43}$$

We note that the total cross-section $\sigma = \int (d\sigma/d\Omega) d\Omega$ diverges because of the large differential cross-section for small-angle scattering:

$$\frac{d\sigma}{d\Omega} \sim \left(\frac{Z_1 Z_2 e^2}{4\pi\epsilon_0 E_k}\right)^2 \frac{1}{\theta^4} \quad (\theta \ll 1) \,. \tag{3.44}$$

This divergence is due to the fact that incident particles of arbitrarily large impact parameters are deflected. The total elastic cross-section for scattering angles greater than θ_{min} is (using $d\Omega \sim 2\pi\theta d\theta$ for $\theta \ll 1$)

$$\sigma(\theta > \theta_{min}) \sim \left(\frac{Z_1 Z_2 e^2}{4\pi\epsilon_0 E_k}\right)^2 \frac{\pi}{\theta_{min}^2} \quad (\theta_{min} \ll 1) \,. \tag{3.45}$$

- Scattering of particles in a Yukawa potential

$$V(r) = \frac{g\hbar c}{r} e^{-r/r_0} \,. \tag{3.46}$$

This potential is identical to the Coulomb potential for $r \ll r_0$ but approaches zero much faster for $r > r_0$. Unlike the case of the Coulomb potential, there is no analytical solution for particle trajectories. It is necessary to integrate numerically the equations of motion to find $\theta(b)$ and $d\sigma/d\Omega$. The result is shown in Fig. 3.10 for an incident particle of energy $E_k = 10g\hbar c/r_0$. We see that for $b < r_0$ (corresponding to $\theta > 0.1$) the scattering angle approaches that for the Coulomb potential, as expected since the two potentials have the same form for $r/r_0 \to 0$. For $b > r_0$, the scattering angle is smaller than the angle for Rutherford scattering since the Yukawa force falls rapidly for $r \gg r_0$. It follows that the differential cross section for small angles is smaller than that for Rutherford scattering, diverging as θ^{-2} rather than as θ^{-4}. The elastic cross section still diverges but only logarithmically, $\sigma(\theta_{min}) \propto \log(\theta_{min})$. We see from the figure that

$$\sigma(\theta > 0.01) = \pi[b(0.01)]^2 \sim 4\pi r_0^2 \,. \tag{3.47}$$

An angle of 0.01 is already quite small and to get a much higher cross-section one has to go to considerably smaller angles. For the Yukawa potential, πr_0^2 therefore gives the order of magnitude of the cross-section for scattering by measurably large angles. We shall see below that in the quantum-mechanical calculation, the cross-section is finite.

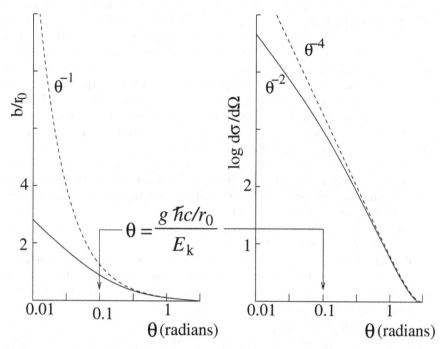

Fig. 3.10. The scattering of a non-relativistic particle in a Yukawa potential $V(r) = g\hbar c\, e^{-r/r_0}/r$. The initial kinetic energy of the particle is taken to be $10g\hbar c/r_0$ so that it can penetrate to about $r = r_0/10$. The left solid curve shows the numerically calculated impact parameter $b(\theta)$ in units of r_0. The right solid curve shows the logarithm of the differential scattering normalized to the backward scattering cross-section ($\theta = \pi$). For comparison, the dashed lines show the same functions for the Coulomb potential $V(r) = g\hbar c/r$. For $\theta > 0.1$, we have $b < r_0$ and the two potentials give nearly the same results. This is to be expected since the two potentials are nearly equal for $r \ll r_0$. For $b \gg r_0$, the Yukawa scattering angle is much less than the Coulomb scattering angle because the force drops of much faster with distance. As a consequence, the cross section is smaller.

Much of our knowledge of nuclear structure comes from the scattering of charged particles (generally electrons) on nuclei. This is because high-energy charged particles penetrate inside the nucleus and their scattering-angle distribution therefore gives information on the distribution of charge in the nucleus. We will see that the correct interpretation of these experiments requires quantum-mechanical calculation of the cross-sections. However, it turns out that the quantum-mechanical calculation of scattering in a $1/r$ potential gives the same result as the classical Rutherford cross-section found above. This means that the Rutherford cross-section can be used to interpret experiments using positively charged particles whose energy is sufficiently low that they cannot penetrate inside the nucleus.

This is how, in 1908, Rutherford discovered that the positive charge inside atoms is contained in a small "nucleus." Rutherford reached this conclusion after hearing of the results of experiments by Geiger and Marsden studying the scattering of α-particles on gold foils. While most α's scattered into the forward direction, they occasionally scatter backward. This was impossible to explain with the then popular "plum pudding" model advocated by J.J. Thomson where the atom consisted of electrons held within a positively charged uniform material. A heavy α-particle cannot be deflected through a significant angle by the much lighter electron. On the other hand, scattering at large angle would be possible in rare nearly head-on collisions with a massive, and therefore immobile, gold nucleus.

After this brilliant insight, Rutherford spent some time (2 weeks [5]) calculating the expected angular distribution which turned out to agree nicely with the observed distribution. Rutherford's model naturally placed the light electrons in orbits around the heavy nucleus.

Another 17 years were necessary to develop the quantum mechanics that explains atomic structure and dynamics.

We expect the Rutherford cross-section calculation to fail if the electron can penetrate inside the nucleus. Classically, this will happen for head-on collisions if the initial kinetic energy of the α particle is greater than the electrostatic potential at the nuclear surface:

$$E_k = \frac{2Ze^2}{4\pi\epsilon_0 R} = \frac{2Z\alpha\hbar c}{R} \sim 1.2 \, A^{2/3} \, \text{MeV} , \tag{3.48}$$

where $2Z$ is the product of the α-particle and nuclear charges, α is the fine structure constant, and we have used $R \sim 1.2A^{1/3}$ fm and $Z \sim A/2$. We expect the backward scattering to be suppressed for energies greater than this value. The naturally occurring α-particles used by Rutherford have energies of order 6 MeV so the effect can only be seen for $Z/A^{1/3} < 3$ corresponding to $A < 11$. Rutherford and collaborators used this effect to perform the first measurement of nuclear radii.

3.3 Quantum mechanical scattering on a fixed potential

Quantum mechanical collision theory is treated at length in many textbooks. [1] In this section, we will present two simple approximate methods applicable to scattering due to weak and electromagnetic interactions. The

---

[1] See for instance M. L. Goldberger and K. M. Watson, *Collision Theory*, John Wiley & Sons, 1964; K. Gottfried, *Quantum Mechanics*, W. A. Benjamin, 1966; J-L. Basdevant and J. Dalibard, *Quantum Mechanics* Chapter 18, Springer-Verlag, 2002.

first uses standard time-dependent perturbation theory applied to momentum eigenstates and the second uses wave packets. The first is an essential part of this chapter because it can be easily generalized to inelastic scattering. The second is mostly a parenthetical section intended to improve our understanding of the physics.

To prepare the ground for the perturbation calculation, we first briefly discuss the concept of *asymptotic states* and their normalization. Other technical ingredients, the limiting forms of the Dirac δ function, and basics results of time-dependent perturbation theory in quantum mechanics are reviewed in Appendix C.

3.3.1 Asymptotic states and their normalization

In studying nuclear, or elementary interactions, we are most of the time not interested in a space-time description of phenomena.[2] Instead, we study processes in which we prepare initial particles with definite momenta and far away from one another so that they are out of reach of their interactions at an initial time t_0 in the "distant past" $t_0 \sim -\infty$. We then study the nature and the momentum distributions of final particles when these are also out of range of the interactions at some later time t in the "distant future" $t \rightarrow +\infty$. (The size of the interaction region is of the order of 1 fm, the measuring devices have sizes of the order of a few meters.)

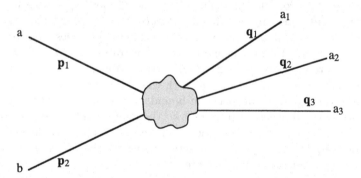

Fig. 3.11. Asymptotic states in a collision

Under these assumptions, the initial and final states of the particles under consideration are *free* particle states. These states are called *asymptotic states*. The decay of an unstable particle is a particular case. We measure the energy and momenta of final particles in asymptotic states.

---

[2] One exception is the case of "neutrino oscillations" discussed in Sect. 4.4.

By definition, the asymptotic states of particles have definite momenta. Therefore, strictly speaking, they are not physical states, and their wave functions $e^{ipx/\hbar}$ are not square integrable. Physically, this means that we are actually interested in wave packets who have a non vanishing but very small extension Δp in momentum, i.e. $\Delta p/|p| \ll 1$.

It is possible to work with plane waves, provided one introduces a proper normalization. A limiting procedure, after all calculations are done, allows to get rid of the intermediate regularizing parameters. This is particularly simple in first order Born approximation, which we will present first. The complete manipulation of wave packets is possible but somewhat complicated. However, it gives interesting physical explanations for various specific problems, and we shall discuss it in Sect. 3.3.5.

We will consider that the particles are confined in a (very large but finite) box of volume L^3. We will let L tend to infinity at the end of the calculation. Besides its simplicity, this procedure allows to incorporate relativistic kinematics of ingoing and outgoing particles in a simple manner.

In such a box of size L, the normalized momentum eigenstates are

$$|\boldsymbol{p}\rangle \to \psi_{\boldsymbol{p}}(\boldsymbol{r}) = L^{-3/2}e^{i\boldsymbol{p}\cdot\boldsymbol{r}/\hbar} \quad \text{inside the box} \quad , \tag{3.49}$$

$$\psi_{\boldsymbol{p}}(\boldsymbol{r}) = 0 \quad \text{outside the box} \quad .$$

These wave functions are normalized in the sense that

$$\int |\psi_{\boldsymbol{p}}(\boldsymbol{r})|^2 \, \mathrm{d}^3\boldsymbol{r} \ = \ 1 \ . \tag{3.50}$$

For convenience, we will define here the Hilbert space with *periodic* boundary conditions : in one dimension $\psi(L/2) = \psi(-L/2)$ and $\psi'(L/2) = \psi'(-L/2)$ (this amounts to quantizing the motion of particles on a large circle of radius $R = L/2\pi$). In such conditions, the operators $\hat{p} = (\hbar/i)\partial/\partial x$ and $\hat{p}^2$ have a discrete spectrum $p_n = 2\pi n\hbar/L$. In three dimensions the quantization of momentum is $\boldsymbol{p} = (2\pi\hbar/L)(n_1, n_2, n_3)$, where the n_i are arbitrary integers.

The advantage of using periodic boundary conditions is that the states (3.49) are normalized eigenstates of *both* the energy and the momentum, as we wish. This is not the case in the usual treatment of a "particle in a box" where one requires that the wave function vanish at the edge of the box. The energy eigenfunctions in this case are

$$\psi_E(\boldsymbol{r}) \ = \ L^{-3/2} \sin n_x \pi x/L \ \sin n_y \pi y/L \ \sin n_z \pi z/L \tag{3.51}$$

where n_x, n_y, n_z are positive integers and $E = \pi^2\hbar^2(n_x^2 + n_y^2 + n_z^2)/2mL^2$. In this case the energy eigenfunctions are not eigenstates of momentum. However, both boundary conditions give the same density of states so we need not worry about which regularization procedure is used.

The orthogonality relation between momentum eigenstates

$$\langle \boldsymbol{p}|\boldsymbol{p}'\rangle = \delta_{n_1 n_1'}\delta_{n_2 n_2'}\delta_{n_3 n_3'}$$

can also be written in the following manner, useful to take limits,

$$\langle p|p'\rangle = (2\pi\hbar/L)^3 \Delta_L^3(\boldsymbol{p} - \boldsymbol{p}') \ , \tag{3.52}$$

where $\Delta_L^3(\boldsymbol{p} - \boldsymbol{p}')$ is a limiting form of the delta function discussed in Appendix (C.0.2).

Since each component of momentum is quantized in steps of $2\pi\hbar/L$, the number of states in a momentum volume $\mathrm{d}^3\boldsymbol{p}$ is

$$\mathrm{d}N(\boldsymbol{p}) = (2s+1)(\frac{L}{2\pi\hbar})^3 \mathrm{d}^3\boldsymbol{p} \equiv \rho(\boldsymbol{p})\mathrm{d}^3\boldsymbol{p} \tag{3.53}$$

where $2s+1$ is the number of spin-states for a particle of spin s. This defines the *density of states* (in momentum space): $\rho(\boldsymbol{p}) = (2s+1)(L/2\pi\hbar)^3$. This corresponds to a density in *phase space* (momentum×real space) of $(2s+1)$ states per elementary volume $(2\pi\hbar)^3$.

In what follows, we will be interested in the number of states in an interval $\mathrm{d}E$. To obtain this, we note that the number of states within a momentum volume $\mathrm{d}^3\boldsymbol{p}$ can be written as

$$\mathrm{d}N(\boldsymbol{p}) = \rho(\boldsymbol{p})\mathrm{d}^3\boldsymbol{p} = (2s+1)(\frac{L}{2\pi\hbar})^3 p^2\mathrm{d}p\mathrm{d}\Omega \ , \tag{3.54}$$

where $\mathrm{d}\Omega$ is the solid angle covered by $\mathrm{d}^3\boldsymbol{p}$. Taking $E\mathrm{d}E = c^2 \, p\mathrm{d}p$ (which holds both in the relativistic and non-relativistic regimes), we find the number of states in the interval $\mathrm{d}E$ and in the solid angle $\mathrm{d}\Omega$ is

$$\mathrm{d}N(E, \mathrm{d}\Omega) = (2s+1)(\frac{L}{2\pi\hbar})^3 \frac{pE}{c^2}\mathrm{d}E\mathrm{d}\Omega \ . \tag{3.55}$$

3.3.2 Cross-sections in quantum perturbation theory

The simplest way to calculate cross-sections in quantum mechanics is to use standard time-dependent perturbation theory (Appendix C). The idea is to describe the system by a Hamiltonian that is the sum of an "unperturbed" H_0 and a perturbation H_1. In the present context, H_0 will represent the kinetic energy of the incoming and outgoing beam particles and the perturbation H_1 will be the interaction potential (which acts for a very short time).

Perturbation theory gives the transition rates between energy eigenstates of the unperturbed Hamiltonian, i.e. between the initial state $|i\rangle$ of energy E_i and one of the possible final states $|f\rangle$ of energy E_f. The first order result is

$$\lambda_{i\to f} = \frac{2\pi}{\hbar} |\langle f|H_1|i\rangle|^2 \, \delta(E_f - E_i) \ , \tag{3.56}$$

where $\delta(E)$ satisfies

$$\int_{-\infty}^{\infty} \delta(E)\mathrm{d}E = 1 \ , \tag{3.57}$$

and is a limiting form of the delta function discussed in (C.0.2). The dimension of $\delta(E)$ is 1/energy so the transition rate λ has dimension of 1/time as expected.

If higher-order perturbation theory is necessary, energy will still be conserved in the transition rate since energy conservation is an exact result due to the time-translation invariance of the Hamiltonian. The more general transition rate including higher-order effects is then written as

$$\lambda_{i \to f} = \frac{2\pi}{\hbar} |\langle f|T|i\rangle|^2 \, \delta(E_f - E_i) \,, \tag{3.58}$$

where T is the "transition matrix element." In the context of scattering theory, the first order result, $T = H_1$, is called the "Born approximation."

For the initial and final states, we choose plane waves of momentum p and p' as defined above

$$\psi_i(r) = \frac{e^{i p \cdot r/\hbar}}{L^{3/2}} \qquad \psi_f(r) = \frac{e^{i p' \cdot r/\hbar}}{L^{3/2}} \,, \tag{3.59}$$

where L^3 is the normalization volume. The classical scattering angle is defined by

$$\cos \theta = \frac{p \cdot p'}{|p||p'|} \,. \tag{3.60}$$

Of more importance for quantum calculations is the momentum transfer

$$q = p' - p \,, \tag{3.61}$$

and its square

$$q^2 = q \cdot q = |p|^2 + |p'|^2 - 2p \cdot p' \,. \tag{3.62}$$

For elastic scattering we have $|p| = |p'| = p$ so

$$q^2 = 2p^2(1 - \cos \theta) = 4p^2 \sin^2 \theta/2 \quad \text{(elastic scattering)} \,. \tag{3.63}$$

The small angle limit is often useful:

$$q^2 \sim p^2 \theta^2 \quad \theta \ll 1 \,, \quad \text{(elastic scattering)} \,. \tag{3.64}$$

The matrix element between initial and final states is then

$$\langle p'|V|p\rangle = \frac{1}{L^3} \int e^{i(p-p')\cdot r/\hbar} V(r) \mathrm{d}^3 r = \frac{\tilde{V}(p-p')}{L^3} \,. \tag{3.65}$$

It is proportional to the Fourier transform of the potential

$$\tilde{V}(q) = \int e^{i q \cdot r/\hbar} V(r) \mathrm{d}^3 r \,. \tag{3.66}$$

Note that the dimensions of $\tilde{V}$ defined here are energy$\times$volume.

The transition rate to the final state is

$$\lambda_{i \to f} = \frac{2\pi}{\hbar} \frac{|\tilde{V}(p-p')|^2}{L^6} \delta(E' - E) \,. \tag{3.67}$$

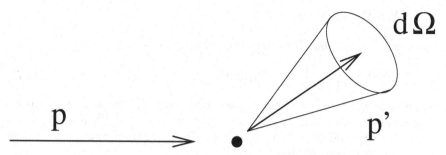

Fig. 3.12. Scattering of a single particle by a fixed potential.

We cannot measure the transition rate to a *single* momentum state so we must sum the transition rate over a group of interesting final states. Within a volume L^3, the number of momentum states in the momentum range d^3p' is given by (3.53). Multiplying (3.67) by this number of states we get the total transition rate into the momentum volume d^3p

$$\lambda(d^3p') = \left(\frac{L}{2\pi\hbar}\right)^3 d^3p' \frac{2\pi}{\hbar} \frac{|\tilde{V}(q)|^2}{L^6} \delta(E'-E) . \qquad (3.68)$$

We carefully drop the factor $(2s+1)$ in the number of states. We do this because it is often the case that only one spin state is produced with high probability in a reaction. When this is not the case, it is then necessary to sum over all possible final spin states.

The number of states in the momentum range dE' and momentum oriented into the solid angle $d\Omega$ at angles (θ, ϕ) with respect to a given direction is given by (3.55) so the total transition rate into these states is then

$$\lambda(dE', d\Omega) = \left(\frac{L}{2\pi\hbar}\right)^3 \frac{p'E'}{c^2} dE' d\Omega \frac{2\pi}{\hbar} \frac{|\tilde{V}(q)|^2}{L^6} \delta(E'-E) . \qquad (3.69)$$

We use the delta function to integrate over energy in order to find the transition rate into (energy-conserving) states within the solid angle $d\Omega$

$$\lambda(d\Omega) = \frac{v'}{L^3} \frac{(E')^2}{4\pi^2\hbar^4 c^4} |\tilde{V}(q)|^2 d\Omega , \qquad (3.70)$$

where $v' = p'c^2/E'$ is the velocity of the final state particle.

We remark, as mentioned above, that the crucial factor in (3.70) is the presence of the modulus squared of the Fourier transform of the potential $|\tilde{V}(q)|^2$. This is the main result of this calculation. Information on the differential cross-section gives us direct access to the potential through its Fourier transform. This result is very concise and elegant. It is basically the same effect that one encounters in diffraction phenomena. If one neglects multiple scattering, the amplitude of the diffraction pattern is the Fourier transform of the diffracting system (crystal, macro-molecule, etc.).

3.3.3 Elastic scattering

For elastic scattering, $E' = E$, so the transition rate is

$$\lambda(d\Omega) = \frac{v}{L^3} \frac{E^2}{4\pi^2\hbar^4 c^4} |\tilde{V}(\boldsymbol{q})|^2 \, d\Omega \,, \qquad (3.71)$$

where v is the velocity of the initial state particle. The transition rate is proportional to the density of scattering centers ($n = 1/L^3$) and to the velocity of the projectile. Using (3.30), we divide by these two factors to get the differential cross-section

$$\frac{d\sigma}{d\Omega} = \frac{E^2}{4\pi^2(\hbar c)^4}|\tilde{V}(\boldsymbol{q})|^2 \,, \qquad (3.72)$$

where $E = \sqrt{p^2 c^2 + m^2 c^4}$ is the energy of the incident particle and $\tilde{V}$ is given by (3.66).

We remark that in the above expression the normalization parameter L cancels off identically. Therefore, we can readily take the limit $L \to \infty$.

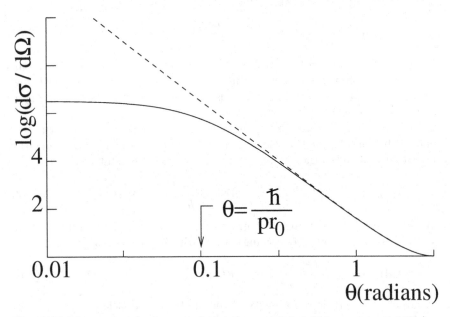

Fig. 3.13. The scattering of a non-relativistic particle in a Yukawa potential $V(r) = g\hbar c e^{-r/r_0}/r$. The momentum of the particle is $p = 10\hbar/r_0$. The solid line shows the quantum mechanical differential cross-section (3.75). For small angles $\theta < \hbar/(pr_0)$ the cross section is flat, avoiding the divergence present in the classical calculation (Fig. 3.10). At large angles $\theta > \hbar/(pr_0)$ the scattering follows the Coulomb cross-section shown by the dashed line.

As an example of potential scattering, we can take the Yukawa potential

$$V(r) = \frac{g\hbar c}{r} e^{-r/r_0} , \qquad (3.73)$$

where the range of the interaction is the Compton wavelength $r_0 = \hbar/Mc$ of the exchanged particle of mass M. The Coulomb potential between particles of charge Z_1 and Z_2 corresponds to $g = Z_1 Z_2 \alpha$ and $r_0 \to \infty$. The Fourier transform $\tilde{V}(q)$ for the Yukawa potential is

$$\tilde{V}(q) = \frac{4\pi g\hbar c \hbar^2}{q^2 + (\hbar/r_0)^2} = \frac{4\pi g(\hbar c)^3}{q^2 c^2 + M^2 c^4} , \qquad (3.74)$$

which gives a differential cross-section

$$\frac{d\sigma}{d\Omega} = 4g^2 (\hbar c)^2 \left(\frac{E}{4p^2 c^2 \sin^2 \theta/2 + M^2 c^4} \right)^2 , \qquad (3.75)$$

where we have used (3.63). The cross section, shown in Fig. 3.13, does not diverge at small angles like the classical cross-section. The total elastic cross-section is therefore finite:

$$\sigma_{el} = 8\pi g^2 (\hbar c)^2 \left(\frac{E}{M^2 c^4} \right)^2 \frac{1}{1 + 2p^2/M^2 c^2} . \qquad (3.76)$$

We remark that in many strong interaction calculations, the Born approximation is not valid. Indeed the dimensionless parameter g is larger than one and perturbation theory does not apply. Nevertheless, the above result bears many qualitatively useful features.

In what follows, we consider cases where the Born approximation is valid. There are two simple limits corresponding to the mass of the exchanged particle Mc^2 being much greater than or much less than pc.

- $Mc^2 \ll pc$, i.e. $r_0 \gg \hbar/p$. As illustrated in Fig. 3.13, the differential cross section is angle-independent for $\theta < \hbar/(pr_0)$ and Rutherford-like for $\theta > \hbar/(pr_0)$. We can then find the cross-section for the Coulomb potential by taking the limit $r_0 \to \infty$ and setting $g = Z_1 Z_2 \alpha$:

$$\frac{d\sigma}{d\Omega} = \left(\frac{Z_1 Z_2 e^2}{4\pi\epsilon_0} \right)^2 \left(\frac{E}{2p^2 c^2} \right)^2 \frac{1}{\sin^4 \theta/2} . \qquad (3.77)$$

In the non-relativistic limit, $E = mc^2$, $E_k = p^2/2m$, and the formula reduces to the classical Rutherford cross-section (3.43).

$$\frac{d\sigma}{d\Omega} = \left(\frac{Z_1 Z_2 e^2}{16\pi\epsilon_0} \right)^2 \left(\frac{1}{p^2/2m} \right)^2 \frac{1}{\sin^4 \theta/2} . \qquad (3.78)$$

This coincidence of the classical and the quantum theory seems, at first, amazing. It is actually quite simple to understand by dimensional analysis. The non-relativistic cross-section (3.78) calculated quantum mechanically turns out to be independent of $\hbar$. It is proportional to the square of the *only* length, $a = e^2/4\pi\epsilon_0(p^2/2m)$, that is linear in $e^2/4\pi\epsilon_0$ and a combination of powers of p, m and $\hbar$. Since this is the only length available, the quantum cross-section must be $\hbar$-independent and can therefore agree with a classical cross-section, also $\hbar$-independent.[3] The same is not true for the Yukawa potential where the differential cross-section derived from (3.74) depends on $\hbar$ and consequently *cannot* agree with the classical calculation, as seen by comparing Figs. (3.10) and (3.13). It is the existence of another length scale, r_0, that allows one to form a cross-section that depends on $\hbar$.

In the case where the incident particle is ultra-relativistic, $E \sim pc$, we have

$$\frac{d\sigma}{d\Omega} = \left(\frac{Z_1 Z_2 \alpha}{2}\right)^2 \left(\frac{\hbar c}{E}\right)^2 \frac{1}{\sin^4 \theta/2} \quad . \tag{3.79}$$

The cross-section is proportional to α^2 and to the square of the only length, $\hbar c/E$, that can be formed from $\hbar$, c, and E. (In the relativistic limit, the cross-section no longer depends on m.)

- $Mc^2 \gg pc$, i.e. $r_0 \ll \hbar/p$. In this case, the differential cross-section is angle-independent for all θ

$$\frac{d\sigma}{d\Omega} = \frac{G^2 E^2}{4\pi^2 (\hbar c)^4} \quad \text{i.e.} \quad \sigma = \frac{G^2 E^2}{\pi (\hbar c)^4} \quad , \tag{3.80}$$

where

$$G = 4\pi \frac{g(\hbar c)^3}{(Mc^2)^2} . \tag{3.81}$$

Not surprisingly, this cross-section results also from the delta potential, i.e. a contact interaction.

$$V(r) = G\delta^3(\boldsymbol{r}) \quad \Rightarrow \tilde{V}(\boldsymbol{q}) = G . \tag{3.82}$$

This potential is a good approximation for neutrino interactions like

$$\nu_e e^- \rightarrow \nu_e e^- . \tag{3.83}$$

From Table 3.1, we see that in this case $G \propto G_F$ where G_F is the Fermi constant.

---

[3] Another puzzle lies in the fact that (3.77), which obtained in perturbation theory, actually coincides with the exact non-relativistic result, which can be calculated analytically with the Schödinger equation (see for instance A. Messiah, *Quantum Mechanics* vol. 1, chap. XI-7). This "miraculous" coincidence comes from the fact that since Coulomb forces are long range forces, one is not allowed, in principle, to make use of plane waves as asymptotic states. One should rather use Coulomb wavefunctions, defined in Messiah, as asymptotic states. The miracle is that the sum of the correct perturbation series gives exactly the simple plane-wave formula. This is again related to the fact that $\hbar$ is absent in the classical result.

3.3.4 Quasi-elastic scattering

Potential scattering most naturally applies to elastic scattering because of the classical limit of a light particle moving through the force field of a fixed heavy particle. However, in the quantum treatment, we saw that the potential simply serves to calculate a matrix element between initial and final states. It is not surprising therefore that the same formalism applies to "quasi-elastic" scattering where the light particle changes its nature (i.e. its mass) when it interacts with a fixed particle. Obvious candidates are the weak interactions of leptons scattering on nucleons, e.g.

$$\bar{\nu}_e \, p \leftrightarrow e^+ \, n \, . \tag{3.84}$$

We note that the reaction going to the right is endothermic and the reaction going to the left is exothermic. Since these two reactions are due to the exchange of W bosons, we can use a delta-potential of the form (3.82) and rely on the fundamental theory of weak interactions (Table 3.1) to provide us with the effective G for each reaction.

The rate calculation proceeds as in the elastic case up to (3.70) at which point we have to take into account the fact the the initial and final state momenta are not equal. Since we will want to factor out the initial state velocity, we write the rate as

$$\lambda(\mathrm{d}\Omega) = \frac{v}{L^3} \frac{v'}{v} \frac{(E')^2}{4\pi^2\hbar^4 c^4} |\tilde{V}(\boldsymbol{q})|^2 \, \mathrm{d}\Omega \, , \tag{3.85}$$

corresponding to a cross-section

$$\frac{\mathrm{d}\sigma}{\mathrm{d}\Omega} = \frac{v'}{v} \frac{(E')^2}{4\pi^2\hbar^4 c^4} |\tilde{V}(\boldsymbol{q})|^2 \, . \tag{3.86}$$

For the delta-potential, the angular distribution is isotropic and the cross-section is

$$\sigma = \frac{v'}{v} \frac{(E')^2}{\pi\hbar^4 c^4} G^2 \, . \tag{3.87}$$

At sufficiently high energy, the initial and final state velocities approach c so the factor v'/v is of no importance. At low energy, this factor generates a very different behavior for the two reactions.

The endothermic reaction $\bar{\nu}_e p \to e^+ n$ has a threshold neutrino energy of $E_{\mathrm{th}} = (m_n + m_e - m_p)c^2 = 1.8 \, \mathrm{MeV}$. Near threshold, the final state positron has an energy $E' \sim m_e c^2$ and a velocity $v' \sim \sqrt{2(E_\nu - E_{\mathrm{th}})/m_e}$. The initial velocity for the nearly massless neutrino is $v \sim c$ so the cross-section is

$$\sigma = \left(\frac{2(E_\nu - E_{\mathrm{th}})}{m_e c^2} \right)^{1/2} \frac{(m_e c^2)^2}{\pi\hbar^4 c^4} G^2 \qquad E_\nu > E_{\mathrm{th}} \, . \tag{3.88}$$

The situation for the exothermic reaction $e^+ n \to \bar{\nu}_e p$ is quite different. As the velocity v of the positron approaches zero, the energy E' of the final state neutrino approaches $(m_n + m_e - m_p)c^2 = 1.8\,\text{MeV}$ so the cross-section approaches

$$\sigma = \frac{c}{v} \frac{(m_n + m_e - m_p)^2 c^4}{\pi \hbar^4 c^4} G^2 \,. \tag{3.89}$$

The cross-section is proportional to the inverse of the velocity, as anticipated in Sect. 3.1.5. The reaction rate, proportional to the product of the velocity and the cross-section is therefore velocity independent.

3.3.5 Scattering of quantum wave packets

The calculations of the last section were very efficient in yielding reaction rates and cross-sections in cases where perturbation theory applies. However, they are not able to elicit various physical properties of interest. In this section, we will provide a more physical description using wave packets, which we shall use later on.

In quantum mechanics, particles are represented by wavefunctions, $\psi(\mathbf{r})$ giving the probability $|\psi(\mathbf{r})|^2 \mathrm{d}^3 r$ to find the particle in a volume $\mathrm{d}^3 r$ near $\mathbf{r}$. If the particle interacts only via a potential $V(r)$, the wavefunction satisfies the Schrödinger equation

$$\mathrm{i}\hbar \frac{\partial \psi}{\partial t} = \frac{-\hbar^2}{2m} \nabla^2 \psi + V(r)\psi \,. \tag{3.90}$$

As illustrated in Fig. 3.14, a scattering experiment on a single target particle with a short range potential corresponds to the situation where $V(r) \sim 0$ except in a small region $r < R$ near the target particle. Initially, the wavefunction is a broad wave packet, ψ_{in}, that propagates freely in the z direction far from the target. The transverse width of the wave packet is taken to be much greater than R so that the entire potential is "sampled" by the wavefunction. When the wave packet reaches the target ($t = 0$), the interaction with the potential generates a scattered wave packet ψ_{sc} which now accompanies the transmitted wave packet.

The essential result of the calculation that follows is that the scattered wave is found by summing spherical waves emanating from each point in the region where $V \neq 0$. This is illustrated in Fig. 3.15. It will turn out that the scattered wave from each point is proportional to the product of the potential and the incident wave at that point. This is physically reasonable since the scattered wave must vanish when either the potential or the incident wave vanishes. When one integrates the waves over the region of non-vanishing potential, the result (3.113) that the scattered wave is proportional to the Fourier transform of the potential will emerge in a natural way. Physically, this comes about since, as illustrated in Fig. 3.15, the waves add coherently in the forward direction but with increasingly random phases away from the forward direction. This leads to a decreasing cross-section with increasing

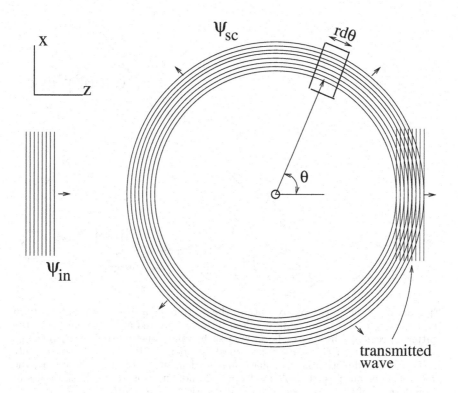

Fig. 3.14. A wave packet that impinges upon a region with $V(r) \neq 0$ will interact in a way that will produce a scattered wave packet and a transmitted wave packet. The probability to find the particle in the box in the scattered wave is proportional to the differential scattering cross-section, $d\sigma/d\Omega$.

angle. Mathematically, this is just what the Fourier transform does since it is maximized at $\boldsymbol{q} = 0$ ($\theta = 0$).

We now start the wave packet calculation of the differential cross-section. This cross-section is related to the rate of particles counted by a detector placed at an angle θ with respect to the beam shown in Fig. 3.14. The rate is given by

$$\frac{dN_{\text{det}}}{dt} = N_{\text{T}} F \frac{d\sigma}{d\Omega} d\Omega \,, \tag{3.91}$$

where N_{T} is the number of target particles and F is the incident particle flux. We average the flux over some arbitrary time T much greater than the time of passage of the wave packet. The mean incident flux is given by the probability to find the incident particle near the z-axis:

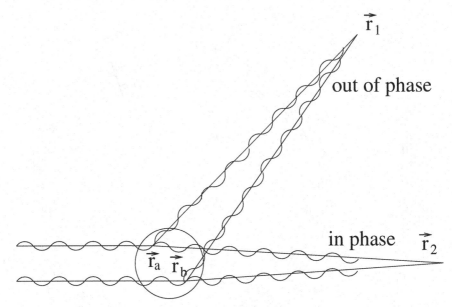

Fig. 3.15. In the Born approximation, the scattered wave at any point far from the region of $V \neq 0$ is the sum of spherical waves emitted at each point r' in the scattering region. The figure shows who such waves, one emitted at r_a and one at r_b. The phase of the scattered wave at the point r is $k(z' + |r - r'|)$. Only in the forward direction is this phase independent of r'. In other directions, the phase depends on r' so the spherical waves do not sum coherently. This results in a diminishing of the cross-section for angles satisfying $|p' - p|R > \hbar$.

$$F(x = y = 0, z) \;=\; \frac{1}{T} \int_{-\infty}^{\infty} \mathrm{d}z |\psi_{\mathrm{in}}(x = y = 0, z, t < 0)|^2 \;. \qquad (3.92)$$

We use a wave packet that is sufficiently broad that this flux is constant over the entire extent of the region with $V \neq 0$.

The detection rate is proportional to the probability to find the particle in the box shown in Fig. 3.14:

$$\frac{\mathrm{d}N_{\mathrm{det}}}{\mathrm{d}t} \;=\; \frac{1}{T}(\mathrm{d}\theta)^2 \int_0^{\infty} \mathrm{d}r \, r^2 |\psi_{\mathrm{sc}}(r, \theta, t \gg 0)|^2 \;. \qquad (3.93)$$

Using (3.91), we find

$$\frac{\mathrm{d}\sigma}{\mathrm{d}\Omega} \;=\; \frac{\int \mathrm{d}r \, |\psi_{\mathrm{sc}}(r, \theta, t \gg 0)|^2 r^2}{\int \mathrm{d}z \, |\psi_{\mathrm{in}}(x = y = 0, z, t < 0|^2} \qquad (3.94)$$

To calculate the differential cross-section we need only calculate ψ_{sc} for a given ψ_{in}. To do this, it is useful to express the wavefunction as a superposition of the energy eigenfunctions $\psi_E(r)$ satisfying the eigenvalue equation

$$-\frac{\hbar^2}{2m}\nabla^2 \psi_E(r) + V(r, t)\psi_E(r) \;=\; E\psi_E(r) \;. \qquad (3.95)$$

For $V = 0$ the eigenfunctions are just the familiar plane waves, $\exp(i\boldsymbol{p} \cdot \boldsymbol{r}/\hbar)$ and a superposition makes a wave packet of the form

$$\psi(\boldsymbol{r}, t) = \frac{1}{(2\pi)^{3/2}} \int d^3k \, \phi(\boldsymbol{k}) \, e^{i(\boldsymbol{k} \cdot \boldsymbol{r} - \omega(k)t)} , \qquad (3.96)$$

where $\boldsymbol{k} = \boldsymbol{p}/\hbar$ and $\omega(k) = E(p)/\hbar$. With $V = V(r) \neq 0$, far from the target, $r \gg R$, the eigenfunctions are sums of plane waves and radial waves emanating from the target:

$$\psi(\boldsymbol{r}, t) = \frac{1}{(2\pi)^{3/2}} \int d^3k \, \phi(\boldsymbol{k}) \left[e^{i\boldsymbol{k} \cdot \boldsymbol{r}} + \frac{f(\theta)}{r} e^{ikr} \right] e^{-i\omega(k)t} . \qquad (3.97)$$

The first term in the integral represents the initial and transmitted wave packet and the second term is the scattered wave. (We will see that the second term integrates to zero for $t \ll 0$ so it does not contribute to the initial wave packet.) The "scattering amplitude" $f(\theta)$ is a function of the angle between the momentum $\boldsymbol{p}$ and the position vector $\boldsymbol{r}$:

$$\cos \theta(\boldsymbol{k}, \boldsymbol{r}) = \frac{\boldsymbol{k} \cdot \boldsymbol{r}}{|\boldsymbol{k}||\boldsymbol{r}|}. \qquad (3.98)$$

Since $f(\theta)$ has the dimensions of length, we can anticipate that

$$\frac{d\sigma}{d\Omega} = |f(\theta)|^2 . \qquad (3.99)$$

To describe a particle impinging on the target along the z direction we take $\phi(\boldsymbol{k})$ to be strongly peaked at $\boldsymbol{k}_0 = (k_x = 0, k_y = 0, k_z = k_0 = p_0/\hbar)$. Therefore only the values of $\boldsymbol{k}$ near $\boldsymbol{k}_0$ will contribute. We therefore expand $\omega(k)$

$$\hbar\omega(k) = E(p_0) + \boldsymbol{\nabla} E(\boldsymbol{p}) \cdot (\boldsymbol{p} - \boldsymbol{p}_0) + \dots$$

$$= E(p_0) + v_0(p_z - p_0) + \dots , \qquad (3.100)$$

where v_0 is the group velocity. For a wave packet representing a massive particle, the group velocity is the classical velocity $v_0 = p_0/m$. Keeping only the first two terms in the expansion, the first term of (3.97) is

$$\psi_{\text{in}}(\boldsymbol{r}, t) = \frac{1}{(2\pi)^{3/2}} e^{i(k_0 z - \omega_0 t)} \psi_{\text{env}}(\boldsymbol{r} - \boldsymbol{v}_0 t) , \qquad (3.101)$$

where the "envelope" function is

$$\psi_{\text{env}}(\boldsymbol{r} - \boldsymbol{v}_0 t) = \int d^3k \, \phi(\boldsymbol{k}) \, e^{i(\boldsymbol{k} - \boldsymbol{k}_0) \cdot (\boldsymbol{r} - \boldsymbol{v}_0 t)} . \qquad (3.102)$$

We see that ψ_{in} is the product of a plane wave and an envelope that is a function only of $\boldsymbol{r} - \boldsymbol{v}_0 t$, i.e. the envelope moves with the group velocity. For example, if $\phi(\boldsymbol{k})$ is a real Gaussian function peaked at $\boldsymbol{k} - \boldsymbol{k}_0 = 0$, then the envelope will be a Gaussian peaked at $\boldsymbol{r} - \boldsymbol{v}_0 t = 0$, i.e. at $\boldsymbol{r} = \boldsymbol{v}_0 t$, with the variances of the Gaussians satisfying the Heisenberg uncertainty relations

$\sigma_x \sigma_{p_x} = \sigma_y \sigma_{p_y} = \sigma_z \sigma_{p_z} \sim \hbar$. Including higher-order terms in the expansion (3.100) leads to spreading of the wave packet at large times.

The scattered wavefunction has a similar structure:

$$\psi_{sc}(r, \theta, t) = \frac{f(\theta)}{r} e^{i(k_0 r - \omega_0 t)} \int d^3 k \, \phi(\boldsymbol{k}) \, e^{i(k_z - k_0)(r - v_0 t)} \qquad (3.103)$$

where we have dropped a factor $\exp(i(k_x^2 + k_y^2)r/k_0)$ which is near unity for a sufficiently wide wave packet. We have also taken $f(\theta)$ out of the integral since $\phi(\boldsymbol{k})$ is strongly peaked around $\boldsymbol{k}_0$ and therefore (3.98) is well approximated by

$$\cos\theta \sim \frac{\boldsymbol{k}_0 \cdot \boldsymbol{r}}{|\boldsymbol{k}_0||\boldsymbol{r}|} . \qquad (3.104)$$

Comparing (3.102) and (3.103), we see that

$$|\psi_{sc}(r - v_0 t, \theta, t)|^2 = \frac{|f(\theta)|^2}{r^2} |\psi_{in}(x = y = 0, z = r - v_0 t)|^2 . \qquad (3.105)$$

This tells us the scattered wavefunction is simply a replica of ψ_{in} that is scaled down by a factor $f(\theta)/r$. Note also that (3.105) implies that the scattered wave vanishes for $t \ll 0$ since it is proportional to the incident wave at $(t \ll 0, z \gg 0)$ which vanishes.

Substituting (3.105) into (3.94) we find the required identification of the differential scattering cross-section and the square of the scattering amplitude (3.99).

We now need to find the relation between $f(\theta)$ and the potential $V(r)$. This is easy to do if the potential is sufficiently weak that the wave packet is only slightly perturbed as it passes through the potential. We rewrite the eigenvalue equation (3.95) as

$$(\nabla^2 - k^2)\psi_k(\boldsymbol{r}) = 2mV(r)\psi_k(\boldsymbol{r})/\hbar^2 \qquad (3.106)$$

where $k = \sqrt{2mE}/\hbar$. We will look for solutions of the form

$$\psi_k = e^{ikz} + \psi_{k\,sc} , \qquad (3.107)$$

where the first term is a solution of the eigenvalue equation with $V = 0$ and the second term is a particular solution to the equation with $V \neq 0$. Since the effect of the potential is assumed to be small, it should be a good approximation to replace the wavefunction of the right-hand side of (3.106) with the incident plane wave:

$$(\nabla^2 - k^2)\psi_k(\boldsymbol{r}) = 4\pi S(\boldsymbol{r}) , \qquad (3.108)$$

where

$$S(\boldsymbol{r}) = \frac{2mV(r)\exp(i\boldsymbol{k} \cdot \boldsymbol{r})}{4\pi\hbar} . \qquad (3.109)$$

For $k = 0$ this is the Poisson equation of electrostatics with the electrostatic potential replaced by $\psi_k(\boldsymbol{r})$ and the charge density replaced by $S(\boldsymbol{r})$. The solution is well-known:

$$\psi_{k\,\text{sc}}(\boldsymbol{r}) = \frac{1}{4\pi} \int d^3 r' \frac{1}{|\boldsymbol{r} - \boldsymbol{r}'|} S(\boldsymbol{r}') \qquad (k = 0) . \qquad (3.110)$$

For $k \neq 0$, the solution is only slightly more complicated:

$$\psi_{k\,\text{sc}}(\boldsymbol{r}) = \frac{1}{4\pi} \int d^3 r' \frac{\exp(ik|\boldsymbol{r} - \boldsymbol{r}'|)}{|\boldsymbol{r} - \boldsymbol{r}'|} S(\boldsymbol{r}') . \qquad (3.111)$$

This formula has a simple physical interpretation: the scattered wave is a sum of spherical waves generated at each point $\boldsymbol{r}'$ in the potential well and having an amplitude proportional to $S(\boldsymbol{r}') \propto V(\boldsymbol{r}')$.

Equation (3.111) can be written as

$$\psi_{k\,\text{sc}}(\boldsymbol{r}) = \frac{2m}{4\pi\hbar^2} \int d^3 r' \frac{\exp(ik|\boldsymbol{r} - \boldsymbol{r}'|)\exp(ikz')}{|\boldsymbol{r} - \boldsymbol{r}'|} V(r) . \qquad (3.112)$$

We are interested in ψ_{sc} far from the scattering center in which case we can approximate $\boldsymbol{r}' = 0$ (in the denominator) and $|\boldsymbol{r} - \boldsymbol{r}'| \sim r - \boldsymbol{r} \cdot \boldsymbol{r}'/r$ (in the numerator). A particle observed at $\boldsymbol{r}$ will be interpreted as having a momentum $\boldsymbol{p}' = p_0 \boldsymbol{r}/r$ implying $|\boldsymbol{r} - \boldsymbol{r}'| \sim r - \boldsymbol{k}' \cdot \boldsymbol{r}'/k_0$ so the scattered wavefunction is

$$\psi_k(\boldsymbol{r}) = \frac{2m e^{ikr}}{4\pi r} \int d^3 r' V(r') \exp(i\boldsymbol{q} \cdot \boldsymbol{r}'/\hbar) , \qquad (3.113)$$

where $\boldsymbol{q} = \boldsymbol{p} - \boldsymbol{p}'$ is the momentum transfer of magnitude $|\boldsymbol{q}|^2 = 2p_0^2(1 - \cos\theta)$. We see that the scattered wave is proportional to the Fourier transform of the potential.

$$\tilde{V}(\boldsymbol{q}) = \int e^{i\boldsymbol{q}\cdot\boldsymbol{r}/\hbar} V(\boldsymbol{r}) d^3 r . \qquad (3.114)$$

The differential cross section is then

$$\frac{d\sigma}{d\Omega} = \frac{m^2}{4\pi^2\hbar^4} |\tilde{V}(\boldsymbol{p} - \boldsymbol{p}')|^2 \qquad (3.115)$$

as found in the previous section.

Equation, (3.115) tells us that the cross-section takes an especially simple form if $\boldsymbol{q} = 0$

$$\frac{d\sigma}{d\Omega}(\boldsymbol{q} = 0) = \frac{m^2}{4\pi^2\hbar^4} \left| \left(\int V(r) d^3 r \right) \right|^2 . \qquad (3.116)$$

Since $q^2 = 2p^2(1 - \cos\theta)$, this condition is met either in the forward direction, $\theta = 0$, or in the low-energy limit where the de Broglie wavelength is much greater than the range of the potential, $\hbar/p \gg R$. For $q^2 \neq 0$, the exponential in the integrand is an oscillating function of $\boldsymbol{r}'$ so the integral is suppressed. This can be intuitively understood by saying that far from the region where $V \neq 0$, the spherical waves generated at different positions are not entirely in phase and therefore partially cancel. As seen in (3.112) and in Fig. 3.15, only for $\theta = 0$ or for $\hbar/p \gg R$ is the phase independent of $\boldsymbol{r}'$ so the spherical waves are entirely in phase at the observer's position $\boldsymbol{r}$.

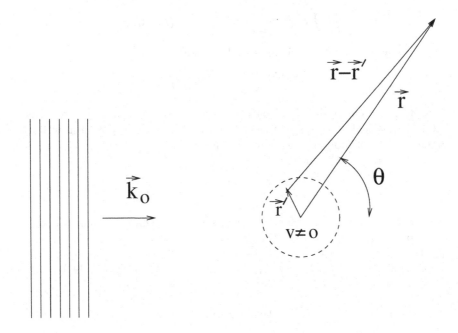

Fig. 3.16. A wave packet of central momentum $\boldsymbol{p}_0 = \hbar \boldsymbol{k}_0$ that impinges upon a region with $V(r) \neq 0$. The scattered wave packet at $\boldsymbol{r}$ is the superposition of spherical waves generated at each point $\boldsymbol{r}'$. A particle observed at $\boldsymbol{r}$ will be interpreted as having a momentum $\boldsymbol{p}' = p_0 \boldsymbol{r}/r$ implying a momentum transfer squared of $|\boldsymbol{q}|^2 = |\boldsymbol{p}' - \boldsymbol{p}|^2 = 2p_0^2(1 - \cos\theta)$.

The suppression of the cross-section for $\theta \neq 0$ because of destructive interference is quite different from the classical case. Here, the large angle cross-section is suppressed simply because the particle trajectory must pass near the center of the potential in order to produce a wide-angle scatter.

While the decline of the cross-section with increasing scattering angle has different origins in quantum and classical mechanics, we saw previously that the classical and quantum calculations may give identical answers as long as $q^2 \neq 0$. In fact, what distinguishes quantum scattering from classical scattering is that in quantum scattering the cross-section must be isotropic for $qR \ll 1$. This condition is met at all scattering angles if the de Broglie wavelength of the incident particle is much greater than the range R of the potential. This is equivalent to the condition

$$\frac{p^2}{2m} < \frac{(\hbar c)^2}{8R^2 mc^2} = 5 \, \text{MeV} \, \frac{1 \, \text{GeV}}{mc^2} \left(\frac{1 \, \text{fm}}{R} \right)^2 . \tag{3.117}$$

For incident energies below this limit, the scattering must be isotropic. For incident energies above this limit, the scattering will still be isotropic at small angles:

$$\theta < \frac{\hbar c}{\sqrt{2(p^2/2m)mc^2 R}} = 2 \left(\frac{5\,\text{MeV}}{p^2/2m}\right)^{1/2} \left(\frac{1\,\text{GeV}}{mc^2}\right)^{1/2} \frac{1\,\text{fm}}{R} ,\tag{3.118}$$

where we have taken the small-angle limit $(1-\cos\theta) = \theta^2/2$. For angles larger than this values, the cross-section decreases, as seen in Fig. 3.6.

3.4 Particle–particle scattering

We now return to the treatment of scattering using time-dependent perturbation theory as in Sect. 3.3.2. In this section, we complicate slightly the scattering problem by taking into account the recoil of the target particle. The immediate result will be that the translation invariance of the Hamiltonian enforces momentum conservation, a fact that was ignored in fixed-potential scattering.

3.4.1 Scattering of two free particles

We consider now the scattering to two particles, 1 and 2, with initial momenta $\boldsymbol{p}_1$ and $\boldsymbol{p}_2$, and final momenta $\boldsymbol{p}_1'$ and $\boldsymbol{p}_2'$. We take the potential energy to be $V(\boldsymbol{r}_1-\boldsymbol{r}_2)$, i.e. a function only of the *relative coordinates* of the two particles. The conservation of momentum will be a consequence of the assumption that the interaction potential $V(\boldsymbol{r}_1 - \boldsymbol{r}_2)$ is translation invariant.

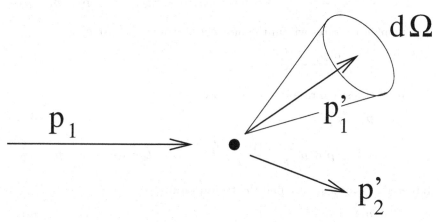

Fig. 3.17. Scattering of two particles with recoil.

The treatment of this problem follows the treatment of scattering on fixed potential starting with the transition rate given by (3.56). The initial and final state wavefunctions are now

$$\psi_i(\boldsymbol{r}_1, \boldsymbol{r}_2) = \frac{e^{i\boldsymbol{p}_1 \cdot \boldsymbol{r}_1}}{L^{3/2}} \frac{e^{i\boldsymbol{p}_2 \cdot \boldsymbol{r}_2}}{L^{3/2}} \qquad \psi_f(\boldsymbol{r}_1, \boldsymbol{r}_2) = \frac{e^{i\boldsymbol{p}_1' \cdot \boldsymbol{r}_1}}{L^{3/2}} \frac{e^{i\boldsymbol{p}_2' \cdot \boldsymbol{r}_2}}{L^{3/2}} . \tag{3.119}$$

The matrix element between initial and final states is

$$\langle f|V|i\rangle = \frac{1}{L^6} \int e^{i(\boldsymbol{p}_1-\boldsymbol{p}'_1)\cdot\boldsymbol{r}_1/\hbar} e^{i(\boldsymbol{p}_2-\boldsymbol{p}'_2)\cdot\boldsymbol{r}_2/\hbar} V(\boldsymbol{r}_1-\boldsymbol{r}_2)\mathrm{d}^3r_1\mathrm{d}^3r_2$$

$$= \frac{1}{L^6} \int e^{i(\boldsymbol{p}_1-\boldsymbol{p}'_1)\cdot(\boldsymbol{r}_1-\boldsymbol{r}_2)/\hbar} e^{i(\boldsymbol{p}_1+\boldsymbol{p}_2-\boldsymbol{p}'_1-\boldsymbol{p}'_2)\cdot\boldsymbol{r}_2/\hbar} V(\boldsymbol{r}_1-\boldsymbol{r}_2)\mathrm{d}^3r_1\mathrm{d}^3r_2 \; .$$

Replacing the integration variable $\boldsymbol{r}_1$ by $\boldsymbol{r} = \boldsymbol{r}_1 - \boldsymbol{r}_2$, we find

$$\langle f|V|i\rangle = \frac{\tilde{V}(\boldsymbol{p}_1-\boldsymbol{p}'_1)}{L^6} (2\pi\hbar)^3 \Delta_L^3(\boldsymbol{p}_1+\boldsymbol{p}_2-\boldsymbol{p}'_1-\boldsymbol{p}'_2) \; , \qquad (3.120)$$

where

$$\Delta_L^3(\boldsymbol{p}) = \prod_{i=x,y,z} \left(\frac{1}{\pi} \frac{\sin p_i L/2\hbar}{p_i} \right) \qquad (3.121)$$

is a limiting form of the three-dimensional delta function (see Appendix C.0.2). The matrix element is the product of the Fourier transform of the potential introduced previously and an oscillating function (3.121) whose role, when squared, is to force momentum conservation:

$$[\Delta_L^3(\boldsymbol{p})]^2 = \frac{L^3\delta^3(\boldsymbol{p})}{(2\pi\hbar)^3} \; . \qquad (3.122)$$

Substituting into (3.56), the transition rate to the final state is

$$\lambda_{i\to f} = \frac{2\pi}{\hbar} \frac{|\tilde{V}(\boldsymbol{p}_1-\boldsymbol{p}'_1)|^2}{L^{12}} \delta(E_f-E_i)L^3(2\pi\hbar)^3\delta^3(\boldsymbol{p}_1+\boldsymbol{p}_2-\boldsymbol{p}'_1-\boldsymbol{p}'_2) \; .$$

The number of states within the momentum volume $\mathrm{d}^3p'_1\mathrm{d}^3p'_2$ is

$$\mathrm{d}N = \frac{L^3\mathrm{d}^3p'_1}{(2\pi\hbar)^3} \frac{L^3\mathrm{d}^3p'_2}{(2\pi\hbar)^3} \; . \qquad (3.123)$$

The total transition rate into these states is then

$$\lambda(\mathrm{d}^3\boldsymbol{p}'_1, \mathrm{d}^3\boldsymbol{p}'_2) = \qquad\qquad\qquad\qquad\qquad\qquad\qquad (3.124)$$

$$\left(\frac{L}{2\pi\hbar}\right)^3 \mathrm{d}^3p'_1\mathrm{d}^3p'_2 \frac{2\pi}{\hbar} \frac{|\tilde{V}(\boldsymbol{p}_1-\boldsymbol{p}'_1)|^2}{L^6}\delta(E'-E)\delta^3(\boldsymbol{p}_1+\boldsymbol{p}_2-\boldsymbol{p}'_1-\boldsymbol{p}'_2) \; .$$

Integrating over $\mathrm{d}^3\boldsymbol{p}'_2$, we find the transition rate

$$\lambda(\mathrm{d}^3\boldsymbol{p}'_1) = \qquad\qquad\qquad\qquad\qquad\qquad\qquad\qquad (3.125)$$

$$\left(\frac{L}{2\pi\hbar}\right)^3 \mathrm{d}^3p'_1 \frac{2\pi}{\hbar} \frac{|\tilde{V}(\boldsymbol{p}_1-\boldsymbol{p}'_1)|^2}{L^6}\delta(E'_1+E'_2-E_1-E_2) \; ,$$

where $E'_2 = E_2(\boldsymbol{p}_1 + \boldsymbol{p}_2 - \boldsymbol{p}'_1)$ is determined by momentum conservation. This is the same transition rate as in the fixed potential case (3.68) except that the energy-conservation delta function now includes the effect of nuclear recoil. If the nuclear recoil is negligible, the reaction rate is identical to that calculated for a fixed potential. In particular, for a heavy target at rest, we have $E_2 = m_2c^2$ and $E_2' = m_2c^2 + (p_2')^2/2m_2$ and

$$\lambda(\mathrm{d}^3\boldsymbol{p}_1') = \left(\frac{L}{2\pi\hbar}\right)^3 \mathrm{d}^3\boldsymbol{p}_1' \frac{2\pi}{\hbar} \frac{|\tilde{V}(\boldsymbol{p}_1 - \boldsymbol{p}_1')|^2}{L^6} \delta(E_1' - E_1 + (p_2')^2/2m_2) \,,$$

which reduces to the fixed potential result when $m_2 \to \infty$.

Another interesting limit is the collision of two ultra-relativistic particles. We treat the problem in the center-of-mass so that $E_1 = E_2 = E_{\mathrm{cm}}/2$. The transition rate is

$$\lambda(\mathrm{d}^3\boldsymbol{p}_1') = \left(\frac{L}{2\pi\hbar}\right)^3 \mathrm{d}^3\boldsymbol{p}_1' \frac{2\pi}{\hbar} \frac{|\tilde{V}(\boldsymbol{p}_1 - \boldsymbol{p}_1')|^2}{L^6} \delta(2E_1' - 2E_1) \,, \qquad (3.126)$$

or

$$\lambda(\mathrm{d}\Omega, \mathrm{d}E_1') = \qquad\qquad\qquad\qquad\qquad\qquad\qquad (3.127)$$

$$\frac{2c}{L^3} \left(\frac{1}{2\pi\hbar}\right)^3 \mathrm{d}\Omega p_1' E_1' \mathrm{d}E_1' \frac{\pi}{\hbar c} |\tilde{V}(\boldsymbol{p}_1 - \boldsymbol{p}_1')|^2 \delta(2E_1' - 2E_1) \,.$$

Dividing by the factor $2c/L^3$ gives the cross-section (where L cancels off identically).

A simple example is high-energy neutrino–electron elastic scattering in which case $\tilde{V} \propto G_{\mathrm{F}}$. This gives isotropic scattering in the center-of-mass

$$\frac{\mathrm{d}\sigma}{\mathrm{d}\Omega} \propto \frac{G_{\mathrm{F}}^2 E_{\mathrm{cm}}^2}{4\pi^2} \,, \qquad\qquad\qquad\qquad\qquad (3.128)$$

with a total cross-section of

$$\sigma \propto \frac{G_{\mathrm{F}}^2 E_{\mathrm{cm}}^2}{\pi} \,. \qquad\qquad\qquad\qquad\qquad (3.129)$$

The correct numerical factors are given in Table 3.1.

By taking into account the recoil of the target, we have introduced that added constraint of momentum conservation into the cross-section. Since momentum conservation is the result of the translation invariance of the Hamiltonian, we can anticipate that it will hold in more general reactions between two (or more) free particles. Consider a reaction

$$a_1 + a_2 \to b_1 + b_2 + \ldots + b_n \,. \qquad\qquad\qquad (3.130)$$

The initial momenta are $\boldsymbol{p}_1$ and $\boldsymbol{p}_2$ with $\boldsymbol{p} = \boldsymbol{p}_1 + \boldsymbol{p}_2$ being the total momentum and $E = E_1 + E_2$ the total energy. The final state momenta and energies are $\boldsymbol{q}_i$, $i = 1, \ldots, n$ and E_i'.

We can anticipate that the cross-section will be of the form

$$d\sigma = \frac{2\pi}{\hbar v_0} G(\boldsymbol{p}_1, \boldsymbol{p}_2; \boldsymbol{q}_1, \ldots, \boldsymbol{q}_n)(2\pi\hbar)^3$$

$$\times\, \delta(\boldsymbol{p} - \Sigma\boldsymbol{q}_i)\delta(E - \Sigma E_i') \frac{\mathrm{d}^3\boldsymbol{q}_1}{(2\pi\hbar)^3} \cdots \frac{\mathrm{d}^3\boldsymbol{q}_n}{(2\pi\hbar)^3} \,, \qquad (3.131)$$

where G is the square of the relevant transition amplitude. This expression only has meaning after we integrate it over four independent variables (for instance one momentum and one energy) in order to remove the delta functions coming from the conservation of energy and momentum.

Finally, we note that, in general, the relative velocity v_0 of the initial particles is

$$v_0 = \left((v_1 - v_2)^2 - (v_1 \wedge v_2)^2/c^2\right)^{1/2} \tag{3.132}$$

This expression must be used if the initial particles are not collinear.

3.4.2 Scattering of a free particle on a bound particle

It is often the case that free particles are scattered on particles that are not free but rather bound in potential wells as illustrated in Fig. 3.18. Consider a particle a of mass m_a that can scatter on a particle b of mass m_b that is bound near the origin by a potential $U(r_b)$ (acting only on particle b). The interaction between a and b is described by another potential $V(r_a - r_b)$. The possible wavefunctions of b in the potential U are called $\{\psi_n(r_b)\}$.

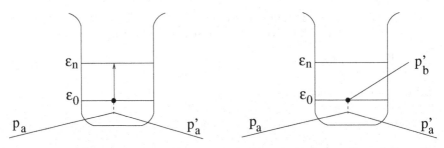

Fig. 3.18. Scattering of particle a on particle b in a bound state. Particle b can be left in bound state (left) or ejected from the potential (right).

An example of such a process is the scattering of electrons on deuterons. The deuteron is a bound state of a proton and neutron interacting through the nucleon–nucleon potential. The electron interacts with the proton via the Coulomb potential. Another example is ν_e-deuteron scattering. In this case the ν_e interacts with the neutron, i.e. $\nu_e n \to e^- p$.

We assume that initially b is in its ground state $\psi_0(r_b)$. The initial state wavefunction is then the product of a plane wave and a bound-state wavefunction, $L^{-3/2} e^{i p \cdot r/\hbar} \psi_0(r_b)$. In the final state, b can either stay in its ground state or be placed in an excited state $\psi_n(r_b)$:

$$|f\rangle \to L^{-3/2} e^{i p' \cdot r/\hbar} \psi_n(r_b) \quad . \tag{3.133}$$

Energy conservation implies

$$E(p') = E(p) - (\varepsilon_n - \varepsilon_0) \,. \tag{3.134}$$

In the Born approximation, the scattering matrix element is

$$\langle f|T|i\rangle \;=\; \frac{1}{L^3}\int e^{i(\boldsymbol{p}-\boldsymbol{p}')\boldsymbol{r}/\hbar}\psi_n^*(\boldsymbol{r}_b)\psi_0(\boldsymbol{r}_b)V(\boldsymbol{r}-\boldsymbol{r}_b)\mathrm{d}^3r\,\mathrm{d}^3r_b \,. \tag{3.135}$$

After changing integration variables $(\boldsymbol{r}_a, \boldsymbol{r}_b) \to (\boldsymbol{r}_a - \boldsymbol{r}_b, \boldsymbol{r}_b)$ this becomes

$$\langle f|T|i\rangle = L^{-3}\,\tilde{V}(\boldsymbol{p}-\boldsymbol{p}')\,F_n(\boldsymbol{p}-\boldsymbol{p}') \tag{3.136}$$

where $\tilde{V}$ is the previously defined Fourier transform

$$\tilde{V}(\boldsymbol{q}) = \int e^{i\boldsymbol{q}\cdot\boldsymbol{r}/\hbar}V(r)\mathrm{d}^3r \tag{3.137}$$

and where F_n is defined as

$$F_n(\boldsymbol{q}) = \int e^{i\boldsymbol{q}\cdot\boldsymbol{r}_b/\hbar}\psi_n^*(\boldsymbol{r}_b)\psi_0(\boldsymbol{r}_b)\mathrm{d}^3r_b \,. \tag{3.138}$$

The amplitude is then the product of the Fourier transforms of the potential $V(\boldsymbol{r}_a - \boldsymbol{r}_b)$ and the Fourier transform of the product of the initial and final state wavefunctions of b. This leads to a factorization of the cross-section for the excitation of the final state n:

$$\frac{\mathrm{d}\sigma_n}{\mathrm{d}\Omega} = \frac{\mathrm{d}\sigma_{\mathrm{f}}}{\mathrm{d}\Omega}\,|F_n(\boldsymbol{p}-\boldsymbol{p}')|^2 \tag{3.139}$$

where $\mathrm{d}\sigma_{\mathrm{f}}/\mathrm{d}\Omega$ is the cross-section on a free particle b. It is given by (3.72) for elastic scattering ($n = 0$) and, in general, by (3.86). The function $|F_n(\boldsymbol{p}-\boldsymbol{p}')|^2$ is called the *form factor* for excitation of the state n.

The target particle can also be ejected from its potential well into a free state of momentum $\boldsymbol{p}_b'$. In this case, the final state is

$$|f\rangle \to e^{i\boldsymbol{p}'\cdot\boldsymbol{r}/\hbar}e^{i\boldsymbol{p}_1\cdot\boldsymbol{r}_b/\hbar}/L^3 \quad. \tag{3.140}$$

Energy conservation now implies

$$E(\boldsymbol{p}') + E(\boldsymbol{p}_b') \;=\; E(\boldsymbol{p}) + \varepsilon_0 \,. \tag{3.141}$$

We introduce the Fourier transform of the initial wavefunction:

$$\tilde{\psi}_0(\boldsymbol{p}_b) = \frac{1}{(2\pi\hbar)^{3/2}}\int e^{-i\boldsymbol{p}_b\cdot\boldsymbol{r}_b/\hbar}\psi_0(\boldsymbol{r}_b)\mathrm{d}^3r_b \,, \tag{3.142}$$

which gives the amplitude for the initial bound particle to have a momentum $\boldsymbol{p}_b$. We then obtain, after a straightforward calculation, the scattering amplitude

$$\langle \boldsymbol{p}_b'\boldsymbol{p}'|T|\boldsymbol{p}\rangle \;=\; L^{-3}\tilde{V}(\boldsymbol{p}-\boldsymbol{p}')\left(\frac{2\pi\hbar}{L}\right)^{3/2}\tilde{\psi}_0(\boldsymbol{p}_b = \boldsymbol{p}' + \boldsymbol{p}_b' - \boldsymbol{p}) \,. \tag{3.143}$$

We see that the scattering amplitude is a product of the amplitude for scattering on free particle, $\boldsymbol{p} \to \boldsymbol{p}'$, and the amplitude for the initial bound particle to have the correct momentum to give momentum conservation. The cross-section for dissociation then factorizes:

$$\frac{d\sigma}{d^3\boldsymbol{p}'d^3\boldsymbol{p}'_b} = \frac{d\sigma}{d^3\boldsymbol{p}'}(\boldsymbol{p}\boldsymbol{p}_b \to \boldsymbol{p}'\boldsymbol{p}'_b)\,|\tilde{\psi}_0(\boldsymbol{p}_b)|^2 \quad . \tag{3.144}$$

In other words, the cross-section is the product of the cross-section on a free particle of momentum $\boldsymbol{p}_b$ times the probability $|\tilde{\psi}_0(\boldsymbol{p}_b)|^2$ that the ejected particle had the momentum $\boldsymbol{p}_b$ before the collision.

The three types of scattering considered here give complementary information on the target.

- Elastic scattering where b is left in its ground state. The form-factor is just the Fourier transform of the square of the ground-state wavefunction

$$F_0(\boldsymbol{q}) = \int e^{i\boldsymbol{q}\cdot\boldsymbol{r}/\hbar}|\phi_0(\boldsymbol{r})|^2 d^3\boldsymbol{r} \quad . \tag{3.145}$$

  We see that if we know the elementary cross-section, $d\sigma_f/d\Omega$, a measurement of the cross section on bound b's yields the (modulus squared of the) Fourier transform of the square of the ground-state wavefunction. Since the ground-state wavefunction has no zeroes and can be taken to be real, this can be inverted to give the wavefunction itself. We will see in the next section how this allows us to determine the charge distribution of nuclei.
- Production of an excited state. This reaction gives information on the wavefunction, quantum numbers and lifetime of the excited state. In fact, the so-called *Coulomb excitation* of nuclear states due to the passage of a charged particle is one of the important methods of deducing lifetimes of low-lying states. Unfortunately, the formalism we have given here is not general enough to completely explain this effect. It is better described as a two-step process: the emission of a virtual photon by the incident particle and the absorption of the photon by the target.
- Dissociation of the bound state. This reaction allows us to deduce the momentum distribution of the target particle in its initial ground state. We will see that this will allow us to deduce the momentum distribution of quarks within nucleons from inelastic electron–nucleon scattering.

We note that $|F_0(\boldsymbol{q} = 0)|^2 = 1$ and that $|F_0(\boldsymbol{q} \neq 0)|^2 < 1$, i.e. that the form factor acts to suppress the elastic cross section at large $\boldsymbol{q}^2$. This is understandable intuitively because we saw in Sect. 3.3.5 that the decline of the cross-section with increasing $\boldsymbol{q}^2$ is due, in the Born approximation, to the fact that the spherical waves emanating from different points in the region of $V \neq 0$ will not be in phase with each other except in the forward direction. If, in addition, the center of the potential is "smeared out" by the wavefunction of b, the phases of the emanating waves are further randomized, leading to a stronger decrease with $\boldsymbol{q}^2$.

Three examples of wavefunctions and their form factors are shown in Table 3.2.

Table 3.2. Three squared wavefunctions and their Fourier transforms. The common mean square radius is $\bar{r}^2 = 4\pi \int r^4 |\psi(r)|^2 \mathrm{d}r$.

$|\psi(r)|^2$	$F(q^2)$
$\bar{r}^{-3}(3^{3/2}/\pi)\exp(-2\sqrt{3}r/\bar{r})$	$(1 + \bar{r}^2 q^2/12\hbar^2)^{-2}$
$\bar{r}^{-3}(3/2\pi)^{3/2}\exp(-3r^2/2\bar{r}^2)$	$\exp(-\bar{r}^2 q^2/6\hbar^2)$
$\bar{r}^{-3}(3/5)^{3/2}(3/4\pi)\quad r < \sqrt{5/3}\bar{r}$ $\qquad\qquad\qquad 0 \quad r > \sqrt{5/3}\bar{r}$	$3\alpha^{-3}(\sin\alpha - \alpha\cos\alpha)\quad \alpha = \sqrt{5/3}|q|\bar{r}/\hbar$

Some care must be taken with regards to the validity and the generalization of these results. For more details, see for instance Mott and Massey *The theory of atomic collisions*, Chap. XII.

These results can be generalized to the case of a bound state of n particles with a wavefunction $\psi(r_1, \ldots, r_n)$. Taking the case of elastic scattering, we call $\langle T_f^i \rangle$ the scattering amplitude on the particle i. The contribution to the elastic amplitude of the particle i is

$$\langle T_b^i \rangle = \langle T_f^i \rangle F_0^i(q) \tag{3.146}$$

with $q = p - p'$, $\langle T_b \rangle$ and $\langle T_f \rangle$ are respectively the matrix elements on bound and free particles and

$$F_0^i(q) = \int e^{iq \cdot r_i/\hbar} |\psi_0(r_1, \ldots r_i, \ldots, r_n)|^2 \mathrm{d}^3 r_1 \ldots \mathrm{d}^3 r_n \quad . \tag{3.147}$$

In the Born approximation, we can ignore multiple scattering so the total amplitude is simply the sum

$$\langle p'|T|p \rangle = \sum_{i=1}^{n} \langle p'|T_b^i|p \rangle \quad . \tag{3.148}$$

The cross section is then proportional to $|\langle p'|T|p \rangle|^2$.

3.4.3 Scattering on a charge distribution

Suppose that $\psi_0(r_1, \ldots r_n)$ is the wavefunction of a set of bound point charges $Z_1, \ldots Z_n$, and consider the elastic scattering of a charge Z on the bound state. The elementary scattering amplitude is simply the Rutherford amplitude that we write as

$$\langle p'|T|p \rangle = ZZ_i\alpha\hbar c \tilde{T}^i(p, p') \tag{3.149}$$

by factoring out the coupling $ZZ_i\alpha\hbar c$. The quantity

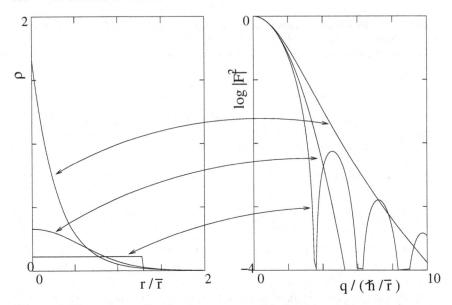

Fig. 3.19. Three charge distributions, exponential, Gaussian, and square (i.e. constant within $r \leq \bar{r}$ and zero otherwise), and their form factors. The three distributions shown on the left all have the same total charge, $4\pi \int r^2 \rho(r) \mathrm{d}r = 1$ and the same mean square radius $\langle r^2 \rangle = 4\pi \int r^4 \rho(r) \mathrm{d}r$. with $\bar{r} \equiv \langle r^2 \rangle^{1/2}$. The equality of their $\langle r^2 \rangle$ requires that at low q^2 their form factors, on the right are all equal, $F \sim 1 - (1/6)q^2 \langle r^2 \rangle / \hbar^2$. Only for $q^2 > 4\hbar^2 / \langle r^2 \rangle$ do the form factors differ significantly.

$$P_i(\boldsymbol{r}) = \int |\psi_0(\boldsymbol{r}_1, \ldots, \boldsymbol{r}_i = \boldsymbol{r}, \boldsymbol{r}_n)|^2 \mathrm{d}^3\boldsymbol{r}_1 \ldots \mathrm{d}^3\boldsymbol{r}_{i-1}\mathrm{d}^3\boldsymbol{r}_{i+1} \ldots \mathrm{d}^3\boldsymbol{r}_n$$

is the probability density to find the particle i in the volume $\mathrm{d}^3\boldsymbol{r}$ at $\boldsymbol{r}$, and $\rho_i(\boldsymbol{r})$, defined by

$$\rho_i(\boldsymbol{r}) = Z_i\sqrt{\alpha\hbar c}P_i(\boldsymbol{r}) \,, \tag{3.150}$$

is the contribution of the charge i to the total charge density $\rho(\boldsymbol{r})$ of the bound state

$$\rho(\boldsymbol{r}) = \sqrt{\alpha\hbar c}\sum_{i=1}^{n} Z_i P_i(\boldsymbol{r}) = \sqrt{\alpha\hbar c}\tilde{\rho}(\boldsymbol{r}) \tag{3.151}$$

with

$$\int \tilde{\rho}(\boldsymbol{r})\mathrm{d}^3\boldsymbol{r} = \sum_{i=1}^{n} Z_i = Z_{\text{tot}} \tag{3.152}$$

where Z_{tot} is the total charge of the system.

In these conditions, the cross-section is

$$\frac{d\sigma_0}{d\Omega} = \frac{d\sigma^{\text{Ruth}}}{d\Omega}(Z\alpha\hbar c)|F_0^{Z_{\text{tot}}}(\boldsymbol{q})|^2 \, , \tag{3.153}$$

where $\frac{d\sigma^{\text{Ruth}}}{d\Omega}(Z\alpha\hbar c)$ is the Rutherford cross-section and

$$F_0^{Z_{\text{tot}}}(\boldsymbol{q}) = \int e^{i\boldsymbol{q}\cdot\boldsymbol{R}/\hbar}\tilde{\rho}(\boldsymbol{r})d^3\boldsymbol{r} \tag{3.154}$$

is the Fourier transform of the charge density (divided by e). Clearly

$$F_0^{Z_{\text{tot}}}(0) = Z_{\text{tot}} \tag{3.155}$$

so for small momentum transfer (3.153) is reduced to the Rutherford cross-section of a particle of charge Z on a *point* particle of charge $Z_{\text{tot}}e$. One says that at low momentum transfer, there is *coherent* scattering on the bound state. The cross-section is proportional to the total charge $(Z_{\text{tot}})^2$ of the composite system.

3.4.4 Electron–nucleus scattering

The nucleus is not a point particle so we can expect the elastic scattering will be suppressed for momentum transfers greater than $\hbar/R$ where $R \sim 1.2A^{1/3}$ fm is the radius of the nucleus. The most efficient way to see this effect is in electron–nucleus scattering. Electrons are insensitive to the nuclear force so the elementary cross-section (the *Mott* cross-section, the relativistic generalization of the Rutherford cross-section) is known very precisely and the deviations at large scattering angle can give information of the charge distribution, i.e. the distribution of protons in the nucleus. For relativistic electrons $E = pc$, the non-vanishing size causes deviations from the Rutherford cross section for

$$\theta > \frac{\hbar c}{ER} \sim \frac{200\,\text{MeV}}{E}A^{-1/3} \, , \tag{3.156}$$

where we use the small angle approximation $q = p\theta$. For electrons of energy $E = 400\,\text{MeV}$ scattering on calcium, $A^{1/3} \sim 4$ so we expect a suppression of the cross-section for $\theta > .1$, i.e. $\theta > 7$ deg.

Figure 3.20 shows the cross-section as a function of angle for electrons scattering on calcium [32]. The deviations from the Mott formula are clear and allow one to derive the nuclear charge distribution. The oscillating pattern of the cross-section indicates clearly that this nucleus has a relatively flat charge distribution with a relatively well-defined edge (Fig. 3.19). This behavior is seen on all nuclei with $A > 10$ except strongly deformed nuclei where the averaging over nuclear orientations leads to a "fuzzier" edge. Light nuclei, $A < 10$ have a more Gaussian charge distribution as indicated by the absence of oscillations in their differential cross-sections.

As emphasized in Sect. 3.4.2, the scattering of an electron on a nucleus can leave the nucleus in one of its discrete excited states. This is illustrated in Fig. 3.21 which shows the energy distribution distribution of electrons scattered

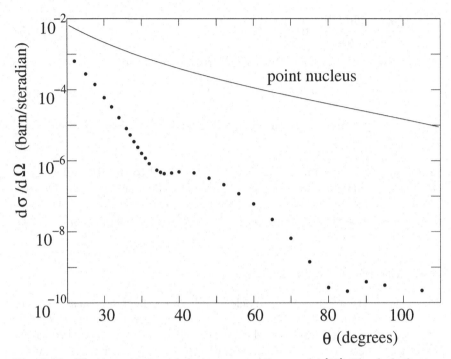

Fig. 3.20. Elastic scattering of electrons on calcium nuclei [32]. The dashed curve shows the Mott scattering cross-section for electrons on a heavy spinless nucleus.

at a fixed angle on ^{12}C. Electrons that leave the nucleus in an excited state must lose energy in order to conserve energy. Peaks corresponding to the ^{12}C ground state (elastic scattering) and to the first three excited states are seen.

Finally, the scattering of an electron on a nucleus can also lead to the ejection of a nucleon and this is also reflected in the energy distribution of the final-state electron. This distribution is shown for electron–^{2}H scattering in Fig. 3.22. Their are no excited states of the deuteron so the distribution shows two peaks, one for elastic scattering and a broader peak at lower energy for nucleon ejection. The second peak is at lower energy because the electron loses energy to eject a nucleon and, more importantly, the electron here scatters on a quasi-free nucleon of smaller mass than the nucleus. The nucleon therefore recoils with more energy than would the nucleus as a whole (Exercise 3.6).

Generally speaking, electron–nucleon scattering dominates over elastic scattering at high q^2. This is because the form-factor suppresses the elastic cross-section at high q^2 and this suppression compensates the fact that the elastic cross-section is proportional to Z^2 while the sum of the cross-sections on individual nucleons if proportion only to Z. Nucleon-ejection then dominates for q^2 greater than the value defined by $|F(q^2)|^2 < Z^{-1}$. From Fig. 3.19 we see that for $Z \sim 30$, the required suppression by a factor ~ 30 occurs for $q^2 \langle r^2 \rangle / \hbar^2 > 10$, i.e. $pc\theta > 600$ MeV.

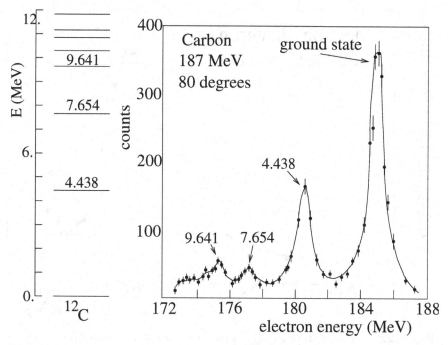

Fig. 3.21. The spectrum of excited states of ^{12}C (left) and the final-state energy spectrum of 187 MeV electrons scattering at 80 deg on ^{12}C (right). The peak at 185 MeV corresponds to elastic scattering. (2 MeV is taken by the recoiling nucleus.) The other peaks correspond to inelastic scattering leaving the ^{12}C nucleus in an excited state. The three lowest excitations are clearly visible.

3.4.5 Electron–nucleon scattering

Electron scattering on nucleons is very similar to electron scattering on nuclei. At low values of q^2 one observes only elastic scattering. The angular distribution is well described by Rutherford scattering modified by two effects:

- The nucleon and electron spins must be taken into account since they lead to magnetic forces that modify the angular distribution.
- As in the case of nuclei, nucleons are not point particles so both the charge and magnetic moments have spatial distributions leading to electric and magnetic form factors.

The first indications of the internal structure of the proton and neutron came from the values of their magnetic moments. According to Dirac's theory, a point like elementary particle of spin 1/2 and charge q should have a magnetic moment

$$\mu_p = \frac{e\hbar}{2m_p} \quad , \quad \text{and} \quad \mu_n = 0 \quad (q_n = 0) , \tag{3.157}$$

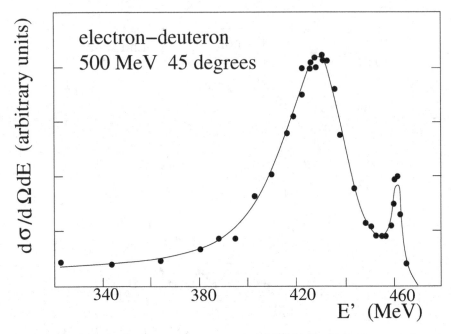

Fig. 3.22. The final-state energy spectrum of 500 MeV electrons scattering at 45 deg on ^{2}H. The peak at 460 MeV corresponds to elastic scattering. (40 MeV is taken by the recoiling nucleus.) The broader peak at 425 MeV corresponds to dissociation of the deuteron. This can be considered as elastic scattering of the electron on either the proton or neutron. The peak is at lower energy because $B(2,1) = 2.2$ MeV must be provided to break the nucleus and because the recoiling nucleon, because of its lower mass, takes more energy than a recoiling deuteron. The width of the inelastic peak reflects the momentum distribution of the nucleons in the nucleus.

i.e. a gyromagnetic ratio of $\gamma = q/m$. The observed values of the proton and neutron moments based on magnetic resonance experiments are

$$\mu_p = 2.7928444 \frac{e\hbar}{2m_p} \qquad \mu_n = -1.91304308 \frac{e\hbar}{2m_p} \ . \tag{3.158}$$

These results suggests that the proton and neutron are not "elementary" particles but have an internal structure. Today, we attribute this structure to the quarks and gluons that form the nucleons and all other hadrons.

The conclusion based on the proton magnetic moment was confirmed by the electron–proton scattering experiments of Hofstadter. Figure 3.23 shows the angular distribution and that expected for a structureless particle. Unlike the case for nuclei (Fig. 3.20) no diffraction minima are seen. This indicates that the proton has a more or less exponential density profile, as indicated in Fig. 3.19. Figure 3.24 shows the proton form factor deduced from the cross-sections for the scattering of electrons of energies between 200 and 500 MeV.

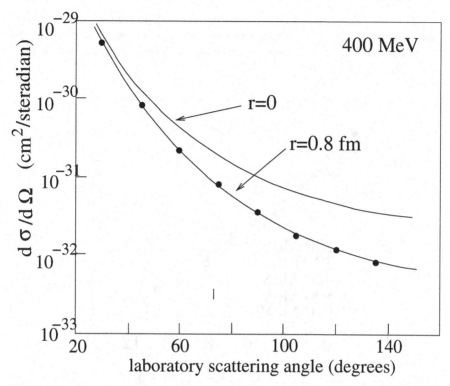

Fig. 3.23. The measured angular distribution of 400 MeV electrons scattered from protons. The two lines show the expected distribution for scattering from a point-like proton and from an exponential charge distribution of mean radius 0.8 fm.

The experimental points can be fitted to a simple empirical charge distribution:

$$\rho_p(r) = \rho_0 e^{-r/a_1} \tag{3.159}$$

which gives a form factor

$$F_p(q) = (1 + q^2 a_1^2/\hbar^2)^{-2} \tag{3.160}$$

where a_1 is the mean square (charge) radius of the proton

$$a = (\langle r^2 \rangle)^{1/2} = \sqrt{12} a_1 . \tag{3.161}$$

The results give $a \sim 0.8$ fm which can be taken as a definition of the "size" of the proton.

The determination of the neutron charge density is more difficult because, being unstable, dense targets cannot be made of neutrons. In practice, one studies electron–deuteron scattering and subtracts off the contribution of the protons.

The charge and magnetic moment densities of protons and neutrons are shown in Fig. 3.24. We note that the neutron has a positively charged core of radius ~ 0.3 fm surrounded by compensating negative charge between 0.3 and 2 fm.

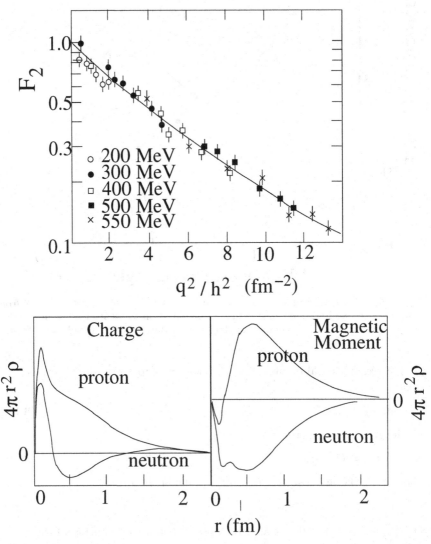

Fig. 3.24. Top panel: the experimental values of the proton form factor $|F(q^2)|^2$ [8]. The curve is the predicted form factor for an exponential charge distribution with a mean charge radius of 0.8 fm. Bottom panel: the derived charge and magnetic moment densities of the proton and neutron [33].

Just as in electron–nucleus scattering, inelastic scattering dominates at high q^2. In electron–nucleus scattering, one see the excitation of excited nuclear states and scattering on individual nucleons followed by their ejection. In electron–nucleon scattering the same sequence is seen. The excited states of nucleons correspond to hadronic resonances. The scattering on constituents corresponds to scattering on individual quarks. The momentum distribution of quarks in the nucleon can be deduced from the energy distribution of scattered electrons just as the momentum distribution of nucleons in nuclei can be deduced (Fig. 3.22).

The difference from electron–nucleus scattering is that while nucleons can be ejected from nuclei and observed, individual quarks are never observed in final states. Rather, one sees events with a multitude of hadrons in addition to a nucleon. The situation is visualized in Fig. 3.25. When the struck quark leaves the nucleon, a quark–antiquark pair is produced which recombines with the spectator quarks so that the final state consists of a nucleon and a pion. The historical sequence of seeing ever more fundamental particles seems to be broken since the fundamental particle in this case is not observed.

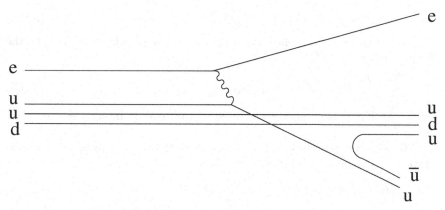

Fig. 3.25. Electron–proton "deep-inelastic scattering." The proton is a bound state uud of three quarks. The electron scatters on an individual quark by exchange of a virtual photon. As the quark leaves the proton, a quark–antiquark pair is created. The pair recombines with the initial quarks to make a final state proton (uud) and π^0 (u$\bar{\text{u}}$).

3.5 Resonances

It is common, especially at low energy, for cross-sections to exhibit resonant behavior with energy dependences of the form

$$\sigma \sim \frac{A}{(E - E_0)^2 + (\Gamma/2)^2} \,, \tag{3.162}$$

where E_0 and $\Gamma/2$ are the energy and width (at half-maximum) of the resonance. In nuclear physics, these resonances are excited states of nuclei that decay rapidly ($\Gamma = \hbar/\tau$) by dissociation or photon-emission. An example is the fourth excited state of ^{7}Li (Fig. 3.5) that leads to a resonance in the n $-^6$ Li cross-section (Fig. 3.4). More spectacular examples are the multitude of highly excited states of heavy nuclei that have sufficient energies to decay by dissociation. The cases of ^{236}U and ^{239}U are shown in Fig. 3.26 where these states appear as resonances in the scattering of neutrons on ^{235}U and ^{238}U. At the hadronic level, almost all the "elementary particles" discovered in the 60's were resonances seen, not as tracks, but simply as maxima in cross-sections or in invariant-mass distributions.

In order to see how resonances come about in quantum mechanics, we examine a simple model due to Wigner and Weisskopf. Consider a discrete state $|a\rangle$ of energy E_a and a set of continuum states $|\alpha\rangle$. The discrete state is coupled to the continuum by a Hamiltonian, H with matrix elements

$$\langle a|H|\alpha\rangle = \langle \alpha|H|a\rangle^* = H(\alpha) \tag{3.163}$$

and

$$\langle a|H|a\rangle = \langle \alpha|H|\alpha\rangle = 0 \quad . \tag{3.164}$$

Because of this coupling, the state $|a\rangle$ is unstable with a lifetime given by the Fermi golden rule (C.15)

$$\frac{1}{\tau} = \frac{2\pi}{\hbar}|H(\alpha \; ; \; E_\alpha = E_0)|^2 \rho_\alpha(E_0) \tag{3.165}$$

where $\rho_\alpha(E_0)$ is the number of states $|\alpha\rangle$ per energy interval evaluated at $E_\alpha = E_0$.

We now calculate the evolution of a system that is initially in a continuum state $|\alpha\rangle$. The most general evolution is

$$|\psi(t)\rangle = \gamma_a(t)e^{-iE_0t/\hbar}|a\rangle + \sum_\beta \gamma_\beta(t)e^{-iE_\beta t/\hbar}|\beta\rangle \; . \tag{3.166}$$

We define

$$\omega_\beta = (E_0 - E_\beta)/\hbar \quad , \tag{3.167}$$

The Schrödinger equation is now two equations

$$i\hbar\dot{\gamma}_a = \sum_\beta \gamma_\beta(t)e^{i\omega_\beta t}H(\beta) \tag{3.168}$$

and

$$i\hbar\dot{\gamma}_\beta = \gamma_a(t)e^{-i\omega_\beta t}H^*(\beta) \quad . \tag{3.169}$$

The initial conditions are

$$\gamma_a(t=0) = 0 \; , \; \gamma_\beta = \delta_{\alpha\beta} \Longleftrightarrow |\psi(0)\rangle = |\alpha\rangle \; . \tag{3.170}$$

Integrating (3.169) we get

$$\gamma_\beta(t) = \delta_{\alpha\beta} + \frac{1}{i\hbar} \int_0^t \gamma_a(t') e^{-i\omega_\beta t'} H^*(\beta) . \tag{3.171}$$

Substituting this into (3.168) we get

$$i\hbar\dot{\gamma}_a = e^{i\omega_a t} H(\alpha) + \frac{1}{i\hbar} \sum_\beta |H(\beta)|^2 \int_0^t e^{i\omega_\beta(t-t')} \gamma_a(t') dt' . \tag{3.172}$$

The Wigner–Weisskopf approximation is to replace $\gamma_a(t')$ in the integral by $\gamma(t)$. This can be justified *a posteriori*. If this is done, the integral can be evaluated exactly. The sum then yields the inverse lifetime of the state (as we will see in Chap. 4). We then have

$$\dot{\gamma}_a(t) = \frac{1}{i\hbar} e^{i\omega_a t} H(\alpha) - \frac{\gamma_a(t)}{2\tau} , \tag{3.173}$$

τ is defined by (3.165).[4] We can now integrate this equation to find

$$\gamma_a(t) = \frac{H(\alpha)[e^{i(E_0 - E_\alpha)t/\hbar} - e^{-t/2\tau}]}{(E_\alpha - E_0) + i\Gamma/2} \tag{3.174}$$

with

$$\Gamma = \hbar/\tau . \tag{3.175}$$

For $t \gg \tau$ this becomes

$$\gamma_a(t) \simeq \frac{H(\alpha) e^{i(E_0 - E_\alpha)t/\hbar}}{(E_\alpha - E_0) + i\Gamma/2} \tag{3.176}$$

giving a steady-state probability to find the system in the state $|a\rangle$

$$P(a) = \frac{|H(\alpha)|^2}{(E_\alpha - E_0)^2 + \Gamma^2/4} . \tag{3.177}$$

Supposing that $|H(\alpha)|^2$ is a slowly varying function of the energy E_α, this probability has a peak of width Γ centered at E_0.

Substituting (3.176) into (3.171 we find

$$\gamma_\beta(t) = \delta_{\alpha\beta} + \frac{1}{i\hbar} \frac{H(\alpha)H^*(\beta)}{(E_\alpha - E_0) + i\Gamma/2} \int_0^t e^{i(E_\beta - E_\alpha)t'/\hbar} dt' . \tag{3.178}$$

For $\beta \neq \alpha$ this is

$$\gamma_\beta(t) = \frac{2\pi}{i} \frac{H(\alpha)H^*(\beta)}{(E_\alpha - E_0) + i\Gamma/2} e^{i(E_\beta - E_\alpha)t/2\hbar} \Delta_t(E_\beta - E_\alpha) . \tag{3.179}$$

The transition probability per unit time is then

---

[4] To simplify things, we neglect the principal value in the integral. This corresponds to the energy-shift of the level in second order perturbation theory.

$$\frac{d}{dt} P_{\alpha \to \beta} = \frac{2\pi}{\hbar} \frac{|H(\alpha)|^2 |H(\beta)|^2}{(E_\alpha - E_0)^2 + \Gamma^2/4} \delta_t(E_\beta - E_\alpha) \quad . \tag{3.180}$$

We see that the probability is proportional to the square of the amplitude to form the resonance from the initial state and to the square of the amplitude for the decay of the resonance to the final state. The delta function conserves energy in the transition between the initial and final states. The energy dependence of the probability reflect the parameters (width and mean energy) of the resonance.

The total cross-section is found by summing over final states in the usual way. The factor $|H(\beta)|^2 \delta_t(E_\beta - E_\alpha)$ when summed gives a factor Γ from (3.165). Equation (3.165) can also be used to replace $|H(\alpha)|^2 \propto \Gamma/\rho(E)$. The density of states is $\rho \propto VpE$ for a normalization volume V and center-of-mass momentum and energy p and E. After dividing by the flux density $(pc^2/E)/V$, one finds

$$\sigma(E) = 4\pi \left(\frac{\hbar}{p}\right)^2 \frac{(\Gamma/2)^2}{(E - E_0)^2 + \Gamma^2/4} \quad , \tag{3.181}$$

where p is the center of mass momentum. While we have found this formula using perturbation theory, it turns out that it holds even when perturbation doesn't apply. Note that the cross-section for $E = E_0$ is the so-called "geometrical" cross-section $4\pi(\hbar/p)^2$.

The calculation can be generalized to include the effects of spin and to allow for the existence of several different continuums $\{|\alpha_1\rangle, \ldots, |\alpha_n\rangle\}$ corresponding to different particles coupled to the same unstable state. Let Γ be the total width of the resonance and the Γ_i, $i = 1, \ldots, n$, the "partial widths" for the channel i defined by

$$\Gamma = \Sigma \Gamma_i \quad , \quad \Gamma_i = B_i \Gamma \quad , \tag{3.182}$$

where B_i is the branching ratio to the channel i We then have a *spin averaged* cross-section

$$\sigma_{i \to f}(E) = \frac{(2J + 1)}{(2S_1 + 1)(2S_2 + 1)} 4\pi \left(\frac{\hbar}{p}\right)^2 \frac{(\Gamma_i/2)(\Gamma_f/2)}{(E - E_0)^2 + \Gamma^2/4} \quad , \tag{3.183}$$

where J is the spin of the resonance and S_1 and S_2 are the spins of the two initial state particles. The factor $(2J + 1)$ is then due to the sum over the possible intermediate resonant states while the factors $(2S_i + 1)$ take into account the fact that the widths Γ_i are due to disintegration to all possible spin states.

An example of a cross-section exhibiting resonance behavior is shown in Fig. 3.26, showing cross-sections for neutrons on ^{235}U and ^{238}U. The peaks correspond to excited states of ^{236}U and ^{239}U. The states of ^{239}U can only decay by neutron or photon emission and therefore contribute to the elastic and (n, γ) cross-sections. The states of ^{236}U can also decay by fission. The dips in the elastic cross-section at energies just below some of the peaks are due to interference between the resonant amplitude and the non-resonant amplitude.

3.6 Nucleon–nucleus and nucleon–nucleon scattering

Up to now, our analysis of scattering has been based on perturbation theory. Here, we develop a method that is useful in calculating the elastic cross-sections on potentials that strongly affect the incoming wavefunction. It will be most easily applied to neutron–nucleus scattering in cases where the de Broglie wavelength, $2\pi\hbar/p$ of the incident neutron of momentum p is much larger than the range R of the potential. This is equivalent to

$$kR \ll 1 \tag{3.184}$$

where $k = p/\hbar$. For neutron–nucleus scattering, this requirement is satisfied if

$$\frac{p^2}{2m_{\mathrm n}} \ll \frac{(\hbar c)^2}{2m_{\mathrm n}c^2 R^2} \sim \frac{13\,\mathrm{MeV}}{A^{2/3}} \tag{3.185}$$

where we used $R = 1.2A^{1/3}$ for the nuclear radius.

3.6.1 Elastic scattering

We return to the problem first considered in Sect. 3.3.5 of finding the eigenfunctions of the Schrödinger equation

$$\left(\frac{\hbar^2}{2m}\nabla^2 + V(r)\right)\psi_k(\boldsymbol{r}) = \frac{\hbar^2 k^2}{2m}\psi_k(\boldsymbol{r})\,, \tag{3.186}$$

where m is the reduced mass of the neutron–nucleus system.[5] We will look for solutions of the form

$$\psi_k(\boldsymbol{r}) = e^{ikz} + \frac{fe^{ikr}}{r} \qquad r > R \tag{3.187}$$

where f is a constant independent of θ. This solution corresponds to a particle incident in the positive z direction followed by *isotropic* scattering. We will see that such a solution exists as long as (3.184) is satisfied. To do this, we write (3.187) as the sum of a function that vanishes for $kr \ll 1$ and a function that depends only on r:

$$\psi_k(\boldsymbol{r}) = \left[e^{ikz} - \frac{\sin kr}{kr}\right] + \left[\frac{\sin kr}{kr} + \frac{fe^{ikr}}{r}\right] \qquad r > R\,. \tag{3.188}$$

---

[5] Again, we intentionally focus on aspects which seem particularly relevant for our discussion. The general discussion can be found in standard textbooks such as M. L. Goldberger and K. M. Watson, *Collision Theory*, John Wiley & Sons, 1964.

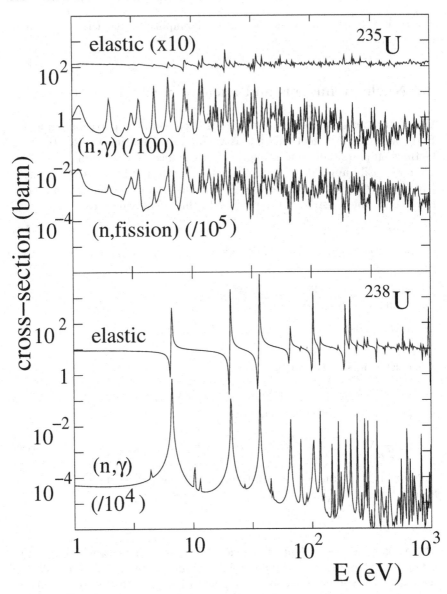

Fig. 3.26. The elastic and inelastic neutron cross-sections on ^{235}U (top) and ^{238}U (bottom). The peaks correspond to excited states of ^{236}U and ^{239}U. The excited states can contribute to the elastic cross-sections by decaying through neutron emission. They contribute to the (n, γ) cross-section by decaying by photon emission to the ground states of ^{236}U and ^{239}U. In the case of ^{236}U the states can also decay by fission so they contribute to the neutron-induced fission cross-section on ^{235}U.

It can readily be verified that the first bracketed term is a solution of (3.186) if $V(r) = 0$. Furthermore, the first bracketed term vanishes in the region $r < R$ where its two terms cancel as long as (3.184) is satisfied. It is therefore a solution of (3.186) even if $V \neq 0$.

The second bracketed function depends only on r:

$$\phi(r) = \frac{\sin kr}{kr} + \frac{f e^{ikr}}{r} = \frac{i}{2kr} \left[e^{-ikr} - (1 + 2ikf)e^{ikr} \right] . \qquad (3.189)$$

This function corresponds to a spherically symmetric wave directed toward the origin (the term $\propto \exp(-ikr)$) and reflected with an amplitude proportional to $(1 + 2ikf)$. We want to find f such that it is a solution of (3.186) for $r > R$. This can be done by matching the solution for $r < R$ with that for $r > R$ at $r = R$, i.e. by requiring that the function and its derivative be continuous at $r = R$.

Finding a solution of the form (3.189) is simpler task than finding a more general solution to (3.186) because the wavefunction now depends only on the radial coordinate r. Defining

$$u_k(r) = r\psi_k(r)$$
$$= e^{-ikr} - (1 + 2ikf)e^{ikr} \qquad r > R , \qquad (3.190)$$

(3.186) becomes

$$\left(\frac{\hbar^2}{2m} \frac{d^2 u_k}{dr^2} + V(r) \right) u_k(r) = \frac{\hbar^2 k^2}{2m} u_k(r) , \qquad (3.191)$$

which is just the one-dimensional Schrödinger eigenvalue equation. This equation can always be solved, numerically if need be.

As a simple example, we consider the spherical square-well potential

$$V(r) = +\infty \quad r < R \qquad V(r) = 0 \quad r > R , \qquad (3.192)$$

corresponding to an impenetrable sphere. In this case, the boundary condition is that the wavefunction vanish at the surface of the sphere, $u(R) = 0$. Equation (3.190) then tells us that

$$(1 + 2ikf)e^{ikr} = e^{-ikr} \quad \Rightarrow \quad f = -R(1 - ikR + \ldots) , \qquad (3.193)$$

where we have expanded in the small parameter $kR \ll 1$. This gives for $k \to 0$

$$\frac{d\sigma}{d\Omega} = R^2 \quad \Rightarrow \quad \sigma = 4\pi R^2 . \qquad (3.194)$$

The cross-section is 4 times the naive expectation, πR^2.

A more realistic example that can be applied to neutron–nucleus scattering is the potential that we used to analyze the deuteron in Sect. 1.4.1:

$$V(r) = -V_0 \quad r < R \qquad V(r) = 0 \quad r > R , \qquad (3.195)$$

where $V_0 > 0$. There are two interesting potentials, one for spin-aligned nucleons ($s = 1$) and one for spin-anti-aligned nucleons ($s = 0$).

The potential is shown in Fig. 3.27a. For $r < R$, the solution for $p^2/2m \ll V_0$ is

$$u(r) = A \sin k'r \quad k' = \sqrt{2m(V_0 + E)}/\hbar \quad r < R, \tag{3.196}$$

where $E = p^2/2m = \hbar^2 k^2/2m$. We discard the $\cos k'r$ solution because it leads to a wavefunction, $u(r)/r$, that is singular at the origin.

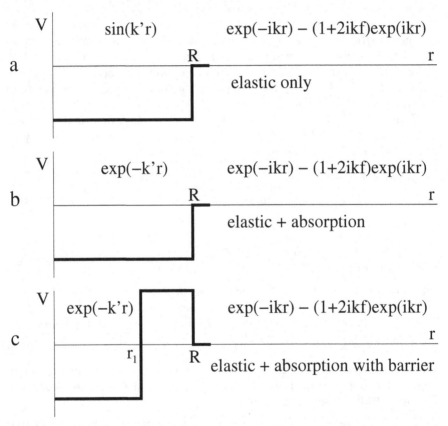

Fig. 3.27. Scattering from a spherical square-well potential. Figures a) and b) differ only in the choice of the wavefunction inside the well, the first leading only to elastic scattering (3.197) and the second to elastic scattering (3.208) and absorption (3.211). The well in Fig. c) has a barrier leading to elastic scattering and absorption proportional to the barrier penetration probability (3.216). In case (a) we require the $u(r)$ vanish at the origin so that the wavefunction $u(r)/r$ is non-singular. In cases b and c the wavefunction is assumed to be absorbed before reaching the origin.

The boundary condition is now that the $r > R$ solution (3.190) and the $r < R$ solution (3.196) have the same value and derivative at $r = R$. This condition is sufficient to determine f. In the low-energy limit, $kR \ll 1$, we have $k'(E = 0) = \sqrt{2mV_0}/\hbar$ and we find

$$f(k = 0) = -R \left[\frac{\tan(\sqrt{2mV_0R^2}/\hbar)}{\sqrt{2mV_0R^2}/\hbar} - 1 \right] . \tag{3.197}$$

The pre-factor $(-R)$ would, by itself, give a total cross-section of $4\pi R^2$. It is multiplied by a factor that depends on the number of neutron wavelengths, $\sqrt{2mV_0R^2}/\hbar$ that fit inside the potential well. As we emphasized in Sect. 1.4.1, the two neutron–proton effective potentials have approximately $1/4$ of a wavelength inside the well, i.e.

$$\frac{\sqrt{2mV_0R^2}}{\hbar} \sim \pi/2 \quad \Rightarrow \quad V_0R^2 \sim 109\,\mathrm{MeV\,fm^2} . \tag{3.198}$$

For the $s = 1$ np system, V_0R^2 is slightly greater than $109\,\mathrm{MeV\,fm^2}$ so the deuteron is bound. For the $s = 0$ system, V_0R^2 is slightly less than $109\,\mathrm{MeV\,fm^2}$ so there is no bound state. In the scattering problem considered here, the quarter wavelength leads to a scattering cross-section that is much larger than $4\pi R^2$ since the tangent in (3.197) is large.

Table 3.3. The low-energy nucleon–nucleon scattering amplitudes and effective ranges taken from the compilation [34]. The last two columns give the potential parameters derived from the deuteron binding energy and the scattering formula (3.197). Note that $f \gg R$ for the $s = 0$ amplitudes.

	f (fm)	R (fm)	V_0 (MeV)	V_0R^2 (MeV fm^2)
n–p (s=1, T=0)	$+5.423 \pm 0.005$	1.73 ± 0.02	46.7	139.6
n–p (s=0, T=1)	-23.715 ± 0.015	2.73 ± 0.03	12.55	93.5
p–p (s=0, T=1)	-17.1 ± 0.2	2.794 ± 0.015	11.6	90.5
n–n (s=0, T=1)	-16.6 ± 0.6	2.84 ± 0.03	11.1	89.5

The cross-section for the scattering on unpolarized neutrons on unpolarized protons is the weighted sum of the cross-section in the $(s = 0)$ and $(s = 1)$ state. Since there are three spin-aligned states and only one anti-aligned state we have

$$\sigma_{\mathrm{n-p}} = (3/4)4\pi|f_{s=1}|^2 + (1/4)4\pi|f_{s=0}|^2 = 20.47\,\mathrm{b} . \tag{3.199}$$

This corresponds to the low-energy limit of the neutron–proton cross-section shown in Fig. 3.4. The contributions from the $(s = 0)$ and $(s = 1)$ amplitudes can be separated by a variety of methods. For instance, the neutron index of refraction (Sect. 3.7) depends on the weighted sum of the *amplitudes* rather than of the on the weighted sum of the squares of the amplitudes. A measurement of the index of refraction combined with the unpolarized cross-section therefore allows one to deduce the amplitudes in the $(s = 1)$ and $(s = 0)$ states.

The measured amplitudes are listed in the first column of Table 3.3. Also listed are the effective ranges R of the potentials. These can be found by considering the energy dependence of the cross-section. One finds

$$\sigma(k) = \frac{4\pi|f(k=0)|^2}{\left|1 - \frac{1}{2}f(k=0)Rk^2\right|^2 + |f(k=0)|^2k^2} \tag{3.200}$$

The cross-section slowly declines with increasing energy, as can be seen in Fig. 3.4. The ranges deduced from the energy dependence are listed in the second column of Table 3.3.

Quite generally, in strong interactions, one calls the quantity

$$a = -f(k=0) \tag{3.201}$$

the *scattering length*. The total low energy cross-section is therefore

$$\sigma(k \simeq 0) = 4\pi a^2 . \tag{3.202}$$

We have seen in equation (3.116) that, in Born approximation, the scattering length is related to the potential by

$$a = \frac{m}{2\pi\hbar^2} \int V(\boldsymbol{r}) \, \mathrm{d}^3 r \quad . \tag{3.203}$$

Here, it is the more complicated relation (3.197)

It should be emphasized that (3.200) applies only to the isotropic component of the elastic cross-section. At energies where one no longer has $kR \ll 1$, our treatment based on isotropic scattering must be modified. When the potential is central (or better the interaction is rotation invariant) the complete treatment of the angular dependence of scattering amplitudes is done by projecting the Schrödinger equation on spherical harmonics. This results in including higher "partial wave amplitudes," each of which corresponds to a given value of the angular momentum ℓ so that the scattering amplitude becomes

$$f(\theta) = \sum_{\ell=0}^{\infty} a_\ell \, P_\ell(\cos\theta) \tag{3.204}$$

where $P_\ell(\cos\theta)$ is a Legendre polynomial and where the amplitudes of the "ℓ^{th}" partial wave is given in terms of the "phase shifts" δ_ℓ

$$a_\ell = (2\ell+1)\frac{e^{2i\delta_\ell} - 1}{2ik} . \tag{3.205}$$

The a_ℓ can be found by solving the partial wave Schrödinger equation for the potential in question. Our original treatment based on (3.187) supposed that f was independent of angle and therefore corresponds to keeping only the $\ell = 0$ part of the wavefunction. The inclusion of $\ell \neq 0$ partial waves leads to a peaking of the cross-section in the forward direction once the assumption $kR < 1$ breaks down. This energy evolution of the differential cross-section is illustrated in Fig. 3.6.

Table 3.3 also shows the amplitudes of proton–proton and neutron–neutron scattering in the $s = 0$ state. (The $s = 1$ state is forbidden by the Pauli principle at low energy.) The proton–proton amplitude can be derived from large-angle scattering where Coulomb scattering is unimportant (Exercise 3.7). Since neutron–neutron scattering is difficult to observe directly, its amplitude must be derived indirectly from other reactions.

We note the similarity in Table 3.3 of the amplitudes and ranges of the three $s = 0$ amplitudes, n–p, p–p, and n–n. This similarity is indicative of the isospin symmetry of the strong interactions. The three $s = 0$ systems form an isospin triplet whose interactions must be identical in the limit of isospin symmetry. There is however a large difference between the iso-triplet potential and the iso-singlet ($s = 1$) potential.

3.6.2 Absorption

While our analysis was intended only for potential scattering, we can include phenomenologically the effects of absorption of the incident particle (e.g. by radiative capture) by appropriately choosing the wavefunction for $r < R$. For example, if a incoming spherical wave is completely absorbed once it enters the nucleus, we can use (Fig. 3.27b)

$$u(r) = A e^{-ik'r} \quad r < R .$$

(3.206)

This corresponds to an ingoing wave and no outgoing wave. (One can think of the ingoing wave as being absorbed near the origin.) Requiring continuity of the wavefunction and its derivative at $r = R$ gives

$$f = -R\left(1 - \frac{2k}{k'}\right) + i\left(\frac{1}{k'} + kR^2 - \frac{k}{k'^2}\right) ,$$

(3.207)

giving an elastic cross-section of

$$\sigma_{el}(k = 0) = 4\pi\left(R^2 + \frac{1}{k'^2}\right) \quad k' = \sqrt{2mV_0}/\hbar .$$

(3.208)

The total (elastic + absorption) cross-section can be found from the optical theorem to be derived in Sect. 3.7:

$$\sigma_{tot} = \frac{4\pi \, \mathrm{Im}(f)}{k}$$

(3.209)

which gives in this case

$$\sigma_{tot} = \frac{4\pi}{kk'} + 4\pi\left(R^2 - \frac{1}{k'^2}\right) .$$

(3.210)

Subtracting off the elastic cross-section we get the absorption cross-section:

$$\sigma_{\text{abs}}(k \to 0) = \frac{4\pi}{kk'} . \tag{3.211}$$

The cross section is proportional to the inverse of the velocity of the incident particle. This $1/v$ behavior is seen if there is a barrier-free exothermic inelastic channel, e.g. radiative absorption (n, γ). Examples are given in Fig. 3.4.

Exothermic reactions between charged particles can only take place if the two particles penetrate their mutual Coulomb barriers. An example is the (p, γ) reaction in ^{6}Li whose cross-section is shown in Fig. 3.4. For such reactions, the cross-section vanishes as $k \to 0$ because the barrier penetration probability vanishes for $k \to 0$.

To see how the barrier penetration factor enters, we treat the third situation in Fig. 3.27c. There is a square potential barrier of height V_0 for $r_1 < r < R$. The solution in this region is

$$u(r) = Ae^{\kappa r} + Be^{-\kappa r} \qquad \kappa = \sqrt{2m(V_0 - E)}/\hbar . \tag{3.212}$$

Imposing the matching conditions at $r = r_1$ and $r = R$ one finds a scattering amplitude

$$f(k \to 0) = -R\left(1 - \frac{1}{\kappa R}\right)\left[1 - ikR\left(1 - \frac{1}{\kappa R}\right) + \frac{2\epsilon}{\kappa R}\right] , \tag{3.213}$$

where the small parameter ϵ is proportional to the barrier penetration probability

$$\epsilon = \frac{k' + ik_1}{k' - ik_1} \exp(-2\kappa(R - r_1)) . \tag{3.214}$$

The elastic and absorption cross-sections are

$$\sigma_{\text{el}}(k = 0) = 4\pi R^2 \left(1 - \frac{1}{\kappa R}\right)^2 \tag{3.215}$$

$$\sigma_{\text{abs}} = \frac{4\pi R}{k}\left(1 - \frac{1}{\kappa R}\right) Im(\epsilon)$$

$$= \frac{4\pi R}{k}\left(1 - \frac{1}{\kappa R}\right)\frac{2k_1 k'}{k_1^2 + k'^2}\exp(-2\kappa(R - r_1)) . \tag{3.216}$$

The absorption cross-section is proportional to the exponential barrier penetration probability, as expected.

The dimensional factor in the absorption cross-section is $R/k \propto R\lambda$. where λ is the de Broglie wavelength. In the more realistic case of a Coulomb barrier, there is no dimensional parameter R so the cross-section must be, by dimensional analysis, proportional to λ^2. The cross-section is generally written as

$$\sigma_{\text{abs}} = \left(\frac{\lambda}{2\pi}\right)^2 S(E)\exp(-2\pi Z_1 Z_2 \alpha/\beta) , \tag{3.217}$$

where $S(E)$ is a slowly varying function of the energy and $\beta = v/c$ for a relative velocity v. The exponential barrier penetration factor, derived previously in Sec. 2.6 vanishes as the velocity approaches zero.

3.7 Coherent scattering and the refractive index

Up to now we have supposed that, in calculating the probability for an interaction in a target, the scattering probability on individual target particles can be added. This implies that the probability for an interaction after passage through a slice of matter of thickness dz is given by $dP = n\sigma dz$. As emphasized in Sect. 3.1, this is only justified if the waves emanating from the different target particles have random phases. Let $A_i = |A_i|e^{i\theta_i}$ be the amplitude for scattering from particle i. The square of the sum of the amplitudes is

$$\left| \sum_i A_i \right|^2 = \sum_i |A_i|^2 + \sum_{i \neq j} |A_i||A_j| e^{i(\theta_i - \theta_j)} \tag{3.218}$$

For random phases, the second term vanishes and we are left with the sum of the squared amplitudes. We will see that in the forward direction it is not justified to assume random phases, so it is necessary to add amplitudes to get correct results. By doing this, we will derive an expression for the index of refraction and for the total cross-section in terms of the forward scattering amplitude (the optical theorem).

As shown in Fig. 3.28, we consider particles incident upon a then slice of material of thickness dz containing a density n of scatterers. Beyond the slice, the wavefunction is the sum of the incident wave, $\psi = \exp(ikz)$ and the scattered wave found by summing the contributions of all scatterers:

$$\psi(z > 0) = \exp(ikz) +$$

$$n dz \int dx \int dy f(\theta) \frac{\exp(ik\sqrt{x^2 + y^2 + z^2})}{(x^2 + y^2 + z^2)^{1/2}}, \tag{3.219}$$

where the scattering angle is

$$\theta(x, y) = \tan^{-1} \frac{\sqrt{x^2 + y^2}}{z}. \tag{3.220}$$

The exponential in (3.219) is a rapidly oscillating function of the integration variables x, y except at the "stationary point" $(x = 0, y = 0)$ where the phase's partial derivatives with respect to x and y vanish. We can anticipate that the integral will be dominated by the region near $x = y = 0$, corresponding to $\theta = 0$. This is just the mathematical equivalent of the physical statement that scattering is generally only coherent in the forward direction. We therefore replace $f(\theta)$ with $f(\theta = 0)$, set $x = y = 0$ in the denominator, and expand the exponential, obtaining:

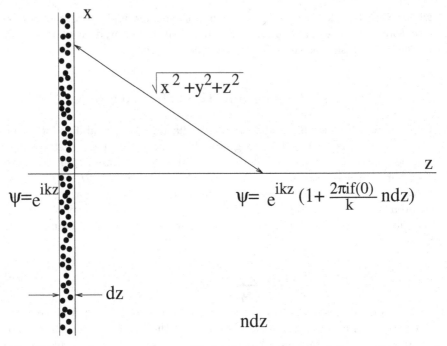

Fig. 3.28. A plane wave, $\psi = \exp(ikz)$ is incident from the left on a thin slice of material of thickness dz containing a density n of scatterers. Beyond the slice, the wave is the sum of the incident wave and the scattered waves from all scatterers in the slice. As shown in the text, the wavefunction takes the form of Equation (3.222).

$$\psi(z > 0) = \exp(ikz)$$

$$\times \left[1 + \frac{nf(0)dz}{z} \int dx \int dy \exp\left(\frac{ik(x^2 + y^2)}{2z}\right)\right] \tag{3.221}$$

Using $\int_{\infty}^{\infty} e^{iu^2} du = \sqrt{2\pi} e^{-i\pi/4}$, we find

$$\psi(z > 0) = \exp(ikz)\left(1 + \frac{2\pi i f(\theta = 0)}{k} n dz\right) \tag{3.222}$$

The magnitude of ψ is

$$|\psi(z > 0)|^2 = 1 - \frac{4\pi Im(f(0))}{k} n dz . \tag{3.223}$$

The total cross-section is defined by the probability $dP = \sigma_{\text{tot}} n dz$ so we deduce the so-called *optical theorem*:

$$\sigma_{\text{tot}} = \frac{4\pi Im(f(0))}{k} . \tag{3.224}$$

The effect of the real part of f can be seen by writing the wavefunction (3.222) just after the slice:

$$\psi(\mathrm{d}z) = \left(1 - \frac{2\pi Im(f(0))}{k}\right) \exp\left[ik\left(1 + \frac{2\pi n Re f(0)}{k^2}\right)\mathrm{d}z\right] . (3.225)$$

This implies that the index of refraction is

$$\text{index of refraction} = 1 + \frac{\lambda^2 n Re(f(0))}{2\pi} , \qquad (3.226)$$

where $\lambda = 2\pi/k$ is the neutron wavelength. Note that for solids $n \sim a_0^{-3}$ so, if $f \sim$ fm then the index differs significantly from unity for $\lambda \sim 100a_0$ corresponding to neutrons with $E \sim 10^{-5}$eV.

The fact that low-energy neutron refraction index is significantly different from unity is due to the fact that the scattering amplitude on nuclei approaches a constant for $k \to 0$. This is unlike the case of photons where the amplitude approaches zero for $k \to 0$ and, consequently, the refraction index approaches a constant, often near unity.

The neutron refraction index can be sufficiently important to permit the construction of neutron guides that use total internal reflection. At sufficiently low energies, it is even possible to construct neutron containers that total reflect neutrons of all scattering angles. While this may seem surprising, it is just equivalent to the mean kinetic energy of the neutrons being less than its mean potential energy within the wall of the container.

Figure 3.29 gives a example of how neutrons can be captured and then stored for study. We will see in Chap. 4 how such stored neutrons can be used to measure the neutron lifetime.

3.8 Bibliography

1. *Collision Theory* M. L. Goldberger and K. M. Watson, John Wiley & Sons, 1964
2. *Electron Scattering and Nuclear and Nucleon structure*, R. Hofstadter, W.A. Benjamin, New-York (1963)

Exercises

3.1 Using the data in Fig. 3.4 and neglecting scattering on oxygen, calculate the mean free path of thermal neutrons $(p^2/2m \sim kT)$ in normal water and in heavy water (where the ^{1}H is replaced with ^{2}H. On average, how many elastic collisions will the neutron suffer before being absorbed by the (n, γ) reaction? How much time will this typically take?

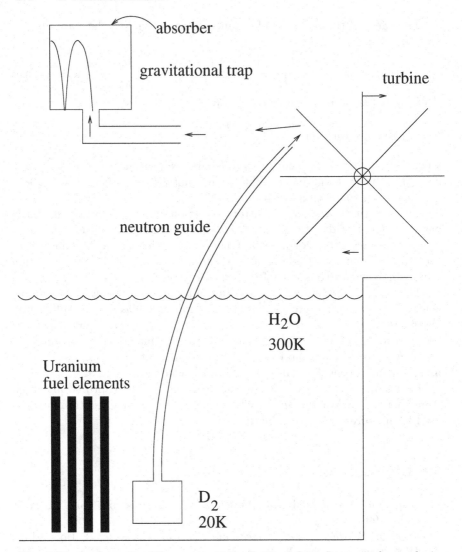

Fig. 3.29. A schematic of the system at the Institute Laue-Langevin for producing ultra-cold neutrons [35]. Neutrons produced by Uranium fission diffuse into the water moderator where they are thermalized to a temperature of $\sim 300\,\mathrm{K}$ by elastic scattering on the protons in water molecules. Some of the neutrons diffuse into a cold deuterium flask where they are further thermalized to $\sim 20\,\mathrm{K}$. (Deuterium is used because of its low radiative capture cross-section; see Fig. 3.4.) Neutrons then escape from the reactor through a neutron guide that uses total internal reflection. They then impinge on a counter rotating turbine, causing them to further loose energy by reflection. The neutrons are then guided to a gravitational trap consisting a container that is reflecting on the bottom and side surfaces but absorbing on the top surface. Neutrons that are trapped must then have energies $< m_n g z \sim 10^{-7}\,\mathrm{eV}$ where $z \sim 1\,\mathrm{m}$ is the vertical size of the container.

3.2 The explosion of the supernova SN1987A, on Feb. 23 1987, released $N \sim 10^{57}$ antineutrinos, some of which were detected by the Japanese detector of Kamiokande. The star that had exploded, was located at a distance of $R \simeq 140,000$ light-years from Earth. The detector contained 2000 tons of water. The particles recorded were positrons produced in the reaction $\bar{\nu}_e p \rightarrow n e^+$, of antineutrinos on the protons of the hydrogen atoms of H_2O. The cross-section of this reaction at the mean energy of 15 MeV of the neutrinos, is $\sigma = 2 \ 10^{-45} \ m^2$. How many events would be expected to be recorded ?

3.3 From the level diagram in Fig. 3.5 and the data in Appendix G. calculate the neutron energy necessary to excite the 7.459 MeV level of ^7Li in a collision with ^6Li.

3.4 The sun has a mean density of $1.4 \, g \, cm^{-3}$. The Thomson cross-section of photons, produced in the solar core, on the electrons of the solar plasma is $\sigma_T(\gamma e^- \rightarrow \gamma e^-) \simeq 0.665 \ 10^{-28} \ m^2$. Calculate the mean free path of photons. In their random walk across the Sun it takes the photons a time of the order of $\tau \simeq (R/l)^2 (l/c) \simeq 10^4$ years to escape and reach us (the radius of the sun is $R \simeq 700.000$ km). Actually, the escape time is larger $\simeq 10^5$–10^6 years due to the higher density by a factor of roughly ~ 200 in the solar core.

3.5 Verify that if (3.37) is satisfied, the maximum target recoil energy is much less than the kinetic energy of the beam particle.

3.6 Verify that the two peak energies in Fig. 3.22 correspond to elastic scattering on protons and on ^2H nuclei.

3.7 By comparing the Rutherford Scattering cross-section with that of nucleon–nucleon scattering (Table 3.3), find the scattering angular range where the nucleon–nucleon amplitude is larger than the Coulomb amplitude for $\sim 2 \, MeV$ proton–proton scattering.

4. Nuclear decays and fundamental interactions

This chapter is primarily concerned with nuclear instability. Generally speaking, there are two types of decays of nuclear species: A- and Z-conserving "dissociative" decays like α-decay and spontaneous fission; and β-decays which transform neutrons to protons or vice versa. Additionally, nuclear excited states decay by emission of photons (γ-decay) or atomic electrons (internal conversion). If their energy is sufficiently high, the excited states can also decay by dissociation, especially nucleon emission.

Our first goal in this chapter is to describe γ- and β-decay of nuclei. The interesting point is that in both cases these decays are due to *fundamental interactions*. The interactions are sufficiently weak that the decays can be treated with standard time-dependent perturbation theory, Appendix C. This is quite different from dissociative decays which are generally viewed as tunneling processes. α-decay is treated in this way in Chap. 2 and spontaneous fission in Chap. 6.

The information gleaned from weak nuclear decays was instrumental in the formulation of the *Standard Model* of fundamental constituents of matter, the families of quarks and leptons, and their interactions. We will end this chapter with a brief introduction to this model.

4.1 Decay rates, generalities

4.1.1 Natural width, branching ratios

Decay rates and mean lifetimes can be defined by the same considerations as lead us to the definition of cross-sections in Chap. 3. An unstable particle has a probability $\mathrm{d}P$ to decay in a time interval $\mathrm{d}t$ that is proportional to $\mathrm{d}t$:

$$\mathrm{d}P = \frac{\mathrm{d}t}{\tau} , \qquad (4.1)$$

where τ clearly has dimensions of time and is called the "mean lifetime" of the particle. This law governs the time dependence of the number $N(t)$ of an unstable state surviving after a time t:

$$N(t + \mathrm{d}t) - N(t) = -N(t)\mathrm{d}P \quad \Rightarrow \quad \frac{\mathrm{d}N}{\mathrm{d}t} = -\frac{N(t)}{\tau} , \qquad (4.2)$$

which has the solution

$$N(t) = N(t = 0)e^{-t/\tau} \, . \tag{4.3}$$

The mean survival time is τ, justifying its name.

The inverse of the mean lifetime is the "decay rate"

$$\lambda = \frac{1}{\tau} \, . \tag{4.4}$$

We saw in Sect. 3.5 that an unstable particle (or more precisely an unstable quantum state) has a rest energy uncertainty or "width" of

$$\Gamma = \hbar\lambda = \frac{\hbar}{\tau} = \frac{6.58 \times 10^{-22} \, \text{MeV sec}}{\tau} \, . \tag{4.5}$$

Since nuclear states are typically separated by energies in the MeV range, the width is small compared to state separations if the lifetime is greater than $\sim 10^{-22}$ sec. This is generally the case for states decaying through the weak or electromagnetic interactions. For decays involving the dissociation of a nucleus, the width can be quite large. Examples are the excited states of ^{7}Li (Fig. 3.5) that decay via neutron emission or dissociation into ^{3}H^{4}He. From the cross-section shown in Fig. 3.4, we see that the fourth excited state (7.459 MeV) has a decay width of $\Gamma \sim 100$ keV.

It is often the case that an unstable state has more than one "decay channel," each channel k having its own "branching ratio" B_k. For example the fourth excited state of ^{7}Li has

$$B_{n^6\text{Li}} = 0.72 \qquad B_{^3\text{H}^4\text{He}} = 0.28 \qquad B_{\gamma^7\text{Li}} \sim 0.0 \, , \tag{4.6}$$

where the third mode is the unlikely radiative decay to the ground state. In general we have

$$\sum_k B_k = 1 \, , \tag{4.7}$$

the sum of the "partial decay rates," $\lambda_k = B_k\lambda$

$$\sum_k \lambda_k = \lambda \, , \tag{4.8}$$

and the sum of the "partial widths," $\Gamma_k = B_k\Gamma$

$$\sum_k \Gamma_k = \Gamma \, . \tag{4.9}$$

4.1.2 Measurement of decay rates

Lifetimes of observed nuclear transitions range from $\sim 10^{-22}$ sec

$$^7\text{Li}\,(7.459\,\text{MeV}) \rightarrow n\,^6\text{Li}, \quad ^3\text{H}\,^4\text{He} \qquad \tau = 6 \times 10^{-21} \, \text{sec} \tag{4.10}$$

to 10^{21} yr

$$^{76}\text{Ge} \rightarrow {}^{76}\text{Se}\,2e^-\,2\bar{\nu}_e \quad t_{1/2} = 1.6 \times 10^{21}\,\text{yr} \tag{4.11}$$

It is not surprising that the techniques for lifetime measurements vary considerably from one end of the scale to the other. Here, we summarize some basic techniques, illustrated in Figs. 4.1- 4.4.

- $\tau > 10^8$ yr (mostly α- and 2β-decay). The nuclei are still present on Earth (whose nuclei were formed about 5×10^9 year ago) and can be chemically and isotopically isolated in macroscopic quantities and their decays detected. The lifetime can then by determined from (4.3) and knowledge of the quantity N in the sample. An illustration of this technique is shown in Fig. 4.1.

- 10 min $< \tau < 10^8$ yr (mostly α- and β-decay). The nuclei are no longer present on Earth in significant quantities and must be produced in nuclear reactions, either artificially or naturally (cosmic rays and natural radioactivity sequences). The lifetimes are long enough for chemical and (with more difficulty) isotopic purification. The decays can then be observed and (4.3) applied to derive τ. The case of ^{170}Tm is illustrated in Fig. 4.2. If the observation time is comparable to τ, knowledge of $N(t = 0)$ is not necessary because τ can be derived from the time variation of the counting rate.

- 10^{-10}s $< \tau < 10^3$s (mostly β-, γ- and α-decay). While chemical and isotopic purification is not possible for such short lifetimes, particles produced in nuclear reactions can be slowed down and stopped in a small amount of material (Sect. 5.3). Decays can be counted and (4.3) applied to derive τ. Examples are shown in Figs. 2.18 and 2.19. The case of the first excited state of ^{170}Yb produced in the β-decay of ^{170}Tm is illustrated in Fig. 4.2.

- 10^{-15}s $< \tau < 10^{-10}$s. (mostly γ-decay). The time interval between production and decay is too short to be measured by standard timing techniques but a variety of ingenious techniques have been devised that apply to this range that covers most of the radiative nuclear decays. One technique uses the fact that the time for a particle to slow down in a material after having been produced in a nuclear reaction can be reliably calculated (Sect. 5.3). For particles with 10^{-15}s $< \tau < 10^{-10}$s, the disposition of material can be chosen so that some particles decay "in flight" and some after coming to rest. For the former, the energies of the decay particles are Doppler shifted and can be distinguished from those due to decays at rest. Measurement of the proportion of the two types and knowledge of the slowing-down time allows one to derive τ. The technique is illustrated in Fig. 4.3.

Another indirect technique for radiative transitions is the *Coulomb excitation* method. The cross-section for the production of an excited state in collisions with a charged particle is measured. As mentioned in Sect. 3.4.2, the cross-section involves the same matrix element between ground- and excited-nuclear states as that involved in the decay of the excited- to ground-state. In fact, the incident charged particle can be considered to be a source of virtual photons that can induce the transition. Knowledge of the cross-section allows one to deduce the radiative lifetime of the state.

- $\tau < 10^{-12}$s i.e. $\Gamma > 6 \times 10^{-10}$ MeV. (mostly γ-decay and dissociation). In this range where direct timing is impossible, the width of the state can be measured and (4.5) applied to derive τ. An example is shown in Fig. 3.4 where the energy dependence of the neutron cross-section on ^{6}Li can be used to derive the widths of excited states. In this example, the state is very wide because it decays by breakup to n^{6}Li or ^{3}H^{4}He. Widths of states that decay radiatively can only be measured with special techniques. An example is the use of the Mössbauer effect, as illustrated in Fig. 4.4.

4.1.3 Calculation of decay rates

Consider a decay

$$a \rightarrow b_1 + b_2 + \ldots + b_N \quad . \tag{4.12}$$

Particle a, assumed to be at rest, has a mass M and an energy $E = Mc^2$. As in scattering theory, we can calculate decay rates by using time-dependent perturbation theory (Appendix C). We suppose that the Hamiltonian consists of two parts. The first, H_0, represents the energies of the initial and final state particles, while the second, H_1, has matrix elements connecting initial and final states. The decay rate, i.e. the probability per unit time that a decays into a state $|f\rangle$ of final particles is

$$\lambda_{a \rightarrow f} = \frac{2\pi}{\hbar} |\langle f|T|a\rangle|^2 \, \delta_t \left(Mc^2 - \sum E_j \right) \tag{4.13}$$

where E_j is the energy of particle b_j. In first order perturbation theory, the transition operator T is just the Hamiltonian responsible for the decay, H_1.

As in the case of nuclear reactions, quantum field theory is the appropriate language to determine which decays are possible and the form of their matrix elements. Lacking this technology, we will usually just give the matrix elements for each process under consideration. However, as in reaction theory where the classical limit of particles moving in a potential was a guide for determining the matrix elements for elastic scattering, certain decay processes have classical analogs that can guide us. This is the case for radiative decays which have the classical limit of a charge distribution generating an oscillating electromagnetic field.

Despite the fact that we will not generally be able to derive rigorously the matrix elements, we can expect that the interaction Hamiltonian is translation invariant. Therefore, the square of the transition matrix element $|\langle f|T|i\rangle|^2$ will be, as in scattering theory, proportional to a momentum conserving delta function. We therefore define the "reduced" transition matrix element $\tilde{T}$ by

$$|\langle f|T|i\rangle|^2 = |\tilde{T}(\mathbf{p}_1 ... \mathbf{p}_N)|^2 V^{-(N+1)} \, V(2\pi\hbar)^3 \delta_L^3(\Sigma \mathbf{p}_j) \,, \tag{4.14}$$

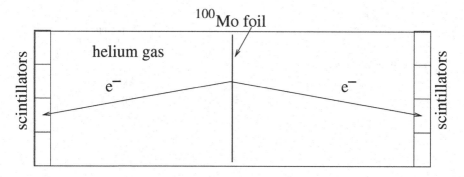

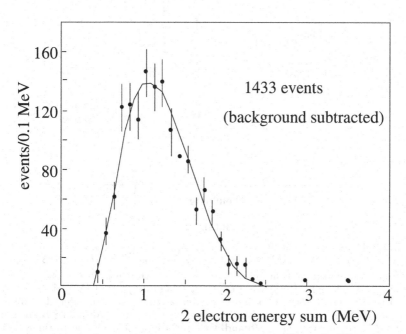

Fig. 4.1. The measurement of the double-β decay of ^{100}Mo $\rightarrow$ ^{100}Ru $2e^-2\bar{\nu}_e$ [36]. The upper figure shows a simplified version of the experiment The source is a 40μm thick foil consisting of 172 g of isotopically enriched ^{100}Mo (98.4% compared to the natural abundance of 9.6%). After a decay, the daughter nucleus stays in the foil but the decay electrons leave the foil (Exercise 4.2) and traverse a volume containing helium gas. The gas is instrumented with high voltage wires that sense the ionization trail left by the passing electrons so as to determine the e^- trajectories. The electrons then stop in plastic scintillators which generate light in proportion to the electron kinetic energy. The bottom figure show the summed kinetic energy of electron pairs measured in this manner. A total of 1433 events were observed over a period of 6140 h, corresponding to a half-life of ^{100}Mo of $(0.95 \pm 0.11) \times 10^{19}$yr.

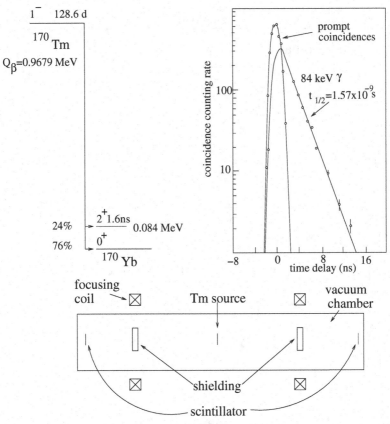

Fig. 4.2. Observation of the decay of ^{170}Tm and measurement of the lifetime of the first excited state of ^{170}Yb [37]. The radioactive isotope ^{170}Tm ($t_{1/2} = 128.6$day) is produced by irradiating a thin foil of stable ^{169}Tm with reactor neutrons. ^{170}Tm is produced through radiative neutron capture, ^{169}Tm(n, γ)^{170}Tm. After irradiation, the foil is placed at a focus of a double-armed magnetic spectrometer. The decay ^{170}Tm $\rightarrow$ ^{170}Yb e$^-$ $\bar{\nu}_e$ proceeds as indicated in the diagram with a 76% branching ratio to the ground state of ^{170}Yb and with at 24% branching ratio to the 84 keV first excited state. The excited state subsequently decays either through γ-emission or by internal conversion where the γ-ray ejects an atomic electron of the Yb. Electrons emerging from the foil are momentum-selected by the magnetic field and focused onto two scintillators. Events with counts in both scintillators are due to a β-electron in one scintillator and to an internal conversion electron in the other. The distribution of time-delay between one count and the other is shown and indicates that the exited state has a lifetime of ~ 1.57 ns.

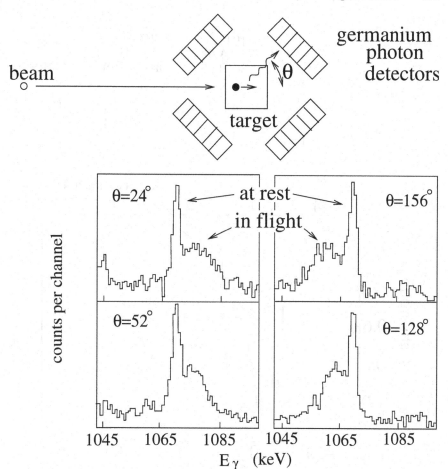

Fig. 4.3. Measurement of radiative-decay lifetimes by the "Doppler-shift attenu-
ation method" [38]. The top figure is a simplified version of the apparatus used to
measure the lifetimes of excited states of ^{74}Br. A beam of 70 MeV ^{19}F ions impinges
upon a ^{58}Ni target, producing a variety of nuclei in a variety of excited states. The
target is sufficiently thick that the produced nuclei stop in the target. Depending
on the lifetime of the produced excited state, the state may decay before stopping
("in-flight" decays) or at rest. The target is surrounded by germanium-diode detectors (the Euroball array) that measure the energy of the photons. The bottom
figure shows the energy distribution of photons corresponding to the 1068 keV line
of ^{74}Br for four germanium diodes at different angles with respect to the beam
direction. Each distribution has two components, a narrow peak corresponding to
decays at rest and a broad tail corresponding to Doppler-shifted in-flight decays.
Note that decays with $\theta > 90$ deg ($\theta > 90$ deg) have Doppler shifts that are positive
(negative). Roughly half the decays are in-flight and half at-rest. Knowledge of the
time necessary to stop a Br ion in the target allowed one to deduce a lifetime of
0.25 ps for the state that decays by emission of the 1068 keV gamma (Exercise 4.4).

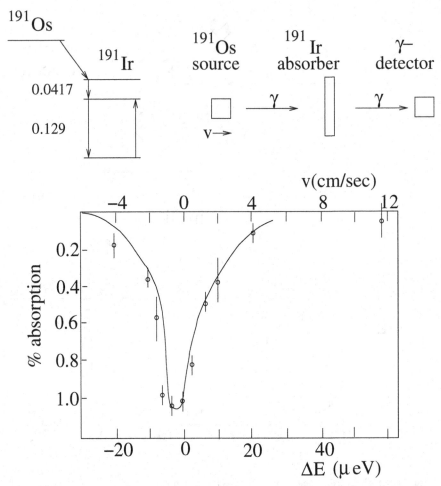

Fig. 4.4. Measurement of the width of the first excited state of ^{191}Ir through Mössbauer spectroscopy [39]. The excited state is produced by the β-decay of ^{191}Os. De-excitation photons can be absorbed by the inverse transition in a ^{191}Ir absorber. This resonant absorption can be prevented by moving the absorber with respect to the source with velocity v so that the photons are Doppler shifted out of the resonance. Scanning in energy then amounts to scanning in velocity with $\Delta E_\gamma / E_\gamma = v/c$. It should be noted that photons from the decay of *free* ^{191}Ir have insufficient energy to excite ^{191}Ir because nuclear recoil takes some of the energy (4.42). Resonant absorption is possible with $v = 0$ only if the ^{191}Ir nuclei is "locked" at a crystal lattice site so the crystal as a whole recoils. The nuclear kinetic energy $p^2/2m_A$ in (4.42) is modified by replacing the mass of the nucleus with the mass of the crystal. The photon then takes all the energy and has sufficient energy to excite the original state. This "Mössbauer effect" is not present for photons with $E > 200\,\mathrm{keV}$ because nuclear recoil is sufficient to excite phonon modes in the crystal which take some of the energy and momentum.

$\tilde{T}$ represents the *dynamics* of the decay, as opposed to the *kinematics* (energy-momentum conservation and state-counting). The factors of the wavefunction normalization volume $V = L^3$ have been added for convenience so that $\tilde{T}$ is V-independent. The factor $V^{-(N+1)}$ comes from the $(N+1)$ wavefunctions in the matrix element $(\exp(ipr)/\sqrt{V})$ while the factor $V(2\pi\hbar)^3$ comes for the square of the integration over the wavefunctions leading to the delta function (C.24). The resulting factor of V^{-N} will be canceled in the sum over final states, as demonstrated explicitly below.

We note that the dimensionality of $\tilde{T}$ depends on the number N of final state particles:

$$[\tilde{T}] = \text{energy} \times \text{length}^{3(N-1)/2} \tag{4.15}$$

For $N = 3$, as in β-decay, it has dimensions of energy$\times$volume and we can anticipate that $\tilde{T} \sim G_{\text{F}}$.

The reaction rate (4.12) is obtained by summing over all possible accessible final states. Just like for cross sections, one first normalizes asymptotic final states in a finite volume $V = L^3$, and one replaces the sum by integrals over the momenta of final particles by making use of the density of states. The normalization volume cancels off and one ends up with the general form

$$\lambda_{a \to b_1 + b_2 \ldots + b_N} \tag{4.16}$$

$$= \frac{(2\pi\hbar)^4}{\hbar^2} \int |\tilde{T}(\boldsymbol{p}_1 \ldots \boldsymbol{p}_N)|^2 \delta^3(\Sigma\boldsymbol{p}_j)\delta(E - \Sigma E_j) \prod_{j=1}^{N} \frac{d^3\boldsymbol{p}_j}{(2\pi\hbar)^3} \,,$$

where $\tilde{T}(\boldsymbol{p}_1 \ldots \boldsymbol{p}_N)$ is the reduced transition matrix element.

In the form (4.16), the transition rate appears as the square of the transition matrix element (divided by $\hbar^2$) integrated over the *phase space* of the accessible final states, i.e. the set of final momenta allowed by energy-momentum conservation. The quantity

$$F = \int (2\pi\hbar)^4 \delta^3(\Sigma\boldsymbol{p}_j)\delta(E - \Sigma E_j) \prod_{j=1}^{N} \frac{d^3\boldsymbol{p}_j}{(2\pi\hbar)^3} \tag{4.17}$$

is called the *volume* of phase space. The larger this volume, the greater is the decay rate (if all other factors are assumed to be equal).

If one is interested in angular distributions, or in energy distributions of final particles, one restricts the integration in (4.16) to the appropriate part of phase space.

4.1.4 Phase space and two-body decays

A simple example is that of two-body decays. Consider the decay of a, of mass m, into a_1 and a_2 of masses m_1 and m_2 respectively. We place ourselves in the rest frame of a, the final momenta are opposite $\boldsymbol{p}_1 = -\boldsymbol{p}_2$ and we set $\boldsymbol{p} \equiv \boldsymbol{p}_1$. The energy of the final state is

$$E = E_1 + E_2 = \sqrt{p^2c^2 + m_1^2c^4} + \sqrt{p^2c^2 + m_2^2c^4} \tag{4.18}$$

from which follows that

$$p\frac{dp}{dE} = \frac{E_1E_2}{Ec^2} \quad . \tag{4.19}$$

The decay rate is

$$\lambda = \frac{1}{4\pi^2\hbar^4} \int |\tilde{T}|^2 d^3p_1 d^3p_2\, \delta^3(\boldsymbol{p}_1 + \boldsymbol{p}_2)\delta(E - E_1 - E_2) \,. \tag{4.20}$$

The integration over $\boldsymbol{p}_2$ fixes $\boldsymbol{p}_2 = -\boldsymbol{p}_1$. Equation (4.19) gives $p_1 dp_1 = (E_1E_2/Ec^2)d(E_1 + E_2)$ which allows a direct integration using the $\delta(E - E_1 - E_2)$ function. If we define the average over angles

$$\langle|\tilde{T}|^2\rangle = \frac{1}{4\pi} \int d\Omega\, |\tilde{T}|^2$$

we obtain

$$\lambda = \frac{1}{\hbar}\frac{pc}{\pi(\hbar c)^3}\frac{E_1E_2}{E}\langle|\tilde{T}|^2\rangle \tag{4.21}$$

where p is the magnitude of the momentum of each final particle.

4.1.5 Detailed balance and thermal equilibrium

In this chapter we mostly concerned with the *irreversible* decays of unstable particles. However, in certain situations it is necessary to consider the inverse of the decay. For instance, in Chap. 8 we will consider the production of ^{8}Be by fusion of two ^{4}He. The ^{8}Be then decays back to the original ^{4}He pair with a lifetime of $\sim 10^{-16}$s ($Q = 92$ keV), so we are led to consider the two directions of the reaction

$$^8\text{Be} \leftrightarrow\, ^4\text{He}\,^4\text{He} \,. \tag{4.22}$$

If Q is comparable to the temperature of the medium, the inverse decay is possible since typical ^{4}He nuclei have enough kinetic energy to fuse to ^{8}Be. This can be the case in stars.

  In situations like this, the relative concentration of the nuclei on the two sides of the reaction may be determined by considerations of chemical equilibrium. For instance, if there is no ^{8}Be originally present, its concentration is built up through fusion until the rate of ^{8}Be decay balances the rate of ^{8}Be fusion. It will turn out that the relative concentrations of nuclear species will be given by a formula (4.32) that is independent of the decay and formation rates.

  We consider a two-body decay a → bc. The decay rate *per particle* into the momentum interval $d^3\boldsymbol{p}_b d^3\boldsymbol{p}_c$ is given by (4.13) multiplied by the number of states in $d^3\boldsymbol{p}_b d^3\boldsymbol{p}_c$

$$\lambda(\boldsymbol{p}_a \to \boldsymbol{p}_b, \boldsymbol{p}_c) = \frac{2\pi}{\hbar} |T(\boldsymbol{p}_a \to \boldsymbol{p}_b, \boldsymbol{p}_c)|^2$$

$$\times \, \delta(E_a - E_b - E_c) \frac{V d^3 \boldsymbol{p}_b}{(2\pi\hbar)^3} \frac{V d^3 \boldsymbol{p}_c}{(2\pi\hbar)^3} \tag{4.23}$$

In the functions λ and T we have not written explicitly the spin variables, s_a, s_b and s_c. To get the decay rate within the volume V from the momentum interval $d^3 \boldsymbol{p}_a$, we must multiply λ by the number of particles in this phase space element:

$$\Lambda(\boldsymbol{p}_a \to \boldsymbol{p}_b, \boldsymbol{p}_c) = \lambda(\boldsymbol{p}_a \to \boldsymbol{p}_b, \boldsymbol{p}_c) \, n_a V f_a(\boldsymbol{p}_a) \, d^3 \boldsymbol{p}_a \,, \tag{4.24}$$

where n_a is the number density and $f_a(\boldsymbol{p}_a)$ is the normalized spin-momentum distribution:

$$\sum_s \int d^3 \boldsymbol{p} f_a(\boldsymbol{p}) = 1 \,. \tag{4.25}$$

Once again, we do not write explicitly the spin variable s in the function f.

We want to compare this decay rate to the formation rate. Per particle pair, this is also given by (4.13)

$$\lambda(\boldsymbol{p}_b, \boldsymbol{p}_c \to \boldsymbol{p}_a) = \frac{2\pi}{\hbar} |T(\boldsymbol{p}_b, \boldsymbol{p}_c \to \boldsymbol{p}_a)|^2 \delta(E_a - E_b - E_c) \frac{V d^3 \boldsymbol{p}_a}{(2\pi\hbar)^3} \,. \tag{4.26}$$

To get the formation rate in the volume V from the momentum interval $d^3 \boldsymbol{p}_b d^3 \boldsymbol{p}_c$, we must multiply λ by the number of particles in this phase space element:

$$\Lambda(\boldsymbol{p}_b, \boldsymbol{p}_c \to \boldsymbol{p}_a) = \lambda(\boldsymbol{p}_b, \boldsymbol{p}_c \to \boldsymbol{p}_a)$$

$$\times \, n_b V f_b(\boldsymbol{p}_b) \, d^3 \boldsymbol{p}_b \, n_c V f_c(\boldsymbol{p}_c) \, d^3 \boldsymbol{p}_c \,, \tag{4.27}$$

where n_a and n_b are the number densities and $f_b(\boldsymbol{p}_b)$ and $f_c(\boldsymbol{p}_c)$ are the normalized momentum distributions.

To find the equilibrium number densities, we need only equate the formation rate (4.27) with the decay rate (4.24). Their equality at equilibrium is due to the *principle of detailed balance* meaning that the rate is balanced with the inverse rate at each point in phase space, not just globally. This is possible if the interaction involved respects time-reversal invariance, in which case the $|T|^2$ is the same for both directions of the reaction. We then get

$$n_b f_b(\boldsymbol{p}_b) \, n_c f_c(\boldsymbol{p}_c) = n_a f_a(\boldsymbol{p}_a) \, (2\pi\hbar)^{-3} \,, \tag{4.28}$$

for all energy conserving combinations of E_a, E_b and E_c. The densities and momentum distributions are then constrained by

$$\frac{n_a}{n_b n_c} = (2\pi\hbar)^3 \frac{f_b(\boldsymbol{p}_b) \, f_c(\boldsymbol{p}_c)}{f_a(\boldsymbol{p}_a)} \,. \tag{4.29}$$

A simple situation occurs when the three species have Maxwell–Boltzmann momentum distributions

$$f_i(p_i) = \frac{1}{2s_i + 1} \frac{1}{(2\pi m_i kT)^{3/2}} \exp[-p_i^2/(2m_i kT)] . \tag{4.30}$$

Using energy conservation

$$m_a c^2 + \frac{p_a^2}{2m_a} = m_b c^2 + \frac{p_b^2}{2m_b} + m_c c^2 + \frac{p_c^2}{2m_c} , \tag{4.31}$$

we find that (4.29) becomes a simple relation between the densities as a function of temperature:

$$\frac{n_a}{n_b n_c} = \frac{(2s_a + 1)}{(2s_a + 1)(2s_a + 1)}$$

$$\times (2\pi\hbar)^3 \left(\frac{m_a}{2\pi m_b m_c kT}\right)^{3/2} \exp(-\Delta m c^2/kT) , \tag{4.32}$$

where $\Delta m = m_a - m_b - m_c$. As could be expected, n_a vanishes for $kT \ll \Delta m c^2$.

Another simple situation occurs when particle c is a photon. In thermal equilibrium, the photon phase-space density is the Planck distribution:

$$n_\gamma f_\gamma(p_\gamma) = \frac{1}{(2\pi\hbar)^3} \frac{1}{\exp(E_\gamma/kT) - 1} . \tag{4.33}$$

Substituting this into the formation rate (4.27) it appears that thermal equilibrium is not possible since, if we equate it with the decay rate (4.24), the Boltzmann factors no longer cancel as they did with the Boltzmann distribution. We are saved since for radiative transitions, the formation and decay matrix elements are no longer equal because the latter is enhanced by *stimulated emission*. This means that the decay matrix element to a given state is multiplied by $(1 + N(p_\gamma))$, where $N(p_\gamma)$ is the number of photons already present in the photon state:

$$|T(p_a \rightarrow p_b, p_\gamma)|^2 = |T(p_b, p_\gamma \rightarrow p_a)|^2 \left[1 + \frac{1}{\exp(E_\gamma/kT) - 1}\right] .$$

Substituting this into (4.24) and equating with the formation rate (4.27) we see that the factors $|T|^2/(\exp(E_\gamma/kT) - 1)$ nicely cancel. We then impose energy conservation

$$m_a c^2 + \frac{p_a^2}{2m_a} = m_b c^2 + \frac{p_b^2}{2m_b} + E_\gamma , \tag{4.34}$$

to find a simple expression for the ratio of the densities

$$\frac{n_a}{n_b} = \frac{2s_a + 1}{2s_b + 1} \exp(-\Delta m c^2/kT) , \tag{4.35}$$

where $\Delta m = m_a - m_b$. Once again, n_a vanishes for $kT \ll \Delta m c^2$. This explains why excited nuclear states are not seen on Earth where $kT \sim 0.025\,\text{eV}$.

Thermal photons can not only excite nuclei through $\gamma(A, Z) \to (A, Z)^*$, but also dissociate them. An example that will turn out to be very important in cosmology is the dissociation of ^{2}H so we consider equilibrium

$$d\gamma \leftrightarrow pn \,. \tag{4.36}$$

Following the same reasoning as above, the dissociation rate is

$$\Lambda(\boldsymbol{p}_d\boldsymbol{p}_\gamma \to \boldsymbol{p}_p, \boldsymbol{p}_n) = \frac{2\pi}{\hbar} |T(\boldsymbol{p}_d\boldsymbol{p}_\gamma \to \boldsymbol{p}_p, \boldsymbol{p}_n)|^2 \delta(E_d + E_\gamma - E_p - E_n)$$

$$\times \frac{V d^3 \boldsymbol{p}_p}{(2\pi\hbar)^3} \frac{V d^3 \boldsymbol{p}_n}{(2\pi\hbar)^3} n_d V f_d(\boldsymbol{p}_d) \, d^3 \boldsymbol{p}_d n_\gamma V f_\gamma(\boldsymbol{p}_\gamma) \, d^3 \boldsymbol{p}_\gamma \,. \tag{4.37}$$

We must then equate this with the rate for the inverse process. Taking into account stimulated emission

$$|T(\boldsymbol{p}_p\boldsymbol{p}_n \to \boldsymbol{p}_d, \boldsymbol{p}_\gamma)|^2 = |T(\boldsymbol{p}_d\boldsymbol{p}_\gamma \to \boldsymbol{p}_p\boldsymbol{p}_n)|^2 \left[1 + \frac{1}{\exp(E_\gamma/kT) - 1}\right] ,$$

and using the Planck relation (4.33), we find the Saha equation:

$$\frac{n_d}{n_p n_n} = \frac{s_d + 1}{(s_p + 1)(s_n + 1)}$$

$$\times (2\pi\hbar)^3 \left(\frac{m_d}{2\pi m_p m_n kT}\right)^{3/2} \exp(-\Delta B/kT) \,, \tag{4.38}$$

where $B = 2.2 \, \text{MeV}$ is the deuteron binding energy. This is basically the same equation as (4.32) except that here the two-body state has higher energy than the one-body state.

4.2 Radiative decays

We now consider the dynamics of *radiative decays*. Radiative decays of atoms, nuclei or elementary particles are those in which photons are emitted. Most common are the decays of excited states

$$A^* \to A + \gamma \,. \tag{4.39}$$

This process is also called the spontaneous emission of a photon by A^*. In radiative decays of nuclei or atoms, the radiator is always much heavier than the mass difference

$$\frac{m_{A*} - m_A}{m_A} \ll 1 \tag{4.40}$$

in which case it is simple to check that practically all the decay energy is taken by the photon. In general we have

$$E_\gamma = (m_{A*} - m_A)c^2 - \frac{p^2}{2m_A} \,. \tag{4.41}$$

In nuclear decays, $E_\gamma = pc \sim$ MeV and $mc^2 \sim A$ GeV giving a nuclear recoil of $p^2/2m \sim 0.5$ keV$/A$, which is indeed much less than E_γ implying

$$E_\gamma \sim (m_{A^*} - m_A)c^2 . \tag{4.42}$$

4.2.1 Electric-dipole transitions

The quantum field theory of photons, i.e. quantum electrodynamics, is necessary to derive the matrix element T for radiative transitions. Fortunately, these transitions have the classical analog of an oscillating charge distribution radiating a classical electromagnetic field and the formula for the radiated power will lead us to the correct formula for the transition rate. The simplest case is that of a single charge q moving in a 1-dimensional harmonic potential so that its position is $x(t) = a \cos \omega t$. This corresponds to an oscillating electric dipole $D_x(t) = qa \cos \omega t$. Classical electrodynamics can be used to calculate the radiated power associated with the oscillating electromagnetic field created by the oscillating dipole. The time averaged (classical) power is given by the Larmor formula

$$P = \frac{2}{3} \frac{\langle \ddot{D}^2 \rangle_t}{4\pi\epsilon_0 c^3} = \frac{1}{3} \frac{q^2}{4\pi\epsilon_0} \frac{a^2 \omega^4}{c^3} , \tag{4.43}$$

where the $\langle \ \rangle_t$ indicates time averaging, replacing the factor $\cos^2 \omega t$ by $1/2$. If the motion is circular, the power is doubled since this corresponds to linear harmonic motion in two dimensions.

A quantum harmonic oscillator in a state n of energy $E_n = \hbar\omega(n + 1/2)$ can decay to the state $n - 1$ by emitting a photon of energy $\hbar\omega$. The time averaged power due to this transition is

$$P = \hbar\omega \, \lambda(n \rightarrow n - 1) , \tag{4.44}$$

where $\lambda(n \rightarrow n - 1)$ is the decay rate of the state n. Equating this with the classical power (4.43) we get

$$\lambda(n \rightarrow n - 1) \sim \frac{1}{\hbar\omega} \frac{1}{3} \frac{q^2}{4\pi\epsilon_0} \frac{2\langle x^2 \rangle_n \omega^4}{c^3} , \tag{4.45}$$

where we have equated the mean of x^2 in the state n with its classical value $a^2/2$. We have written $\sim$ instead of $=$ because we expect the classical formula to apply only in the limit $n \rightarrow \infty$. Introducing the fine-structure constant $\alpha = e^2/4\pi\epsilon_0\hbar c$ we get a more elegant formula

$$\lambda(n \rightarrow n - 1) \sim \frac{2}{3} \frac{q^2}{e^2} \alpha \frac{\langle x^2 \rangle_n \omega^3}{c^2} = \frac{2}{3} \frac{q^2}{e^2} \alpha \frac{\langle E \rangle_n}{mc^2} \omega , \tag{4.46}$$

where in the second form we use $\langle x^2 \rangle_n = E_n/(m\omega^2)$. We see that for a non-relativistic elementary particle ($q = e$), a state lives for many oscillation periods, of order $\alpha^{-1} mc^2/E_n$.

While (4.46) turns out to be (nearly) the correct rate as calculated using quantum electrodynamics, it is not in a form that can be generalized to other problems. Since classically the radiation is due to the oscillating position of the particle, we would expect that the decay rate is related to the matrix element of x taken between initial and final states, rather than the the expectation value of x^2 in the initial state. This suggests replacing $\langle x^2 \rangle_n$ with

$$2|\langle n-1|x|n\rangle|^2 \;=\; \frac{\hbar n}{m\omega} \;=\; \langle n|x^2|n\rangle \frac{n}{n+1/2} \,, \tag{4.47}$$

which has the same large n limit but makes sense quantum mechanically. Substituting this into (4.46) we get

$$\lambda(n \to n-1) \;=\; \frac{4}{3} \frac{q^2}{e^2} \alpha \frac{\omega^3}{c^2} |\langle n-1|x|n\rangle|^2 \,. \tag{4.48}$$

This turns out to be correct and generalizable to any transition involving a single charged particle:

$$\lambda_{i \to f} \;=\; \frac{4\alpha}{3} \frac{q^2}{e^2} \frac{1}{\hbar} \frac{E_\gamma^3}{(\hbar c)^2} \langle f|\boldsymbol{r}|i\rangle \cdot \langle i|\boldsymbol{r}|f\rangle \,. \tag{4.49}$$

If we drop the assumption of a single charged particle involved in the transition, we simply replace the position vector $\boldsymbol{r}$ with $\boldsymbol{D}/q$ where $\boldsymbol{D}$ is an appropriate dipole operator. Transitions governed by (4.49) are called *electric-dipole transitions* or "E1" transitions for short.

Equation (4.49) is used in atomic and nuclear physics when transitions involve, to good approximation, a single particle in which case we have $q = e$. This is the case in hydrogen and alkali atoms where excited states correspond to excitations of the valence electron that moves in the field of an inert core that remains unaffected by the transition. In nuclear physics, the formula applies best to even–odd nuclei where the single unpaired nucleon moves in the potential of the paired nucleons. This is an especially good picture when the paired nucleons form a closed shell corresponding to a magic number of protons or neutrons.

In these cases, we need only take into account the wavefunctions, $\psi(\boldsymbol{r})$ of the initial and final state valence particles

$$\langle f|\boldsymbol{r}|i\rangle \;=\; \int \mathrm{d}^3 r\, \psi_f^*(\boldsymbol{r}) \boldsymbol{r} \psi_i^*(\boldsymbol{r}) \,. \tag{4.50}$$

Equation (4.49) is not very useful for precise calculations because the nuclear matrix element of $\boldsymbol{r}$ is not easily calculable. However, it can be used to estimate the lifetimes of atomic and nuclear excited states. We note the presence of the fine structure constant $\alpha = e^2/4\pi\epsilon_0\hbar c$, which represents the square of the coupling constant of the electromagnetic field with the (charge of the) electron. Note also the fact that the decay rate varies as the *third power* of the photon energy, and as the *square* of the size of the system. For atomic transitions, $\hbar\omega \sim$ eV and $\langle r \rangle \sim 10^{-10}$ m which gives rates of the order of 10^7 to $10^9 \mathrm{s}^{-1}$, i.e. lifetimes of the order of 10^{-7} to 10^{-9}s. The corresponding width, $\Gamma = \hbar/\tau \sim 10^{-7}$eV is much less than the photon energy ($\sim$ eV) but much greater than the atomic recoil energy $E_\gamma^2/2m_A c^2 \sim 10^{-9}$eV.

For nuclear transitions, $\langle r \rangle \sim A^{1/3}10^{-15}$ m so

$$\lambda(\text{E1}) \sim \hbar^{-1}\alpha E_\gamma^3 \left(\frac{A^{1/3}\text{fm}}{\hbar c}\right)^2 . \tag{4.51}$$

For $E_\gamma \sim$ MeV, this gives rates of the order of 10^{15} to 10^{17}s^{-1}, i.e. lifetimes of the order of 10^{-17} to 10^{-15}s. The corresponding width, $\Gamma = \hbar/\tau \sim 10$ eV is much less than the photon energy.

The width is also much less than the the the nuclear recoil energy $E_\gamma^2/2m_A c^2 \sim 10^3$ eV. This has the interesting consequence that, unlike photons emitted by atoms, photons emitted in a nuclear transition do not generally have sufficient energy to re-excite another nucleus through the inverse transition:

$$E_\gamma = (m^* - m_A)c^2 - p^2/2m_A < (m^* - m_A)c^2 - \hbar/\tau . \tag{4.52}$$

The inverse transition can only be induced in special cases where nuclear recoil is suppressed in a crystal by the Mössbauer effect (Fig. 4.4) or when the excited nucleus decays in flight with the Doppler effect compensating the nuclear recoil (Fig. 4.19).

4.2.2 Higher multi-pole transitions

It is often the case that electric-dipole decay of an excited state is forbidden because there are no lower-lying states for which the matrix element $\langle f|r|i\rangle$ is non-zero. A famous example of this is the $n = 2, l = 0$ state of atomic hydrogen that cannot decay to the $n = 1$ state

$$\langle 2\,0\,0|r|1\,0\,0\rangle = \int \mathrm{d}^3 r \psi_1^*(r)r\psi_2(r) = 0 . \tag{4.53}$$

The integral vanishes because the two wavefunctions are spherically symmetric whereas r averages to zero when integrated over all directions. This illustrates a *selection rule* requiring that for E1 transitions, the parity of the initial and final wavefunctions be different and that the angular momentum differ by $\leq$ one unit.

If an E1 transition is forbidden, a state may still decay radiatively by the action of operators that are the quantum analogs of the higher order classical radiation processes: oscillating magnetic dipoles (M1), electric quadrupoles (E2), etc. Each type of radiation has its own selection rules that are given in Table 4.1. Classically, the radiated power for an oscillating l-pole is proportional to ω^{2l+1} so we expect the quantum transition rates to be proportional to E_γ^{2l+1}. Conventionally, one therefore writes the transition rate in the form

$$\lambda(l) = \frac{8\pi(l+1)}{l[(2l+1)!!]^2} \frac{1}{\hbar} \frac{E_\gamma^{2l+1}}{(\hbar c)^{2l}} \alpha B(l) , \tag{4.54}$$

where $(2l+1)!! = (2l+1)(2l-1)(2l-3)\ldots$. The *reduced transition rate $B(l)$* contains all the (difficult) nuclear physics and has dimension of length^{2l}. From (4.49) we see that for E1 transitions we have

$$B(\mathrm{E}1) = (\langle f|\boldsymbol{r}|i\rangle \cdot \langle i|\boldsymbol{r}|f\rangle) \sim R^2 . \tag{4.55}$$

Dimensional analysis suggest that higher order elements are of order

$$B(\mathrm{E}l) \sim R^{2l} . \tag{4.56}$$

It turns out that this formula generally overestimates transition rates, with the exception of electric quadrupole transitions (E2) as discussed below.

Just as higher classical multi-poles are less efficient radiators than classical electric dipoles, the quantum radiative rates decrease with increasing pole number:

$$\frac{\lambda(\mathrm{E}l)}{\lambda(\mathrm{E}1)} \sim \left(\frac{E_\gamma R}{\hbar c} \right)^{2l} \sim \left(\frac{E_\gamma(\mathrm{MeV})A^{1/3}}{200} \right)^{2l} , \tag{4.57}$$

i.e. about 2 orders of magnitude per pole.

Magnetic l-pole radiation is weaker than the corresponding electric l-pole radiation because fields generated by oscillating currents are smaller than fields generated by oscillating charges by a factor v/c where v is the velocity of the radiating charge. The uncertainty principle suggests that the velocity of nucleons in nuclei is of order $\hbar/(Rm_\mathrm{p})$ so we expect

$$B(\mathrm{M}l) \sim \left(\frac{\hbar c}{m_\mathrm{p}c^2 R} \right)^2 B(\mathrm{E}l) = \left(\frac{1}{5A^{1/3}} \right)^2 B(\mathrm{E}l) . \tag{4.58}$$

This implies that Ml transitions have rates between those of El and E$(l+1)$ transitions.

Table 4.1. Selection rules for radiative transitions

type	symbol	angular momentum change $|\Delta J| \leq$	parity change
electric dipole	E1	1	yes
magnetic dipole	M1	1	no
electric quadrupole	E2	2	no
magnetic quadrupole	M2	2	yes
electric octopole	E3	3	yes
magnetic octopole	M3	3	no
electric 16-pole	E4	4	no
magnetic 16-pole	M4	4	yes

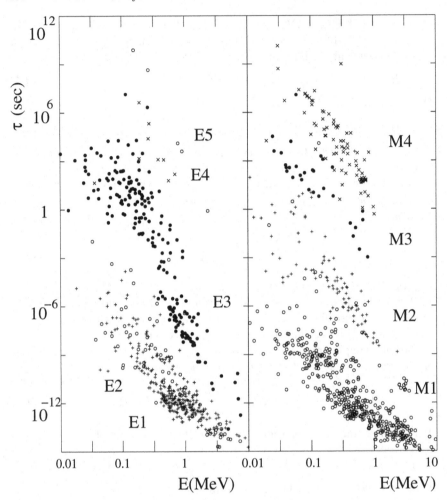

Fig. 4.5. Lifetimes of excited nuclear states as a function of E_γ for various electric and magnetic multipoles.. The various multipoles separate relatively well except for the E1 (open circles) and E2 (crosses) transitions that have similar lifetimes. (For clarity, only 10% of the available E1 and E2 transitions appear in the plot.) The surprising strength of the E2 transitions is because they are generally due to collective quadrupole motions of several nucleons, whereas E1 transitions can often be viewed as single nucleon transitions.

Figure 4.5 shows the lifetimes of radiative transitions as a function of E_γ and of multipole. The expected rate decrease with increasing multipole and decreasing E_γ is evident. The longest lived states appearing in the plot are the M4 transition

$$^{108}\text{Ag}(6^+\ 109.44\,\text{keV}) \rightarrow\ ^{108}\text{Ag}(2^-\ 79.13\,\text{keV}) \quad \text{t}_{1/2} = 418\,\text{yr}\ ,$$

and the E5 transition

$$^{192}\text{Ir}(11^-\ 168.14\,\text{keV}) \rightarrow\ ^{192}\text{Ir}(6^+\ 12.98\,\text{keV}) \quad \text{t}_{1/2} = 241\,\text{yr}\ .$$

The existence of such long-lived isomeric states is only possible because the nucleus is isolated from its environment by its atomic electrons. Long-lived *atomic* states are not possible because they de-excite during frequent collisions with other atoms.

There also exist a few exceptionally short-lived nuclear states. An example is the E1 transition

$$^{11}\text{B}(1/2^-\ 320\,\text{keV}) \rightarrow\ ^{11}\text{Be}(1/2^+)\gamma(320\,\text{keV}) \quad \tau = 166\,\text{fs}\ . \quad (4.59)$$

The lifetime is much shorter than those shown in Fig. 4.5. This is explained by the fact the ^{11}B is a *halo nucleus* consisting of a single neutron orbiting far from a ^{10}Be core [7]. This loosely bound nucleus has only two states, the ground state and the 320 keV state. The short lifetime is due to the large radius of the nucleus [see eq. (4.56)] and to the large matrix element for the simple one-particle wavefunctions.

Except for the case of E2 transitions, (4.56) tends to overestimate the rates. E2 transition rates are underestimated because these transitions are often due to the collective motions of deformed nuclei with permanent quadrupole moments, so the effective charge involved in the transition is large, $q > e$. In this case, the reduced matrix element for an E2 transition for a state of angular momentum j to a state $j - 2$ is

$$\alpha\,B(\text{E2}) = \frac{15}{8\pi}\,Q_0^2\,\frac{j(j-1)}{(2j-1)(2j+1)}\ , \quad (4.60)$$

where Q_0 is the permanent electric quadrupole moment. This relation combined with lifetime measurements can be used to estimate nuclear deformations (Fig. 1.8).

4.2.3 Internal conversion

Whereas an isolated excited nucleus decays by photon emission, an excited nucleus surrounded by atomic electrons can also decay by transferring energy to an atomic electron which is ejected from the atom. This process, called *internal conversion*, can be thought of as the two-step process illustrated in Fig. 4.6 where the photon emitted by the excited nucleus is absorbed by an atomic electron which is subsequently ejected. The energy of the ejected electron is the photon energy minus the binding energy of the electron. One

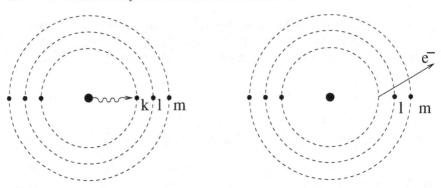

Fig. 4.6. An excited nucleus can transfer its energy to an atomic electron which is subsequently ejected from the atom. The process is called "internal conversion." The ejected electron can come from any of the atomic orbitals. In the figure, an electron from the deepest orbital is ejected, so-called K-conversion. Ejection of electrons in higher orbitals (L-, M- ... conversion) are generally less probable.

denotes K-, L-, or M-conversion as ejection of an electron from orbital 1,2, or 3 (beginning with the inner most). Generally speaking K conversion is the most likely.

While it is intuitive to think of internal conversion as a two-step process, it must be remembered that it is, in fact, a single quantum process whose amplitude can be calculated by standard perturbation theory.

The amplitude for internal conversion is proportional to the same nuclear matrix element responsible for radiative decay. The factor of proportionality depends on the multipolarity of the transition. An approximate expression for the probability for K-conversion compared to that for γ-emission is

$$\alpha_K \sim Z^3 \alpha^4 \frac{l}{l+1} \left(\frac{2m_e c^2}{E_\gamma} \right)^{l+5/2} . \tag{4.61}$$

This formula applies in the limit $\alpha_K \ll 1$ and only if the atomic-electron binding energy is negligible compared to E_γ. It implies that internal conversion dominates over γ-emission for low-energy transitions:

$$E_\gamma < (Z^3 \alpha^4)^{1/(l+5/2)} m_e c^2 . \tag{4.62}$$

Since we always have $Z^3 \alpha^4 < 1$ this means that internal conversions is negligible for $E_\gamma > m_e c^2$. For E1 transitions, internal conversion is almost always small but for large l it becomes increasingly dominant for $E_\gamma < m_e c^2$. In all circumstances, numerical values α_K can be derived. These estimates are sufficiently accurate that the multipolarity of a transition can usually be determined if the conversion factor is measured. This is an important element in the assignment of spins and parities to nuclear states .

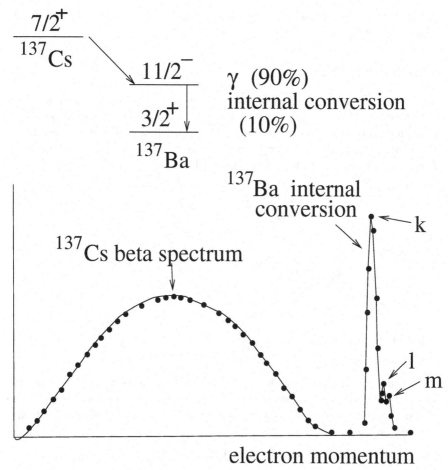

Fig. 4.7. The β-spectrum of ^{137}Cs and the internal conversion lines from the decay of the first excited state of ^{137}Ba [40]. Captures from the K, L and M orbitals are seen.

4.3 Weak interactions

Whereas the electromagnetic interactions responsible for the radiative decays conserve the number of protons and the number of neutrons, the weak interactions transform protons to neutrons or vice versa as well as the numbers of charged leptons and numbers of neutrinos. The archetype of a weak decay is nuclear β-decay

$$(A, Z) \rightarrow (A, Z \pm 1) \, e^-(e^+) \, \nu_e(\bar{\nu}_e) \, . \tag{4.63}$$

Fermi gave a remarkably efficient theory of this process as soon as 1933. The structure of this theory became more profound in 1968 with the advent of the unified theory of weak and electromagnetic interactions due to Glashow, Salam and Weinberg.

4.3.1 Neutron decay

In the Fermi theory, neutron decay n → pe⁻v̄ₑ is a point-like process. This is similar to what we discussed in Chap. 3, concerning the small range of weak interactions owing to the large masses of intermediate bosons. Here, the neutron transforms into a proton and a virtual W⁻ boson, which itself decays into e⁻v̄ₑ. This process is shown schematically in Fig. 4.8.

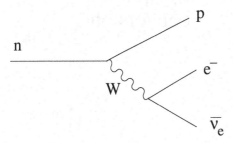

Fig. 4.8. Neutron decay.

To find the matrix element for neutron decay we first recall the matrix element for scattering of two free particles, as discussed in Sect. 3.4.1. If the two particles 1 and 2 interact via a potential $V(r_1 - r_2)$, then the scattering element is

$$\langle f|V|i \rangle = \frac{1}{L^6} \int e^{i(p_1 - p_1')\cdot r_1/\hbar} e^{i(p_2 - p_2')\cdot r_2/\hbar} V(r_1 - r_2) d^3 r_1 d^3 r_2 \,,$$

where p_1 and p_1' are the initial and final momenta for particle 1, and likewise for particle 2. The weak interactions can be described by a delta-function potential $V \sim G\delta(r_1 - r_2)$ so the matrix element is

$$\langle f|V|i \rangle = \frac{G}{L^6} \int e^{i(p_1 - p_1')\cdot r/\hbar} e^{i(p_2 - p_2')\cdot r/\hbar} d^3 r \,. \tag{4.64}$$

There is a factor $\exp(i p \cdot r)$ for each initial-state particle and a factor $\exp(-i p \cdot r)$ for each final-state particle. This suggests that for neutron decay we use the matrix element

$$\langle pe^-\bar{v}_e|H_1|n \rangle = \frac{2.4G_F}{L^6} \int d^3 r \, \exp[i(p_n - p_p - p_e - p_v) \cdot r/\hbar] \,. \tag{4.65}$$

The factor $2.4G_F$ is the effective G for neutron decay and will be discussed below. Since we will not always want to use plane waves, we also write the more general matrix element as

$$\langle \mathrm{pe^-\bar{v}_e}|H_1|\mathrm{n}\rangle = 2.4 G_F \int \mathrm{d}^3 r \psi_n^*(\boldsymbol{r})\psi_p(\boldsymbol{r})\psi_e(\boldsymbol{r})\psi_v(\boldsymbol{r}) . \tag{4.66}$$

The integral in (4.65) will give a momentum conserving delta function so we find that the transition amplitude defined by (4.14) is constant

$$|\tilde{T}|^2 = (2.4 G_F)^2 . \tag{4.67}$$

The fact that $\tilde{T}$ is constant means that the Hamiltonian has no preference for particular final states as long as they conserve energy and momentum. Using (4.16), the differential decay rate of the neutron is

$$\mathrm{d}\lambda = \frac{(2.4 G_F)^2}{8\pi^5 \hbar^7} \mathrm{d}^3 \boldsymbol{p}_p \mathrm{d}^3 \boldsymbol{p}_e \mathrm{d}^3 \boldsymbol{p}_v \delta(\boldsymbol{p}_p + \boldsymbol{p}_e + \boldsymbol{p}_v)\delta(m_n c^2 - E_p - E_e - E_v) .$$

We integrate over the proton momentum which eliminates the momentum-conserving delta function

$$\mathrm{d}\lambda = \frac{c^6}{8\pi^5 \hbar}\left(\frac{2.4 G_F}{(\hbar c)^3}\right)^2 \mathrm{d}^3 \boldsymbol{p}_e \mathrm{d}^3 \boldsymbol{p}_v \delta\left((m_n - m_p)c^2 - p_v c - E_e\right) . \tag{4.68}$$

Here, we have neglected the recoil of the proton, i.e. $E_p \sim m_p c^2$. The momenta $\boldsymbol{p}_e$ and $\boldsymbol{p}_v$ can have all values compatible with energy conservation.

We now use spherical coordinates

$$\mathrm{d}^3 \boldsymbol{p}_e = p_e^2 \mathrm{d}p_e \mathrm{d}\Omega_e \qquad \mathrm{d}^3 \boldsymbol{p}_v = p_v^2 \mathrm{d}p_v \mathrm{d}\Omega_v , \tag{4.69}$$

where the angles refer to the direction of the electron and neutrino momenta. The integration over angles is straightforward and it gives a factor $(4\pi)^2$. Integration over the neutrino momentum eliminates the final delta function and introduces a factor $p_v^2 = ((m_n + m_p)c^2 - E_e)^2/c^2$. We therefore obtain

$$\mathrm{d}\lambda = \frac{2}{\pi^3 \hbar}\left(\frac{2.4 G_F}{(\hbar c)^3}\right)^2 ((m_n - m_p)c^2 - E)^2 \sqrt{E^2 - m_e^2 c^4}\, E\, \mathrm{d}E \tag{4.70}$$

where E is the electron energy.

The energy distribution of the electron is then

$$\frac{\mathrm{d}\lambda}{\mathrm{d}E} \propto E((m_n - m_p)c^2 - E)^2 \sqrt{E^2 - m_e^2 c^4} \tag{4.71}$$

which reproduces the experimental data as shown in Fig. 4.9.

It is the existence of such an energy spectrum which led Pauli in 1930 to the idea of the neutrino. Indeed, the β decays could not be two-body decays into, e.g. p e$^-$, otherwise the electron would be mono-energetic. A third particle had to be present in the final state in order to account for the energy balance. This idea was taken up and formalized by Fermi.

The neutron lifetime is obtained by calculating the integral (4.70). the result is

$$\lambda = \frac{1}{2\pi^3 \hbar}\left(\frac{2.4 G_F}{(\hbar c)^3}\right)^2 (m_e c^2)^5$$

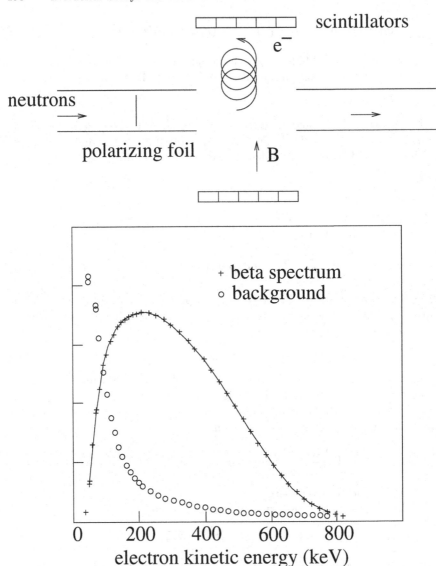

Fig. 4.9. Measurement of the energy spectrum of electrons from neutron β-decay, n → pe⁻v̄ₑ. The top figure shows the apparatus of [41]. Cold neutrons for the Institut Laue-Langevin nuclear reactor enter the apparatus from the left (see Fig. 3.29). The neutrons pass through a magnetized foil that reflect neutrons with spin aligned in the direction of the magnetization. The polarized neutrons then enter a region containing a magnetic field where a certain fraction decay. The decay electrons spiral in the field until stopping in a plastic scintillator. The light output of the scintillator gives a measurement of the electron kinetic energy. Spectrum of the measured electron kinetic energy is shown in the bottom panel. The curve shows the theoretical spectrum.

$$\times \left[\frac{1}{15}(2x^4 - 9x^2 - 8)\sqrt{x^2 - 1} + x\log(x + \sqrt{x^2 - 1})\right] \quad (4.72)$$

$$x = (m_n - m_p)/m_e \quad (4.73)$$

this gives a lifetime in agreement with the experimental lifetime $\tau_{exp} = 886.7 \pm 1.9$ s. The good agreement should not be considered a triumph: the factor 2.4 was, in fact, derived from the neutron lifetime.

The neutron lifetime is best most accurately measured with ultra-cold neutrons [42] as illustrated in Fig. 3.29. The ultra-cold neutrons are stored in a box for a time T after which they are released through a shutter into a pipe that leads them to a neutron counter. The count rate as a function of storage time allows one to deduce the neutron lifetime after small corrections for absorption losses on the walls of the containing vessels.

We notice that a quick order of magnitude of (4.70) can be obtained by neglecting m_e compared to $\Delta m = m_n - m_p$ (a rather crude approximation). One then obtains

$$\lambda \simeq \frac{2c^5}{\pi^3 \hbar} \left(\frac{2.4 G_F}{(\hbar c)^3}\right) \int_0^{p_{max}} (\Delta m c - p)^2 p^2 dp \quad (4.74)$$

with $p_{max} = \sqrt{\Delta m^2 c^2 - m_e^2 c^2}$. By replacing in this expression $\Delta m c$ by p_{max}, we obtain

$$\lambda \sim \frac{1}{15\pi^3 \hbar} (p_{max} c)^5 \left(\frac{2.4 G_F}{(\hbar c)^3}\right)^2 \quad \text{i.e. } \tau \sim 951 \text{ s.} \quad (4.75)$$

The interesting thing about this approximate formula is to show that it is the volume of phase space and the *fifth power of the maximum momentum of the electron* which is the decisive factor. This is a direct consequence of the dimensionality of the Fermi constant.

Our simple treatment has lead to a transition matrix T (4.67) that is "democratic," i.e. it is independent of the final state as long a momentum and energy are conserved. More realistic Hamiltonians introduce non-trivial dynamics that induce correlations between particles.

The first complication that we would like to introduce is by taking into account the spins of the particles. A non-relativistic neutron is described by two wavefunctions

$$|n\rangle = \begin{pmatrix} \psi_{+n}(\boldsymbol{r}) \\ \psi_{-n}(\boldsymbol{r}) \end{pmatrix}, \quad (4.76)$$

and

$$\langle n| = \left(\psi_{+n}^*(\boldsymbol{r}), \psi_{-n}^*(\boldsymbol{r})\right), \quad (4.77)$$

where $\psi_{+n}(\boldsymbol{r})$ and $\psi_{-n}(\boldsymbol{r})$ gives the amplitude to find the neutron at $\boldsymbol{r}$ with spin up or down. The analogous definitions hold for non-relativistic protons, electrons and neutrinos.

It turns out that the Hamiltonian appropriate for neutron β-decay in the (unrealistic) limit where all participating particles are non-relativistic is approximately

$$\langle \mathrm{pe^- \bar{v}_e}|H_1|\mathrm{n}\rangle =$$

$$G_\mathrm{F} \int \mathrm{d}^3 r \left[\langle \mathrm{p}|I|\mathrm{n}\rangle\langle \mathrm{e}|I|\bar{\mathrm{v}}_\mathrm{e}\rangle + g_A \langle \mathrm{p}|\boldsymbol{\sigma}|\mathrm{n}\rangle \cdot \langle \mathrm{e}^\mathrm{T}|\boldsymbol{\sigma}|\bar{\mathrm{v}}_\mathrm{e}\rangle \right], \quad (4.78)$$

where I is the unit matrix and $\boldsymbol{\sigma}$ are the Pauli spin matrices. The constant $g_A \sim 1.25$ has a value that can be derived in principle from the underlying theory of quark decay. In practice, it is fixed empirically by the value of the neutron lifetime. The transposed spinor for the electron is

$$\langle \mathrm{e}^\mathrm{T}| = \left(\psi^*_{-\mathrm{e}}(\boldsymbol{r}), \; \psi^*_{+\mathrm{e}}(\boldsymbol{r}) \right). \quad (4.79)$$

The matrix element (4.78) contains no surprises. It is the sum of four terms. The first two, $I \cdot I$ and $\sigma_z \sigma_z$, yield a proton with the same spin as the neutron and opposite spins for the e$^-$ and $\bar{\mathrm{v}}_\mathrm{e}$. The last two, $\sigma_x \sigma_x$ and $\sigma_y \sigma_y$, flip the nucleon spin and yield e$^-$ and $\bar{\mathrm{v}}_\mathrm{e}$ with the same spin. We see that all four terms guarantee angular momentum conservation in the zero-velocity limit for all particles where there is no orbital angular momentum. In fact, the conservation of angular momentum is forced by the rotational invariance of each term, II and the scalar product $\boldsymbol{\sigma} \cdot \boldsymbol{\sigma}$.

While the non-relativistic limit (4.78) gives no new physics, the relativistic generalization does. Such matrix elements use 4-component Dirac spinors rather than 2-component Pauli spinors to describe the two spin states of particles in addition to the two spin states of their antiparticles. The formalism has sufficient flexibility to reproduce the following correlations observed in β-decay:

- The directions of the e$^-$ and $\bar{\mathrm{v}}_\mathrm{e}$ momenta are correlated. For an angle θ between the two momenta, the distribution of $\cos\theta$ is proportional to $1 + a\cos\theta$ where $a = (1 + g_A^2)/(1 + 3g_A^2) \sim 0.5$.
- A correlation of the same form exists between the spin of the neutron and the direction of the e$^-$ and $\bar{\mathrm{v}}_\mathrm{e}$ momenta. The measured correlation for electrons is shown in Fig. 4.10. We see that of order 5% more electrons are emitted in the direction of the neutron spin than opposed to the neutron spin. This is of profound importance because in indicates that parity conservation is violated in β-decay since the correlation would be opposite if the experiment were observed in a mirror (Fig. 4.10).
- The $\bar{\mathrm{v}}_\mathrm{e}$ is always emitted with its spin aligned with its momentum, i.e. it has positive *helicity*:

$$\frac{\boldsymbol{p} \cdot \boldsymbol{s}}{|\boldsymbol{p}|} = +\hbar/2 \quad (\bar{\mathrm{v}}_\mathrm{e}) \quad (4.80)$$

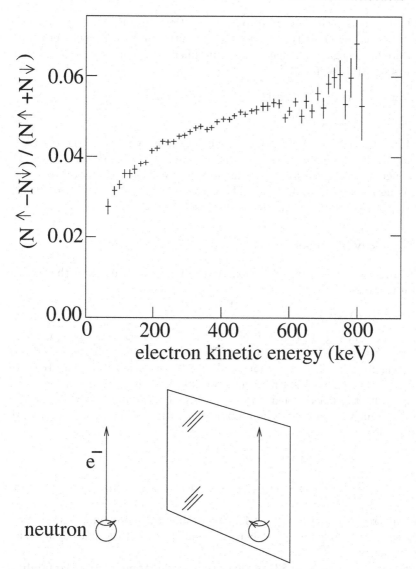

Fig. 4.10. The neutron β decay asymmetry for polarized neutrons. About 5% more neutrons are emitted in the direction of the neutron spin than opposite the direction of the neutron spin. This indicates that parity is violated in β-decay. This is demonstrated in the bottom figure where a spinning neutron decays with the electron emitted in the direction of the neutron (spin) angular momentum. Viewed in the mirror, the spin is reversed but the direction of the electron is not. The excess of electrons emitted in the direction of the neutron spin becomes, viewed in the mirror, an excess of electrons emitted opposite the direction of the neutron spin. What is viewed in the mirror does not correspond to the real world, indicating that physics in the real world does not respect parity symmetry.

This is, again, an indication of parity violation since helicity is reversed if viewed in a mirror. It turns out that ν_e emitted in β^+−decay or electron-capture have negative helicity. The experimental demonstration of this will be discussed in Sect. 4.3.4.

Finally, we mention the origin of the factor 2.4:

$$2.4G_F \sim \cos\theta_c G_F(1 + 3g_A)/2 \,. \tag{4.81}$$

This formula can only be understood within the framework of the complete relativistic theory but we see that the constant $g_A \sim 1.25$ comes from the spin-dependent couplings. Its non-integer value can be understood from the underlying theory of quark decay. The *Cabibbo angle*, $\cos\theta_c = 0.975 \pm 0.001$, comes from the mixing of quarks as discussed in Sect. 4.4.3.

4.3.2 β-decay of nuclei

As we already emphasized, β-radioactivity of nuclei stems from the fundamental processes

$$n \to p\,e^-\,\bar{\nu}_e, \quad p \to n\,e^+\,\nu_e \,. \tag{4.82}$$

As illustrated in Fig. 4.11, it is useful to think of nuclear β-decay as neutron or proton β-decay to empty orbitals. The decaying nucleons are considered as moving in a fixed nuclear potential due to the other nucleons. If there is only one Pauli-unblocked nucleon that is in a position to decay (e.g. ^{13}N in Fig. 4.11), the matrix element is the same as (4.66) except that the nucleon wavefunctions are not plane waves but rather normalized bound state wavefunctions:

$$\langle (A, Z+1)e^-\bar{\nu}_e|H_1|(A, Z)\rangle =$$

$$\frac{2.4G_F}{L^3} \int d^3r\, \psi_p^*(r)\psi_n(r) \exp[-i(p_e + p_v)\cdot r/\hbar] \,. \tag{4.83}$$

As in neutron decay, we write $2.4G_F$ to simulate a more complicated spin-dependent matrix element that can be calculated from the underlying relativistic theory.

The matrix element (4.83) involves the Fourier transform of the product of the initial and final nucleon wavefunctions. A useful approximation comes about by noting that the typical lepton wavelengths, $2\pi\hbar c/1\,\mathrm{MeV} \sim 10^3\,\mathrm{fm}$ are much greater than nuclear sizes, $A^{1/3}\,\mathrm{fm}$, i.e. much greater than the extent of the nucleon wavefunctions in (4.83). We can therefore replace the lepton wavefunctions by their value at $r = 0$ and we have

$$\langle (A, Z+1)e^-\bar{\nu}_e|H_1|(A, Z)\rangle = \frac{2.4G_F}{L^3} \int d^3r\, \psi_p^*(r)\psi_n(r) \,. \tag{4.84}$$

The matrix element is proportional to the overlap of the initial and final state nucleons.

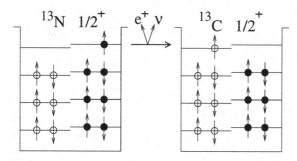

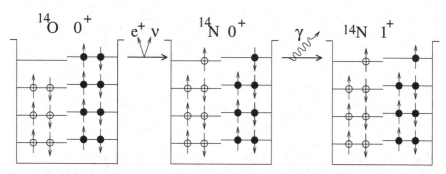

Fig. 4.11. Independent particle picture of nuclear β-decay. The top panel shows the decay ^{13}N $\rightarrow$ ^{13}Ce^{+}ν$_{e}$. The decay can be considered to be the β-decay of the valence proton that moves in the field of an inert ^{12}C core. The decay is shown with no spin-flip but both flip and non-flip processes are possible. The bottom panel shows the decay ^{14}O $\rightarrow$ ^{14}Ne^{+}ν$_{e}$. This decay is predominantly to the first excited state (0^{+}) of ^{14}N followed by the radiative decay to the ground state (1^{+}) (see Fig. 4.12) . For the decay of ^{14}O, there are two valence protons, either of which can decay to a neutron.

The calculation of the decay rate then proceeds in the same manner as for free neutron decay except that momentum conservation no longer holds. This is because the initial and final nuclei are represented by a fixed potential in which the decaying nucleon is confined. The fixed nuclear potential therefore has effectively an infinite mass. This is just as in single-particle scattering in a fixed potential where there was also no momentum conservation. The differential decay rate is

$$d\lambda = \frac{c^6}{8\pi^5\hbar}\left(\frac{2.4G_{\mathrm{F}}}{(\hbar c)^3}\right)^2 |M|^2\delta(\Delta mc^2 - E_\nu - E_{\mathrm{e}})\mathrm{d}^3p_\nu\mathrm{d}^3p_{\mathrm{e}}\,, \qquad (4.85)$$

where Δm is the difference in nuclear masses and the matrix element M is

$$M = \int \psi_{\mathrm{n}}^*(\boldsymbol{r})\psi_{\mathrm{p}}(\boldsymbol{r})\mathrm{d}^3\boldsymbol{r}\,. \qquad (4.86)$$

The differential rate (4.85) is the same as the neutron rate (4.68) except the $m_{\mathrm{n}} - m_{\mathrm{p}}$ is replaced with $m_{A,Z} - m_{A,Z\pm1} = \Delta m$ and a factor $|M|^2$ is added.

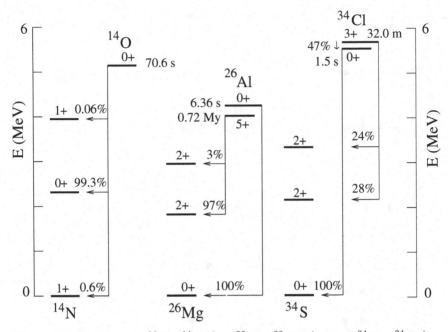

Fig. 4.12. The β-decays ^{14}O $\rightarrow^{14}$ Ne$^+\nu_e$, ^{26}Al $\rightarrow^{26}$ Mge$^+\nu_e$, and ^{34}Cl $\rightarrow^{34}$ Se$^+\nu_e$. ^{14}O has three allowed decays, to the ground state ($Q_{ec} = 5.14$ MeV) and to the 2.31 MeV and 3.95 MeV excited states. The highest branching fraction is for the super-allowed decay to the 0^+ excited state. ^{26}Al has two β-decaying states, the 5^+ ground state and the 0+ ($E = 0.228$ MeV) isomeric state. The later decays via a super-allowed decay to the ground state of ^{26}Mg ($Q_{ec} = 4.233$ MeV). The ground state has no allowed decays, explaining its long lifetime, $t_{1/2} = 7.2 \times 10^5$ yr. For ^{34}Cl, the roles of the isomer ($E = 0.146$ MeV) and ground states are reversed with the ground state decaying through the super-allowed mode ($Q_{ec} = 5.49$ MeV). The isomer has two allowed β-modes to the 2^+ excited states of ^{34}S. The radiative transition to the ground state of ^{34}Cl has a 47% branching fraction making this a mixed β-radiative-decay.

Integrating over the neutrino momenta, we find the electron energy spectrum

$$\frac{d\lambda}{dE_e} = \frac{2}{\pi^3\hbar}\left(\frac{2.4G_F}{(\hbar c)^3}\right)^2 |M|^2 (\Delta mc^2 - E)^2 \sqrt{E^2 - m_e^2 c^4} E . \qquad (4.87)$$

The spectrum is the same as that in neutron decay with the replacement $m_n - m_p \rightarrow \Delta m$.

By integrating over the electron energy we get the total decay rate.

$$\lambda = \frac{2}{\pi^3\hbar}\left(\frac{2.4G_F}{(\hbar c)^3}\right)^2 |M|^2$$

$$\times \int_{m_e c^2}^{\Delta mc^2} (\Delta mc^2 - E)^2 \sqrt{E^2 - m_e^2 c^4}\, E dE . \qquad (4.88)$$

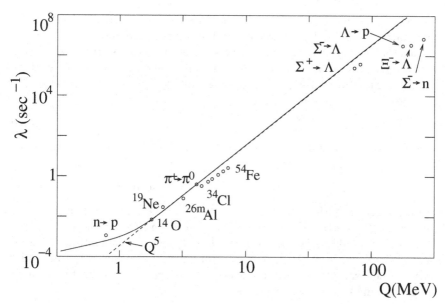

Fig. 4.13. β transition rates for super-allowed β-decays as a function of the max-imum kinetic energy of the final electron. The decays include elementary particles as well has nuclei where symmetry requires the integral (4.86) to be unity. The solid line shows the phase-space integral with the dashed line corresponding to the high-energy value Q_β^5. The super-allowed β^+ emitters, ^{14}O, ^{26m}Al, ^{34}Cl, ^{38m}K, ^{42}Sc, ^{46}V, ^{50}Mn, and ^{54}Fe fall slightly below the line because Coulomb corrections re-duce the rate. The *strangeness-changing* decays like $\Lambda \to p$ fall rather far below the extrapolation because these decays are suppressed by the Cabibbo factor $\sin^2 \theta_c$, as discussed in Sect. 4.4.3.

The rate as a function of $Q_\beta = (\Delta m - m_e)c^2$ is shown in is shown in Fig. 4.13 in the case $|M|^2 \sim 1$. For $Q_\beta \gg m_ec^2$, the rate goes like the fifth power of the decay energy:

$$\lambda \sim |M|^2 G_F^2 Q_\beta^5 . \tag{4.89}$$

Generally, one has $|M|^2 \ll 1$, reducing considerably the rate below the rate calculated by extrapolating from free-neutron decay. However, in certain circumstances, isospin symmetry requires that the initial and final wavefunc-tion overlap nearly perfectly so that $|M|^2 \sim 1$. Such decays are called *super-allowed* decays. Three examples are shown in Fig. 4.12 and their rates are shown in Fig. 4.13. Note that for β^+ (β^-) decays, the Coulomb modifications discussed below make the $|M|^2 = 1$ rates a bit below (above) that expected by extrapolating from neutron decay.

In order to calculate more reliable decay rates, it is necessary to modify (4.84) to take into account the spin of the nucleons and leptons. As in the case of neutron decay, there are two important terms, the *Fermi* term proportional to the overlap integral (4.84), and the *Gamow–Teller* term proportional to the matrix element of σ between initial and final state nuclei.

Just as in radiative decays, there are selection rules governing which combinations of initial J_i and final J_f spins are possible. The Fermi term will vanishes if the angular dependences of the initial and final wavefunctions are orthogonal so we require

$$\text{Fermi}: \quad J_i = J_f \,. \tag{4.90}$$

The Gamow–Teller term can change the spin but vanishes if the initial and final angular momenta are zero:

$$\text{GT}: \quad J_i = J_f, J_f \pm 1 \quad J_i = J_f = 0 \text{ forbidden}\,. \tag{4.91}$$

Additionally, in both cases, the parity of the initial and final nuclei must be the same. Transitions that respect the selection rules are called "Allowed" decays. "Forbidden" decays are possible only if one takes into account the spatial dependence of the lepton wavefunctions, i.e. using (4.83) instead of (4.84) The examples of forbidden decays in Fig. 4.12 illustrate the much longer lifetimes for such transitions.

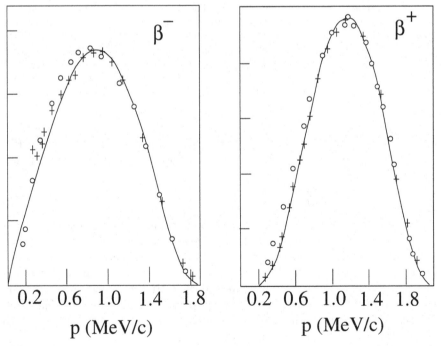

Fig. 4.14. The β^- and β^+ spectra of ^{64}Cu [44]. The suppression the of the β^+ spectrum and enhancement of the β^- at low energy due to the Coulomb effect is seen.

Finally, it is necessary to take into account the fact that in the presence of a charged nucleus, the energy eigenfunctions of the final state $e^{\pm}$ are not plane waves but are suppressed (enhanced) near the nucleus for e^+ (e^-). This has the effect of suppressing β^+ decays at low positron energy and of enhancing β^- decays at low electron energy. This effect is clearly seen in Figs. 4.13 and 4.14. The spectrum (4.87) is modified as

$$\frac{\mathrm{d}\lambda}{\mathrm{d}E_e} \propto F(Z', E_e)\, E_e \sqrt{E_e^2 - m_e^2 c^4}(\Delta mc^2 - E_e)^2 \,, \qquad (4.92)$$

where $F(Z', E_e)$ the "Coulomb correction." Tabulated values can be found in [43].

4.3.3 Electron-capture

All atomic nuclei that can decay by β^+ emission can also capture an atomic electron thereby decaying via

$$(A, Z)\,e^- \;\rightarrow\; (A, Z-1)\,v_e \,. \qquad (4.93)$$

This process is illustrated in Fig. 4.15. The neutrino energy is

$$Q_{ec} \;=\; (m(A, Z) - m(A, Z-1)]c^2 + 2m_e c^2 \;=\; Q_{\beta+} + 2m_e c^2 \,. \quad (4.94)$$

Electron capture is the only decay mode possible if $Q_{\beta+} < 0$ and $Q_{\beta-} < 0$, i.e. if the nuclear masses of neighboring isobars differ by less the the electron mass.

The effective Hamiltonian is

$$\langle (A, Z-1)v_e|H_1|(A, Z)e^-\rangle \;=$$

$$\frac{2.4 G_F}{L^{3/2}} \int \mathrm{d}^3 r \, \psi_p^*(r)\psi_n(r)\psi_e(r) \exp\mathrm{i}[p_v \cdot r/\hbar] \,. \qquad (4.95)$$

As in the case of nuclear β-decay, we can usually make the approximation that the neutrino and electron wavefunctions are constant over the nucleus so that

$$\langle (A, Z+1)v_e|H_1|(A, Z)e^-\rangle \;=\; \frac{2.4 G_F}{L^{3/2}}\psi_e(r = 0) \int \mathrm{d}^3 r \, \psi_p^*(r)\psi_n(r) \,.$$

This gives a decay rate

$$\lambda \;=\; \frac{c}{\pi(\hbar c)^4}(2.4 G_F)^2 |\psi_e(0)|^2 |M|^2 Q_{ec}^2 \,. \qquad (4.96)$$

The inner-most atomic electrons experience only the nuclear electric field so their wavefunctions are hydrogen-like except that the potential is $\sim Ze^2/r$ instead of $\sim e^2/r$. The effective Bohr radius is then a factor Z smaller implying that the wavefunction at the origin is a factor Z^3 larger:

$$|\psi_e(0)|^2 \;\sim\; \frac{Z^3}{a_0^3} \,. \qquad (4.97)$$

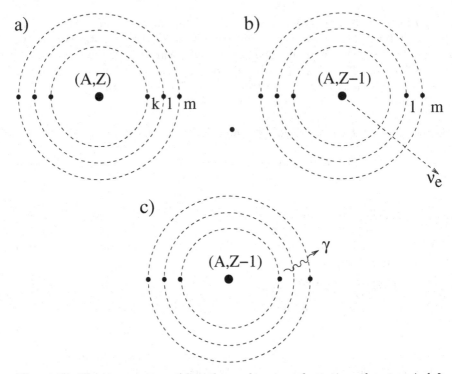

Fig. 4.15. Electron capture. After the nuclear transformation, the atom is left with an unfilled orbital, which is subsequently filled by another electron with the emission of photons (X-rays). As in the case of nuclear radiative decay, the X-ray can transfer its energy to another atomic electron which is then ejected from the atom. Such an electron is called an Auger electron.

The decay rate is then

$$\lambda = \frac{c}{\pi(\hbar c)^4} \frac{(2.4 G_{\rm F})^2 Z^3}{a_0^3} |M|^2 Q_{\rm ec}^2 \ . \tag{4.98}$$

Compared with nuclear β-decay, the Q dependence is weak, $Q_{\rm ec}^2$ rather than Q_β^5. This means that for small Q_β, electron-capture dominates over β^+ decay, as can be seen in Fig. 2.13. The strong Z dependence coming from the decreasing electron orbital radius with increasing Z means that electron-capture becomes more and more important with increasing Z.

Finally, we note that nuclear decay by electron capture leaves the atom with an unfilled atomic orbital. This orbital is filled by other atomic electrons falling into it and radiating photons. The photons are in the keV (X-ray) range since the binding energy of the inner most electron of an atom of atomic number Z is

$$E \sim 0.5 Z^2 \alpha^2 m_e c^2 = 0.01 \, Z^2 \, {\rm keV} \ . \tag{4.99}$$

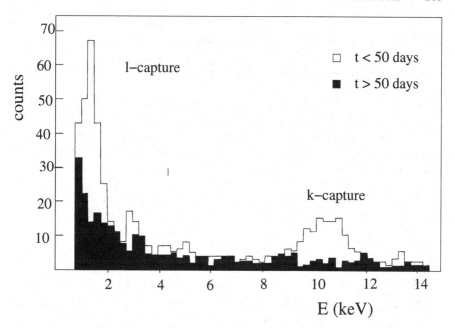

Fig. 4.16. The spectrum of Auger electrons from the electron-capture decay of ^{71}Ge as measured by the Gallium Neutrino Observatory (continuation of the GALLEX project). The spectrum is measured with a small gas-proportional counter filled with a $Xe - GeH_4$ mixture. A small number of the germanium nuclei are radioactive ^{71}Ge nuclei ($t_{1/2} = 11.4$day) produced by solar neutrinos as described in Sect. 8.4.1. Two peaks, corresponding to K- and L-capture, are observed in the first 50 days of counting. After 50 days, most of the ^{71}Ge has decayed and the counts are due to ambient radioactivity due to impurities in the counter.

As in the case of nuclear radiative decays, these photons can transfer their energy to another atomic electron that is then ejected from the atom. Such an electron is called an "Auger electron." Since the only high-energy particle emitted in electron capture is a neutrino, X-rays and Auger electrons are generally the only sure way to signal a decay be electron capture. A typical spectrum is shown in Fig. 4.16

4.3.4 Neutrino mass and helicity

Studies of nuclear β-decay have revealed two interesting properties of neutrinos. The most obvious is its small mass. This is seen in the fact the the energy spectrum of electrons is consistent with (4.87) which was calculated assuming $m_\nu = 0$. A non-zero neutrino mass modifies this distribution in two ways. First, the maximum electron energy is lowered

$$E_e(\text{max}) = (m_{A,Z} - m_{A,Z+1} - m_\nu)c^2 . \tag{4.100}$$

Second, the shape of the electron energy spectrum (4.92) is modified, especially near the "end point," i.e. the maximum electron energy:

$$\frac{1}{FE_e\sqrt{E_e^2 - m_e^2 c^4}}\frac{d\lambda}{dE_e} \propto (E_0 - E_e)\sqrt{(E_0 - E_e)^2 - m_\nu^2 c^4}\,,\qquad(4.101)$$

where $E_0 = (m(A, Z) - m(A, Z'))c^2$ is the maximum electron energy in the case $m_\nu = 0$. A plot of the square root of the left-hand side vs. E_e is, for $m_\nu = 0$, a straight line intersecting zero at $E_e = E_0$. Such a plot is called a *Kurie plot*. As shown in Fig. 4.17, this causes the Kurie plot to curve down near the endpoint.

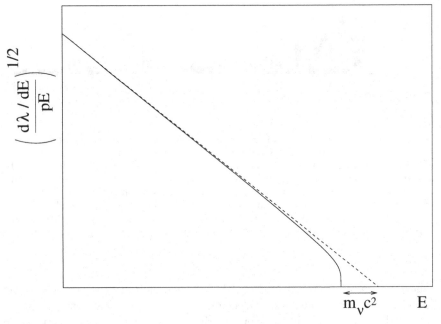

Fig. 4.17. The Kurie plot near the electron maximum energy in the case of $m_\nu = 0$ (dashed line) and $m_\nu \neq 0$ (solid line).

Very precise measurements of the electron energy spectrum have been made for tritium β-decay

$$^3\mathrm{H} \rightarrow {}^3\mathrm{He}\, e^- \bar{\nu}_e \qquad Q_\beta = 18.54\,\mathrm{keV}\,.\qquad(4.102)$$

Tritium is chosen since the low value of Q_β facilitates accurate measurements of E_e near the end point.

The most precise experiment to date [45] is shown in Fig. 4.18 which measures the electron spectrum with an electrostatic spectrometer. The agreement of the observed spectrum with that expected for massless neutrinos,

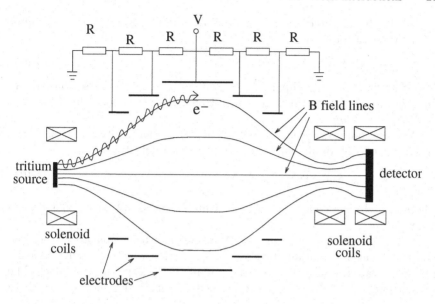

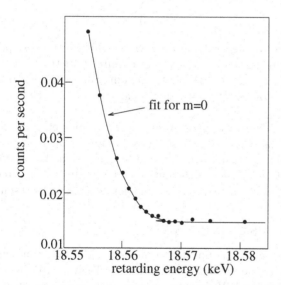

Fig. 4.18. The Mainz tritium β-decay spectrometer [45]. Electrons emerging from a thin tritium source enter a region with an electric and magnetic field. Electrons spiral around the magnetic field lines and the electric field reflects all but the most energetic ones. Those passing this barrier then pass through a symmetric electric field to recover their initial kinetic energy before being detected. The counting rate a a function of the retarding potential reflects the neutrino energy spectrum near the end-point. The curve shows the prediction for vanishing neutrino mass (after correction for experimental resolution).

appropriately corrected for experimental resolution, results in an upper limit on the mass of the $\bar{\nu}_e$ of

$$m_{\bar{\nu}_e} c^2 \; < \; 3\,\text{eV} \, . \tag{4.103}$$

Note that the limit is comparable to *atomic* binding energies so to correctly analyze the experiment, it is necessary to consider the spectrum of the produced helium *atom*.

A second important characteristic of neutrinos (antineutrinos) is that they are produced in β-decay with negative (positive) helicity:

$$\frac{\boldsymbol{s} \cdot \boldsymbol{p}}{|\boldsymbol{p}|} \; = \; +\hbar/2 \quad (\bar{\nu}) \qquad = \; -\hbar/2 \quad (\nu) \, , \tag{4.104}$$

where $\boldsymbol{s}$ is the neutrino spin. This is a manifestation of parity violation in the weak interactions.

There is only one experiment where one has been able to deduce the neutrino helicity, essentially by creating a situation where the neutrino helicity is correlated with a photon helicity and then measuring the latter. The experiment [46] uses the electron-capture decay of ^{152m}Eu shown in Fig. 4.19

$$e^- \; {}^{152m}\text{Eu}(0^-) \; \to {}^{152}\text{Sm}^*(1^-)\,\nu_e \qquad Q_{ec} = 840\,\text{keV}$$

$$^{152}\text{Sm}^*(1^-) \; \to \; {}^{152}\text{Sm}(0^+)\,\gamma\,(961\,\text{keV}) \, . \tag{4.105}$$

As shown in the figure, if the neutrino and photon are emitted in opposite directions, conservation of angular momentum requires that the photon spin have the opposite sign of that of the neutrino spin. Since they go in opposite directions, the photon helicity must have the same sign as the neutrino helicity.

Because the directions of the photon and neutrino momenta are nearly uncorrelated, the problem is to select only those photons that happen to be emitted opposite to the neutrino. This is done by scattering the photons in a sample of ^{152}Sm before they enter the detector. Only photons emitted opposite to the direction of the neutrino have sufficient energy to be resonantly scattered

$$\gamma \, {}^{152}\text{Sm}(0^+) \; \to \; {}^{152}\text{Sm}^* (1^-) \; \to \; \gamma \, {}^{152}\text{Sm}(0^+) \, . \tag{4.106}$$

The fact that they were emitted opposite to the neutrino nearly compensates for the energy taken by the ^{152}Sm in the radiative decay (4.105) (Exercise 4.9). Photons of energy 961 keV detected in the NaI scintillator must then have been emitted opposite to the neutrino. (All photons can Compton scatter but this lowers their energy below 961 keV.)

Once one has selected the photons emitted opposite to the neutrinos, it is only necessary to measure their helicity. This can be done by placing magnetized iron in their path. Because the Compton cross-section depends on the relative orientations of the photon and electron spins, photons are scattered out of their initial direction at rates that depend on the relative

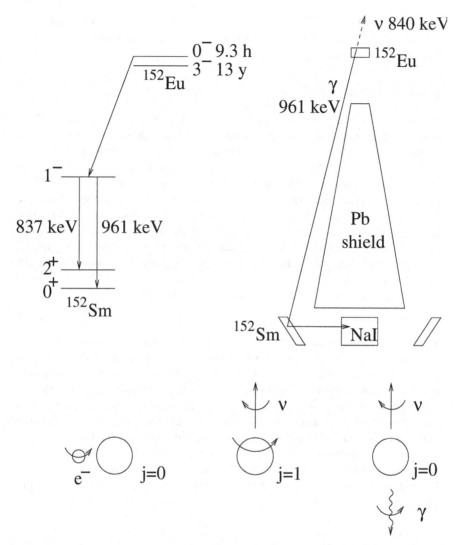

Fig. 4.19. Experiment used to measure the helicity of the neutrino [46]. The experiment uses the electron capture decay of the 0^- isomer of ^{152}Eu decaying with a half-life of 9.3 h to the 1^- excited state of ^{152}Sm. In the experimental apparatus shown on the upper right, the 961 keV de-excitation photons can be scattered in ^{152}Sm and be detected in a NaI scintillator. The de-excitation photons are emitted in directions that are uncorrelated with the neutrino direction, but only photons emitted opposite to the neutrinos have sufficient energy to excite the 1^- state of ^{152}Sm and therefore be resonantly scattered into the NaI. This is because in this case, the recoil energy of the 1^- nucleus compensates for the recoil of the 0^+ nucleus so the photon has the entire 961 keV transition energy (Exercise 4.9). As shown on the bottom of the figure, in order to conserve angular momentum, photons emitted opposite to the neutrino must have the same helicity as the neutrino. The neutrino helicity can therefore be deduced by measuring the helicity of the photon.

orientation of their spins and the magnetization vector. Measurement of the photon counting rate for two magnetizations then allows one to deduce the photon polarization.

4.3.5 Neutrino detection

Because of their tiny cross-sections, neutrino detection presents the challenge of using detectors that are large enough to have a sufficient counting rate yet intelligent enough to distinguish neutrino events from much more common backgrounds coming from cosmic-rays and natural radioactivity. These difficulties explain the long delay between Pauli's neutrino hypothesis in 1934 and their experimental detection by Reines and Cowan in 1956. Since their pioneering experiment, observations of solar neutrinos (Sect. 8.4.1), supernova neutrinos (Sect. 8.4.2) and of "artificial" neutrinos from reactors and particle accelerators have lead not only to important advances in astrophysics but also the surprising observation of "neutrino oscillations" caused by neutrino masses (Sect. 4.4).

The reaction used by Reines and Cowan was

$$\bar{\nu}_e \, p \rightarrow e^+ n \, . \tag{4.107}$$

This reaction was chosen for several reasons. First, nuclear-fission reactors are intense sources of $\bar{\nu}_e$ because most fission products are short-lived β^--emitters (Chap. 6). Second, massive targets can be made since ordinary organic compounds contain large numbers of protons. Finally, the reaction gives a distinct "signature" of a positron produced in the neutrino reactions followed, some time later, by the capture of the thermalized neutron. (The neutron is thermalized by elastic scatters on nuclei, especially hydrogen nuclei.)

The $\bar{\nu}_e$ production rate of a nuclear reactor is easily estimated since each fission yields about 200 MeV of thermal energy and, on average, 5 $\bar{\nu}_e$ coming from the cascade of β-decays of the two fission products. The neutrino production rate rate is therefore simply related to the thermal power P of the reactor

$$\frac{dN_{\bar{\nu}_e}}{dt} = \frac{5P}{200 \, \text{MeV}} \sim 1.5 \times 10^{20} \text{s}^{-1} \left(\frac{P}{1 \, \text{GW}} \right) \, . \tag{4.108}$$

The mean $\bar{\nu}_e$ energy is $\sim 4 \, \text{MeV}$ so the mean $\bar{\nu}_e$ cross-section, from Table 3.1 is $\sigma \sim 10^{-47} \, \text{m}^2$. At a distance R from the reactor, this gives a rate per proton of

$$\lambda(\bar{\nu}_e p \rightarrow e^+ n) = \frac{\sigma}{4\pi R^2} \frac{dN_{\bar{\nu}_e}}{dt} = 10^{-30} \, \text{sec}^{-1} \left(\frac{R}{10 \text{m}} \right)^2 \, . \tag{4.109}$$

To get a rate of 1 event per hour at a distance of 10 m, we need about 2×10^{26} protons corresponding to about 2 kg of organic material (CH).

A typical realization of such a detector is shown schematically in Fig. 4.20. The organic target is in the form of a liquid scintillator that emits light

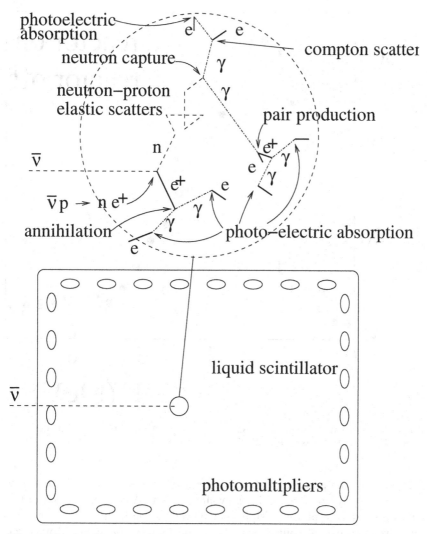

Fig. 4.20. A schematic of the standard method of detecting $\bar{\nu}_e$ through the re-action $\bar{\nu}_e\,p \rightarrow e^+\,n$. The detector consists of liquid scintillator instrumented with photomultipliers. The $\bar{\nu}_e$ scatters on a proton contained in the scintillator, an or-ganic compound. The positron stops through ionization loss (Sect. 5.3) and then annihilates, $e^+e^- \rightarrow \gamma\gamma$. The neutron thermalizes though elastic scatters on pro-tons and is eventually captured on a nucleus, $n(A, Z) \rightarrow \gamma\gamma(A + 1, Z)$. The photons produced in the capture and in the annihilation either convert to e^+e^- pairs or lose energy through Compton scattering, eventually being absorbed photoelectri-cally. Scintillation light is produced by the electrons and positrons slowing down in the scintillator. The scintillation light is detected by the photomultipliers. The light comes in two flashes, the first from the positron produced in the original interaction, and the second from the Compton and photo-electrons after the thermalization and capture of the neutron.

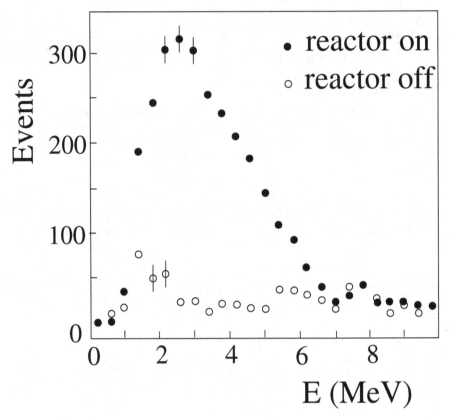

Fig. 4.21. Spectrum of positrons created by the reaction $\bar{\nu}_e\, p \rightarrow n\, e^+$ as observed by the Chooz neutrino experiment [47].

in response to the passage of charged particles. A flash of light is then produced by the positron created in the $\bar{\nu}_e$ interaction. A certain fraction of this light is detected by photomultipliers deployed in the liquid. The neutron is thermalized by elastic collisions with the protons in the scintillator (Exercise 4.10). It is eventually captured, either by a proton or a nucleus with a high capture cross-section that is added to the scintillator. The photons produced in the capture then interact in the scintillator via Compton scattering, electron–positron pair production, and photoelectric absorption (Sect. 5.3.4, Exercise 4.11). Each of these processes create electrons and their total effect is to create a second pulse of scintillation light. The time between the first and second pulse is determined by the neutron-capture rate and is generally ~ 1 ms. The double pulse then distinguishes neutrino events from more common events due to natural radioactivity.

Figure 4.21 shows the positron energy spectrum deduced from the quantity of scintillation light produced in the initial pulse. The positron energy is just equal to the neutrino energy minus the reaction threshold, $(m_n + m_e - m_p)c^2 = 1.8\,\text{MeV}$. The distribution of neutrino energies agrees with that expected from the β-decays of the fission products.

The detection of ν_e from β^+-decay is more difficult than detection of $\bar{\nu}_e$ because the analogous reaction $\nu_e n \rightarrow e^- p$ requires a target consisting of free neutrons. A good approximation to a free neutron target uses deuterons:

$$\nu_e\, {}^2\text{H} \rightarrow e^-\, p\, p\,. \tag{4.110}$$

Sufficiently massive neutrino detectors can be constructed from heavy water and have been used to detect solar neutrinos (Sect. 8.4.1). In these detectors, the final-state electron is seen through its emission of "Cherenkov" light that is detected by photomultipliers. Because there are no neutrons produced in the event, there is no distinctive double pulse to signal a neutrino interaction so the reaction can only be used for $E_\nu > 5\,\text{MeV}$ where there is little background coming from radioactivity.

The SNO experiment using heavy water also observes the deuteron breakup reaction

$$\nu\, {}^2\text{H} \rightarrow \nu\, p\, n\,. \tag{4.111}$$

This reaction has a rather nondistinctive signature, the liberation of a neutron and its subsequent capture. It has the advantage of being a neutral current reaction so it is equally sensitive to all neutrino species, ν_e, ν_μ and ν_τ.

Heavy water (and light water) detectors are also sensitive to neutrino-electron elastic scattering

$$\nu\, e^- \rightarrow \nu\, e^-\,. \tag{4.112}$$

This is a mixed neutral-current/charged reaction so it is sensitive to all species but with a larger cross-section for ν_e.

Low-energy neutrinos can be also detected through capture reactions

$$\nu_e\, (A, Z) \rightarrow e^-\, (A, Z + 1)\,. \tag{4.113}$$

Important examples that have been used for solar neutrino detection are

$$\nu_e\, {}^{37}\text{Cl} \rightarrow e^-\, {}^{37}\text{Ar} \qquad Q = -0.81569\,\text{MeV}\,, \tag{4.114}$$

and

$$\nu_e\, {}^{71}\text{Ga} \rightarrow e^-\, {}^{71}\text{Ge} \qquad Q = -0.22194\,\text{MeV}\,. \tag{4.115}$$

The magnitudes of the (negative) Q values give the threshold neutrino energy. In both cases *radiochemical* techniques are used where the final state nucleus is extracted from a multi-ton target and then observed to decay in a small counter.

Finally, we mention that high-energy neutrinos are copiously produced at high-energy accelerators and by cosmic rays in the Earth's atmosphere. The most important production mechanism is the decay of pions produced in nucleon-nucleon interactions

$$\pi^{\pm} \rightarrow \mu^{\pm} \nu_{\mu}(\bar{\nu}_{\mu}) \ . \tag{4.116}$$

Mostly ν_{μ} are produced in pion decay but some ν_e and ν_{τ} are produced by the decays of heavier particles. While still requiring massive detectors, their observation is simplified by the fact that their energy is much higher than that of natural radioactivity.

4.3.6 Muon decay

We have already presented the μ lepton, or muon, in Sect. 1.8, when we studied muonic atoms.

The muon is elementary in the same sense as the electron. It has the same charge, the same spin, but it is 200 times heavier, $m_{\mu} = 206.8 m_e$, and it is unstable. It decays into an electron and two neutrinos : $\mu^- \rightarrow e^- \bar{\nu}_e \nu_{\mu}$ with a lifetime $\tau = 2 \times 10^{-6}$ s.

The existence of the muon was an enigma for nearly 40 years. When it was discovered, Rabi said "Who ordered that?" Why a heavy electron? All the matter we know around us can be built with protons, neutrons, electrons and neutrinos, or, in terms of fundamental constituents, with the family of quarks and leptons $\{u, d, e, \nu\}$. Why should there be a heavy electron, with which one can achieve the dreams of Gulliver? Because the size of atomic systems are inversely proportional to the system's reduced mass, one could imagine atoms, molecules, a chemistry, a biology 200 times smaller but 200 more energetic than the beings we know! We have found many applications of the muon as probes of nuclei, of crystal structure, and of pyramids,[1] but why does it exist? What is its use in nature?

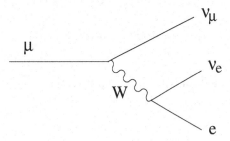

Fig. 4.22. Decay process of the muon.

A first clue comes from the calculation of the muon lifetime $\mu \rightarrow e\bar{\nu}\nu$ (we do not specify the charges or lepton numbers since one of them determines all the others). This is *also* a weak decay governed by the Fermi constant! For the first time, we are facing the *universality* of weak interactions.

---

[1] Muons are by far the most penetrating charged particles since they have no strong interactions and, because of their large mass, radiate (bremsstrahlung) much less efficiently than electrons.

This reaction is represented on Fig. 4.22. It occurs via the virtual production of a W boson, i.e. to a very good approximation it is a point-like interaction.

We denote as p_μ, p_{ν_μ}, p_e, p_{ν_e} the four-momenta of the 4 particles. The square of the matrix element for this reaction is

$$|\tilde{T}|^2 = 4\,G_F^2\,c^4\frac{(p_\mu \cdot p_{\nu_e})(p_{\nu_\mu} \cdot p_e)}{E_\mu\,E_{\nu_e}\,E_{\nu_\mu}\,E_e}\,. \tag{4.117}$$

We can make a few comments on this result:

- By inserting this into (4.16) we notice that the decay rate is written in an explicitly Lorentz invariant form: besides scalar products of four-vectors there is a factor d^3p/E for each final-state particle. It is readily shown that this factor is Lorentz invariant. This leaves only the factor $1/E$ for the initial state particle which simply reflects the fact that the mean lifetime is affected by relativistic time dilation.
- One can check that with the same notations, if we change the muon into a neutron and the ν_μ into a proton, and if we neglect the recoil of the final proton, this expression boils down to the same as in (4.67) for neutron decay except that the factor 2 becomes 2.4. This factor is due to the fact that the neutron is not an elementary particle (as opposed to a muon).

To simplify the calculation, we shall neglect the electron mass (it is easy to take it into account, which is necessary in an accurate calculation) so that $E_e \simeq p_e c$ where p_e is the electron momentum.

In order to simplify the notations, we set:

$$p_1 \equiv p_e\,, \quad p_2 \equiv p_{\nu_e}\,, \quad p_3 \equiv p_{\nu_\mu}\,,$$

and we place ourselves in the rest frame of the muon. By energy-momentum conservation, we have $p_\mu = p_{\nu_\mu} + p_e + p_{\nu_e}$, i.e. $p_\mu - p_{\nu_e} = p_{\nu_\mu} + p_e$, and, by squaring

$$p_{\nu_\mu} \cdot p_e = \frac{1}{2}(p_\mu - p_{\nu_e})^2 = \frac{1}{2}m_\mu c\,(m_\mu c - 2p_{\nu_e})$$

always in the approximation $m_e \simeq 0$.

Altogether, the muon decay rate is

$$\lambda_\mu = \frac{2G_F^2 m_\mu}{\hbar^2(2\pi\hbar)^5}\,I \tag{4.118}$$

with

$$I = \int \frac{p_2(m_\mu c - 2p_2)}{p_1 p_2 p_3}\delta(\boldsymbol{p}_1 + \boldsymbol{p}_2 + \boldsymbol{p}_3)$$

$$\times\;\delta(m_\mu c^2 - p_1 c - p_2 c - p_3 c)\mathrm{d}^3\boldsymbol{p}_1\mathrm{d}^3\boldsymbol{p}_2\mathrm{d}^3\boldsymbol{p}_3 \tag{4.119}$$

We use the momentum conservation δ function to integrate over $\boldsymbol{p}_3$ ($\boldsymbol{p}_3 = -(\boldsymbol{p}_1 + \boldsymbol{p}_2)$).

Transforming to spherical coordinates for p_1 and p_2, we obtain ($d^3p = p^2dpd\Omega$)

$$I = \int \frac{p_2(m_\mu c - 2p_2)}{p_3}\delta(m_\mu c^2 - p_1 c - p_2 c - p_3 c)p_1 p_2 d\Omega_1 d\Omega_2 dp_1 dp_2 \quad .$$

Let θ be the angle between p_1 and p_2, the integration on the other angles gives a factor of $4\pi \times 2\pi = 8\pi^2$. We now use the energy conservation δ function $\delta(m_\mu c^2 - p_1 c - p_2 c - p_3 c) = (1/c)\delta(m_\mu c - p_1 - p_2 - p_3)$ in order to integrate over $\cos\theta$, thanks to the relation $\delta(f(x)) = \frac{1}{|f'(x_0)|}\delta(x - x_0)$, $f(x_0) = 0$.

We have $p_3(\cos\theta) = (p_1^2 + p_2^2 + 2p_1 p_2 \cos\theta)^{1/2}$ therefore

$$\left| \frac{dp_3}{d\cos\theta} \right| = \frac{p_1 p_2}{p_3}$$

i.e.

$$I = \frac{8\pi^2}{c} \int p_2(m_\mu c - 2p_2) dp_2 dp_1 \quad . \tag{4.120}$$

The integration bounds are obtained by noticing that in a three-body system of zero mass particles of zero total momentum and of total energy Mc^2 :

- the maximum and minimum values of the energy of any of these particles are obtained when the three momenta are aligned along the same axis;
- the maximum energy of one of the particles is $Mc^2/2$;
- the minimum invariant mass m_{12} of a two-body subsystem ($m_{12}^2 = (p_1 + p_2)^2/c^2$) is $M/2$.

For a fixed value of p_1, which therefore varies between 0 and $m_\mu c/2$, the integration bounds in p_2 are $m_\mu c/2 - p_1$ and $m_\mu c/2$. A simple calculation leads to

$$I = \frac{8\pi^2}{6c}(\frac{m_\mu c}{2})^4 , \tag{4.121}$$

and to a decay rate

$$\lambda_\mu = \frac{1}{\tau_\mu} = \left(\frac{G_F}{(\hbar c)^3}\right)^2 \frac{(m_\mu c^2)^5}{192\pi^3\hbar} \quad . \tag{4.122}$$

Numerically, this gives a lifetime $\tau_\mu = 2.187 \ 10^{-6}$ s, in excellent agreement with the experimental value $\tau_\mu = 2.197 \ 10^{-6}$ s. Note that we recover in this formula the fifth power of the maximum energy of the final electron.

Actually, muon decay is the cleanest way to determine the value of the Fermi constant. The muon and electron are both point-like particles, so that there are no wavefunction corrections to care about, and there are no final state interactions. The further corrections to the matrix element and to the above result are well understood.

The importance of this result is that it is a first indication of the *universality of weak interactions* of leptons. The pair (e, ν_e) couples to the intermediate bosons of weak interactions in strictly the same way as the couple (μ, ν_μ).

If we compare with the calculation of the neutron lifetime, we are tempted to think that universality extends to protons and neutrons. However, protons and neutrons are not elementary point-like particles and we must reconsider the problem from the point of view of quarks.

4.4 Families of quarks and leptons

The electromagnetic interactions of observed elementary particles are universal in the sense that all charges are multiples of the fundamental charge e of the electron. At the quark level, this universality persists though the fundamental charge is $e/3$.

Universality of the weak interactions is more subtle. First, the Fermi constant governing weak decays, G_F, is not a fundamental coupling constant but rather an effective coupling proportional to m_W^{-2}. Second, the fact that neutrons and protons are not fundamental particles means that fundamental constants are renormalized by non-trivial factors when going from the quark level to the nucleon level. This is the origin of the strange factor $g_A \sim 1.25$ in the effective Fermi constant in β-decay (4.81). Finally, will see that even at the quark level the fundamental couplings are not to quarks and leptons of definite mass but rather to *mixtures* of quarks defined by a unitary matrix. This effect is the origin of the Cabibbo angle in the effective Fermi constant (4.81).

In this section, we will go into some of the details of these problems.

4.4.1 Neutrino mixing and weak interactions

For many years, it was believed that there where three conserved "Lepton numbers," i.e. electron-number, muon-number, and tauon-number. The apparent conservation of these three numbers meant that it was useful to classify leptons in three *generations*:

$$\begin{pmatrix} \nu_e \\ e^- \end{pmatrix} \quad \begin{pmatrix} \nu_\mu \\ \mu^- \end{pmatrix} \quad \begin{pmatrix} \nu_\tau \\ \tau^- \end{pmatrix}. \tag{4.123}$$

These three groups (and their antiparticles) are ordered in increasing masses of the charged lepton:

$$m_e(0.511\,\mathrm{MeV}/c^2) \ll m_\mu(105.6\,\mathrm{MeV}/c^2) \ll m_\tau(1777.0\mathrm{MeV}/c^2).$$

It was believed that the weak interactions do not mix these families, so that for instance, in β-decay a e^- (e^+) must be created in coincidence with a $\bar{\nu}_e$ (ν_e) so as to conserve electron number.

In this simple picture, the weak interactions of leptons are *universal*. The decay rates of the μ and τ can be calculated in a analogous manner as that of the neutron, in terms of the Fermi constant. The decay rate of the τ lepton can be deduced from (4.122). By changing the values of masses, we obtain the rates for $\tau^- \to e^- \bar{\nu}_e \nu_\tau$ and for $\tau^- \to \mu^- \bar{\nu}_\mu \nu_\tau$. These reactions are interpreted as originating in the universal coupling of the intermediate bosons $W^\pm$ to leptons in the processes $W^- \to e^- \bar{\nu}_e$, $W^- \to \mu^- \bar{\nu}_\mu$, $W^- \to \tau^- \bar{\nu}_\tau$ and all corresponding crossed processes.

While this picture is consistent with most experimental results, it is now generally believed that the ν_e, ν_μ and ν_τ are in fact mixtures of three leptons ν_1, ν_2 and ν_3 of masses m_1, m_2 and m_3:

$$\begin{pmatrix} \nu_e \\ \nu_\mu \\ \nu_\tau \end{pmatrix} = \begin{pmatrix} U_{e1} & U_{e2} & U_{e3} \\ U_{\mu1} & U_{\mu2} & U_{\mu3} \\ U_{\tau1} & U_{\tau2} & U_{\tau3} \end{pmatrix} \begin{pmatrix} \nu_1 \\ \nu_2 \\ \nu_3 \end{pmatrix} . \tag{4.124}$$

This fancy notation means only that the amplitude to produce a ν_1 paired with a e^+ is proportional to U_{e1}, the amplitude to produce a ν_2 paired with a μ^+ is proportional to $U_{\mu2}$, etc. Generally, it is supposed that the U matrix is unitary

$$\sum_{l=e,\mu,\tau} U_{li} U_{lk}^* = \delta_{ik} . \tag{4.125}$$

A 3×3 unitary matrix can be parameterized by 3 mixing angles, θ_{ij}, $i \neq j = 1, 2, 3$ and one phase angle δ [1]:

$$\begin{pmatrix} c_{12}c_{23} & s_{12}c_{13} & s_{13}e^{-i\delta} \\ -s_{12}c_{23} - c_{12}s_{23}s_{13}e^{i\delta} & c_{12}c_{23} - s_{12}s_{23}s_{13}\,e^{i\delta} & s_{23}c_{13} \\ s_{12}s_{23} - c_{12}c_{23}s_{13}e^{i\delta} & -c_{12}s_{23} - s_{12}c_{23}s_{13}\,e^{i\delta} & c_{23}c_{13} \end{pmatrix} \tag{4.126}$$

where $c_{ij} = \cos\theta_{ij}$ and $s_{ij} = \sin\theta_{ij}$.

To simplify things, we will first consider the mixing of only two neutrino species, e.g.

$$\begin{pmatrix} \nu_e \\ \nu_\mu \end{pmatrix} = \begin{pmatrix} \cos\theta & \sin\theta \\ -\sin\theta & \cos\theta \end{pmatrix} \begin{pmatrix} \nu_1 \\ \nu_2 \end{pmatrix} . \tag{4.127}$$

The unitary mixing matrix is now a function of a single parameter, the mixing angle θ.

In the case of muon and electron coupling to two neutrinos, the one muon decay mode in Fig. 4.22 becomes the four modes shown in Fig. 4.23

$$\mu^- \to \nu_i e^- \bar{\nu}_j \quad i = 1, 2 \quad j = 1, 2$$

The mixing angle enters into the four amplitudes as indicated in the figure. For example, the amplitude for $\mu \to e\nu_2\bar{\nu}_2$ is proportional to $\cos\theta \sin\theta$. Strangely enough, as long as all neutrino masses are much less than the charged lepton masses so that phase-space factors for all modes are equal, the total decay rate of the muon is unaffected:

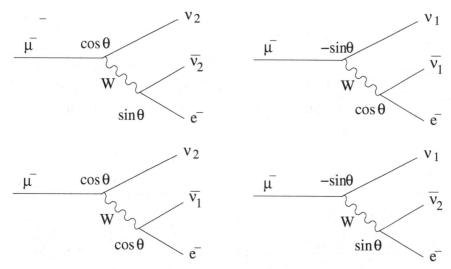

Fig. 4.23. The four decay modes of the muon in the case of mixing of the ν_μ and ν_e. If the neutrinos all have masses that are negligible compared to the m_μ, the total decay rate is unchanged from the case of no mixing shown in Fig. 4.22.

$$\tau^{-1} \propto (\cos^2 \theta)^2 + (\cos \theta \sin \theta)^2 + (\cos \theta \sin \theta)^2 + (\sin^2 \theta)^2 = 1 .$$

The unitarity of the mixing matrix guarantees that this is the case for any number of neutrinos.

In the case of the mixing of two neutrinos, the neutron now has two decay modes shown in Fig. 4.24

$$n \to pe^- \bar{\nu}_i \quad i = 1, 2 .$$

As with the muon, the neutron lifetime is unaffected

$$\tau^{-1} \propto \cos^2 \theta + \sin^2 \theta = 1 .$$

In order to see the effect of the existence of multiple neutrinos, it is necessary to observe the neutrinos themselves. A neutrino or antineutrino source using nuclear β-decay produces a ν_1 (ν_2) flux proportional to $\cos^2 \theta$ ($\sin^2 \theta$). The ν_1 (ν_2) then interact as in Fig. 4.24 with a cross-section proportional the same factor $\cos^2 \theta$ ($\sin^2 \theta$). The total neutrino interaction rate is then

$$\text{Rate} \propto \cos^4 \theta + \sin^4 \theta = 1 - \frac{1}{2} \sin^2 2\theta . \tag{4.128}$$

This is less than the rate that would be observed if there were only one neutrino species produced in β-decay.

If this were the whole story, the existence of neutrino mixing would have been discovered long ago. In fact, the situation is much more interesting because the neutrino masses are so small that an argument based on the uncertainty principle shows that it is generally impossible to determine whether

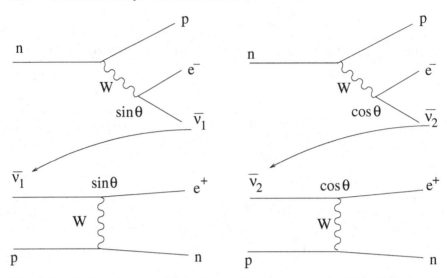

Fig. 4.24. The two decay modes of the neutron in the case of mixing of the ν_μ and ν_e. The total decay rate is unchanged from the case of no mixing. The $\bar\nu_1$ and $\bar\nu_2$ produced in the decay can later interact with a proton target through the reaction $\bar\nu p \to n e^+$. The total scattering rate per decay is proportional to $\cos^4\theta + \sin^4\theta$, i.e. different from the case of no mixing.

it was a ν_1 or a ν_2 that interacted. In this case, one should not sum the interaction *rates* but rather the interaction *amplitudes*.

The situation is illustrated in Fig. 4.25. A nucleus (A, Z) confined to a region of size Δx decays producing an electron of energy E_{e^-}, a recoiling daughter nucleus of energy $E_{(A,Z+1)}$ and a neutrino of energy $E_\nu = E_{(A,Z)} - E_{(A,Z+1)} - E_{e^-}$. The neutrino momentum for $m_\nu c^2 \ll E_\nu$ is

$$p_\nu c \sim E_\nu - \frac{m_\nu^2 c^4}{2E_\nu} . \tag{4.129}$$

The two neutrino species have different momenta but this is allowed as long as the difference in momenta is less than the uncertainty in the momentum of the initial nucleus

$$\frac{(m_i^2 - m_j^2)c^4}{2E_\nu} < c\Delta p \sim \frac{\hbar c}{\Delta x} . \tag{4.130}$$

If this condition is respected, both ν_1 and ν_2 emission is possible with the identical energies for all final-state particles. For $\Delta x \sim 10^5$ fm (corresponding to the size of an atom) and $E \sim 1$ MeV, (4.130) is respected if all neutrinos have masses $\ll 60$ keV, which is the case.

The neutrino then interacts on a proton producing a neutron of energy E_n and a positron of energy E_{e^+}. Once again, if (4.130) is respected, ν_1 and ν_2 scattering can lead to final state particles of identical energies. Since the

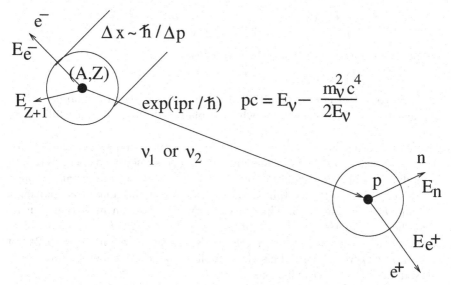

Fig. 4.25. Neutrino production and interaction. A nucleus confined to a region of size $\Delta x \sim \hbar/\Delta p$ β-decays to one of two neutrinos, ν_1 or ν_2. If the mass difference is sufficiently small, $(m_1^2 - m_2^2)/2E_\nu < \Delta p$, then either neutrino can be produced with the same energies for the decay products, E_{Z+1} and E_{e^-}. The two neutrinos can later interact and produce the same energy neutron and positron. The global final state, i.e. of particles produced in the decay and scatter, are identical and the amplitudes of ν_1 scattering and ν_2 scattering must be added.

two neutrinos scatter into identical final states, we must add the scattering amplitudes before squaring. The amplitude for ν_1 is

$$M_1 \propto \cos\theta \exp(ip_\nu r/\hbar) \cos\theta \,. \tag{4.131}$$

The first $\cos\theta$ comes from the amplitude for the production of a ν_1 while the second comes from the absorption. The propagation factor $\exp(ipr/\hbar)$ comes from the fact that the neutrino wavefunction enters into the absorption amplitude. The amplitude for ν_2 is the same except that $\cos\theta$ is replaced by $\sin\theta$ and that the neutrino momentum (4.129) is different because of the different mass. The total rate is found be summing amplitudes and squaring:

$$\lambda(\nu \to e^+) \propto \left| \cos^2\theta \exp\left(-i\frac{m_1^2 c^3 r}{2E_\nu \hbar} \right) + \sin^2\theta \exp\left(-i\frac{m_2^2 c^3 r}{2E_\nu \hbar} \right) \right|^2$$

$$= 1 - \sin^2 2\theta \sin^2\left(\frac{(m_1^2 - m_2^2)c^4 r}{4E_\nu \hbar c} \right) \,. \tag{4.132}$$

As a function of the distance between the decay and scatter, the rate oscillates about the "incoherent rate" given by (4.128). The oscillation length corresponding to the distance between rate maxima or minima is

$$L_{\text{osc}} = \frac{4\pi E_{\text{v}} \hbar c}{(m_1^2 - m_2^2)c^4} = 2.5\,\text{m}\,\frac{E_{\text{v}}}{1\,\text{MeV}}\frac{(m_1^2 - m_2^2)c^4}{1\,\text{eV}^2}. \tag{4.133}$$

For $r \to 0$ the rate reduces to the one neutrino case

$$\lambda(\nu \to e^+) \propto 1 \qquad r \ll L_{\text{osc}}. \tag{4.134}$$

In order to see the effect of multiple neutrinos, it is necessary to observe them sufficiently far from the decay, which explains why the effect is difficult to see.

On the other hand, sufficiently far from the neutrino source, the oscillations are washed out because one must average over neutrino energies (corresponding to different oscillation phases). This averaging is necessary because of the experimental uncertainty in the energies of the final state particles, E_n and E_{e^+}, in Fig. 4.25. This uncertainty generates an uncertainty in E_{v} and therefore in the oscillation phase $\phi = \Delta m^2 c^4 r / 2 E_{\text{v}} \hbar c$. If one averages over a neutrino energy interval ΔE_{v}, ϕ is averaged over more than 2π for $r > 4\pi \hbar c E_{\text{v}}^2 / \Delta E_{\text{v}} \Delta m^2 c^4$. This then implies that the incoherent rate is observed after $E_{\text{v}}/\Delta E_{\text{v}}$ oscillations periods:

$$\langle \lambda(\nu \to e^+) \rangle \propto 1 - \frac{1}{2}\sin^2 2\theta \qquad r \gg \frac{\langle E_{\text{v}} \rangle}{\Delta E_{\text{v}}} L_{\text{osc}}. \tag{4.135}$$

It is also interesting to consider the production of muons by the β-decay neutrinos:

$$(A, Z) \to (A, Z+1)e^- \bar{\text{v}} \quad \text{followed by} \quad \bar{\text{v}}\,\text{p} \to \text{n}\,\mu^+. \tag{4.136}$$

[This sequence would require the acceleration of the (A, Z) to an energy sufficiently high to produce $\bar{\text{v}}$ of sufficient energy to produce muons, $m_\mu c^2 = 105\,\text{MeV}$.] The $\bar{\text{v}}_1 \to \mu^+$ amplitude has a factor $-\sin\theta$ and the $\bar{\text{v}}_2 \to \mu^+$ amplitude a factor $+\cos\theta$ so we get

$$\lambda(\bar{\text{v}} \to \mu^+) \propto$$

$$\left| -\cos\theta\sin\theta\exp\left(-\text{i}\frac{m_1^2 c^3 r}{2 E_{\text{v}} \hbar}\right) + \sin\theta\cos\theta\exp\left(-\text{i}\frac{m_2^2 c^3 r}{2 E_{\text{v}} \hbar}\right) \right|^2$$

$$= \sin^2 2\theta \sin^2\left(\frac{(m_1^2 - m_2^2)c^4 r}{4 E_{\text{v}} \hbar c}\right) \tag{4.137}$$

For $r \to 0$, the rate vanishes because the amplitudes for the two neutrinos cancel. This explains the "myth" of separate conservation of electron number and muon number. Near the decay point, antineutrinos produced in decays with electrons can only produce positrons in subsequent interactions. It was thus natural to call these antineutrinos "$\bar{\text{v}}_e$." Further from the decay point, the antineutrinos can also produce muons. One says that the original $\bar{\text{v}}_e$ have partially "oscillated" to $\bar{\text{v}}_\mu$.

Equation (4.132) can be used to determine $\sin^2 2\theta$ and $m_2^2 - m_1^2$ by measuring the rate as a function of the distance from the neutrino source. Experiments using reactor $\bar{\text{v}}$ found results consistent with no mixing for $r < 1\,\text{km}$

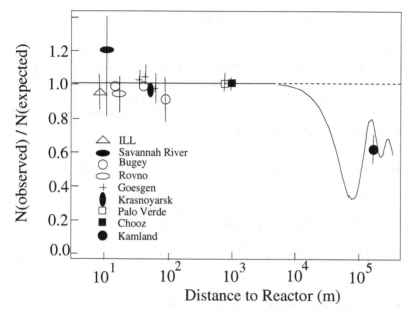

Fig. 4.26. The $\bar{\nu}_e$ detection rate vs. distance from nuclear reactors.

while the Kamland experiment [48] at $r \sim 180\,\mathrm{km}$ gave a rate of about 0.6 ± 0.1 times the no-mixing rate with an average antineutrino energy of $\sim 5\,\mathrm{MeV}$ (Fig. 4.26). Interpreting this result as a mixing between two neutrinos suggests

$$\sin^2 2\theta_{12} > 0.5 \qquad 10^{-3}\,\mathrm{eV}^2 > (m_2^2 - m_1^2)c^4 > 10^{-5}\,\mathrm{eV}^2 , \qquad (4.138)$$

where θ_{12} parametrizes the mixing between ν_e and ν_μ. Much more precise constraints come from experiments using solar neutrinos (Chap. 8). In order to correctly interpret these results, one needs to include the effects on the neutrino phase of the index of refraction of the solar medium. The totality of the results imply [48]

$$\sin^2 2\theta_{12} = 0.84 \pm 0.06 \qquad (m_2^2 - m_1^2)c^4 = (7 \pm 2) \times 10^{-5}\,\mathrm{eV}^2 . (4.139)$$

Evidence for mixing between ν_μ and ν_τ has been found using neutrinos from the decays of pions produced by cosmic rays in the Earth's atmosphere (Fig. 5.4). The dominant decay mode is $\pi^\pm \to \mu^\pm \bar{\nu}(\nu)$ so, in the absence of mixing, these neutrinos would be expected to produce muons through sequences like

$$\pi^- \to \mu^- \bar{\nu} \qquad \text{followed by} \qquad \bar{\nu}p \to n\mu^+ . \qquad (4.140)$$

The muons produced by these neutrinos scattering on protons have been observed by the Kamiokande experiment consisting of 30 kton water instrumented by photomultipliers. An example of a similar detector is illustrated in Fig. 8.14. The muons produced Cherenkov light in the water which is detected by the photomultipliers.

The Superkamiokande experiment [49] observes that neutrinos produced by cosmic rays in the atmosphere *above* the detector produce muons at the rate expected for no mixing. On the other hand, neutrinos produced in the atmosphere on the far side of the Earth interact at about half the expected rate indicating maximal mixing, $\theta \sim \pi/4$. The rate deficit is not compensated by an increased production of electrons so this result is believed to indicate mixing of ν_μ and ν_τ between ν_2 and ν_3. The mixing angles and masses are

$$\sin^2 2\theta_{23} > 0.86 \qquad (m_3^2 - m_2^2)c^4 = (3 \pm 1) \times 10^{-3}\,\mathrm{eV}^2\,, \qquad (4.141)$$

where θ_{23} parametrizes the mixing between ν_μ and ν_τ.

The two mixing angles θ_{12} and θ_{23} are both nearly 45 deg indicating that the mixing matrix (4.126) is approximately

$$U \sim \begin{pmatrix} 1/\sqrt{2} & 1/\sqrt{2} & s_{13}e^{-i\delta} \\ -1/2 & 1/2 & 1/\sqrt{2} \\ 1/2 & -1/2 & 1/\sqrt{2} \end{pmatrix}, \qquad (4.142)$$

where the third mixing angle is small: $s_{13} \ll 1$ and $c_{13} \sim 1$. It can be measured by observing $\nu_\mu \to \nu_e$ oscillations. A non-vanishing phase δ would induce CP violation so that $\nu_\mu \to \nu_e$ oscillations would not have the same rate as $\bar{\nu}_\mu \to \bar{\nu}_e$ oscillations. Efforts are underway to determine these last two parameters of the neutrino mixing matrix.

4.4.2 Quarks

The quark model. The quark model was imagined by Gell-Mann [51] in 1964. Since then, it has received considerable experimental confirmation. This model consists in constructing the hundreds of hadrons which were discovered in the years 1960–1980 with a simple set of point-like fermions, the quarks.

Just as for leptons, there exist two series of three quarks which can be classified by the mass hierarchy. For leptons we have

$q = -1:$	e^-	μ^-	τ^-	$q = +1:$	e^+	μ^+	τ^+
$q = 0:$	ν_e	ν_μ	ν_τ	$q = 0:$	$\bar{\nu}_e$	$\bar{\nu}_\mu$	$\bar{\nu}_\tau$

where ν_e, ν_μ and ν_τ are actually mixtures of ν_1, ν_2 and ν_3. For quarks we have

$q = -1/3:$	d	s	b	$q = +1/3:$	$\bar{d}$	$\bar{s}$	$\bar{b}$
$q = +2/3:$	u	c	t	$q = -2/3:$	$\bar{u}$	$\bar{c}$	$\bar{t}$

For the moment, we ignore the possibility of quark mixing so that we pair the d-quark with only the u-quark, etc.

Two types of hadrons have been observed[2]:

---

[2] Recently, evidence has been presented for the production of a particle consisting of three quarks and a quark–antiquark pair [50].

- The *baryons*, are *fermions* composed of three quarks,
- The *mesons* are *bosons* composed of a quark and an antiquark.

For example, the proton and the neutron are bound states of "ordinary," or light quarks of the type:

$$|p\rangle = |uud\rangle, \qquad |n\rangle = |udd\rangle \quad . \tag{4.143}$$

The $\pi^\pm$ mesons, of spin zero and mass $m_\pi c^2 \simeq 140$ MeV, are states :

$$|\pi^+\rangle = |u\bar{d}\rangle \quad |\pi^0\rangle = \frac{1}{\sqrt{2}}\left[|u\bar{u}\rangle + |d\bar{d}\rangle\right] \quad |\pi^-\rangle = |d\bar{u}\rangle . \tag{4.144}$$

"Strange " particles are obtained by substituting a strange quark s for a u or d quark in an ordinary hadron. For instance the Λ^0 hyperon, of mass $m_\Lambda = 1115$ MeV$/c^2$ is a state $|\Lambda^0\rangle = |uds\rangle$, to be compared with the neutron $|n\rangle = |udd\rangle$; the strange K$^-$ meson is a state $|K^-\rangle = |s\bar{u}\rangle$ to be compared with $|\bar{\pi}^-\rangle = |d\bar{u}\rangle$. Similarly, one constructs charm particles by substituting a charm quark c to the quark u. One observes, for instance, particles which have both charm and strangeness such as $|D_s^+\rangle = |c\bar{s}\rangle$, and so on.

Quarks at first appeared simply to provide a handy classification scheme for the observed hadrons. The experimental indication of the actual "existence" of quarks was given in 1969. In "deep inelastic scattering " of electrons on nucleons, we saw in Sect. 3.4.2 that the scattering cross-section is directly related to the elementary cross-section on the constituent particles, i.e. the quarks. As it turns out, the observed electron–quark scattering cross-section corresponds to that of scattering on point-like spin 1/2 particles of charges 2/3 and −1/3.

With our present methods of investigation, quarks present neither an internal structure nor excited states. The charm quark was discovered in 1974. The bottom quark b was discovered shortly after in 1976. The top quark t was only discovered in 1995, because of its high mass.

The original idea of Gell-Mann was to put some order in the hundreds of hadrons that were continually discovered. It is remarkable that this is possible with such a minimal set of elementary constituents.

Basic Structure of the Model.

1. Quarks have spin 1/2. This is necessary since some hadrons have half-integer spins. One cannot construct a half-integer spin starting from integer spins, but the reverse is possible.
2. Baryons are bound states of three quarks $|q_1 q_2 q_3\rangle$ (antibaryons of three antiquarks). Therefore, the baryon number of quarks is $B = 1/3$ (the baryon number of antiquarks is -1/3). Baryonic spectra correspond to the three-body spectra of $(q_1 q_2 q_3)$ systems.
3. Mesons are quark–antiquark states $|q_1 \bar{q}_2\rangle$. The meson spectrum is the two-body spectrum of $(q_1 \bar{q}_2)$ systems.
4. Quark electric charges are fractional: $+2/3$ or $-1/3$ (antiquarks: $-2/3$ or $+1/3$).

5. In addition to electric charge and baryon number, other additive quantum numbers are conserved in strong interactions: strangeness s, charm c, beauty b and top t. One therefore attributes a *flavor* to each quark. This is an additive quantum number which is conserved in strong interactions (an antiquark has the opposite flavor). The flavor of a hadron is the sum of flavors of its constituent quarks.

6. The u and d quarks are called "usual " quarks. They are the constituents of the protons and neutrons and are therefore the only quarks present in matter surrounding us. One can convince oneself that the conservation of the u and d flavors amounts simply to the conservation of electric charge.

7. Quarks are *confined* in hadrons. The observable free (asymptotic) states are baryons (3 quarks states) and mesons (quark–antiquark states). In fact, the interaction that binds quarks corresponds to a potential which increases linearly at large distances. If one attempts to extract a quark from a hadron, it is necessary to provide an amount of energy which, at a certain point, will be transformed into quark–antiquark pairs. These pairs rearrange themselves with the initial quarks to materialize the energy in the form of new hadrons. (In some sense, this phenomenon is similar to what happens if one tries to separate the north and south poles of a magnet.)

8. "Bare " masses [3] of quarks are

$$m_{\rm d} \simeq 7.5\,{\rm MeV} \sim m_{\rm u} \simeq 4.2\,{\rm MeV} \ll m_{\rm s} \simeq 150\,{\rm MeV}$$

$$\ll m_{\rm c} \simeq 1.5\,{\rm GeV} \ll m_{\rm b} \simeq 4.2\,GeV \ll m_{\rm t} \simeq 175\,{\rm GeV}.$$

Color . Among the excited states of the proton and neutron, a "resonance" has been known since the 1950's. it is called the Δ of mass $m = 1232$ MeV and spin-parity $J^P = 3/2^+$. It is an isospin quadruplet $T = 3/2$, the Δ^{++}, the Δ^+, the Δ^0 and the Δ^- of similar masses (between 1230 and 1234 MeV).

In the quark model, the quark content of these particles is simple:

$$|\Delta^{++}\rangle = |{\rm uuu}\rangle, \quad |\Delta^+\rangle = |{\rm uud}\rangle \quad |\Delta^0\rangle = |{\rm udd}\rangle \quad |\Delta^-\rangle = |{\rm ddd}\rangle \quad .$$

But this is a catastrophe! In this construction, the Pauli principle is violated.

In fact, the states Δ are spin excitations of the nucleons. They are the ground states of three u or d quarks in the total spin state $J^P = 3/2^+$. Unless a serious pathology occurs in the dynamics, the ground state of a three particle system has a zero total orbital angular momentum, and, likewise, all the relative two-body orbital angular momenta are zero. Therefore, the spatial wavefunction of the three quarks is *symmetric*, and in order to ensure that the spin of the three quark system be $J^P = 3/2^+$, the three spins must be aligned, i.e. the total spin state must be symmetric.

---

[3] Since quarks are permanently confined in hadrons, it is not possible to observe the mass of a free quark. We use a terminology of solid-state physics which corresponds to the fact that the particles are permanently interacting with their environment.

Consider the states $|\Delta^{++}\rangle = |uuu\rangle$ or $|\Delta^{-}\rangle = |ddd\rangle$. These are identical particle states whose spatial wavefunctions and spin wavefunctions are both symmetric. This violates the Pauli Principle.

Faced with such a situation, there were three possibilities:

- forget about the quark model because it was inconsistent,
- forget about the Pauli principle,
- give new degrees of freedom to quarks so as to make the wavefunctions antisymmetric.

The third possibility was realized in 1964 by Greenberg. His solution to the problem consists in assuming that there exists a new quantum number exactly conserved by all interactions, called *color*. At first, this hypothesis could appear as an artificial way to satisfy the Pauli Principle, but it turned out to be the fundamental concept of strong interactions and it lead to the theory of *Quantum Chromodynamics*.

One assumes that each quark exists in three forms of color, say blue, green and red. (An antiquark has the complementary color of the corresponding quark.) The color state of any three quark state is *totally antisymmetric*, i.e.

$$|\psi_{qqq}\rangle = \frac{1}{\sqrt{6}} \left[|bgr\rangle + |grb\rangle + |rbg\rangle - |brg\rangle - |rgb\rangle - |gbr\rangle\right] \qquad (4.145)$$

Returning to the Δ^{++} and Δ^{-}, it is clear that the Pauli Principle is now satisfied.

As a general fact, a fundamental property of color is that it is not observable in physical states. Nevertheless, its effects are observable, as we shall see.

The formal theory of color is performed within an *exact* symmetry of fundamental interactions, based on the Lie group $SU(3)$.

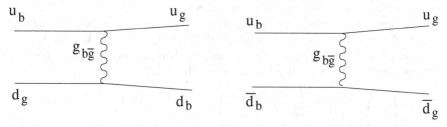

Fig. 4.27. An example of gluon exchange between two quarks (left) and between a quark and antiquark (right).

Quarks interact by the exchange of *gluons*, massless particles that are chargeless but carry one color and one anticolor, e.g. $b\bar{r}$. Figure 4.27 shows schematically how gluons are exchanged.

In quantum field theory, it is possible to show that it is the fact that the gluons carry a charge, i.e. color, that leads to confinement, i.e. to a long-range force that that grows linearly with separation. In electromagnetism, the exchanged photons do not themselves carry charge and the force is not confining. It is this confinement that prevents a free quark from emerging in electron deep-inelastic scattering as shown in Fig. 3.25.

4.4.3 Quark mixing and weak interactions

It is tempting to think that the classifications of leptons and of quarks are parallel from the point of view of weak interactions. One observes a similar mass hierarchy with two sequences of three fermions separated by one unit of charge. In analogy with the lepton families (4.123), we then write three quark families as

$$\begin{pmatrix} u \\ d' \end{pmatrix} \quad \begin{pmatrix} c \\ s' \end{pmatrix} \quad \begin{pmatrix} t \\ b' \end{pmatrix} \tag{4.146}$$

In the absence of mixing, we would just have $d' = d$, $s' = s$ and $b' = b$. The W could induce only the transmutations $u \leftrightarrow d$, $c \leftrightarrow s$ or $t \leftrightarrow b$. But then, strange particles would be *stable*, since no $s \to u$ would exist. Unlike lepton mixing which is manifested only in subtle effects in neutrino oscillations, quark mixing is responsible for the instability of otherwise stable particles.

If the strange and/or charm particles were stable, then we would be facing several coexisting nuclear worlds. In addition to usual nuclei, we would see a whole series of other nuclear species, similar but heavier. For instance, in any nucleus, one could replace the neutron by the Λ^0 hyperon, which is a (u, d, s) state of mass $m_{\Lambda^0} = 1116 \, \text{MeV}/c^2$. Furthermore, since the Pauli principle does not constrain the neutron and the Λ^0, the usual nuclei would possess a larger number of heavy isotopes, since the Λ^0's can sit on the same shells as the neutrons (in a shell model for instance).

In order to destabilize strange particles, we must have quark mixing such that d', s' and b' are linear combinations of d, s and b. Generally, it is supposed that the transformation is unitary (like the corresponding neutrino transformation) so we write in analogy with (4.124)

$$\begin{pmatrix} d' \\ s' \\ b' \end{pmatrix} = \begin{pmatrix} U_{dd} & U_{ds} & U_{db} \\ U_{sd} & U_{ss} & U_{sb} \\ U_{bd} & U_{bs} & U_{bb} \end{pmatrix} \begin{pmatrix} d \\ s \\ b \end{pmatrix} ,$$

where the unitary matrix can be put in the form (4.126). (The three quark mixing angles and phase have, *a priori*, nothing to do with the neutrino mixing angles.)

The elements of the mixing matrix can be determined from decay rates. Consider the decays

$$\text{n} \to \text{p} \, \text{e}^- \, \bar{\nu}_i \quad (i = 1, 2, 3) \qquad \lambda = 1.1 \times 10^{-3} \, \text{s}^{-1} , \tag{4.147}$$

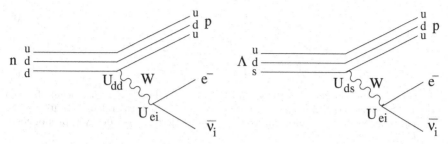

Fig. 4.28. Neutron decay and Λ^0 decay. Both can be thought of as the decay of a quark in the presence of two spectator quarks. Both decays contain elements of the quark and neutrino mixing matrices.

$$\Lambda^0 \rightarrow pe^-\bar{\nu}_i \quad (i = 1, 2, 3) \qquad \lambda = 3.2 \times 10^6\,s^{-1}. \tag{4.148}$$

Just as in nuclear physics where the β-decay of nuclei can be interpreted in terms of the β-decay on the constituent nucleons, the decay of hadrons can be interpreted in term of quark decay, as shown in Fig. 4.28. The quark decays corresponding to neutron and Λ^0 decay are

$$d \rightarrow u e^-\bar{\nu}_i \quad (i = 1, 2, 3), \tag{4.149}$$

$$s \rightarrow u e^-\bar{\nu}_i \quad (i = 1, 2, 3). \tag{4.150}$$

The decay rates, summed over the three neutrino species, are proportional to the squares of elements of the mixing matrix

$$\lambda(n \rightarrow pe^-\bar{\nu}) \propto \sum_{i=1}^{3} |U_{dd}U_{ei}|^2 = |U_{dd}|^2, \tag{4.151}$$

$$\lambda(\Lambda^0 \rightarrow pe^-\bar{\nu}) \propto \sum_{i=1}^{3} |U_{ds}U_{ei}|^2 = |U_{ds}|^2, \tag{4.152}$$

From the ratio of the two decay rates, it is thus possible to determine the ratio $U_{ds}/U_{dd} = \tan\theta_{12}$. Of course, we cannot simply equate this to the rate ratio because of other factors in the rates. In studying neutron decay, we saw that phase space acts as the fifth power of the maximum energy of the final electron. Now, for Λ^0 decay we have $p_{max} = 163\,MeV/c$, whereas for neutron decay $p_{max} = 1.2\,MeV/c$. If we define $a = \lambda/p_{max}^5$, we obtain

$$\frac{a(\Lambda^0 \rightarrow p + e^- + \bar{\nu}_e)}{a(n \rightarrow p + e^- + \bar{\nu}_e)} \sim 0.06 \sim \tan^2\theta_{12}. \tag{4.153}$$

We see that the s quark transforms much more weakly to a u than a d quark transforms to a u quark. Confirmation of this comes by comparing the rates of $K^+ \rightarrow \mu^+\nu$ or $K^- \rightarrow \mu^-\bar{\nu}$ with $\pi^+ \rightarrow \mu^+\nu$ or $\pi^- \rightarrow \mu^-\bar{\nu}$, Fig. 4.29. The rates of these two weak decays should also be in the ratio $\tan^2\theta_{12}$ once phase space is taken into account.

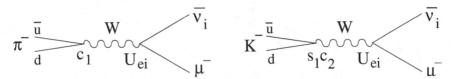

Fig. 4.29. The decays $\pi^- \to \mu^- \nu_i$ and $K^- \to \mu^- \bar{\nu}_i$. The two decays occur through quark–antiquark annihilation by creation of a W^- which then decays to $\mu^- \bar{\nu}_i$. The rate of pion decay is proportional to $|U_{dd}|^2$ while that of kaon decay is proportional to $|U_{ds}|^2$.

The idea of quark mixing was proposed in 1964 by N. Cabibbo. At that time, only normal and strange hadron where known to exist. Cabibbo suggested that the W couple the u quark to a linear combination of the d and s quarks:

$$|d'\rangle = \cos\theta_c\, |d\rangle + \sin\theta_c\, |s\rangle \qquad (4.154)$$

where the value of the angle θ_c, now called the Cabibbo angle, is $\theta_c \simeq 13.1°$. The Cabibbo theory correctly predicts the decay rates of the neutron, Λ^0, pions and kaons as well as the "hyperons" like Σ^+ (uus, $m_\Sigma = 1189\,\mathrm{MeV}/c^2$). The rates of these decays are indicated on Figs. 4.13.

With the discovery of the c, b, and t quarks, we now know that the quark mixing is a bit more complicated with the three families (4.146) mixed according to the Kobayashi–Maskawa–Cabibbo scheme (4.4.3).

The moduli of the coefficients of this matrix are now known to be in the ranges [1]:

$$\begin{pmatrix} 0.9742 \text{ to } 0.9757 & 0.219 \text{ to } 0.226 & 0.002 \text{ to } 0.005 \\ 0.219 \text{ to } 0.225 & 0.9734 \text{ to } 0.9749 & 0.037 \text{ to } 0.043 \\ 0.004 \text{ to } 0.014 & 0.035 \text{ to } 0.043 & 0.9990 \text{ to } 0.9993 \end{pmatrix} \qquad (4.155)$$

Note that, unlike the case of neutrino mixing, the diagonal elements are all near unity so the mixing angles are all small.

It is important that the non-zero value of the phase δ implies automatically the violation of CP symmetry (product of a charge conjugation and a parity reversal). This is not of just academic interest since we will see in Chap. 9 that, in the absence of CP violation, one would expect equal amounts of matter and antimatter in the Universe, in conflict with the observation that there is very little antimatter in the observable Universe. Note that a general 2-dimensional unitary mixing matrix has no complex phase that cannot be eliminated by a suitable redefinition of the base states. Only with at least three families is it possible to induce CP violation.

Unfortunately, it does not appear that CP violation in the quark mixing matrix is directly connected with the CP violation that generated the cosmological matter-antimatter asymmetry. Nevertheless, the 3-family-CP-violation connection emphasizes that important effects can come from the existence of particles that do not normally appear in nature. To this extent,

CP violation provides a clue to the answer to Rabi's question about the third generation, "Who ordered that?"

The second question (and there the theory is silent at present) is why this hierarchy of masses? In each class one has $m_b \gg m_s \gg m_d$, $m_t \gg m_c \gg m_u$, $m_\tau \gg m_\mu \gg m_e$. This is one of the big enigmas of present theoretical physics, as is, in general, the question of the origin of mass of particles. This is directly related to the third question: why these values of the Cabibbo–Kobayashi–Maskawa angles?

4.4.4 Electro-weak unification

Nuclear β-decay is an example of a reaction due to *charged-current* weak interactions, by which we mean they are due to exchange of the charged $W^\pm$. We summarize the $W^\pm$ couplings symbolically by

$$|W^-\rangle \to g\left(|e\bar{\nu}_e\rangle + |\mu\bar{\nu}_\mu\rangle + |d'\bar{u}\rangle + |s'\bar{c}\rangle\right) \tag{4.156}$$

Each of the terms of the right-hand-side is a decay channel of the W^- but they are meant to be more general than that. For instance the W^- decays to $e^-\bar{\nu}_e$ but it can also be emitted by a $e^- \to \nu_e$ transition. The ν_e, ν_μ, d' and s' are the linear combinations of the physical particles defined by (4.124) and (4.4.3). (To streamline the formulas, in this section we will ignore the third generation of leptons and quarks.) The dimensionless coupling constant g gives the strength of the coupling and is related to the W mass and to the Fermi constant by

$$\left(\frac{g}{m_W c^2}\right)^2 = 4\sqrt{2}\,\frac{G_F}{(\hbar c)^3} \quad \Rightarrow \quad \frac{g^2}{4\pi} = \frac{1}{29.46} \quad . \tag{4.157}$$

Similarly for the W^+ if we replace fermions by their antiparticles.

In the same manner, in electromagnetic interactions, the photon is universally coupled to the electric charge and, using the same notation,

$$|\gamma\rangle \to \sqrt{4\pi\alpha}\left(-|e^-e^+\rangle + (2/3)|u\bar{u}\rangle - (1/3)|d\bar{d}\rangle\right) \tag{4.158}$$

where α is the fine structure constant. We see from (4.156) that $g^2/4\pi$ is the weak charged current equivalent of the electromagnetic fine-structure constant. The fact that the two fine-structure constants are of the same order of magnitude suggests they have something to do with each other, as indeed they do.

Until 1973, one had only observed charged current weak interactions due to the exchange of the massive $W^\pm$ and the electromagnetic interactions due to the exchange of the massless photon. In 1973 it was discovered that there were also *neutral current* weak interactions due to the exchange of the massive Z^0. As we will now see, the existence of such interactions had been expected from symmetry considerations.

Weak Isospin. From the point of view of weak interactions, quarks as well as leptons appear as *weak-isospin* doublets that are mathematically equivalent to the strong-isospin doublet formed from the u and d quarks, seen in Chap. 1. The doublets (v_e, e), (u, d'), $(e^+, \bar{v}_e)$, (v_μ, μ) etc.

Consider, for instance, the doublets (v_e, e) and $(e^+, \bar{v}_e)$. Within this assumption, we can construct a weak isospin triplet $|\psi_m^1\rangle$ and a singlet $|\psi_0^0\rangle$ as follows [4]

$$\begin{cases} |\psi_{+1}^1\rangle = |v_e e^+\rangle \\ |\psi_0^1\rangle = (|v_e \bar{v}_e\rangle - |e^- e^+\rangle)/\sqrt{2} \\ |\psi_{-1}^1\rangle = |e^- \bar{v}_e\rangle \end{cases} \tag{4.159}$$

$$|\psi_0^0\rangle = (|v_e \bar{v}_e\rangle + |e^- e^+\rangle)/\sqrt{2} \ . \tag{4.160}$$

Note from (4.156) that the W^- couples to $|\psi_{-1}^1\rangle$ and that the W^+ couples to $|\psi_{+1}^1\rangle$. This suggests that we introduce a W^0 couples to $|\psi_0^1\rangle$ so that the situation is completely symmetric:

$$|W^m\rangle \to g|\psi_m^1\rangle \qquad m = +1, 0, -1 \ . \tag{4.161}$$

This relation can be extended to the other weak-isospin doublets: $(v_\mu \mu^+)$, $(d'u)$ and $(s'c)$.

Because of the triplet of W bosons, its relativistic gauge theory is said to be based on a non-abelian group SU(2), the weak isospin.

Weak Hypercharge. One might have thought that the photon could be identified with the W^0 but this is not possible, if only because it couples to the chargeless neutrino. To find the photon, we must first introduce another neutral particle, the B^0 that couples to the, as yet unused $|\psi_0^0\rangle$ in (4.160). We choose to make the coupling proportional to a new universal coupling constant g' and to the *weak hypercharge* of the doublet in question:

$$Y = 2(Q - T_3) \ . \tag{4.162}$$

The weak-hypercharge is equal to -1 for leptons and to 2/3 for quarks (opposite values for the antiparticles).

The B^0 couplings are then

$$|B^0\rangle \to g'[-\frac{1}{\sqrt{2}}(|v_e \bar{v}_e\rangle + |e^- e^+\rangle) + (2/3)\frac{1}{\sqrt{2}}(|u\bar{u}\rangle + |d\bar{d}\rangle) + \ldots] \tag{4.163}$$

Because there is only one B^0, the quantum field theory of the B^0 field is a gauge theory on an abelian group $U(1)$ analogous to electromagnetism.

---

[4] The difference in signs of the $m = 0$ combinations compared to previous expressions comes from the fact that we are dealing with fermion–antifermion systems instead of fermion–fermion systems.

The Glashow–Weinberg–Salam Mechanism. Like the W^0, the B^0 cannot be the photon since the coupling (4.163) is not proportional to the charge. In particular, it couples to the neutrino. It is, however, simple to find a linear combination of the two neutral bosons that *does* have couplings proportional to the charge: Let us therefore define :

$$\left\{ \begin{aligned} |Z^0\rangle &= \cos\theta_w |W^0\rangle - \sin\theta_w |B^0\rangle \\ |\gamma\rangle &= \sin\theta_w |W^0\rangle + \sin\theta_w |B^0\rangle \end{aligned} \right. \tag{4.164}$$

where θ_w is called the weak-mixing angle or also the "Weinberg angle."

The photon's couplings are proportional to charge if we take

$$\sin\theta_w = \frac{g'}{\sqrt{g^2 + g'^2}} \qquad \cos\theta_w = \frac{g}{\sqrt{g^2 + g'^2}} . \tag{4.165}$$

One can check easily, restricting to the doublet (e, ν_e) doublet that

$$|\gamma\rangle \rightarrow -\frac{gg'}{\sqrt{g^2 + g'^2}}(|\psi_0^1\rangle - |\psi_0^0\rangle)/\sqrt{2} = -\frac{gg'}{\sqrt{g^2 + g'^2}}|e^- e^+\rangle \quad , \tag{4.166}$$

i.e. the coupling to $\nu_e \bar{\nu}_e$ vanishes as expected for a neutral particle. This gives us a relation with the fine structure constant

$$\frac{gg'}{\sqrt{g^2 + g'^2}} = g\sin\theta_w = \sqrt{4\pi\alpha} , \qquad \Rightarrow \frac{g'^2}{4\pi} = \frac{1}{107.5} . \tag{4.167}$$

It can be verified that the couplings of the photon to the quarks are proportional to their charges.

We also need a way to ensure that the photon mass vanishes. This can be done by considering mass-squared matrix [5] in the basis $\{|W^0\rangle, |B^0\rangle\}$

$$\mathcal{M} = \begin{pmatrix} m_W^2 & -A \\ -A & m_B^2 \end{pmatrix} \tag{4.168}$$

where A characterizes the $W^0 \rightleftharpoons B^0$ transition which is responsible for the mixing.

The diagonalization of this matrix is straightforward. If one wants the photon mass to vanish, we need to take $A = m_W m_B$ which yields by identifying the eigenvectors of $\mathcal{M}$ with the expressions (4.164):

$$m_\gamma = 0, \qquad m_Z^2 = m_W^2 + m_B^2 = \frac{m_W^2}{\cos^2\theta_w} , \qquad \cos\theta_w = \frac{m_W}{m_Z} . \tag{4.169}$$

---

[5] Why the masses *squared* and not the masses themselves? This is because bosons, such as the W and B, satisfy the Klein–Gordon equation which we mentioned in Sect. 1.4 (1.63), which involves the squares of the masses and not the masses themselves. If one incorporates an interaction, this feature will remain. For fermions, on the contrary, the Dirac equation is of first order and it is directly on the mass that one operates.

Bringing together all these results and using the values of the fine-structure constant and G_F, we obtain the following masses as a function of the weak-mixing angle θ_w[6]

$$m_W c^2 = \frac{38.7 \text{ GeV}}{\sin \theta_w} \qquad m_{Z_0} = \frac{m_W}{\cos \theta_w} \qquad (4.170)$$

$$\cos^2 \theta_w + 4\pi\alpha/g^2 = 1 \quad . \qquad (4.171)$$

Originally, θ_w was a free parameter of the Glashow, Weinberg and Salam model. It is now known to good precision:

$$\sin^2 \theta_w = 0.2237 \pm 0.0008 \quad , \qquad (4.172)$$

$$m_W c^2 = 80.33 \pm 0.15 \text{ GeV} , \qquad (4.173)$$

$$m_Z c^2 = 91.187 \pm 0.007 \text{ GeV} \quad . \qquad (4.174)$$

We should emphasize that the question of the mass of the W and Z is very delicate. In the relativistic formulation of the theory of Glashow, Weinberg and Salam, a consistent quantum field gauge theory is constructed on a non-abelian group. This can only be done if the initial masses of the four states $|W^+\rangle$, $|W^0\rangle$, $|W^-\rangle$ and $|B^0\rangle$ are all identically zero. They acquire non-vanishing masses m_W and m_B by a spontaneous symmetry breaking mechanism where *Higgs bosons* play a crucial role. The fact that, in the end, the photon is a zero-mass particle is a *natural* automatic consequence of the theory, and not a condition imposed on the values of parameters as we have done above. The Higgs bosons is the last particles of the standard model that is yet to be observed. Its couplings to the other particles are responsible for the particle masses. An intense experimental activity is underway in order to identify them.

We have ignored the question of helicity in our discussion of electro-weak theory. In fact the theory is constructed so that the charged current interactions maximally violate parity (as required by the observation that neutrinos are 100% polarized) while the electromagnetic interactions conserve parity.

We see that the theory unifies electromagnetic and weak interactions. It predicts the existence of weak interactions which do not change the electric charge. These interactions are mediated by the neutral Z^0 boson. An example of such a reaction is

$$\nu\, p \rightarrow \nu\, p \, . \qquad (4.175)$$

It is illustrated in Fig. 4.30

The theory of neutral currents also had a role in the prediction of the charm quark. At the time when the electroweak theory appeared, neutral currents were unknown and only 3 quarks had been discovered, i.e. the u

---

[6] The first relation is obtained after taking into account various radiative corrections. Otherwise one obtains $m_W c^2 = 37.3 \text{ GeV}/\sin \theta_w$.

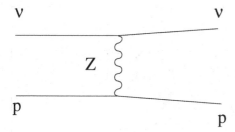

ν   ν

Z

p

    p

Fig. 4.30. Neutrino-proton elastic scattering mediated by the exchange of the Z^0 boson.

the d and the strange quark s. In 1970, in investigating neutral current reactions, Glashow, Iliopoulos and Maiani were led to invent a mechanism which predicted the existence of charm more than 4 years before its discovery.

As we have written the theory with the weak mixing of the quarks via a unitary matrix (4.4.3), the Z^0 does not have "non-diagonal" couplings to things like $\bar{d}s$. In the absence of the unitary mixing, it is readily shown that the Z^0 would have such couplings. In this case, the K^0 would decay to $\mu^+\mu^-$ through $\bar{d}s \to Z^0 \to \mu^+\mu^-$. However, the upper experimental limits on the rate of this reaction are so low that one is sure that it is excluded up to second order perturbation theory (to which the exchange of charged currents W^+ and W^- contributes in principle).

One should keep in mind the following aspects of neutral current weak interactions of quarks.

1. The couplings of the Z^0 are diagonal. This boson couples to $(u\bar{u})$, $(d\bar{d})$, $(s\bar{s})$, but not to $(d\bar{s})$, etc.
2. The coupling constants have a form similar to the lepton couplings.
3. The partial decay width of the Z^0 into 2 fermions $f\bar{f}$ is given by

$$\Gamma(Z^0 \to f\bar{f}) = \frac{g_Z^2 m_Z c^2}{48\pi}(|c_V|^2 + |c_A|^2) .$$

4. The constants $|c_V|^2$ and $|c_A|^2$ are such that one has
   - for neutrinos : $|c_V|^2 + |c_A|^2 = 1/2$
   - for e, μ, τ : $|c_V|^2 + |c_A|^2 = (4\sin^4\theta_w + (2\sin^2\theta_w - 1)^2)/2 \sim 1/4$
   - for u, c, t : $|c_V|^2 + |c_A|^2 = 1/4 + (1/2 - 4/3\sin^2\theta_w)^2 \sim 0.29$
   - for d, s, b : $|c_V|^2 + |c_A|^2 = 1/4 + (1/2 - 2/3\sin^2\theta_w)^2 \sim 0.37$
   where θ_w is the Weinberg angle.

This means that the branching ratio is 6% for each neutrino species , it is 14% for d,s,b quarks (taking into account the color of quarks which multiplies by a factor of 3 the contribution of each quark flavor).

Altogether, one obtains the estimate

$$\Gamma_{Z^0} \sim 2.5\,\text{GeV} . \tag{4.176}$$

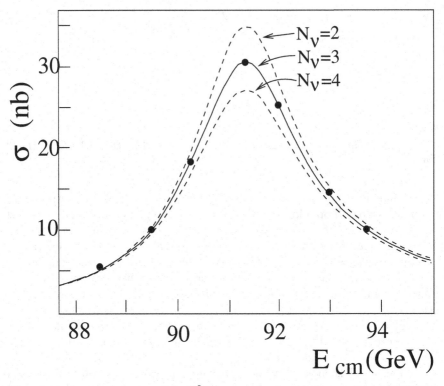

Fig. 4.31. Resonance curve of the Z^0 measured by the Aleph experiment at CERN in e^+e^- collisions at the LEP colliding ring. The three curves represent the theoretical predictions according to the number of different neutrino species. (Courtesy Aleph Collaboration.)

We note that the existence of another, new, neutrino would increase this width by 167 MeV, that of a new family would increase it by a factor of 4/3, that of a neutrino and its lepton, assuming the quarks are too heavy, by 264 MeV.

We therefore understand the interest of measuring the Z^0 width in order to count the number of different fermions whose masses are smaller than $m_Z/2$.

The answer obtained in CERN with steadily increasing accuracy since 1989 is

$$\Gamma_{Z^0} = 2.4963 \pm 0.0032 \,\text{GeV} \quad \text{i.e.} \quad N_\nu = 2.991 \pm 0.016.$$

In other words, there are three families of quarks and leptons and *only* three (neutrinos lighter that the Z^0, of course).

Originally, quarks were a simple way to classify hundreds of hadrons. The idea got enriched with the construction of quantitative dynamical models of hadron spectroscopy.

With deep inelastic scattering, jets in e + e− collisions, the explanation of the magnetic moments of hadrons, i.e. by studying the electromagnetic interactions of quarks, these acquired a physical reality and their elementarity was established: they are point-like particles. Then, using weak interactions, one discovers a fascinating symmetry between quarks and leptons which is suggestive of a possible underlying unification, which would be more general and more profound, of the constituents of matter and their fundamental interactions. It is one of the great hopes of present theoretical physics to construct a complete and unified theory of all interactions including gravitation. Superstring theory and the reduction of all corresponding theories to a single unique "M-Theory " is a very active field of research at present.

4.5 Bibliography

1. F. Halzen and A.D. Martin: *Quarks and Leptons* (Wiley, New York, 1984)
2. W. Greiner and B. Mïler:*Gauge Theory of Weak Interactions* (Springer, Berlin, 2000)
3. U. Mosel: *Fields, Symmetries, and Quarks* (Springer, Berlin, 1999)
4. Q. Ho-Kim and X.-Y. Pham:*Elementary Particles and Their Interactions* (Springer, Berlin, 1998)
5. *Introduction to Elementary Particles*, D. Griffiths, (Harper & Row, New York, 1987).

Exercises

4.1 Discuss the electron energy spectrum from the 2β-decay of ^{100}Mo shown in Fig. 4.1. The average total energy taken by the 2 electrons in ~ 1.2 MeV. How much energy on average is taken by the two neutrinos?

4.2 Charged particles lose energy in when traversing a material by ionization. The minimum rate of energy loss (applicable to relativistic particles) is ~ 2 MeV g^{-1}cm^2. (Multiplying by the density of the medium gives the energy loss per unit distance in MeV cm^{-1}.) Consider the experiment shown in Fig. 4.1. Estimate the energy loss by the decay electrons before leaving the foil. Does this energy loss significantly affect the energy measurement by the scintillators?

4.3 Discuss the decay scheme shown in Fig. 4.2. What factors determine the branching ratios of β-decay to the ground and first excited states of ^{170}Yb? Estimate the internal conversion fraction for the decay of the 84 keV state by using (4.61). Is the use of the formula justified?

4.4 Consider the photon spectra shown in Fig. 4.3. From the maximum Doppler shift of forward going photons, estimate the velocity and energy of the decaying ^{74}Br nuclei. Use the Bethe–Bloch formula (5.32) to estimate the energy-loss rate of Br ions in nickel and the time for the nuclei to come to rest. Estimate the lifetime of the decaying state. What are the possible criticisms of this estimate?

4.5 A commonly used γ-source is ^{60}Co which β-decays to an excited state of ^{60}Ni as in Fig. 4.32. Use the indicated spins and parities to explain why ^{60}Co does not decay directly to the ground state of ^{60}Ni. Explain why two photons are emitted in the de-excitation of ^{60}Ni. Estimate the time delay between the β-decay and the photon emissions.

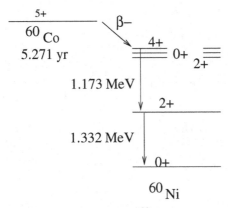

Fig. 4.32. The decay of ^{60}Co.

4.6 What energy photons are emitted after each of the five β-decays shown in Fig. 4.12?

4.7 Estimate the energy of the k-capture Auger electrons emitted after the electron-capture decay of ^{71}Ge (Fig. 4.16) by assuming that the inner most electrons occupy hydrogen-like orbitals.

4.8 Verify the β spectrum in the case of a non-zero neutrino mass (4.101).

4.9 Consider the decay scheme in Fig. 4.19. Explain the large difference in $t_{1/2}$ for ^{152}Eu and the isomer ^{152m}Eu. Calculate the kinetic energy of the ^{152}Sm nucleus after the electron-capture decay. What is the energy of the

"961 keV" photons that happen to be emitted in the direction opposite to that of the neutrino? Do they have sufficient energy to re-excite ^{152}Sm nuclei in the ground state?

4.10 Consider the neutrino event illustrated in Fig. 4.20. The final-state neutron takes only a small amount of the neutrino energy. Estimate the maximum neutron energy corresponding to positrons emitted opposite to the direction of the neutrino. What is the cross-section for such neutrons to elastically scatter on protons (Fig. 3.4). What is the neutron mean-free path in the scintillator (CH, $\rho \sim 1\,\mathrm{g\,cm}^{-2}$)? Using the fact that a neutron loses on average half of its kinetic energy in a collision with a proton, estimate the number of collisions necessary for the neutron to be thermalized. After thermalization, what is the neutron capture cross-section on hydrogen (Fig. 3.4)? What is the mean time be for the neutron is captured?

4.11 In Fig. 4.20, what is the mean-free path of the photons in the scintillator (CH) assuming that the photons interact mostly via Compton scattering on electrons, $\sigma \sim \sigma_\mathrm{T}$.

5. Radioactivity and all that

Our galaxy's stock of heavy elements was mostly produced in supernovae explosions that eject stable and unstable nuclei into the interstellar medium. The interstellar clouds containing these elements may later condense to form stellar systems. By the time this happens, most of the short-lived nuclei would have decayed, so planets contain, at their birth, only nuclei with lifetimes greater than, say, 10^6 yr. On Earth, $\sim 4.5 \times 10^9$ yr after its formation, most of the original unstable nuclei have decayed leaving only those with $t_{1/2} > 10^8$ yr. There is also continuous production of unstable nuclei by cosmic-rays entering the Earth's atmosphere. The α and β decays of these nuclei constitute the natural *radioactivity* that was discovered by Becquerel at the end of the nineteenth century.

All living species have evolved in this bath of radiation. What is new on Earth is the existence of an artificial radioactivity due to the development of nuclear technologies, most importantly energy production by nuclear reactors. There are now also numerous uses of radioactive nuclei, most notably in medical treatment and dating.

With these developments, public health concerns have naturally arisen because of the effect of ionizing radiation on living tissue. Since natural radioactivity is present everywhere, it should be emphasized that the important questions concern keeping the ambient radioactivity at a level not much above its natural level. This obviously requires confinement of the large amounts of radioactive elements produced in nuclear reactors.

In this chapter, we will first discuss the natural sources of radioactivity and the production of radioactive elements in neutron and charged-particle beams. We will then discuss how high-energy particles lose energy while traversing ordinary matter. Particle detectors based on the energy-loss mechanisms will be briefly discussed. Finally, we discuss a few applications of radioactivity.

5.1 Generalities

Most nuclei are inherently unstable, being capable of decaying to other nuclei by β-decay, α-decay, or by spontaneous fission. The set of all possible decays is referred to globally as *radioactivity*.

Let τ be the (mean) lifetime of a nuclear species. The *half-life* $t_{1/2}$ is defined by

$$t_{1/2} \equiv \tau \log 2 \sim 0.693\tau \ . \tag{5.1}$$

The name "half-life" refers to the fact that the probability that the nucleus has decayed after $t_{1/2}$ is equal to $1/2$. Nuclear physicists traditionally use half-lives, whereas particle physicists use (mean) lifetimes. The two definitions differ by a factor of $\ln 2 = 0.693$. (We hope that the trivial mistakes in this book that come from forgetting this factor will disappear in the editing process.)

The *activity* of a radioactive sample is the number of decays per unit time. For a source containing N particles the activity is

$$A = N\lambda = N/\tau \ . \tag{5.2}$$

If one knows A and τ, one can deduce the concentration N of the radioactive product in the sample.

Activities are measured in Becquerel:

$$1\,\mathrm{Bq} \ = \ 1\,(\mathrm{decay})\,\mathrm{s}^{-1} \ . \tag{5.3}$$

The previously used unit was the Curie : 1 Curie $= 3.7 \times 10^{10}\,\mathrm{Bq}$. This corresponds to the activity of $1\,\mathrm{g}$ of the radium isotope $^{226}\mathrm{Ra}$.

According to the nature of emitted particles, α particles, $e^{\pm}$, or photons, one distinguishes three types of radioactivity: α, β and γ. They were previously described theoretically in Chap. 4. The types are summarized in Table. 5.1.

5.2 Sources of radioactivity

At the end of the 1930s, one made the distinction between two classes of radioactivity, natural and artificial. Natural radioactivity was discovered accidentally by Henri Becquerel in 1896 who observed that photographic plates (unexposed to light) are blackened when placed next to uranium sulfide crystals. After some time, the primary particles responsible for the blackening where identified as $^4\mathrm{He}$ ions from a chain of decays starting with

$$^{238}\mathrm{U} \ \rightarrow \ ^{234}\mathrm{Th}\,\alpha \qquad t_{1/2} \simeq 4.468 \times 10^9\,\mathrm{yr} \ . \tag{5.4}$$

Artificial radioactivity coming from elements produced in laboratory reactions was discovered by Frédéric and Irène Joliot-Curie in 1934 who observed

$$\alpha\,^{27}\mathrm{Al} \ \rightarrow \ \mathrm{n}\,^{30}\mathrm{P} \tag{5.5}$$

followed by

$$^{30}\mathrm{P} \ \rightarrow \ ^{30}\mathrm{Si}\,\mathrm{e}^+\,\nu_{\mathrm{e}} \qquad t_{1/2} = 150.0\,\mathrm{sec} \ . \tag{5.6}$$

Table 5.1. Main types of radioactivity. The first four, $\beta^{\pm}$, electron-capture, and 2β are A-conserving decays that lead to the one β-stable isobar for the given A. The next two, α and (spontaneous) fission change A. All these decays may leave the daughter nuclei in excited states that decay (generally rapidly) to the ground states by the last two forms of radioactivity, γ and internal conversion. Two types of radioactivity, e^--capture and internal conversion occur only if the nucleus is surrounded by atomic electrons.

name	reaction	energy spectrum
β^-	$(A,Z) \to (A, Z+1)e^- \bar{\nu}_e$	continuous
β^+	$(A,Z) \to (A, Z-1)e^+ \nu_e$	continuous
e^--capture	$e^-(A,Z) \to (A, Z-1)\nu_e$	discrete
	$+$atomic X-rays/Auger e^-'s	
2β	$(A,Z) \to (A, Z \pm 2)2e^{\pm}2\bar{\nu}_e(2\nu_e)$	continuous
α	$(A,Z) \to (A-4, Z-2)^4\mathrm{He}$	discrete
fission	$(A,Z) \to (A_1, Z_1)(A_2, Z_2)$ ($+$neutrons)	continuous
γ	$(A,Z)^* \to (A,Z)\gamma$	discrete
internal conversion	$e^-(A,Z)^* \to (A,Z)e^-$	discrete
	$+$atomic X-rays/Auger e^-'s	

One can divide natural radioactivity into "fossil" radioactivity due to elements present at the formation of the Earth, and "cosmogenic" radioactivity due to elements continually produced in the atmosphere by cosmic-rays.

Techniques for producing artificial radioactivity can be divided into those that use neutrons (neutron activation) and those that use charged particle beams. An example of the latter is reaction (5.5) where the Joliot-Curies used naturally occurring α radiation ($E_\alpha \sim 5\,\mathrm{MeV}$) to induce the transmutation of aluminum to phosphorus. Modern alchemists use artificially accelerated ions of arbitrary energies, greatly increasing the possibilities, as discussed in Sect. 3.1.5.

5.2.1 Fossil radioactivity

Most elements present on Earth condensed from a interstellar cloud about $4.5 \times 10^9\,\mathrm{yr}$ ago. This cloud consisted mostly of "primordial" $^1\mathrm{H}$ and $^4\mathrm{He}$ that was produced in the first minutes after the "Big Bang" (see Chap. 9). Additionally, the cloud contained heavier elements that had been produced in earlier generations of stars and dispersed into the interstellar medium by various processes (see Chap. 8). Most radioactive nuclei were produced in supernovae that generate a mix of neutron-rich nuclei in a period of time of order a few seconds. When, some millions of years later, the cloud condensed to form the solar system, a few radioactive nuclear species were still present. At the present epoch, only nuclei with mean lives greater than, say, 10^8 yr, are still present in significant numbers.

The nuclei with 10^8 yr $< \tau < 10^{12}$ yr are listed in Table 5.2. These long-lived nuclei involve either highly forbidden β decays (large spin changes) or are α-decays that happen to have Q-values that place the half-lives in this range.

Table 5.2. Nuclei with 10^8 yr $< t_{1/2} < 10^{12}$ yr. The "isotopic abundance" is for the terrestrial mix and the activity for the purified element corresponds to the terrestrial isotopic mix. For the activity in the Earth's crust, the uranium and thorium activities include the activities of the daughters. Note that three nuclides, ^{40}K, ^{147}Sm and ^{235}U have lifetimes much less than the age of the Earth ($\sim 4.5 \times 10^9$ yr) and therefore have very small isotopic abundances.

decay	half-life (years)	isotopic abundance (percent)	activity $(Bq\,kg^{-1})$ (element)	activity $(Bq\,kg^{-1})$ (crust)
^{40}K $\rightarrow$ ^{40}Ca $e^- \bar{\nu}_e$ 89% $\rightarrow$ ^{40}Ar ν_e 11%	1.28×10^9	0.0117	3.0×10^4	6.3×10^2
^{87}Rb $\rightarrow$ ^{87}Sr $e^- \bar{\nu}_e$	4.75×10^{10}	27.83	8.8×10^5	8.0×10^1
^{146}Sm $\rightarrow$ ^{142}Nd α	1.03×10^8	$< 10^{-7}$	< 1	$< 10^{-4}$
^{147}Sm $\rightarrow$ ^{143}Nd α	1.06×10^{11}	15.1	1.3×10^5	9×10^{-1}
^{176}Lu $\rightarrow$ ^{176}Hf e^-	3.78×10^{10}	2.61	5.5×10^4	4×10^{-2}
^{187}Re $\rightarrow$ ^{187}Os $e^- \bar{\nu}_e$	4.15×10^{10}	62.6	1.1×10^6	8×10^{-4}
^{232}Th $\rightarrow$ ^{228}Ra α	1.405×10^{10}	100	4.05×10^6	3.5×10^2
^{235}U $\rightarrow$ ^{231}Th α	7.038×10^8	0.72	5.7×10^5	1.7×10^1
^{238}U $\rightarrow$ ^{234}Th α	4.468×10^9	99.275	1.2×10^7	4.7×10^2

Most of the nuclei in Table 5.2 are isotopes of rare elements with the exception of ^{40}K. The decay modes of this isotope, shown in Fig. 5.1, yields a β^- with $E_{max} = 1.31$ MeV (BR=89.3%) or a γ with $E_\gamma = 1.46$ MeV (BR=10.5%). Potassium has a chemistry similar to that of sodium and is plentiful in biological material where it is the dominant source of radioactivity.

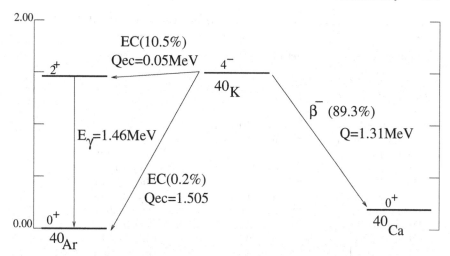

Fig. 5.1. The β-decay of ^{40}K. Being an N-odd,Z-odd nucleus, it has two modes: electron capture to ^{40}Ar and β$^-$ to ^{40}Ca. The long half life $(1.28 \times 10^9 \text{yr})$ is explained by the large angular momentum changes in the decays to the ground states (making the decays highly forbidden) and the small value of Q_{ec} to the 2^+ state of ^{40}Ar.

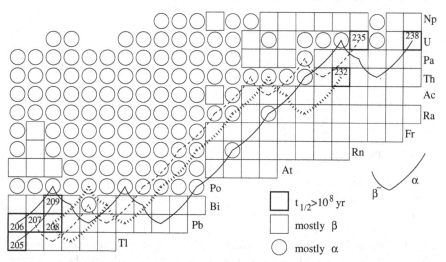

Fig. 5.2. The three chains of natural radioactivity: ^{238}U $\rightarrow^{206}$ Pb; ^{235}U $\rightarrow^{207}$ Pb; and ^{232}Th $\rightarrow^{208}$ Pb. Each chain consists of a series of α$-$ and β$^-$-decays. In the case of the ^{232}Th chain, a branch occurs at ^{212}Bi that has a 64% branching ratio for β$^-$-decay and a 36% ratio for α-decay. Similar, though less balanced, branchings occur throughout the chains and only the primary routes are shown. Also shown is the recently discovered ^{209}Bi $\rightarrow^{205}$ Tl chain consisting of a single α decay [52].

Table 5.3. Members of the three natural radioactivity chains and their half-lives.

uranium series	actinium series	thorium series
^{238}U$_{92}$ 4.468 Gyr		
^{234}Th$_{90}$ 24.10 d		
^{234}Pa$_{91}$ 1.17 m		
^{234}U$_{92}$ 245.5 kyr	^{235}U$_{92}$ 0.7038 Gyr	
^{230}Th$_{90}$ 75.38 kyr	^{231}Th$_{90}$ 25.52 h	^{232}Th$_{90}$ 14.05 Gyr
^{226}Ra$_{88}$ 1600 y	^{231}Pa$_{91}$ 32760 y	^{228}Ra$_{88}$ 5.75 y
^{222}Rn$_{86}$ 3.8235 d	^{227}Ac$_{89}$ 21.773 y	^{228}Ac$_{89}$ 6.15 h
^{218}Po$_{84}$ 3.10 m	^{227}Th$_{90}$ 18.72 d	^{228}Th$_{90}$ 1.9116 y
^{214}Pb$_{82}$ 26.8 m	^{223}Ra$_{88}$ 11.435 d	^{224}Ra$_{88}$ 3.66 d
^{214}Bi$_{83}$ 19.9 m	^{219}Rn$_{86}$ 3.96 s	^{220}Rn$_{86}$ 55.6 s
^{214}Po$_{84}$ 164.3 μs	^{215}Po$_{84}$ 1.781 ms	^{216}Po$_{84}$ 0.145 s
^{210}Pb$_{82}$ 22.3 y	^{211}Pb$_{82}$ 36.1 m	^{212}Pb$_{82}$ 10.64 h
^{210}Bi$_{83}$ 5.013 d	^{211}Bi$_{83}$ 2.14 m	^{212}Bi$_{83}$ 60.55 m
^{210}Po$_{84}$ 138.376 d	^{207}Tl$_{81}$ 4.77 m	64% ^{212}Po$_{84}$ 0.299μs
		36% ^{208}Tl$_{81}$ 3.053 m
^{206}Pb$_{82}$ > 10^{20} yr	^{207}Pb$_{82}$ > 10^{20} yr	^{208}Pb$_{82}$ > 10^{20} yr

The three heavy nuclides, ^{238}U, ^{235}U and ^{232}Th, are the origins of the three natural radioactivity chains illustrated in Fig. 5.2 and Table 5.3. The chains proceed through a series of α and β decays that end at three isotopes of lead

$$^{238}\mathrm{U} \;\rightarrow\; ^{206}\mathrm{Pb} \quad (A = 4n + 2)\,,$$

$$^{235}\mathrm{U} \;\rightarrow\; ^{207}\mathrm{Pb} \quad (A = 4n + 3)\,,$$

$$^{232}\mathrm{Th} \;\rightarrow\; ^{208}\mathrm{Pb} \quad (A = 4n)\,,$$

The three chains end with an isotope of Z-magic lead. (^{208}Pb is doubly magic, $N = 126$.) The fourth chain with $A = 4n + 1$ has no long-lived nuclei heavier than ^{209}Bi. The chain thus consists of a single decay ^{209}Bi $\rightarrow \alpha\, ^{205}$Tl ($t_{1/2} = 1.9 \times 10^{19}$ yr) [52].

As seen in Table 5.3, the lifetimes of the daughter nuclei are all short on geological timescales so in many applications the cascade can be considered to pass "instantaneously" to the final lead isotope. The fact that ^{238}U, ^{235}U, and ^{232}Th are long-lived whereas the elements with $209 < A < 232$ have much shorter half-lives can be traced to the shell structure of the nucleus which makes the doubly magic nucleus ^{208}Pb especially bound. As a result, the Q-values of α-decay are relatively large for $208 < A < 232$ (Fig. 2.14) making the α rates high.

The uranium and thorium chains and ^{40}K dominate the radioactivity inside the Earth. The activity of the lower atmosphere is mostly due to three isotopes of the noble gas radon that are daughters of uranium and thorium and can diffuse out of rocks into the air. It is typically $\sim 20\,\mathrm{Bq\,m^{-3}}$ inside reasonably ventilated buildings [53]. It should be also noted that the Earth's stock of helium originates mostly in the α particles emitted in the uranium–thorium chains. (Inert primordial helium was not retained during the formation of the Earth.)

Minerals that contain uranium and thorium also contain their daughters. Neglecting leakage out of the mineral, all species in the chain

$$1 \rightarrow 2 \rightarrow 3 \rightarrow 4 \ldots n$$

will generate quantities of each nucleus governed by the coupled differential equations

$$\dot{N}_1 = -N_1/\tau_1 \tag{5.7}$$

$$\dot{N}_i = N_{i-1}/\tau_{i-1} - N_i/\tau_i \quad i \neq 1 . \tag{5.8}$$

In the case of the three natural chains, the first nucleus is by far the longest lived $\tau_1 \gg \tau_{i\neq1,n}$. In this case, the solution is

$$N_1 = N_1(t=0)e^{-t/\tau_1} \tag{5.9}$$

$$N_i = N_1\frac{\tau_i}{\tau_1} \quad i \neq 1,n \tag{5.10}$$

$$N_n = N_1(t=0)(1 - e^{-t/\tau_1}) . \tag{5.11}$$

The unstable daughters thus reach an equilibrium abundance proportional to their half-lives where the activity of each daughter is equal to the activity of the parent.

It was Marie and Pierre Curie who chemically isolated the daughter nuclei present in uranium ore. Once uranium ore is refined only the parents ^{238}U and ^{235}U and the daughter ^{234}U remain in the chemically pure uranium.

Natural lead ore contains the natural isotopes, ^{204}Pb, ^{206}Pb, ^{207}Pb and ^{208}Pb, as well as the uranium–thorium daughters due to the decay of any uranium or thorium present in the lead ore. After refining resulting in the removal of the uranium–thorium and non-lead daughters, the purified lead contains the four natural isotopes as well as the the β-emitting daughters ^{210}Pb ($t_{1/2} = 22.37\,\mathrm{yr}$), ^{211}Pb ($t_{1/2} = 36.1\,\mathrm{min}$), and ^{212}Pb ($t_{1/2} = 10.6\,\mathrm{hr}$) (see Fig. 5.2). Ignoring the two very short-lived daughters, we see that recently mined lead contains a ^{210}Pb activity that depends on the amount of ^{238}U present in the ore. It ranges from $\sim 1\,\mathrm{Bq\,kg^{-1}}$ to $2500\,\mathrm{Bq\,kg^{-1}}$ [53] In ancient lead recovered, for example, from sunken Roman galleys, the ^{210}Pb has mostly decayed away and activities as small as $0.02\,\mathrm{Bq\,kg^{-1}}$ have been reported [53]. Such ancient lead is used as shielding in measurements where very low radioactivity is necessary (Exercise 9.7).

5.2.2 Cosmogenic radioactivity

Most interstellar matter is in the form of a thermalized gas (mostly hydrogen and helium). There exists, however a non-thermal component of *cosmic-rays* of similar chemical composition but with an energy spectrum (Fig. 5.3) peaking at a kinetic energy of $\sim 300\,\text{MeV}$ and then falling like $\sim E^{-3}$. While the origin of this component is not entirely established, it is believed that it results from acceleration of particles by time-varying magnetic fields produced by pulsars (neutron stars) and supernova remnants.

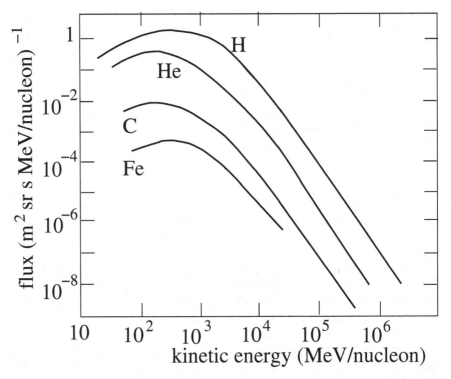

Fig. 5.3. The flux of cosmic radiation outside the Earth's atmosphere [54]. Most particles are protons or ^{4}He nuclei with smaller numbers of heavy nuclei. Carbon and Iron are important examples.

When cosmic-rays enter the Earth's atmosphere, they lose energy through ionization and nuclear reactions. The Earth's atmosphere is sufficiently thick that most of the primary cosmic-radiation stops in the atmosphere. Most cosmic radiation that reaches the Earth's surface consists of muons and neutrinos from the decays of pions produced in these collisions (Fig. 5.4). A small nuclear component consisting mostly of neutrons reaches the surface but is quickly absorbed in the first few meters of the Earth's crust.

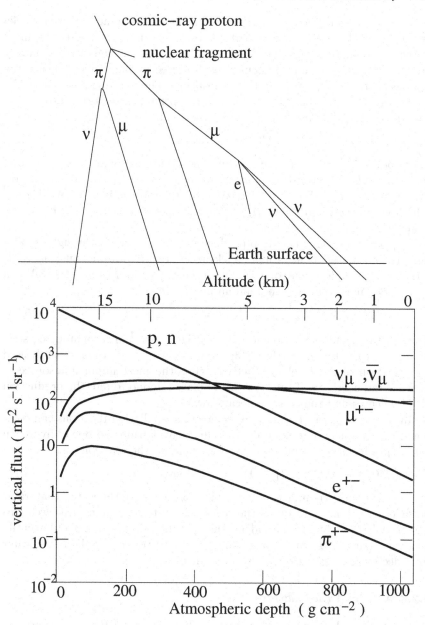

Fig. 5.4. An example of a "shower" induced by a cosmic-ray proton in the upper atmosphere. Two pions and a nuclear fragment are produced when the proton strikes a nucleus. The two pions decay $\pi \to \mu\nu$. and one of the muons decay $\mu \to e\nu\bar{\nu}$ The undecayed muon reaches the Earth's surface where it stops because of ionization energy loss. The lower panel shows the fluxes as a function of depth in the atmosphere [54].

Cosmic-rays produce radioactive nuclei via their interactions with nuclei in the atmosphere and in the Earth's crust. In the atmosphere, the most common radioactivity produced is that of ^{14}C. This nucleus is a secondary product of neutrons who are themselves produced by high-energy cosmic-ray protons that breakup nuclei in the atmosphere. The neutrons then either decay or are absorbed. The most common absorption process is the *exothermic* (n,p) reaction on abundant atmospheric nitrogen

$$n\,^{14}N \rightarrow\,^{14}C\,p \quad t_{1/2}(^{14}C) = 5730\,\text{yr}. \tag{5.12}$$

The produced ^{14}C is mixed throughout the atmosphere and enters the food chain through CO_2 ingesting plants. This results in a ^{14}C abundance in live organic material of about 10^{-12} relative to non-radioactive carbon. The resulting activity is about $250\,\text{Bq}\,\text{kg}^{-1}$. As we will see in Sect. 5.5.2, this allows the estimation of ages of dead organic material.

High energy cosmic rays also produce radioactive nuclei through "spallation" reactions where one or more nucleons are removed from a nucleus (Sect. 3.1.5). For example a neutron with kinetic energy greater than $\sim 20\,\text{MeV}$ can remove two nucleons for a germanium nucleus, e.g.

$$n\,^{70}Ge \rightarrow 3n\,^{68}Ge \quad t_{1/2}(^{68}Ge) = 270.7\,\text{day}. \tag{5.13}$$

This reaction results in a radioactivity of $0.3\,\text{mBq}\,\text{kg}^{-1}$ in germanium crystals produced at the Earth's surface [53]. In high-purity germanium crystals used for detection of low-level radioactivity, it is the most important source of intrinsic radioactivity. If the crystals are placed underground, the production of ^{68}Ge ceases and the activity decays away.

In rare circumstances, nucleons removed in spallation reactions can combine to form nuclei. For example the radioactive tritium isotope of hydrogen can be produced by cosmic rays by (for example) the reaction

$$n\,^{16}O \rightarrow\,^{3}H\,^{14}N \quad t_{1/2}(^{3}H) = 12.35\,\text{yr}. \tag{5.14}$$

The atmospheric tritium combines with oxygen to form water that rains down on the Earth. Prior to the atmospheric testing of nuclear weapons and the Chernobyl reactor accident, this was the primary source of naturally occurring tritium. Because of its short half-life, tritium is absent in water from deep water reserves and also in crude oil.

5.2.3 Artificial radioactivity

Radioactive nuclei can be created in the laboratory by the same reactions that are induced by cosmic-rays, though in the laboratory we can choose beams and targets that maximize production rates. Two general techniques are used, those based on charged particle beams, and those using (thermal) neutrons produced by nuclear reactors.

To produce a radioactive nucleus, it is generally necessary to take a stable nucleus and add or subtract nucleons, or to transmute protons to neutrons

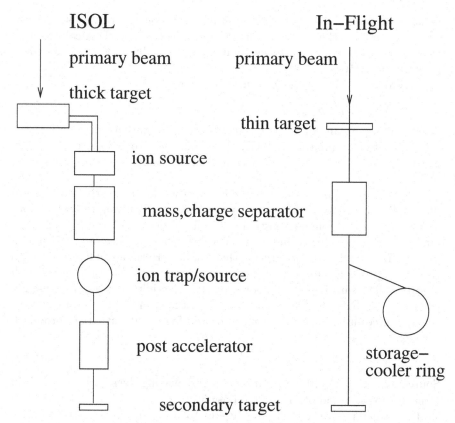

Fig. 5.5. The two primary methods of producing beams of radioactive nuclei [31]. In the ISOL method ("Isotope Separator On-Line") a primary ion beam is incident upon a thick target. Nuclei produced in inelastic reactions are stopped in the target but eventually diffuse into a system that collects, purifies, and accelerates ions. In the "In-Flight" technique, the primary ions are incident upon a target that is sufficiently thin that produced nuclei are not stopped. Nuclei emerging from the target can then be mass selected.

or vice versa. High energy ion collisions (Sect. 3.1.5) can be used to produce a wide variety of nuclides by fragmentation or fusion reactions. If a specific nuclide is desired, it is more efficient to use low-energy ions. For example, in order to produce the radioactive germanium isotope ^{71}Ge ($t_{1/2} = 11.44\,$day), two simple reactions come immediately to mind. The first uses an accelerated proton beam:

$$\text{p}\,^{71}\text{Ga} \rightarrow \text{n}\,^{71}\text{Ge}\,. \tag{5.15}$$

The proton kinetic should be about $10\,$MeV so that the Coulomb barrier between the proton and gallium nucleus is insufficient to prevent the reaction, while the energy is too small to have appreciable cross-sections for other inelastic reactions involving the ejection of nucleons. A target of enriched ^{71}Ga is exposed to the proton beam and then dissolved and chemically treated to extract any germanium. If the gallium is isotopically pure, the germanium is nearly pure ^{71}Ge.

A second possibility uses the (n, γ) reaction

$$n\,^{70}\text{Ge} \rightarrow \gamma\,^{71}\text{Ge} . \tag{5.16}$$

A sample of germanium is placed in a neutron beam. After irradiation, the sample contains a mixture of ^{70}Ge, ^{71}Ge and ^{71}Ga from the decay of ^{71}Ge. Chemical treatment can then yield a mixture of stable ^{70}Ge and radioactive ^{71}Ge.

Both charged-particle- and neutron-activation are currently used to produce radioactive nuclei. Charged particles have the disadvantage in that ions lose their energy through ionization when they enter the sample, so that they quickly become unable to induce nuclear reactions because of the Coulomb barrier. Thermal neutrons, on the other hand, perform a random walk until they are absorbed. One then gets more activity per incident particle with neutrons than with ions. For this reason, large activity sources are generally produced in intense neutron fluxes available at nuclear reactors. For example, cobalt sources used for medical purposes and food sterilization are produced through the reaction

$$n\,^{59}\text{Co} \rightarrow \gamma\,^{60}\text{Co} . \tag{5.17}$$

Sources of activity $> 10^{16}$ Bq can be made by placing the sample of cobalt in a reactor for a period of weeks (Exercise 5.2).

In nuclear research, it is often useful to produce beams of radioactive nuclei that can be accelerated to energies necessary to study their reactions. In recent years, much progress has been made in the production of radioactive beams. The two generic methods of production are illustrated in Fig. 5.5. Examples of experiments using radioactive beams are illustrated in Figs. 1.4, 1.5 and 2.18.

5.3 Passage of particles through matter

Particles produced in nuclear reactions or decays interact with matter in ways that depend on their nature. We can distinguish the following cases:

- Charged nuclei and particles. These particles lose their energy by ionizing the atoms in the medium and eventually come to rest. This process is described in Sect. 5.3.1 for particles with masses $\gg m_e$. The special case of electrons and positrons is studied in Sect. 5.3.3.
- Photons. γ-rays generally lose energy in material through Compton scattering on atomic electrons

$$\gamma e^- \rightarrow \gamma e^- . \tag{5.18}$$

The secondary electrons then deposit their energy in the medium through ionization. The photon continues to Compton scatter until it is photoelectrically absorbed,

$$\gamma \, \text{atom} \rightarrow e^- \text{atom}^+ \,. \tag{5.19}$$

Photons with energies greater than $2m_e$ can be directly absorbed by production of electron–positron pairs

$$\gamma(A, Z) \rightarrow e^+ e^- (A, Z) \,. \tag{5.20}$$

These processes are studied in Sect. 5.3.4.

- Neutrons. Neutrons lose energy by elastic scattering on nuclei until they thermalize. They are eventually absorbed, generally by the (n, γ) reaction $n(A, Z) \rightarrow \gamma(A + 1, Z)$. These processes are described in Sect. 5.3.5.

In this section, we describe these physical process. Their biological effects will be briefly described in Sect. 5.4.

5.3.1 Heavy charged particles

When a charged particle traverses a medium, it progressively loses its energy by transferring it to the electrons of the atoms of the medium. The rate of energy loss can be estimated by considering an ion of mass m_{ion} and charge $z_{ion}e$ that passes near a *free* electron, as illustrated in Fig. 5.6. To simplify the calculation, we first suppose that the ion is non-relativistic, $v \ll c$, and that $m_{ion} \gg m_e$. Since $m_{ion} \gg m_e$, the ion's movement is nearly unaffected by the close encounter with the electron so that its trajectory is, to first approximation a straight line with impact parameter b.

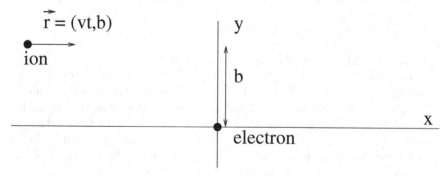

Fig. 5.6. Passage of a charged particle in the vicinity of an atom.

The electron feels a Coulomb force due to the the presence of the ion and therefore recoils after the ion's passage. The electron's momentum can be calculated by integrating the force. The integral is non-zero only in the direction perpendicular to the trajectory:

$$p_e(b, v) = \int F_y dt = \frac{z_{ion} e^2}{4\pi\epsilon_0} \int_{-\infty}^{\infty} \frac{b\, dx/v}{(x^2 + b^2)^{3/2}} = \frac{z_{ion} e^2}{2\pi\epsilon_0\, vb}. \qquad (5.21)$$

This formula is valid for values of b that are sufficiently large that during the passage, the electron recoils through a distance that is small compared to b. The energy loss of the ion, ΔE, is the kinetic energy of the recoiling electron:

$$\Delta E(b, v) = \frac{p_e^2}{2m_e} = \left(\frac{z_{ion} e^2}{4\pi\epsilon_0}\right)^2 \frac{2}{v^2 b^2 m_e}. \qquad (5.22)$$

The energy loss is proportional to v^{-2} because the slower the ion, the longer the time that the electron feels the electric field of the ion.

The energy loss is proportional to b^{-2} so we need to average over impact parameters. The procedure follows precisely what we did in Chap. 3 when we calculated reaction probabilities in terms of cross-sections. We consider a box of volume L^3 containing one electron. The mean energy loss for random impact parameters is

$$\Delta E(v) = \frac{1}{L^2} \int_{b_{min}}^{b_{max}} 2\pi b\, db\, \frac{2}{v^2 b^2 m_e} \left(\frac{z_{ion} e^2}{4\pi\epsilon_0}\right)^2$$

$$= \frac{1}{L^2} \frac{4\pi}{v^2 m_e} \left(\frac{z_{ion} e^2}{4\pi\epsilon_0}\right)^2 \ln(b_{max}/b_{min})$$

$$= \frac{(\hbar c)^2}{L^2 m_e c^2} \frac{4\pi}{\beta^2} (z_{ion}\alpha)^2 \ln(b_{max}/b_{min}), \qquad (5.23)$$

where $\beta = v_{ion}/c$ and α is the fine structure constant. For N_e electrons in the box, the total energy loss is found simply by multiplying by N_e. The rate of energy loss, dE/dx, is then found by dividing by the length L of the box

$$\frac{dE}{dx} = \frac{(\hbar c)^2 n_e}{m_e c^2} \frac{4\pi}{\beta^2} (z_{ion}\alpha)^2 \ln(b_{max}/b_{min}), \qquad (5.24)$$

where $n_e = N_e/L^3$ is the density of electrons in the box.

The energy-loss rate (5.24) has a logarithmic dependence on b_{max}/b_{min} that must be estimated. The naive expectation $b_{min} = 0$ obviously won't do. The source of the problem is that our method for calculating the energy loss $\Delta E(v, b)$ gives an infinite energy loss (5.22) for $b = 0$. In fact, a head-on collision gives an energy loss of only $\Delta E_{max} = 4E_{ion}(m_e/m_{ion})$ where E_{ion} is the kinetic energy of the ion. We can then take for an effective value of b_{min} that value of b for which (5.22) gives $\Delta E(v, b) = \Delta E_{max}$, i.e.

$$b_{min} \sim \frac{z_{ion} e^2}{4\pi\epsilon_0} \left(\frac{m_{ion}}{8E_{ion} v^2 m_e^2}\right)^{1/2} = z_{ion} \frac{\alpha^2}{\beta^2} a_0 \qquad (5.25)$$

where a_0 is the Bohr radius. Since the dependence dE/dx on b_{min} is only logarithmic, we can expect that this estimate will give reasonable results.

The naive result $b_{max} = L$ also is incorrect but for more subtle reasons concerning the fact that the electron is bound to an atom rather than free as we have assumed. In order for the ion to lose energy, the perturbation on the electron due to the passage of the ion must excite the electron from its ground state to a higher energy state. This is only possible in quantum mechanics if the perturbation varies over a time τ that is short compared to the inverse of the Bohr frequency of the transition, $\omega_f - \omega_i$ where f and i refer to the initial and final states. For atomic systems, $\omega_f - \omega_i \sim \alpha c/a_0$. The characteristic time for variations of the perturbation is $V/\dot{V}$ where $V = e^2/4\pi\epsilon_0\sqrt{x^2 + b^2}$ is the perturbing potential. The condition is most stringent by taking $x = 0$:

$$\frac{b_{max}}{v} < \frac{a_0}{c\alpha} , \tag{5.26}$$

i.e.

$$b_{max} \sim \frac{\beta}{\alpha} a_0 . \tag{5.27}$$

This gives our estimate of the energy-loss rate

$$\frac{dE}{dx} = \frac{(\hbar c)^2 n_e}{m_e c^2} \frac{4\pi}{\beta^2} (z_{ion}\alpha)^2 \ln[\beta^3/(z_{ion}\alpha^3)] \qquad \beta > z_{ion}^{1/3}\alpha . \tag{5.28}$$

The condition $\beta > z_{ion}^{1/3}\alpha$ is just $b_{max} > b_{min}$. For slow ions, $\beta < z_{ion}^{1/3}\alpha$, we expect little energy loss because the perturbation is not fast enough to excite atoms.

For $\beta > z_{ion}^{1/3}\alpha$, the energy loss is proportional to the inverse square of the velocity and to the electron density. Eliminating the electron density in favor of the mass density $\rho \sim m_p(A/Z)n_e$ we have

$$\frac{dE}{dx} = \rho z_{ion}^2 \frac{Z}{A} \left[\frac{4\pi(\hbar c)^2\alpha^2}{m_p m_e c^2}\right] \frac{1}{\beta^2} \ln[\beta^3/(z_{ion}\alpha^3)] , \tag{5.29}$$

An improved treatment due to H. Bethe and F. Bloch differs only in the logarithmic term

$$\frac{dE}{dx} = \rho z_{ion}^2 \frac{Z}{A} \left[\frac{4\pi(\hbar c)^2\alpha^2}{m_p m_e c^2}\right] \frac{1}{\beta^2} \left[\ln\left(\frac{2m_e c^2\beta^2\gamma^2}{I}\right) - \beta^2\right] , \tag{5.30}$$

where $I \sim Z\alpha^2 m_e c^2$, is the mean ionization energy of the electrons in the atom. Compared with (5.29), the argument of the logarithm is now $\sim \beta^2/\alpha^2$.

The Bethe–Bloch formula (5.30) applies as long as $\beta > z_{ion}\alpha$. At slower speeds, the perturbation does not excite the atoms of the medium and energy-loss is suppressed. In fact, the ion attaches electrons from the medium so that the effective value of z_{ion} is less than the charge of the naked ion [55].

The order of magnitude of the energy-loss rate is given by the bracketed combination of fundamental constants in (5.30):

$$\frac{4\pi(\hbar c)^2\alpha^2}{m_p m_e c^2} = 0.313\,\mathrm{MeV\,(g\,cm^{-2})^{-1}} . \tag{5.31}$$

Multiplied by the density and dividing by β^2, this gives the order of magnitude of dE/dx. The logarithmic factor increases this by a factor ~ 10 so that for most materials we have

$$\rho^{-1}\frac{dE}{dx} \sim \frac{1\,\text{MeV}\,(\text{g}\,\text{cm}^{-2})^{-1}}{\beta^2}\frac{z_{\text{ion}}^2\,(Z/A)}{0.5}. \tag{5.32}$$

This quantity, $\rho^{-1}dE/dx$, evaluated for $z_{\text{ion}} = 1$ is called the *stopping power* of a material. For $\rho \sim 1\,\text{g}\,\text{cm}^{-3}$ and $Z/A \sim 1/2$, we see that the energy loss rate is of order $1\,\text{MeV}\,\text{cm}^{-1}/\beta^2$.

For $\beta \sim 1$, the Bethe–Bloch formula predicts a roughly constant stopping power, rising only logarithmically with energy. Particles that have energies giving an energy-loss near the minimum value, $\sim 2\text{MeV}\,(\text{g}\,\text{cm}^{-2})^{-1}$, are so-called *minimum ionizing particles*. Most cosmic ray muons reaching the Earth's surface are roughly minimum ionizing and these particles are often used to quickly calibrate energy-loss detectors (see below).

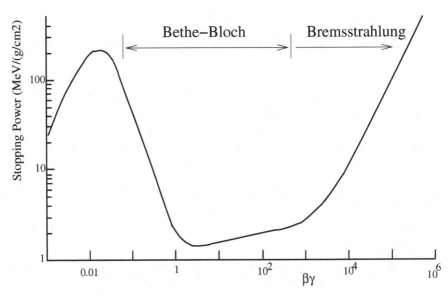

Fig. 5.7. The stopping power for positive muons in copper as a function of $\beta\gamma = v/c/\sqrt{1 - v^2/c^2}$ [1]. For $0.04 < \beta\gamma < 400$ the stopping power follows the Bethe–Bloch formula (5.30). At higher energies the energy loss is dominated by bremsstrahlung. At low energy, the stopping power is less than predicted by Bethe–Bloch because positive ions attach electrons, so their effective charge is less than the naked charge. Nuclei follow the same Bethe–Bloch formula as muons though the limits of its validity are different. At low energy, the formula works for $\beta > z_{\text{ion}}\alpha$ while for $\beta\gamma \gg 1$ the energy loss is dominated by nuclear inelastic collisions rather than by bremsstrahlung.

Figure 5.7 shows the calculated stopping power [1]. The Bethe–Bloch formula (5.30) is applicable for $0.05 < \beta\gamma < 500$. The stopping power falls like β^{-2} until $\beta \sim 1$ and then rises logarithmically. For $\beta < \alpha z_{ion}$ the formula fails because the slowly moving ions capture atomic electrons from the medium, lowering the effective value of z_{ion}. For $\gamma \gg 1$, radiation (bremsstrahlung) eventually becomes important. In Fig. 5.7, this effect is calculated for positive muons where the effect is important for $\gamma > 1000$, i.e. $E > 100\,GeV$. The muon is the only particle other than the electron for which bremsstrahlung is important. Energy loss for hadrons and nuclei with $E > GeV$ is dominated by discrete inelastic scatters on nuclei (Sect. 3.1.5), liberating nucleons and creating hadrons.

The energy-loss can be integrated to give the *range* of a particle, i.e. the distance traveled before stopping.

$$R(E) = \int_0^E \frac{dE}{dE/dx} .$$

(5.33)

For $dE/dx \propto \beta^{-2}$ the integral takes a simple form:

$$R(E) = E \left(\frac{dE}{dx}(E)\right)^{-1} \propto E^2 .$$

(5.34)

An α-particle ($z_{ion}^2 = 4$) of $E \sim 5\,MeV$ has $\beta^2 = 2E/m_\alpha c^2 \sim 2.5 \times 10^{-3}$ giving a stopping power of $\sim 3000\,MeV\,(g\,cm^{-2})^{-1}$. The α-particle will therefore penetrate only $\sim 0.03\,mm$ of a light material like plastic.

Charged particle detectors. Detection of the ionization caused by the passage of charged particles is the basis of most types of particle detectors used in nuclear and high-energy physics. The simplest class consists of gas-filled ionization chambers as illustrated in Fig. 5.8. The liberated electrons and ions drift toward charged electrodes where they create an electric pulse.

Other types of energy-loss detectors are listed in Table 5.4. *Semiconductor silicon* and germanium ionization detectors are very useful because, being denser than gasses, they can stop charged particles, yielding a measurement of their total kinetic energies. *Scintillators* (Fig. 5.9) where a small portion of the energy loss is transformed to visible photons are often used because of their simplicity and low cost.

Cherenkov radiation is created by charged particles moving at a velocity greater than the light velocity c/n in a medium of refraction index n. While of great use in high-energy particle physics, they are rarely used in nuclear physics because of most particles are non-relativistic. One of their most important uses is in massive detectors constructed to detect solar neutrinos (Sect. 8.4.1).

Under special conditions, the energy loss of charged particles can create visible tracks in a medium, as in nuclear emulsions, cloud chambers, and bubble chambers.

Table 5.4. Some charged-particle detectors and the signal produced for 1 keV of total energy loss. The largest intrinsic signal is in silicon semiconductor detectors but that in gas-ionization chambers can be effectively much higher since additional electrons will be liberated when the primary electrons collide with atoms in the high electric fields near collection wires (proportional chambers and Geiger counters). The signal from scintillators and Cherenkov counters will be reduced because the photons must be detected, generally by photomultiplier tubes (Figs. 5.14 and 8.14). The primary advantage of optical detectors is their excellent time resolution, as good as 0.1 ns.

Detector	yield
silicon semiconductors	$\sim 270 e^-$ keV^{-1}
gas-ionization chambers	~ 50 e$^-$ keV^{-1}
organic scintillators	< 15 visible photons keV^{-1}
H$_2$O Cherenkov counters	~ 0.1 visible photons keV^{-1} $(\beta > 1/n \ \ n = 1.33)$

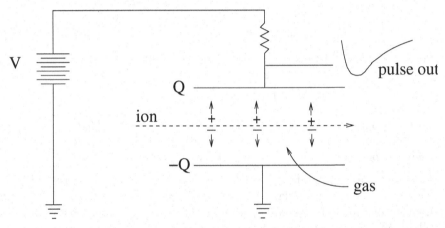

Fig. 5.8. The principle of operation of a gas-filled ionization chamber. The charge liberated by the passage of the particle drifts in the electric field toward the electrodes. If the high-voltage supply is removed after charging the electrodes, the passage of ionizing radiation would simply discharge the chamber. In the early days of radioactivity, this rate of discharge was a standard monitor of radioactivity and cosmic radiation. If the high-voltage supply is left in place, after the passage of the charged particle, the potential difference between the electrodes first drops and then returns to its nominal value. The resulting pulse signals the passage of a single particle. This is the principle of *proportional counters* and *Geiger counters*. If the time of passage of the particle is known (e.g. by an external scintillator counter), the position of the particle can be deduced from the time delay between this time and the arrival of the ions at the electrodes. This is the principle of the operation of *drift chambers*.

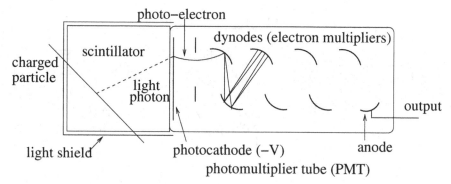

Fig. 5.9. The principle of a scintillation detector. A charged particle creates a small amount of visible light while losing energy in the scintillating material. The photon may be detected by a photomultiplier tube (PMT) where it may liberate an electron on a photocathode through the photo-electric effect. The photocathode is held at high (negative) potential so the photoelectron is directed toward a series of dynodes where each electron creates a certain number of further electrons. The electrons are finally collected at the anode.

5.3.2 Particle identification

Many nuclear species can be produced in nuclear collisions and we are now in a position to understand how the identity of individual nuclei can be determined. A species is characterized by its charge, q, and mass, m, and individual nuclei are additionally characterized by their velocity, v, or equivalently by their momentum p or kinetic energy $p^2/2m$. Charged particle detectors give information on these quantities:

- p/q from the trajectory in a magnetic field;
- v from the flight-time t between detector elements separated by a distance d;
- q^2/v^2 from dE/dx in a thin ionization counter;
- $p^2/2m$ from the total energy deposit in a thick ionization counter.

Clearly, a single type of measurement is insufficient to identify the species. The charge-to-mass ratio q/m can be deduced by combining two measurements, for instance p/q from the magnetic trajectory and v from a time-of-flight measurement. With perfect precision, this would be sufficient to identify the particle since no two species have exactly the same q/m.[1] In real experiments, it is generally necessary to have a third type of measurement. Fig. 5.10 shows how dE/dx measurements where combined with magnetic trajectory and time-of-flight measurements to identify the doubly-magic nuclide ^{100}Sn.

---

[1] This is effectively the case in mass-spectrometers (Fig. 1.3) where the magnetic trajectory is combined with knowledge $p^2/2m$ determined by the accelerating potential. A high precision of $\Delta(m/q)/(m/q) \sim 10^{-8}$ generally allows one to unambiguously identify ionic species.

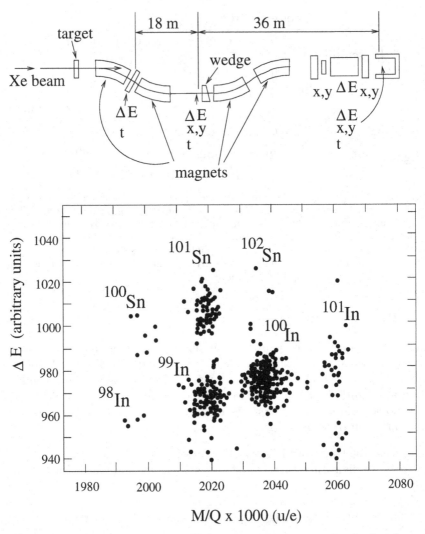

Fig. 5.10. The apparatus used by [56] at the GSI laboratory for the first detection of doubly-magic ^{100}Sn. A ^{124}Xe beam of kinetic energy $\sim 1\,\text{GeV/nucleon}$ is incident on a thin beryllium target. Particles emerging from the target enter a spectrometer consisting of four dipole magnets. The magnetic field is set so that particles of $A \sim 100$ with kinetic energy $\sim 1\,\text{GeV/nucleon}$ pass through the spectrometer. As indicated in the figure, the spectrometer is equipped with ionization counters that measure energy loss (ΔE), position (x,y), and time of passage (t) for individual particles. Additionally, a wedge-shaped aluminum block placed at the midpoint of the spectrometer produced an energy loss that depends on position and therefore on A. Combining the (x,y) information with the time-of-flight information allows one to deduce m/q. A scatter plot (bottom panel) of m/q vs. energy loss then identifies clearly the nuclei. Finally, particles are stopped in a silicon counter where their decays can be observed.

5.3.3 Electrons and positrons

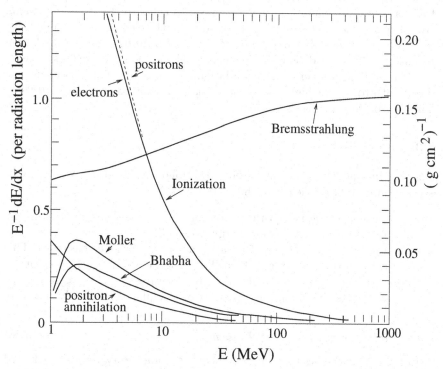

Fig. 5.11. Energy loss of electrons and of positrons in lead as a function of the incident energy [1]. The left scale uses the radiation length of lead (0.56 cm) as the unit of length. Ionization dominates at low energies while bremsstrahlung dominates above 10 MeV. Subdominant contributions come from wide-angle elastic scattering on electrons (*Moller* for electrons and *Bhabba* for positrons) and positron annihilation.

Our derivation of the Bethe–Bloch equation (5.30) does not apply to electrons because we assumed that the ions were sufficiently massive that their trajectories are not affected by close encounters with atomic electrons. In spite of this, for $E_e < 10$ MeV, the stopping power for electrons is to good approximation given by the Bethe–Bloch equation. Electrons and positrons at low energy therefore come to rest in a manner similar to that of heavy ions, though their trajectories are much less straight since they scatter at large angles from nuclei.

After coming to rest, positrons annihilate with atomic electrons yielding two photons of energy $E_\gamma = m_e c^2$

$$e^+ e^- \rightarrow \gamma\gamma. \tag{5.35}$$

For $E_e > 10\,\mathrm{MeV}$, electrons and positrons lose most of their energy by radiation of photons (bremsstrahlung) that results from their acceleration in the electron field of nuclei. The distance over which an electron loses a fraction $1/e$ of its energy through this process is called the *radiation length*, X_0. . Because the effect is due to acceleration in the electric field of a nucleus, $\rho^{-1}X_0$ has a strong dependence on Z, ranging from $61\,\mathrm{g\,cm^{-2}}$ in hydrogen to $6.37\,\mathrm{g\,cm^{-2}}$ in lead. For the low-energy electrons and positrons from β-decay, this effect is generally of secondary importance but rapidly becomes the dominant process for high-energy electrons and positrons with $E > 10\,\mathrm{MeV}$. This can be seen in Fig. 5.11 for lead.

5.3.4 Photons

Photons interact with matter through the following reactions whose cross-section on carbon and lead are shown in Fig. 5.12:

- Rayleigh-scattering, i.e. elastic scattering from atoms. This process has the only cross-section that has no energy threshold.
- Compton scattering, i.e. elastic scattering from quasi-free electrons. This process has a threshold for each shell of bound electrons corresponding to the ionization energy, I, of the shell. In the range $I \ll E_\gamma \ll m_e c^2$, the cross-section is the Thomson cross-section calculated classically in Chap. 3

$$\sigma_T = (8\pi/3)r_e^2 \qquad I \ll E_\gamma \ll m_e c^2 . \tag{5.36}$$

In this energy range, the differential cross-section is forward and backward peaked

$$\frac{d\sigma}{d\Omega} = \frac{\sigma_T \cos^2\theta}{2\pi} , \tag{5.37}$$

where θ is the scattering angle. For $E_\gamma \gg m_e c^2$, the cross-section falls like E_γ^{-2} and is increasingly peaked in the forward direction $\theta < m_e/E_\gamma$.
- Photoelectric absorption by atoms

$$\gamma\,\mathrm{atom} \rightarrow \mathrm{e^-\,atom^+} . \tag{5.38}$$

The threshold for ejection of an electron of a given shell is just the ionization energy of the shell.
- Pair production , i.e.

$$\gamma(A, Z) \rightarrow \mathrm{e^+ e^-}\,(A, Z) . \tag{5.39}$$

The threshold is $2m_e c^2$. This is by far the dominant effect at high energy.

Photons also have a small probability to be absorbed by breaking up nuclei (photo-nuclear absorption) as described in Sect. 3.1.5.

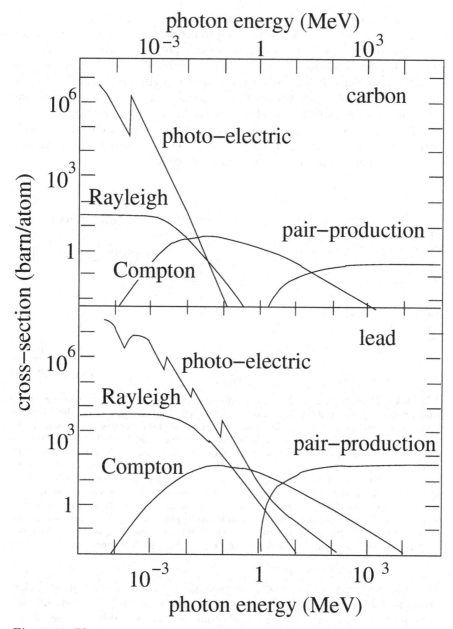

Fig. 5.12. Photon cross-sections on carbon and lead [1] as explained in the text. At low energy, $1\,\text{keV} < E < 100\,\text{keV}$, photo-electric absorption dominates while electron–positron pair production dominates for $E \gg 2m_e c^2$. Compton scattering dominates at intermediate energies. Photo-nuclear absorption (Fig. 3.8) is of minor importance.

Photon Detectors. Nuclear physics is mostly concerned with photons in the range $100\,\text{keV} - 1\,\text{MeV}$ where Compton scattering and photoelectric absorption are the most important processes. A photon entering a massive detector therefore generates electrons by the sequences shown in Fig. 5.13. In "photon detectors" it is actually these electrons that are detected. For instance, in germanium diode detectors, a germanium crystal is polarized and the ionization liberated by the electrons can be sensed as an electric pulse of amplitude proportional to the deposited electron energy. If the photon is absorbed, the detector gives a signal proportional to the photon energy. Various detectors using this principle are listed in Table 5.5.

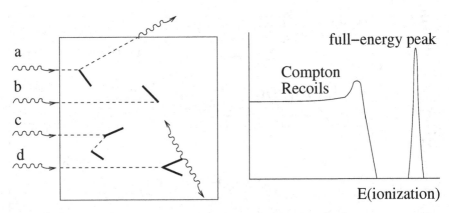

Fig. 5.13. Photons in the MeV range interact with matter by Compton scattering and photoelectric absorption, as illustrated by photons a, b and c on the left. The spectrum of ionization energy deposited by the secondary electrons is shown on the right. Photon a deposits only a fraction of its energy and contributes to the continuous Compton spectrum. Photons b (photoelectric) and c (Compton followed by photoelectric) deposit all their energy and lead to the full-energy peak in the energy spectrum. Photons with $E_\gamma > 2m_e$ can convert to an electron–positron pair (photon d). The electron and positron then lose their kinetic energy by ionization and the positron then annihilates, generally at rest, with another electron. Depending on whether both, one, or neither of the annihilation photons is photo-absorbed in the detector, pair production contributes to the full-energy peak or to "single-escape" or "double escape" peaks $511\,\text{keV}$ or $1022\,\text{keV}$ below the full-energy peak.

At high energy, $E \gg 2m_ec^2$, the passage in matter of photons and electrons is deeply connected. Photons create electrons and positrons by pair production, which in turn create photons by bremsstrahlung. The resulting cascade of particles is called an "electromagnetic shower" of electrons, positrons and photons.

Table 5.5. Some γ- and X-ray detectors and their signal yield. Photons with $E < 2m_ec^2$ interact in the detectors mostly by photo-electric absorption and Compton scattering as shown in Fig. 5.13, and the signal is generated by the secondary electrons. γ-detectors require the use of high-Z elements to give a high probability of photoelectric absorption. The effective signal from scintillators is smaller than that shown below by a factor of order 10 because of limited photon collection efficiency and quantum efficiency of photomultipliers. The intrinsic energy resolution of scintillators is then given by Poisson statistics on the observed number N of photons, $\Delta N = \sqrt{N}$ yielding a resolution of no better than 10 keV for a 1 MeV γ-ray. This resolution is reflected in the observed width of the full-energy peak in Figs. 5.13 and 7.2. The signal from silicon and germanium diodes is so large that statistical fluctuations are generally small compared to other sources, e.g. amplifier noise. This results in γ-ray resolutions of typically ~ 1 keV for germanium diodes, making them by far the best detectors when high resolution is required. On the other hand, scintillators are much less expensive and are therefore favored when large detectors covering a large solid angle are required.

Detector	yield
silicon semiconductor diodes (x-rays)	$270\,e^-\,\text{keV}^{-1}$
germanium semiconductor diodes (γ-rays)	$340\,e^-\,\text{keV}^{-1}$
NaI scintillators (γ-rays)	< 40 visible photons keV^{-1}
BGO scint. $((Bi_2O_3)_2(GeO_2)_3$ (γ-rays)	< 6 visible photons keV^{-1}

5.3.5 Neutrons

Neutrons in the MeV range interact with matter mostly by elastic scattering on nuclei. This results in a progressive loss of the neutron kinetic energy until they are thermalized with a mean energy, $\sim kT$, given by the temperature of the medium (Fig. 5.14). The neutron then continues to perform a random walk with velocity $v \sim 2000\,\text{m s}^{-1}$ until they are absorbed, usually by a (n, γ) reactions. In a homogeneous medium containing nuclei of mass number A, the mean time for absorption after thermalization is

$$\tau = \frac{1}{n\langle \sigma v \rangle} \sim 6\,\mu\text{s}\frac{A\,\text{g cm}^{-3}}{\rho}\frac{1\,\text{b}}{\sigma}\,, \tag{5.40}$$

where n and ρ are the number and mass density of nuclei and where σ is the thermally averaged cross-section at $T = 300$K. Note that the absorption time is substantially sorter than the mean lifetime of a free neutron, ~ 886.7 s.

Neutron detectors. In nuclear physics, neutrons are generally detected by first thermalizing them and then by observing their absorption. Some commonly used absorption reactions are listed in Table 5.6. Absorption on ^{3}He, ^{6}Li and ^{10}B yielding charged particles are preferred because the final state particles are easily detected through their ionization. Radiative absorption on ^{157}Gd is sometimes used because of its enormous cross-section for thermal neutron capture, 2.55×10^5 b.

Table 5.6. Some thermal neutron detectors.

Reaction	cross-section	detector type
$n\,^3He \to p\,^3H$	$5.33 \times 10^3 b$	^3He-filled proportional chambers
$n\,^6Li \to\,^3H\,^4He$	$9.42 \times 10^2 b$	^6Li-doped scintillators
$n\,^{10}B \to\,^4He\,^7Li$	$3.8 \times 10^3 b$	BF_3-filled proportional chambers
$n\,^{157}Gd \to \gamma\gamma\,^{158}Gd$	$2.55 \times 10^5 b$	gadolinium-doped scintillators

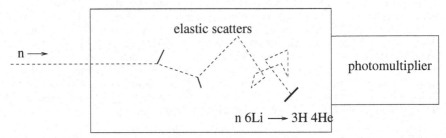

Fig. 5.14. A neutron detector based on the reaction $n^6Li \to\,^3H^4He$. The neutron enters a ^6Li-doped scintillator. The neutron is thermalized by elastic scatters, after which it performs a random walk through the material. Eventually, it is absorbed by a nucleus in the material, in this case by a ^6Li nucleus. The final state particle ^3H and ^4He are charged an therefore stop in the medium by depositing their kinetic energy, mostly by ionization. Scintillation light is detected by the photomultiplier. If the original neutron is sufficiently energetic for the first few nuclear recoils to produce a significant amount of scintillation light, the event consists of two light pulses, the first for the elastic scatters and the second for the capture.

5.4 Radiation dosimetry

We saw in the previous section that all non-thermalized particles, with the practical exception of neutrinos, deposit their energy in the medium through which the propagate. The total energy deposited per unit mass (of the medium) is called the *dose*. It has units of Gy (Gray)

$$1\,Gy \;=\; 1\,J\,kg^{-1} \;=\; 6.25 \times 10^{12}\,MeV\,kg^{-1}\,. \tag{5.41}$$

The dose clearly refers to a flux of particles over a specified period of time or to a specified event (e.g. a nuclear accident).

Biological effects depend not only on the total energy deposited but also on the density of the energy deposit. The *equivalent dose* is therefore used to have a better estimate of biological damage caused by the disruption in cells due to ionization and the resultant breaking of molecular and chemical bonds. The most important risks involve mutations that can cause cancer. The unit is the Sievert (Sv)

$$\text{equivalent dose (Sv)} \;=\; \sum\,[w_R \times \text{dose (Gy)}]\,, \tag{5.42}$$

where the factor w_R takes into account the long term risks in regular exposi-
tions to weak doses of each type of radiation mentioned in Table 5.7 and the

sum is over the different types of radiation. Heavily ionizing radiation involving particles with $\beta < 1$ have risk factors greater than unity. The fact that w_R are multiples of 5 reflects the precision of the biology, not the physics.

Table 5.7. Risk factors in various radiations

Radiation	w_r
X and γ rays, any energy	1
Electrons and muons, any energy	1
Neutrons, $E < 10$ keV	5
Neutrons, $10 < E < 100$ keV	10
Neutrons, $100\ keV < E < 2$ MeV	20
Neutrons, $2 < E < 20$ MeV	10
Neutrons, $E > 20$ MeV	5
Protons, $E > 20$ MeV	5
α particles, fission fragments, heavy nuclei	20

Table 5.8 shows the typical contribution of various sources of radiation to the mean equivalent dose we receive annually (in the absence of a local nuclear event). The primary sources are cosmic rays, 0.26 mSv at sea level, and natural radioactivity, ~ 2 mSv. The $\alpha-$ and $\beta-$particles from natural activity are mostly confined to the material containing the radioactivity and are therefore not dangerous unless ingested. However, $\gamma-$rays from β-decays near the surface of materials radiate us continually. The most energetic of the photons are the 1460 keV ^{40}K and 2620 keV ^{208}Tl γ-rays. They are often seen in the background spectra of photon detectors (Fig. 7.2).

More important than the photon emitters is the α-emitting noble gas Radon produced in the uranium–thorium chains. It diffuses out of the ground and building materials, generating an effective dose of $\sim 1-2$ mSv for building dwellers. This can be an order of magnitude smaller for people living outside and an order of magnitude higher in poorly ventilated buildings or in granite-rock areas.

The effective dose derived from ingesting or inhaling radioactive materials (Table 5.9) is difficult to estimate since it depends critically on where and for how long the body stores the particular material. The *radiotoxicity*, R, associated with a given nuclide is the effective dose (in Sv) resulting from ingesting or inhaling an activity A (in Bq). It can be written as

$$R = \frac{A\, w_R\, \langle E \rangle\, \tau_{\text{eff}}}{m_{\text{eff}}}, \tag{5.43}$$

where $\langle E \rangle$ is the mean energy per decay deposited in the body and w_R is the associated risk factor. Depending on the nucleus in question, the effective retention time τ_{eff} can be the nuclear mean life for short-lived nuclei, the biological retention time for elements that are eliminated from the body (e.g. tritiated water), or the lifetime of the organism itself for long-lived nuclei that can be permanently attached to the body parts, e.g. ^{239}Pu in bones.

Table 5.8. Typical annual effective doses [57].

source	dose (mSv)
cosmic radiation	
sea level	0.26
2000m	0.40
air travel (per 1600 km)	0.01
ground γ–rays	0.46
air (radon)	2.0
Weapons test fallout	0.01
Dwelling (stone/brick/concrete) γ–rays	0.07
Food and drink	0.3
Television	0.01
medical x-rays	0.40
total	3.6

Table 5.9. The whole-body effective dose (radiotoxicity) due to selected radionu-clides if taken internally [59]. Note the similar doses of the three fission products ^{90}Sr, ^{131}I and ^{137}Cs in spite of their very different lifetimes, indicating that their values of τ_{eff} are similar, due to short biological retention times.

Isotope	$t_{1/2}$	$\langle E \rangle$ (MeV)	dose (μSv kBq^{-1})	principal organs affected
^{90}Sr	28.79 yr	1.3 (β^-)	30 (28)	bone marrow, lungs
^{131}I	8.02 day	0.5 (β^-)	11 (22)	thyroid
^{137}Cs	30.07 yr	0.8 (β^-, γ)	6.7 (13)	whole body

The effective mass of the organism, m_{eff}, is the body mass if one calculates the full-body dose or the mass of the organ in which the radioactive material is attached if one wishes to calculate the dose received by that organ.

The radiotoxicity is mostly discussed in the context of nuclear accidents and nuclear-waste storage. The evaluation of the expected dose from with a given nuclide must also take into account the relative risk that the element can be introduced into the food chain. One can also protect oneself from certain nuclides by saturating the body with non-radioactive isotopes. This is how one protects the thyroid from radioactive iodine after a nuclear incident.

The medical risks associated with radiation can be divided into short-term and long-term risks. The lethal dose defined as a 50 % risk of death within 30 days is 3 Sv if it is absorbed in a short time over the whole body. The spinal chord is the most critically affected. The long-term risks of cancer are estimated to be about $0.02\,\mathrm{Sv}^{-1}$ meaning that 2% of people exposed to 1 Sv will develop a cancer due to this exposure before dying of unrelated causes. This very broad estimate can be refined to take into account the source of radiation and the parts of the body that are exposed or contaminated. Note however that in the approximation that cancer rates are proportional to the dose and independent of the organ exposed, the total cancer rate does not depend on these considerations. It should also be mentioned that there is an active and unresolved debate about whether or not these estimates apply to very low levels of radiation.

The recommendation for workers or medical patients exposed to radioactivity in a regular manner is that the equivalent dose absorbed by someone should be less than 20 mSv per year averaged over a period of 5 years, with a maximum of 50 mSv in a single year. For the public, the absorbed dose must be smaller than 5 mSv per year in addition to the natural dose and possible medical doses.

5.5 Applications of radioactivity

5.5.1 Medical applications

The disrupting effect on cells of ionizing radiation can be used to human advantage by killing unwanted, e.g. cancerous cells. The use of radiation from external radioactive sources or accelerated particle beams in cancer therapy has a long history.

It is interesting to examine the cases of protons or heavy ions. Figure 5.15 shows the energy deposition in the medium in terms of the penetration distance. We see that practically *all* the energy is deposited in a very localized region near the stopping point. This comes from the $1/\beta^2$ factor in equation (5.30). Also shown in the figure is a comparison between the effect of ion beams and of photons. One can see the great advantage, from the medical point of view, of heavy ion beams which attack and destroy tumors in a very accurate and localized manner, as opposed to γ-rays which produce damage all around the point of interest.

It is also useful to sometimes inject, intravenously or orally, radioactive substances into the body in a chemical form that accumulates preferentially in particular organs or cancerous tissue. The subsequent decays can then irradiate the tumor or betray its position by emitting externally detectable γ-rays. Pure β^- emitters are preferred for tumor irradiation since all energy is deposited within a few millimeters of the decay. Positron emitters (see below) are preferred for position measurements.

Much progress has been made in the production of radioactive *tracers* that give a physical manifestation of certain biological functions in human bodies. One can make tracers with given chemical or biological properties, so that they become fixed in definite organs or biological functions. For example,

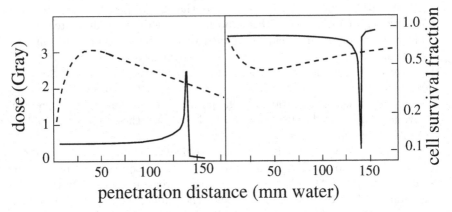

Fig. 5.15. Energy loss of ions (left) and survival rate of cells (right) as a function of the penetration depth. Because of the v^{-2} factor in the Bethe–Bloch formula (5.30), most of the energy is deposited near the stopping point. The dashed curve corresponds to the same quantities for photons. We can see the considerable medical advantage to use heavy ion beams.

glucose accumulates in tumors that have a high metabolic rate. Iron is used to study the various phases of hemoglobin, Iodine is used for the thyroid gland, rare gases (xenon and krypton) in pneumology; phosphorus for the metabolism, etc.

For position measurements, of special interest are β^+-emitters which yield pairs of 511 keV photons when the positrons annihilate. The emission of two particles allows one to more easily constrain the origin of the emission. The principle of this "positron emission tomography" (PET) is illustrated in Fig. 5.16. A commonly used positron emitter is ^{18}F that is produced and decays via

$$p\,^{18}O \to n\,^{18}F\,.$$

$$^{18}F \to\,^{18}O\,e^+\,\nu_e \qquad t_{1/2}\,=\,109.8\,m\,. \tag{5.44}$$

Protons of energy $\sim 15\,$MeV are used and the 109.8 m half-life of ^{18}F allows one to chemically separate and prepare a fluorine containing compound. For instance, ^{18}F-containing glucose can be prepared and then ingested sufficiently rapidly for it to accumulate in high-metabolic tumors.

5.5.2 Nuclear dating

One of the most intellectually interesting applications of radioactivity has been to provide accurate estimates of ancient objects, ranging from archaeological artifacts to stars. Important examples are the use of ^{14}C to date archaeological organic material and ^{87}Rb to give an accurate age of the solar system.

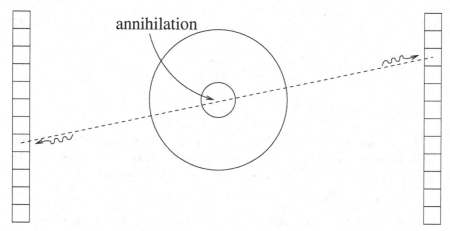

Fig. 5.16. The principle of positron-emission tomography. A β^+-emitter is injected into a body in a chemical form such that it accumulates in the organ or tissue that one wishes to study. Positrons from β^+-decay stop in the material and then annihilate, $e^+e^- \rightarrow \gamma\gamma$. The detection of the two "back-to-back" photons constrains the position of the decay to lie along the line connecting the two detector elements. Accumulation of many events allows one to reconstruct the geometry of the decay region.

To understand the principle of nuclear dating, we consider a closed system containing initially N_{A0} nuclei A and N_{B0} nuclei B. By closed, we mean that there are no exchanges of A and B nuclei with the exterior. If A decays to B, then their numbers at some time t are

$$N_A(t) = N_{A0}\, e^{-t/\tau} \tag{5.45}$$

$$N_B(t) = N_{A0}\,(1 - e^{-t/\tau}) + N_{B0}$$

$$= N_A(t)\, e^{t/\tau}(1 - e^{-t/\tau}) + N_{B0}\,, \tag{5.46}$$

where τ is the (known) mean lifetime of A. If, at time t, N_A and/or N_B are measured and if N_{A0} and/or N_{B0} are known, we may be able to deduce the age t by solving one or both of the equations. We note the following cases:

1. N_A is measured. We must then know N_{A0} to deduce t from (5.45).
2. N_B is measured. We must know both N_{A0} and N_{B0} to deduce t from (5.46).
3. N_B and N_A are measured. We need only N_{A0} or N_{B0}.

Generally neither N_{A0} nor N_{B0} are known but are rather deduced from the measured amount of another species A' or B' for which the ratio $N_{A0}/N_{A'}$ or $N_{B0}/N_{B'}$ is known. Usually, A' (B') is another isotope of A (B). In many applications, A' and B' are stable so that $N_{A'}$ and $N_{B'}$ are time-independent. In this case we write (5.45) and (5.46) as

$$\frac{N_A(t)}{N_{A'}} = \frac{N_{A0}}{N_{A'}} e^{-t/\tau} \tag{5.47}$$

$$\frac{N_B(t)}{N_{B'}} = \frac{N_{A0}}{N_{B'}} (1 - e^{-t/\tau}) + \frac{N_{B0}}{N_{B'}}$$

$$= \frac{N_A(t)}{N_{B'}} e^{t/\tau} (1 - e^{-t/\tau}) + \frac{N_{B0}}{N_{B'}} , \tag{5.48}$$

^{14}C dating uses (5.47) with $A = {}^{14}$C and $A' = $ natural carbon (^{12}C and ^{13}C). Radioactive ^{14}C is produced continuously in the atmosphere, mostly via the exothermic (n,p) reaction induced by cosmic-ray neutrons

$$\text{n}\,{}^{14}\text{N} \rightarrow \text{p}\,{}^{14}\text{C} \quad Q = 0.63\,\text{MeV} \tag{5.49}$$

The produced ^{14}C is mixed in the atmosphere and then fixed in carbon-breathing plants. This leads to an isotopic abundance of ^{14}C in organic material of

$$\frac{{}^{14}\text{C}}{{}^{12}\text{C} + {}^{13}\text{C}} = 1.2 \times 10^{-12} , \tag{5.50}$$

corresponding to a specific activity of $250\,\text{Bq\,kg}^{-1}$. There are good reasons to believe that the cosmic-ray flux has been constant over many ^{14}C lifetimes so that this ratio has remained constant in time, at least until atmospheric nuclear weapon testing between 1955 and 1963. The neutron flux during these tests resulted in a doubling of the ^{14}C content of objects living at that time. Since the end of the tests, the ^{14}C content has returned to within a few percent of the pretest value, and may even have decreased due to the increased burning of ^{14}C-free fossil fuels.

When a living organism dies, it stops fixing carbon so the ^{14}C contained in the organism decays with $t_{1/2} = 5730\,\text{yr}$. The amount of ^{14}C relative to natural carbon measured in an object allows one to date the object through (5.47). The amount of ^{14}C is generally determined by measuring the β-activity of carbon chemically extracted from the sample, though recently mass spectrometer techniques have been become sufficiently precise to directly measure the small amounts of ^{14}C.

Figure 5.17 shows some early data that was used to verify the technique. The test uses objects whose age can be determined by independent means. The activity of the samples decrease with the known age of ^{14}C. This indicates that the production rate of ^{14}C has, indeed, been roughly constant over the last few millenniums.

This method is usefully applied to objects with ages between, say, 500 and 50000 years. Older objects have very little of the original ^{14}C, making measurements difficult.

The ages of minerals can sometimes be dated using potassium-argon dating

$$\text{e}^{-}\,{}^{40}\text{K} \rightarrow {}^{40}\text{Ar}\,\nu_\text{e} \quad t_{1/2} = 1.28 \times 10^{9}\,\text{yr} . \tag{5.51}$$

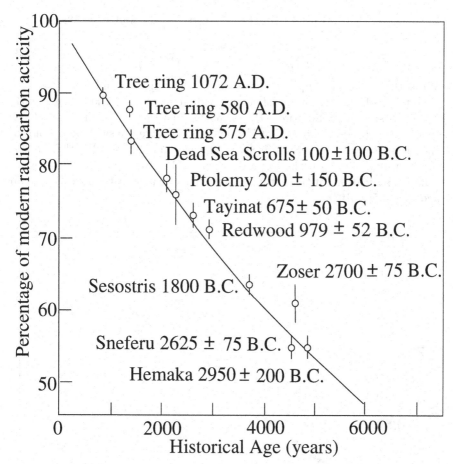

Fig. 5.17. Calibration of the ^{14}C dating method with objects of known age [58].

In this method, (5.46) can be used directly assuming that the initial abundance of the noble gas argon was very small in the mineral. The age is then deduced from the measured ^{40}K and ^{40}Ar abundances. It should be noted that the measured age refers generally to the last solidification of the mineral since melting would generally lead to loss of the argon.

In a similar technique, uranium ores can be dated by measuring the quantities of the descendants ^{206}Pb (from ^{238}U) or ^{207}Pb (from ^{235}U). In this case, it is not reasonable to assume that the initial abundances of the lead isotopes vanished. It is therefore necessary to use (5.48) with $B' = {}^{204}$Pb which has no radiogenic source.

The ages of meteorites have been most accurately measured using the decay

$$^{87}\text{Rb} \rightarrow {}^{87}\text{Sr}\, e^- \, \bar{\nu}_e \, . \tag{5.52}$$

In this case, (5.48) is used with $A = {}^{87}\text{Rb}$, $B = {}^{87}\text{Sr}$ and $B' = {}^{86}\text{Sr}$:

$$\left[\frac{{}^{87}\text{Sr}}{{}^{86}\text{Sr}}\right]_t = \left[\frac{{}^{87}\text{Rb}}{{}^{86}\text{Sr}}\right]_t e^{t/\tau}(1 - e^{-t/\tau}) + \left[\frac{{}^{87}\text{Sr}}{{}^{86}\text{Sr}}\right]_{t=0}, \qquad (5.53)$$

While the chemical composition, i.e. the Rb/Sr ratio, depends on the precise conditions of the formation of an individual meteorite, the isotopic ratio ${}^{87}\text{Sr}/{}^{86}\text{Sr}$ should be uniform at the formation in a well-mixed pre-solar cloud. In this case, (5.48) predicts that if all meteorites have the same age, a plot of the measured value of ${}^{87}\text{Sr}/{}^{86}\text{Sr}$ vs. the measured value of ${}^{87}\text{Rb}/{}^{86}\text{Sr}$ should be a straight line with the slope of $\exp(t/\tau)(1 - \exp(-t/\tau)) \sim t/\tau$. Figure 5.18 [60] indicates that this is the case and that the meteorites have a common age of $\sim 4.5 \times 10^9$ yr.

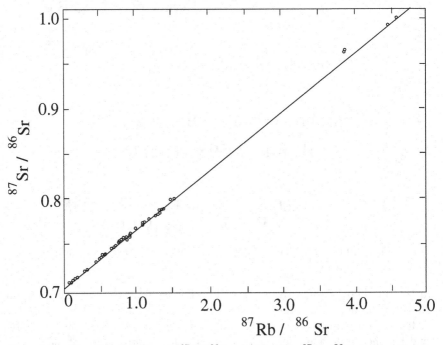

Fig. 5.18. The abundance ratio ${}^{87}\text{Sr}/{}^{86}\text{Sr}$ vs. the ratio ${}^{87}\text{Rb}/{}^{86}\text{Sr}$ for a collection of meteorites [60]. The fact that the points lie on a straight line indicate that the meteorites have a common age of $\sim 4.53 \times 10^9$ yr. At the epoch of formation, the meteorites had a common isotopic ratio ${}^{87}\text{Sr}/{}^{86}\text{Sr} = 0.7003$.

The mean time since the formation of terrestrial heavy elements can be estimated from the abundances of two isotopes of uranium, observed to be ${}^{235}\text{U}/{}^{238}\text{U} = 0.00714 \pm 0.00008$. One uses (5.47) with $A = {}^{235}\text{U}$ and $A' = {}^{238}\text{U}$ but modified to take into account the finite lifetime of ${}^{238}\text{U}$.

$$\left[\frac{N_{235}}{N_{238}}\right]_t = \left[\frac{N_{235}}{N_{238}}\right]_{t=0} e^{-t/\tau_{235}} e^{t/\tau_{238}} , \tag{5.54}$$

with $t_{1/2}(235) = 7.07 \times 10^8$ yr and $t_{1/2}(238) = 4.5 \times 10^9$ yr. The two isotopes are believed to have been formed in nearly equal amounts (Sect. 8.3), so (5.54) implies $t \sim 6 \times 10^9$ yr. The age of the Earth or of meteorites is noticeably less (4.5×10^9 yr) so most the the uranium on Earth had spent a considerable amount of time in interstellar space before condensing on the Earth.

The ages of very old stars have been measured by comparing their atmospheric abundances of uranium and thorium. One uses

$$\left[\frac{N_U}{N_{Th}}\right]_t = \left[\frac{N_U}{N_{Th}}\right]_{t=0} e^{-t/\tau_{238}} e^{t/\tau_{232}} \tag{5.55}$$

where the half-life of ^{232}Th is 1.405×10^{10} yr and for very old stars one can assume that all the uranium is ^{238}U. The oldest stars [61] so far found have a uranium–thorium ratio of about 0.2 implying ages of about 1.2×10^{10}yr (assuming equal initial abundances). This is of the order of the estimated age of the Universe, 1.3×10^{10} yr, indicating that stars formed within the first $\sim 10^9$ yr after the big bang (Chap. 9).

Finally, while (5.47) and (5.48) are generally used to derive the *ages* of samples using a known lifetime, they can also be used to measure the *lifetime* of a nuclide using a sample of known age. While this may seem like a strange way to measure a nuclear lifetime, a comparison with the laboratory-measured lifetime amounts to a comparison of the present-day lifetime with the lifetime in the past. If the two lifetimes are consistent, this verifies the hypothesis that laws and constants governing the decays have not varied.

Very restrictive bounds on such variations have been found using the decay

$$^{187}Re \rightarrow {}^{187}Os\, e^- \bar{\nu}_e \qquad t_{1/2} = 4.35 \times 10^{10} \text{ yr} \quad \text{(laboratory)} . \tag{5.56}$$

The age has also been measured [62] using meteorites whose age was determined by U-Pb or Rb-Sr dating. The ^{187}Re half-life can then be determined by using ^{186}Os as the standard in (5.48). The derived value is $t_{1/2} = (4.16 \pm 0.04) \times 10^{10}$yr is in reasonable agreement with the laboratory measurement. This indicates that the ^{187}Re half-life has changed little over the past 4×10^{10} yr.

The particular interest of using ^{187}Re $\rightarrow {}^{187}$Os is that the tiny energy release, $Q_\beta = 2.64$ keV, makes the decay rate extremely sensitive the the fundamental constants. For example, the value of Q_β is sensitive to the value of the fundamental charge through its effect on the Coulomb energy in the Bethe–Weisäcker formula (2.13), $(0.7103Z^2/A^{1/3})$ MeV. The contribution of this term to the binding energy of ^{187}Re is 698 MeV while that for ^{187}Os is 717 MeV for a difference of 19 MeV. A relative change in the square of the fundamental charge of only 10^{-4} would be sufficient to make ^{187}Os lighter than ^{187}Re so that the former is stable rather than the latter. Smaller changes to the electric charge would change the lifetime of ^{187}Re through the Q_β dependence of β-decay rates calculated in Chap. 4 (4.98). It is therefore simple to transform the limit on the change in the ^{187}Re lifetime into a limit on the time variation of the electric charge, or equivalently on the time variation of the fine-structure constant, proportional to the square of the fundamental charge. Following Dyson [63] a limit $|\Delta\alpha/\alpha| < 2 \times 10^{-6}$ for the variation of α over the last ^{187}Re lifetime was recently derived [64].

5.5.3 Other uses of radioactivity

We mention here a few of the other uses of radiation and radioactivity.

- **Sterilization of foodstuffs**
  This consists of exposing foodstuffs to ionizing radiations in order to destroy insects or micro-organisms and delay the deterioration without altering the edibility. Sources of ^{60}Co emitting ~ 1 MeV photons are most commonly used. This technique generates a loss of the germinal potential. Vegetables are treated this way, such as potatoes, fruits, onions etc. The treatment is simple and produces less alteration of the nutritious properties and taste, compared to classical treatments such as sterilization or chemical treatments. This method has other advantages: it is efficient, non-toxic and has a low cost. It is believed to be "danger-free" and it is under considerable development.
- **Creation of genetically modified plants**
  The irradiation by γ rays of genes of certain plants (wheat, barley, rice, sugar cane, cotton, ...) gives them new properties which can be selected to give better resistance to diseases, to heat, to winter conditions, to unfavorable soils. It also allows one to control the ripening, be it sooner or later, and to improve yields. Radioactive mutation techniques have been known since the 1960's. They have been used in Europe and in the former USSR for the culture of wheat, in the United States for the culture of barley, of beans and of grapefruits, in Pakistan for the culture of rice, in India for cotton and cane sugar, etc. This can be viewed as a primitive form of the more systematic genetic modification now practiced by biologists.
- **Sterilization of insects**
  The same method consists in exposing male insects born in a laboratory to sufficient doses of radiation in order to sterilize them. They are then released in large numbers in infected zones. Female insects who mate with these insects have no descendants and the population of harmful insects decreases progressively. This method has the big advantage that it does not bring chemical pollution, unlike pesticides. It has been successfully used in Japan against the melon fly, in Mexico, in Peru, and in Egypt against the fruit fly, and in Africa against the tsetse fly.
- **Gammagraphy**
  X rays which are used in radiography of the body or materials of low den-

sity, are not penetrating enough to be used for dense or thick materials. In that case, one can use γ rays. The principle is the same: the γ rays irradiate a sample of material and the outgoing rays are recorded on a photographic plate. this reveals possible defects coming from the manufacturing or wearing effects.

- **Radioactive tracers**
If a radio-element is introduced into a body, it is possible to follow its trajectory through the body. By measuring the emitted radiation, tracers allow for instance to follow the displacements of products in circuits of chemical factories, the detection of leaks in dams or in buried pipes. This technique is used in particular to monitor oil pipelines.

- **Fire detectors**
A radioactive source (^{241}Am) ionizes the air permanently. The ionization is modified if smoke particles are present. This modification triggers a warning signal. Such detectors are sensitive to very small amounts of smoke. They are widely used in stores, in factories and in offices.

- **Nuclear batteries**
Radioactive sources such as ^{238}Pu, ^{60}Co and ^{90}Sr are used to construct batteries of several hundred Watts. The heat produced by radioactivity is converted into electricity. Such batteries are used in satellites and in distant meteorological stations. They can function for several years without any maintenance. For example, the Voyager spacecraft was powered by three ^{238}Pu generators. It was launched in 1977, reached Neptune in august 1989, and is now beyond the solar system. Nuclear Batteries are also used in heart pacemakers.

- **Conservation of the artistic objects**
The exposition of works of art and of archaeological documents to radiations can destroy insects, microorganisms and mold that they contain and ensures an excellent sterility. This technique has been used in particular to treat the mummy of Ramses II.
The impregnation of wooden or stone objects by a polymer under gamma irradiation is the principle of the "Nucleart " process. It allows one to treat and to recover pieces of buried water-logged wood.

5.6 Bibliography

1. James E. Turner, Atoms, Radiation, and Radiation protection, John Wiley, New York, 1995.
2. J.E. Coggle, Biological effects of radiation, 1983.
3. Edward Pochin, Nuclear radiation : risks and benefits, Oxford Science Publication, 1983.
4. R. E. Taylor, Radiocarbon Dating: An Archaeological Perspective, Academic Press, Orlando, 1987.

Exercises

5.1 Assume that a given radioactive nuclide has a decay rate λ_1 and that it decays into a daughter nuclide which is itself radioactive with decay rate λ_2. Assume further that at time $t = 0$ there are N_1 parent nuclei and $N_2 = 0$ daughter nuclei. Show that at time $t \geq 0$ the number of daughter nuclei is

$$N_2(t) = N_1 \left(1 - e^{-(\lambda_1 - \lambda_2)t}\right) e^{-\lambda_2 t} \frac{\lambda_1}{(\lambda_1 - \lambda_2)} \, . \tag{5.57}$$

Verify that the limits $t \to 0$, $t \to \infty$, $\lambda_2 \ll \lambda_1$ and $\lambda_2 \gg \lambda_1$ are reasonable.

5.2 One gram of cobalt is placed near the core of a nuclear reactor where it is exposed to a neutron flux of $5 \times 10^{14} \mathrm{cm}^{-2}\mathrm{s}^{-1}$. The cross-section for thermal neutron capture on ^{59}Co is $\sim 40\,$b. Calculate the ^{60}Co activity after 1 week of irradiation.

5.3 About 0.3% of the mass of a human is potassium of which 0.012% is ^{40}K ($t_{1/2} = 1.25 \times 10^9\,$yr). Use the decay scheme shown in Fig 5.1 to discuss the annual dose received by a body from the internal ^{40}K. Argue that almost all of the β^- kinetic energy is absorbed in the body. Use the photon absorption cross-section in Fig. 5.12 to argue that a significant fraction of the 1.46 MeV γ-rays are also absorbed in the body. Estimate then the annual dose received from ^{40}K decay.

5.4 In 1990, archaeologists found, in Pedra Furada in Brazil, human remains whose activity in carbon 14 was $19.6 \times 10^{-4}\,$Bq per gram of carbon. Given that the lifetime of ^{14}C is $\tau = 8200$ years (half life $t_{1/2} = 5700$ years) and that the activity of ^{14}C in living tissues and in atmospheric CO_2 is 0.233 Bq (per gram of C), determine the age of these human remains. Compare the results with the date when the first Neanthropians (i.e. homo sapiens, the species to which you and all other living human beings on this planet belong) appeared in Europe, about 35 000 years ago. See Bahn, Paul G.; "50,000-Year-Old Americans of Pedra Furada," Nature, 362:114, 1993.

5.5 The half-lives of uranium 234,235 and 238 are respectively 2.5×10^5 years, $7.1 \; 10^8$ years and $4.5 \; 10^9$ years. Their relative natural abundances are respectively 0.0057%, 0.72% and 99.27%. Are these data consistent with the assumption that these nuclei were formed in equal amounts at the same time? If not, can one explain the discrepancy knowing that the unstable isotopes $^{234}_{90}$Th and $^{234}_{91}$Pa exist?

5.6 A simple way of making a source of neutrons is to mix an α-emitter with beryllium. The α-particles can then produce neutrons via the exothermic reaction

$$\alpha\,^9\text{Be} \;\rightarrow\; \text{n}\,^{12}\text{C} + 5.7\,\text{MeV}\,. \tag{5.58}$$

The cross-section for this reactions is $\sim 0.4\,\text{b}$ for $E_\alpha \sim 5\,\text{MeV}$. For $E_\alpha < 4\,\text{MeV}$ the cross-section is much smaller because of the Coulomb barrier.

Consider a nucleus emitting $5\,\text{MeV}$ α-particles placed in pure beryllium, $(\rho \sim 1.8\text{g}\,\text{cm}^{-3})$. What is the initial energy-loss rate of the α-particles? How far does an α-particle travel in the beryllium before losing $1\,\text{MeV}$ of energy? What is the probability that a neutron is created by the (α, n) reaction over this distance. What α activity is required for $1\,\text{Bq}$ of neutron activity?

6. Fission

Fission is the spontaneous or induced breakup of a nucleus. While it is "just another reaction" from the physics point of view, it has a special role in public affairs because of its enormous applications in energy production and weaponry.

In this chapter, we start with a general discussion of nuclear energy production. In Sec. 6.2 we catalog the various types of fission reactions and in Sec. 6.3 we show how fission rates can be explained qualitatively by considering fission as a barrier penetration process.

The remaining sections consider the basic practical problems in fission technology. Section 6.4 lists the nuclei that can be used as nuclear fuel. Section 6.5 presents the general conditions for setting up a fission chain reaction. Section 6.6 discusses problems associated with neutrons moderators used in thermal neutron reactors. Section 6.7 introduces the Boltzmann transport equation used to calculate the performance of reactors. Finally, in Sect. 6.8 we describe the operation of the major types of fission reactors.

6.1 Nuclear energy

Nuclear energy is extracted through exothermic nuclear reactions, the two principal types being fission and fusion. Of course, we should remember that radioactivity itself liberates a large amount of energy as first recognized by Pierre Curie and Henri Becquerel. For example, the decay of ^{238}U

$$^{238}\mathrm{U} \rightarrow {}^{234}\mathrm{Th}\,\alpha \qquad t_{1/2} = 4.468 \times 10^9 \mathrm{yr}\,, \qquad Q_\alpha = 4.262\,\mathrm{MeV}\,,$$

generates a power of

$$P = \frac{N_A}{238}\frac{Q_\alpha \ln 2}{t_{1/2}} = 8 \times 10^{-9}\,\mathrm{W\,g^{-1}}\,. \tag{6.1}$$

This tiny amount of power is increased in nuclear reactors by ten orders of magnitude through neutron-induced fission of ^{235}U. This increases Q from 4.668 MeV to $\sim 200\,\mathrm{MeV}$ and decreases the effective lifetime to a few months.

Fission and fusion of nuclei are the main reactions which produce nuclear energy in large amounts.

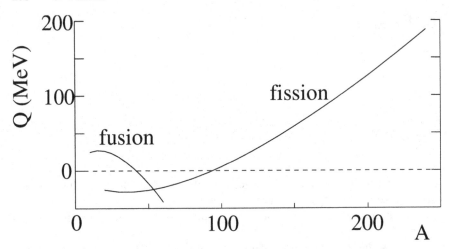

Fig. 6.1. The energy release in fission and self-fusion as predicted by the Bethe–Weizsäcker formula (2.13) for β-stable nuclei. Only nuclei with $40 < A < 95$ are stable against both fission and self-fusion. In this figure, $Q_{\text{fis}}(A, Z)$ is calculated for symmetric fission, $A_1 = A_2 = A/2$ and $Z_1 = Z_2 = Z/2$. $Q_{\text{fus}}(A, Z)$ is calculated for the production of a single nucleus of $A' = 2A$ and $Z' = 2Z$.

A nucleus $^A_Z X$ can breakup, i.e. *fission*, in many ways. In the simplest case, there are two *fragments*

$$^{(A_1+A_2)}_{(Z_1+Z_2)} X \;\rightarrow\; ^{A_1}_{Z_1} X \;+\; ^{A_2}_{Z_2} X \;+ Q_{\text{fis}} \tag{6.2}$$

The *energy release* is

$$Q_{\text{fis}} \;=\; m(A_1 + A_2, Z_1 + Z_2)c^2 - [m(A_1, Z_1) + m(A_2, Z_2)]c^2$$

$$= B(A_1, Z_1) + B(A_2, Z_2) - B(A_1 + A_2, Z_1 + Z_2) \,. \tag{6.3}$$

In the rest frame of the initial nucleus, the sum of the kinetic energies of the fission fragments is Q_{fis}. The difference in binding energies $B(A, Z)$ of the initial and final nuclei is thus transformed into kinetic energy which is eventually transfered to the medium, heating it up.

Symmetrically, two nuclei can undergo fusion if, by a nuclear reaction, they produce a heavier nucleus (plus a light particle y)

$$^{A_1} X_{Z_1} + {}^{A_2} X_{Z_2} \rightarrow {}^A X_Z + y + Q_{\text{fus}} \,. \tag{6.4}$$

If Q_{fus} is positive, the light particle y is necessary to conserve momentum. In the simplest case, y is a photon so $A = A_1 + A_2$, $Z = Z_1 + Z_2$ and

$$Q_{\text{fus}} \;=\; B(A_1 + A_2, Z_1 + Z_2) - B(A_1, Z_1) - B(A_2, Z_2) \,. \tag{6.5}$$

If $Q_{\text{fus}} > 0$, the reaction can take place with the two initial nuclei at rest, in which case Q_{fus} is the energy of the final state photon. As in fission, the difference in binding energies of the initial and final nuclei is transformed to kinetic energy.

The curve of binding energies (Fig. 1.2) shows that the binding energy is maximum for $A \sim 56$ implying

- *heavy* nuclei can *fission* ;
- *light* nuclei can *fuse.*

Figure 6.1 shows Q for symmetric ($A_1 = A_2$) fusion and fission as a function of A of the initial nucleus. We see that only nuclei with $40 < A < 95$ are intrinsically stable against both spontaneous fission and self-fusion. Fortunately for us, both fusion and fission are very slow processes under normal conditions. As we will see in Chap. 7, fusion is inhibited by the Coulomb barrier between positively charged nuclei. As we will see in the Sect. 6.3, spontaneous fission is strongly inhibited by the "fission barrier" resulting from the surface term in the Bethe–Weisäcker formula (2.13).

In practice, we will mostly be concerned with the fission of heavy nuclei, $A \simeq 240$. For two fission fragments with $A \simeq 120$ we can estimate Q_{fis} from Fig. 1.2. For the initial nuclei we have $B/A(240) \simeq 7.6$ MeV, while for the fission fragments we have $B/A(120) \sim 8.5$ MeV. This gives

$$Q_{\text{fis}} \sim 240 \times (8.5 - 7.6) \sim 220 \text{ MeV} \quad . \tag{6.6}$$

(This energy is somewhat greater than that shown in Fig. 6.1 because the figure shows the Q of the initial fission while our estimate includes the energy generated by β-decays of the fission fragments down to β-stable nuclei.) In actual fission processes, there is always a certain number of free *neutrons* produced. Let $\overline{\nu}$ be the average number of such neutrons. Since, by definition, they have no binding energy, we have

$$Q_{\text{fis}} \sim 220 \,\text{MeV} - \overline{\nu} \times 8.5 \,\text{MeV} \quad , \tag{6.7}$$

i.e. for $\overline{\nu} \sim 2.5$, $Q_{\text{fis}} \sim 200 \,\text{MeV}$.

6.2 Fission products

A given nucleus can fission in many ways. For ^{236}U, one possibility is

$$^{236}_{92}\text{U} \rightarrow {}^{137}_{53}\text{I} + {}^{96}_{39}\text{Y} + 3\text{n} \tag{6.8}$$

$$^{137}\text{I} \rightarrow {}^{137}\text{Xe} \, \text{e}^- \, \overline{\nu}_e \quad t_{1/2} = 24.5 \,\text{s}$$

$$^{137}\text{Xe} \rightarrow {}^{137}\text{Cs} \, \text{e}^- \, \overline{\nu}_e \quad t_{1/2} = 3.818 \,\text{m}$$

$$^{137}\text{Cs} \rightarrow {}^{137}\text{Ba} \, \text{e}^- \, \overline{\nu}_e \quad t_{1/2} = 30.07 \,\text{yr}$$

$$^{96}\text{Y} \rightarrow {}^{96}\text{Zr} \, \text{e}^- \, \overline{\nu}_e \quad t_{1/2} = 5.34 \,\text{s} ,$$

or, globally

$$^{236}_{92}\text{U} \rightarrow {}^{137}_{56}\text{Ba} + {}^{96}_{40}\text{Zr} + 3\text{n} + 4\text{e}^- + 4\overline{\nu}_e . \tag{6.9}$$

The original *fission fragments* ^{137}I and ^{96}Y are transformed through a series of β-decays to the β-stable ^{137}Ba and ^{96}Zr. Note the long half-life of ^{137}Cs that makes it effectively stable over a fuel cycle of a nuclear reactor. It is an example of radioactive waste from a fission reactor.

Of course the reaction (6.8) is only one of many possible fission modes. It is necessary to consider the problem statistically. The observed distribution of fragments of neutron-induced fission of ^{235}U, corresponding roughly to spontaneous fission of ^{236}U, is shown in Fig. 6.2. One observes that fission is mainly *binary* : the probability distribution of final products is a two-peak curve. The fission fragments have, statistically, different masses. This is called *asymmetric fission*. In the case of neutron induced fission of uranium 235, one observes:

- A fragment in the "group " $A \sim 95$, $Z \sim 36$ (Br, Kr, Sr, Zr)
- A fragment in the "group " $A \sim 140$, $Z \sim 54$ (I, Xe, Ba).

The two groups are near the magic neutron numbers $N = 50$ and $N = 82$

Because of the large neutron excess of heavy nuclei, the two fission fragments generally are below the line of β-stability. They therefore β$^-$-decay to the bottom of the stability valley. This process is usually quite rapid, though in the above example (6.8) the last β-decay of ^{137}Cs is rather slow. This is an example of the long-lived radioactivity that is important for the storage of waste from nuclear reactors.

We see that the fission process results in the production of a large variety of particles. They can be classified as

- Two fission fragments that are β$^-$-unstable.
- Other "prompt" particles, mostly neutrons emitted in the fission process and photons emitted by the primary fission fragments produced in highly excited states.
- "Delayed" particles mostly e$^-$, $\bar{\nu}_e$, and γ emitted in the β$^-$-decays of the primary fission fragments fragments and their daughters.

Most of the released energy is contained in the initial kinetic energies of the two fission fragments. The kinetic energy of each heavy fragment at the time of fragmentation is of the order of 75 MeV, with initial velocities of roughly 10^7 m s^{-1}. Given their large masses, their ranges are very small $\sim 10^{-6}$ m. The stopping process transforms the kinetic energy to thermal energy.

For a given fissile nucleus we call $\bar{\nu}$ the average number of free neutrons produced, $\bar{\mu}$ the number of β-decays, and $\bar{\kappa}$ the number of photons. The total *energy balance* of a fission reaction is then

$$A \rightarrow B + C + \bar{\nu}n + \bar{\mu}e^- + \bar{\mu}\bar{\nu}_e + \bar{\kappa}\gamma.$$

On average, the various components take the following energies

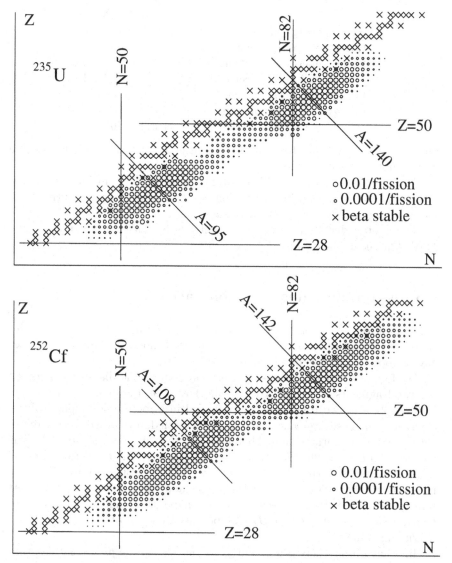

Fig. 6.2. The distribution of fission fragments for neutron induced fission of ^{235}U and for spontaneous fission of ^{252}Cf. The distribution for ^{235}U is dominated by asymmetric fission into a light nucleus ($A \sim 95$) and a heavier nucleus ($A \sim 140$), reasonably near the magic neutron numbers $N = 50$ and $N = 82$. The distribution for ^{252}Cf is broader but still dominated by asymmetric fission. Because of the large neutron excess in nuclei with $A > 230$, almost all fission fragments are below the line of β-stability and therefore decay by β$^-$-emission.

	MeV
Kinetic energy of fragments	165 ± 5
Energy of prompt photons	7 ± 1
Kinetic energy of neutrons	5 ± 0.5
Energy of β decay electrons	7 ± 1
Energy of β decay antineutrinos	10
Energy of γ decay photons	6 ± 1
Total	200 ± 6

Of the ~ 200 MeV, only 190 ± 6 MeV are useful, since the neutrinos escape and do not heat the medium.

The neutrons produced will maintain chain reactions in induced fission processes. For ^{236}U (i.e. for the fission of ^{235}U induced by a neutron) the mean number of produced neutrons is $\bar{\nu} = 2.47$.

The energy distribution of these neutrons is peaked around $E_n = 0.7$ MeV. The mean energy is $< E_n > \sim 2$ MeV.

6.3 Fission mechanism, fission barrier

The complexity of fission seems out of reach of a detailed theoretical description. However, as early as 1939, the liquid droplet model of Bohr and Wheeler gave a good qualitative description.

We follow the potential energy of the system as the distance r between the two fragments A and B varies, as illustrated on Fig. 6.3.

Initially, the two fragments are coalesced in a more or less spherical nucleus. As r increases, the nucleus becomes deformed, its surface area increases compared to the initial shape. Therefore, this deformation *increases* the surface tension energy. On the other hand, the increase in the distance $A - B$ means a decrease of the *Coulomb repulsion* energy between A and B. There is a competition between the nuclear forces and the Coulomb repulsion. At some point, as r varies between r_0 (initial shape of the nucleus) and infinity (separated fragments A and B), the potential energy of the system has a maximum value.

In other words, there is a potential barrier, called the *fission barrier*, that must be crossed in order for the process to occur. One calls *activation energy* E_A, the difference between the maximum of the barrier and the energy of the initial nucleus in its ground state. For nuclei with $A \sim 240$, this energy turns out to be of the order of 6 to 7 MeV.

In order for the original nucleus to decay by spontaneous fission, the barrier must be crossed by the quantum tunnel effect. To get an idea of the factors that determine the difficulty of tunneling through the barrier, it is interesting to compare the surface and Coulomb energies of a spherical nucleus as predicted by the Bethe–Weizsäcker semi-empirical mass formula (2.13)

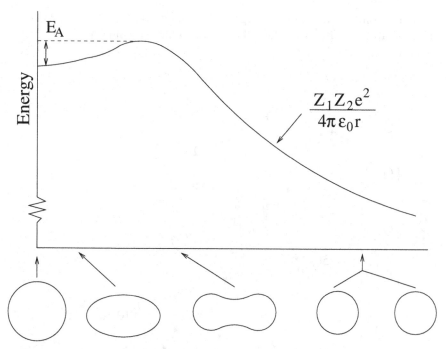

Fig. 6.3. Variation of the energy of a deformed nucleus as a function of the distortion as sketched. For small distortions, the energy increases with increasing distortion because of the increasing surface area. When the two fragments are separated, the energy falls with increasing separation because of the decreasing Coulomb energy. An energy barrier E_A must be crossed for fission to occur.

$$\frac{E_c}{E_s} = \frac{0.7103\,\text{MeV}\; Z^2 A^{-1/3}}{17.804\,\text{MeV}\; A^{2/3}} = \frac{Z^2/A}{25.06}. \qquad (6.10)$$

Nuclei with $Z^2/A > 25$ would be expected to have small barriers because the Coulomb energy (decreasing function of separation) dominates the surface energy (increasing function of separation). In fact, by calculating the surface area and Coulomb energies of a nucleus in the shape of an ellipsoid, it can be shown that the surface area varies twice as fast as the Coulomb energy as the nucleus is deformed while keeping the volume constant. This means that we expect fission to be instantaneous for

$$E_c > 2E_s \quad \Rightarrow \quad \frac{Z^2}{A} > 50. \qquad (6.11)$$

Super-heavy nuclei have $Z/A \sim 1/3$ implying $Z > 150$ for instantaneous fusion. This is an absolute upper limit on the size of nuclei.

Figure 6.4 shows the inverse of the rate of spontaneous fission of selected nuclei as a function of Z^2/A. As expected, the inverse rate decreases rapidly for increasing Z^2/A. For nuclei with $Z < 92$, the lifetime for spontaneous fission becomes immeasurably large making nuclei with $100 < A < 230$ effectively stable in spite of the fact that $Q_{\text{fis}} > 0$ for these nuclei.

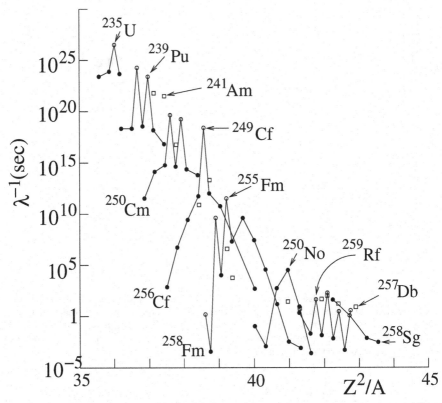

Fig. 6.4. Spontaneous fission lifetimes as a function of the fission parameter Z^2/A for selected nuclei. Circles are for even-Z nuclei, filled circles for even-even nuclei and open circles for even-odd nuclei. Squares are for odd-Z nuclei.

The fission rate can be increased by placing the nucleus in an excited state so as to reduce the fission barrier. This is most simply done by photon absorption:

$$\gamma {}^A_Z X \;\rightarrow\; {}^A_Z X^* \;\rightarrow\; \text{fission} \quad , \tag{6.12}$$

where ${}^A X^*$ is any excited state of the nucleus ${}^A X$. This process is called *photo-fission*.

Figure 6.5 shows the total photo-fission cross-section on ^{236}U as a function of photon energy. The cross-section is negligible for $E_\gamma < 5\,\text{MeV}$, rapidly increases in the range $5\,\text{MeV} < E_\gamma < 6\,\text{MeV}$ and then slowly increases to its maximum value of $0.2\,\text{b}$ at $E_\gamma \sim 14\,\text{MeV}$. Somewhat arbitrarily we can define $\Delta E_S = 5.7\,\text{MeV}$ as the energy that must be added to the ground state

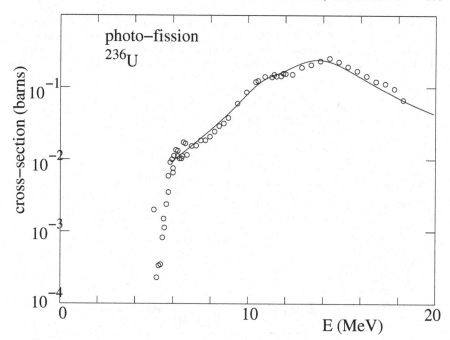

Fig. 6.5. Cross-section for $\gamma\,^{236}\text{U} \to$ fission [30].

of ^{236}U to make the probability for fission reasonably large. This energy is expected to be somewhat less than the height of the barrier E_A

$$\Delta E_S \sim E_A - 1\,\text{MeV}\,. \tag{6.13}$$

This is because it is not necessary to add enough energy to erase the barrier but only enough to make the tunneling rapid.

Table 6.1 lists the values of ΔE_S for selected nuclei in the region $230 < A < 240$. They are all of the order $\Delta E_S \sim 6\,\text{MeV}$.

A second way to induce fission of a nucleus (A, Z) is through neutron absorption by the nucleus $(A - 1, Z)$:

$$\text{n}\,^{A-1}\text{X}_Z \to {}^A\text{X}_Z^* \to \text{fission}\,. \tag{6.14}$$

The effective threshold for neutron-induced fission, i.e. the minimum neutron kinetic energy necessary to give a large probability for inducing fission, is

$$T_\text{n}(A - 1) = \Delta E_S(A) - S_\text{n}(A)\,, \tag{6.15}$$

where S_n is the neutron-separation energy of the nucleus A, i.e. the energy necessary to remove a neutron. The reasoning behind this formula is illustrated in Fig. 6.6 which shows the levels of the systems $A = 236$ and $A = 239$ involved in the fission of ^{236}U and ^{239}U. The ground states of these two nuclei can be transformed to a fissionable state by adding an energy (by photon

Table 6.1. Fission threshold energy ΔE_S and neutron separation energy S_n for selected nuclei (A, Z). ΔE_S gives the effective threshold for photo fission. The effective threshold for neutron-induced fission of the nucleus $(A-1)$ is $T_n = \Delta E_S - S_n$. For the three odd-$(A-1)$ nuclei, $T_n < 0$ so fission can be induced by thermal neutrons.

Fissioning nucleus (A, Z)	ΔE_S (MeV) (A, Z)	S_n (MeV) (A,Z)	T_n(threshold) (MeV) $(A-1, Z)$	neutron target $(A-1, Z)$
$^{234}_{92}$U	5.4	6.9		$^{233}_{92}$U
$^{236}_{92}$U	5.7	6.3		$^{235}_{92}$U
$^{240}_{94}$Pu	5.5	7.3		$^{239}_{94}$Pu
$^{233}_{90}$Th	6.4	5.1	1.3	$^{232}_{90}$Th
$^{235}_{92}$U	5.8	5.3	0.5	$^{234}_{92}$U
$^{239}_{92}$U	6.0	4.8	1.2	$^{238}_{92}$U

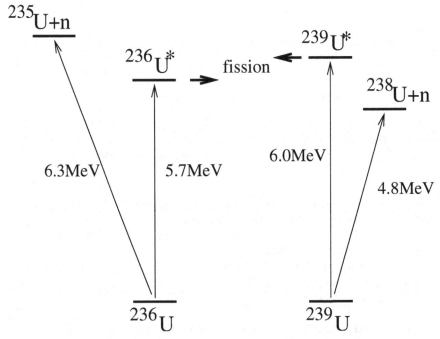

Fig. 6.6. Levels of the systems $A = 236$ and $A = 239$ involved in the fission of ^{236}U and ^{239}U. The addition of a motionless (or thermal) neutron to ^{235}U can lead to the fission of ^{236}U. On the other hand, fission of ^{239}U requires the addition of a neutron of kinetic energy $T_n = 6.0 - 4.8 = 1.2\,\text{MeV}$.

absorption) ΔE_S equal to 5.7 MeV (^{236}U) or 6.0 MeV (^{239}U). A neutron can be removed from the ground states of these two nuclei by adding an energy (by photon absorption) S_n equal to 6.3 MeV (^{236}U) or 4.8 MeV (^{239}U). We see that a ^{235}U nucleus with a free neutron is at a higher energy than the lowest fissionable state of ^{236}U. The addition of a motionless (or thermal) neutron to ^{235}U can thus lead to the fission of ^{236}U. On the other hand, a ^{238}U nucleus with a free neutron is at a lower energy than the lowest fissionable state of ^{239}U, so the addition of a motionless (or thermal) neutron to ^{238}U cannot lead to the fission of ^{239}U but only to the radiative capture of the neutron. Fission of ^{239}U requires the addition of a neutron of kinetic energy $T_n = 6.0 - 4.8 = 1.2$ MeV.

The last column of Table 6.1 gives the values of the neutron-induced fission threshold for selected nuclei. Odd-N target nuclei are fissionable by thermal neutrons ($T_n < 0$), whereas even-N nuclei have a threshold for the kinetic energies of incident neutrons. As illustrated in Fig. 6.6, this is because the last neutron of an odd-N fissioning nucleus is less bound then the last neutron of an even-N fissioning nucleus, as reflected in the pairing term $[\delta(A) = 34 A^{-3/4}]$ in the Bethe–Weizsäcker semi-empirical mass formula.

Figure 6.7 shows the neutron-induced fission cross-sections for ^{235}U and ^{238}U. The cross-section for ^{238}U exhibits the expected effective threshold at $E \sim 1.2$ MeV. The threshold-less cross-section on ^{235}U exhibits the characteristic $1/v$ behavior for exothermic reactions at low energy.

6.4 Fissile materials and fertile materials

The nuclei which are most easily used as fuel in fission reactors are the three even-odd nuclei ^{233}U, ^{235}U and ^{239}Pu which fission rapidly by *thermal* neutron capture.

Of the three fissile nuclides, only ^{235}U exists in significant quantities on Earth, which explains its historical importance in the development of nuclear technology. Terrestrial uranium is (at present) a mixture of isotopes containing 0.72% ^{235}U and 99.3% ^{238}U.

On the other hand, ^{239}Pu and ^{233}U have α-decay lifetimes too short to be present in terrestrial ores. They are produced artificially by neutron capture starting from the *fertile materials* ^{238}U and ^{232}Th:

$$n \, {}^{238}_{92}\text{U} \rightarrow {}^{239}_{92}\text{U} \, \gamma \tag{6.16}$$

$${}^{239}_{92}\text{U} \rightarrow {}^{239}_{93}\text{Np} \, e^- \, \bar{\nu}_e \quad t_{1/2} = 23.45 \, \text{m}$$

$${}^{239}_{93}\text{Np} \rightarrow {}^{239}_{94}\text{Pu} \, e^- \, \bar{\nu}_e \quad t_{1/2} = 2.3565 \, \text{day}$$

and

$$n \, {}^{232}_{90}\text{Th} \rightarrow {}^{233}_{90}\text{Th} \, \gamma \tag{6.17}$$

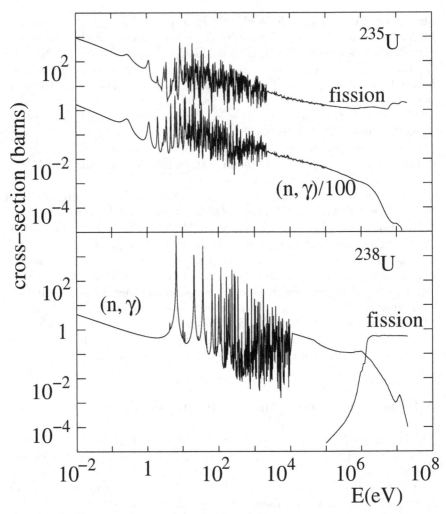

Fig. 6.7. Neutron-induced fission and radiative-capture cross-sections for ^{235}U and ^{238}U as a function of the incident neutron energy. The fission cross-section on ^{238}U has an effective threshold of $\sim 1.2\,\text{MeV}$ while the cross-section on ^{235}U is proportional, at low energy, to the inverse neutron velocity, as expected for exothermic reactions. Both fission and absorption cross-sections have resonances in the range $1\,\text{eV} < E < 10\,\text{keV}$.

$$^{233}_{90}\text{Th} \rightarrow {}^{233}_{91}\text{Pa}\, \text{e}^-\, \bar{\nu}_\text{e} \quad t_{1/2} = 22.3\,\text{m}$$

$$^{233}_{91}\text{Pa} \rightarrow {}^{233}_{92}\text{U}\, \text{e}^-\, \bar{\nu}_\text{e} \quad t_{1/2} = 26.967\,\text{day}$$

In particular, a reactor which burns ^{239}Pu and which contains ^{238}U rods, can produce more Plutonium than it actually consumes owing to the chain (6.16). This is the principle of fast breeder reactors.

6.5 Chain reactions

The induced fission of ^{235}U:

$$\text{n}\,^{235}\text{U} \rightarrow \text{A} + \text{B} + \bar{\nu}\text{n} , \tag{6.18}$$

creates on average $\bar{\nu} \sim 2.5$ neutrons. These secondary neutrons can induce the fission of other ^{235}U nuclei. When they are emitted in a fission reaction, the neutrons have a large kinetic energy, 2 MeV on the average. They can be brought back to thermal energies by exchanging energy with nuclei in the medium via elastic scatters.

Since $\bar{\nu} > 1$, a multiplicative effect can occur. The number of neutrons will be multiplied from one generation to the next and the reaction rate increases accordingly. This is called a chain reaction. In order to evaluate the possibility for a chain reaction to occur, we must know the number k of fission neutrons which will *effectively* induce another fission. The number k is less than $\bar{\nu}$ because a certain fraction of the neutrons will be absorbed by non-fission reactions or diffuse out of the region containing the ^{235}U.

If $k > 1$, the chain reaction occurs. This case is called the supercritical regime. If $k < 1$, the reaction does not develop, i.e. the sub-critical regime. The limit $k = 1$ is called the critical regime.

The only inherent neutron-loss mechanism is radiative capture on the nucleus constituting the fuel

$$\text{n}\,^{A}\text{U} \rightarrow \gamma\,^{A+1}\text{U} \qquad \sigma \equiv \sigma_{(\text{n},\gamma)} . \tag{6.19}$$

If this is the only loss mechanism, then the number of neutrons that induce fission will be

$$\bar{\nu}' = \bar{\nu}\,\frac{\sigma_\text{fis}}{\sigma_\text{fis} + \sigma_{(\text{n},\gamma)}} . \tag{6.20}$$

Table 6.2 gives the values of $\bar{\nu}'$ for pure ^{235}U, ^{238}U, and ^{239}Pu under conditions where the neutrons are "fast" ($E_\text{n} \sim 2\,\text{MeV}$) and thermalized ($E_\text{n} \sim 0.025\,\text{eV}$). For fast neutrons, absorption on ^{235}U and ^{239}Pu is unimportant but for ^{238}U the large absorption cross-section reduces the number of available neutrons from $\bar{\nu} = 2.88$ to $\bar{\nu}' = 0.52$, i.e. sub-critical. For thermal neutrons, absorption reduces the number of available neutrons by $\sim 25\%$ for ^{235}U and ^{239}Pu while, as already noted, there are no fissions of ^{238}U.

Table 6.2. Comparison of selected configurations for nuclear reactors with the last column giving the number k of fission-produced neutrons available to induce further fissions. It is necessary to have $k \geq 1$ for a chain reaction to occur. The neutron energy $E_n \sim 2\,\mathrm{MeV}$ corresponds to "fast" neutron reactors while $E_n \sim 0.025\,\mathrm{eV}$ corresponds to "thermal" neutron reactors. The fuels shown are pure isotopes of uranium and plutonium as well as the natural terrestrial mixture of uranium ($0.7\%\,^{235}\mathrm{U}$) and a commonly used enriched mixture ($2.5\%\,^{235}\mathrm{U}$). σ_{fis} and $\sigma_{(n,\gamma)}$ are the cross-sections (in barns) for neutron induced fission and radiative neutron capture (appropriately weighted for the isotopic mixtures). $\bar{\nu}$ is the mean number of neutrons produced per fission and $\bar{\nu}'$ is the mean number after correction for radiative capture on the fuel mixture. Finally, for thermal neutrons we show, in the final column, the number of neutrons k after multiplying by δ (Table 6.3) to account for neutron losses from radiative capture on the thermalizing medium (moderator). The three thermalizers are normal water, heavy water, and carbon.

E_n	fuel	σ_{fis}	$\sigma_{(n,\gamma)}$	$\bar{\nu}$	$\bar{\nu}'$	k
$\sim 2\,\mathrm{MeV}$	$^{235}\mathrm{U}$	1.27	0.10	2.46	2.28	$= \bar{\nu}'$
	$^{238}\mathrm{U}$	0.52	2.36	2.88	0.52	$= \bar{\nu}'$
	$^{239}\mathrm{Pu}$	2	0.10	2.88	2.74	$= \bar{\nu}'$
$\sim 0.025\,\mathrm{eV}$	$^{233}\mathrm{U}$	524	69	2.51	2.29	1.72 ($^1\mathrm{H_2O}$)
						2.2 ($^2\mathrm{H_2O}$)
						2.0 (C)
	$^{235}\mathrm{U}$	582	108	2.47	2.08	1.56 ($^1\mathrm{H_2O}$)
						2.0 ($^2\mathrm{H_2O}$)
						1.8 (C)
	$^{238}\mathrm{U}$	0	2.7	0	0	0
	$^{239}\mathrm{Pu}$	750	300	2.91	2.08	1.56 ($^1\mathrm{H_2O}$)
						2.0 ($^2\mathrm{H_2O}$)
						1.8 (C)
	$0.7\%\,^{235}\mathrm{U}$	4.07	3.5	2.47	1.33	0.99 ($^1\mathrm{H_2O}$)
						1.3 ($^2\mathrm{H_2O}$)
						1.16 (C)
	$2.5\%\,^{235}\mathrm{U}$	14.5	5.4	2.47	1.8	1.37 ($^1\mathrm{H_2O}$)
						1.8 ($^2\mathrm{H_2O}$)
						1.6 (C)

Reactors using uranium as fuel generally have mixtures of the two isotopes ^{238}U and ^{235}U. It is therefore necessary to take into account fission and absorption by both isotopes. For mixtures not to far from the natural terrestrial mixture, $f_{235} = 0.007$, fast neutron fission and absorption is dominated by the primary isotope ^{238}U. On the other hand, for thermal neutrons fission is due entirely to ^{235}U while absorption is due to both ^{235}U and ^{238}U so for thermal neutrons

$$\bar{\nu}' = \bar{\nu}_{235} \frac{f_{235}\sigma_{\text{fis},235}}{f_{235}(\sigma_{\text{fis},235} + \sigma_{(n,\gamma),235}) + (1 - f_{235})\sigma_{(n,\gamma),238}} . \tag{6.21}$$

As shown in Table 6.2, the natural mixture gives a number of available neutrons $\bar{\nu}' = 1.33$ while increasing f_{235} to 0.025 increases the number to $\bar{\nu}' = 1.8$.

At this point, the values of $\bar{\nu}'$ in Table 6.2 tell us that fast neutron reactors can work with ^{239}Pu or, with somewhat less efficiency, with pure ^{235}U. For reactors using thermal neutrons, the values of $\bar{\nu}'$ indicate that a variety of fuels can yield chain reactions. However, before concluding, we must calculate the number of neutrons lost in the thermalization process. This is done in the next section.

6.6 Moderators, neutron thermalization

The cooling of fission neutrons is achieved through elastic collisions with nuclei of mass $\sim Am_n$ in a moderating medium, as represented in Fig. 6.8. In such a collision, the ratio of final to initial neutron energies as a function of center-of-mass scattering angle θ is

$$E'/E = (A^2 + 2A\cos\theta + 1)/(A + 1)^2 . \tag{6.22}$$

Assuming isotropic scattering in the center-of-mass, a good approximation for neutron energies less than $\sim 1\,\text{MeV}$, one has on the average :

$$\langle E'/E \rangle = (A^2 + 1)/(A + 1)^2 . \tag{6.23}$$

The energy exchange is most efficient for ^{1}H ($A = 1$) where half the energy is lost in a collision, and becomes very inefficient for $A \gg 1$. We will see shortly that a number that is more useful than the mean of E'/E is the mean of its logarithm:

$$\langle \log(E/E') \rangle = -\frac{1}{2} \int_{-1}^{1} \log[E'/E]\,d\cos\theta$$

$$= 1 - \frac{(A-1)^2}{2A} \log \frac{(A+1)}{A-1} . \tag{6.24}$$

(For $A = 1$ this expression reduces to $\langle \log(E/E') \rangle = 1$.)

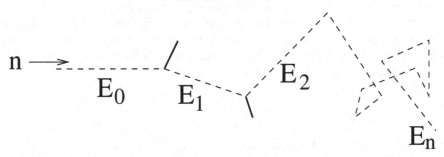

Fig. 6.8. A series of neutron–nucleus elastic scatters leading to the thermalization of the neutron.

Consider a series of collisions as represented in Fig. (6.8). The center-of-mass scattering angles are $\theta_1, \theta_2, \cdots \theta_n$. After n collisions, the mean neutron energy E_n is given by

$$E_n/E_0 = \prod_{i=1}^{n} E_i/E_{i-1} \quad \Rightarrow \quad \log E_n/E_0 = \sum_{i=1}^{n} \log(E_i/E_{i-1}), \quad (6.25)$$

and, in a series of random collisions, there will be after n collisions :

$$\langle \log(E_n/E_0) \rangle = n \langle \log(E'/E) \rangle . \quad (6.26)$$

The average number of collisions N_{col} which are necessary in order to reduce the energy of fission neutrons from $E_{\text{fis}} \sim 2\,\text{MeV}$ to the thermal energy $E_{\text{th}} \sim 0.025\,\text{eV}$, is given by :

$$N_{\text{col}} = \frac{\log(E_{\text{fis}}/E_{\text{th}})}{\langle \log(E/E') \rangle} , \quad (6.27)$$

with the denominator as a function of A given by (6.24) For the three moderators ^1H (light water $^1\text{H}_2\text{O}$) $A = 1$, ^2H (heavy water $^2\text{H}_2\text{O}$) $A = 2$, and C (graphite or CO_2) $A = 12$, the values of N_{col}, are given in Table. 6.3. As expected, hydrogen is the most efficient thermalizer, requiring only ~ 18 collisions, while carbon requires 115.

Neutrons may be lost during the thermalization process by radiative capture on the thermalizing nuclei. Per collision, the probability is

$$p = \frac{\sigma_{(\text{n},\gamma)}}{\sigma_{\text{el}} + \sigma_{(\text{n},\gamma)}} , \quad (6.28)$$

where $\sigma_{(\text{n},\gamma)}$ and σ_{el} are the cross-sections for radiative capture and elastic scattering. The value of p is given for the three moderators in Table 6.3.

The probability that the neutron not be absorbed during the thermalization process,

$$\delta = (1 - p)^{N_{\text{col}}} , \quad (6.29)$$

Table 6.3. Comparison of the three most commonly used neutron moderators in nuclear reactors, water, heavy water and graphite. The cross-sections per molecule for elastic scattering and radiative absorption are σ_{el} and $\sigma_{(n,\gamma)}$. The probability p for absorption per collision is given by the ratio of the elastic cross-section and the total cross-section. The number of elastic collisions N_{col} necessary to thermalize a neutron with $E_n \sim 2\,\mathrm{MeV}$ is given by (6.27). The last column gives the probability of neutron survival during thermalization.

	σ_{el}	$\sigma_{(n,\gamma)}$	$p = \sigma_{(n,\gamma)}/\sigma_{tot}$	N_{col}	$\delta = (1-p)^{N_{col}}$
$^1\mathrm{H_2O}$	44.8	0.664	1.5×10^{-2}	18	0.76
$^2\mathrm{H_2O}$	10.4	10^{-3}	9.6×10^{-5}	25	0.998
C	4.7	4.5×10^{-3}	9.6×10^{-4}	115	0.895

is given in the last column to Table 6.3.

For nuclear reactors using thermal neutrons, the final number of available neutrons for fission is found by multiplying $\bar{\nu}'$ in Table 6.2 by δ from Table 6.3 to get k as the last column in Table 6.2.

From the last column of Table 6.2, we see that there are three main types of theoretically feasible reactors:

- Natural uranium reactors using heavy water or carbon as moderators.
- Enriched uranium reactors. A 2.5% enrichment in $^{235}\mathrm{U}$ allows the use of light water as the moderator.
- Fast neutron reactors work without moderators. The most efficient fuel is $^{239}\mathrm{Pu}$ with $k = 2.74$. The neutron flux is sufficiently high that one often adds a mixture of uranium (generally depleted in $^{235}\mathrm{U}$ after previous use as nuclear fuel) that results in a production of $^{239}\mathrm{Pu}$ through neutron capture on $^{238}\mathrm{U}$ (6.16). Such *breeder* reactors can actually produce more fuel ($^{239}\mathrm{Pu}$) than they consume. Breeder reactors are more complicated than those using thermal neutrons because, in order to avoid thermalizing the neutrons, a liquid containing only heavy nuclei (usually sodium) must be used to evacuate heat from the reactor core.

6.7 Neutron transport in matter

In the previous two sections, we evaluated the possibility for creating nuclear chain reactions by considering the number of neutrons produced in a fission event, and the number of neutrons lost through radiative capture on fuel and moderating nuclei. Here we will consider additional losses due to neutrons escaping outside the sides of the reactor. Roughly speaking the fuel must have a size at least as large as the neutron mean-free path so that the neutrons have a reasonable probability of creating further fissions before escaping.

To go beyond this rough estimate requires a very detailed and complicated analysis. More generally, the construction and the operating of a nuclear reactor require the mastery of the distribution of neutrons both in energy and in space. This is called *neutron transport* in the reactor. It is a very involved problem which necessitates the elaboration of complex computer codes. Several processes occur in the history of an individual neutron; its formation in a fission, its elastic collisions with the various nuclei which are present inside the medium, in particular its slowing down by the nuclei of the moderator, its radiative capture, and finally the new fission that it can induce. Besides that, in a finite medium, one must also consider the number of neutrons that will be lost because they diffuse out of the region containing the fuel. This constraint corresponds to the concept of a "critical mass" of fuel, below which geometric losses necessarily lead to a sub-critical situation. A glance at Fig. 6.9, which shows an actual fuel element (which is itself plunged into the water-moderator) illustrates why the neutron transport is a complicated problem, although all basic ingredients, i.e. the elementary cross-sections and the geometrical architecture of all materials are known.

A detailed study of neutron transport is far beyond the scope of this text. It is both fundamental in nuclear technologies and very complicated to solve. The transport equation is an integro-differential equation whose numerical treatment is in itself an artistry which has been steadily developed for decades in all nuclear research centers. Its complexity comes in part from the fact that it treats the behavior of neutrons both as a function of energy (they can lose energy in collisions) and in space (they scatter). All R&D organizations involved in this problem possess their own "secrets" to address it. Calculations of neutron transport use the Boltzmann equation formalism. In Appendix D we give some indications about how this equation appears in the specific case of neutron transport.

Here, we will consider the problem in a very simple approximation, in order to exemplify why and how the concept of a "critical mass" emerges. The problem is quite simpler if we make the, not totally absurd, assumptions that the neutrons all have the same time-independent energy, and that the medium is homogeneous, though finite in extent.

6.7.1 The transport equation in a simple uniform spherically symmetric medium

We treat a simple system consisting a pure ^{239}Pu fuel with no moderator. The lack of light nuclei in the medium allows us to make the approximation that neutrons do not loose energy in elastic collisions, which simplifies things considerably.

The calculation will result in an estimation of the smallest sphere of ^{239}Pu that will support a nuclear chain reaction. The calculated "critical" radius of this sphere turns out to be

$$R_c = 0.056 \, \text{m} , \tag{6.30}$$

corresponding to a calculated critical mass

$$M_c = 14.7 \, \text{kg} . \tag{6.31}$$

The parameters of the calculation and its result are shown in Table 6.4.

In spheres of radius less than the critical radius, neutrons produced in fission of ^{239}Pu generally leave the sphere before producing a further fission, resulting in $k < 1$. For spheres of greater radius, the neutrons generally produce further fissions (and further neutrons). The number of neutrons increases exponentially.

The size of the critical radius is just somewhat larger than the mean free path of neutrons in ^{239}Pu. This is as expected since it is the mean free path that determines whether neutron escape freely or remain to create further fissions.

Table 6.4. Characteristics of ^{239}Pu needed in the calculation of the critical radius and mass. All cross-sections are given for 2 MeV neutrons.

Elastic scattering cross-section	$\sigma_{el} = 3.45 \, \text{b}$
Neutron-induced fission cross-section	$\sigma_{fis} = 1.96 \, \text{b}$
Radiative capture cross-section	$\sigma_{n,\gamma} = 0.080 \, \text{b}$
Total absorption cross-section	$\sigma_{abs} = \sigma_{n,\gamma} + \sigma_{fis} = 2.04 \, \text{b}$
Total cross-section	$\sigma_{tot} = \sigma_{el} + \sigma_{abs} = 5.87 \, \text{b}$
Neutrons produced per fission	$\bar{\nu} = 2.88$
Neutrons not radiatively absorbed	$k = \bar{\nu}' = 2.74$
Density	$\rho_{239} = 19.74 \times 10^3 \, \text{kg m}^{-3}$
neutron mean free path	$l = (\sigma_{tot}\rho_{239}/m_{239})^{-1} = 0.0343 \, \text{m}$
Critical radius	$R_c = 0.056 \, \text{m}$
Critical mass	$M_c = (4/3)\pi R_c^3 \rho_{239} = 14.7 \, \text{kg}$

We consider neutron transport under the following assumptions.

- The medium is static (i.e. we neglect any small thermal motions); it has a constant density, and it is spherically symmetric, centered at the origin $r = 0$.
- Neutron–neutron scattering is negligible (since the density of neutrons is much smaller than the density of the medium) .
- Neutron decay is negligible, i.e. the neutron lifetime is very large compared to the typical time differences between two interactions.

- The nuclei of the medium are of only one species, (^{239}Pu to be specific) they have a much larger mass than the neutron mass. Therefore, in neutron–nucleus collisions, the neutron kinetic energy is unchanged. In such a collision, the direction on the neutron velocity can change, but not its magnitude.

The neutron distribution is characterized by their density in phase space

$$\frac{\mathrm{d}N}{\mathrm{d}^3p\,\mathrm{d}^3r} = f(\boldsymbol{r},\boldsymbol{p},t) \quad , \tag{6.32}$$

where $\mathrm{d}N$ is the number of neutrons in the phase space element $\mathrm{d}^3p\,\mathrm{d}^3r$. The space density of neutrons and the current describing the spatial flow of neutrons are the integrals over the momentum

$$n(\boldsymbol{r},t) = \int f(\boldsymbol{r},\boldsymbol{p},t)\mathrm{d}^3p \,, \tag{6.33}$$

$$\boldsymbol{J}(\boldsymbol{r},t) = \int \boldsymbol{v} f(\boldsymbol{r},\boldsymbol{p},t)\mathrm{d}^3p \,. \tag{6.34}$$

In the absence of collisions, neutron momenta are time-independent and the flow of particles in phase space is generated by the motion of particles at velocities $\boldsymbol{v} = \boldsymbol{p}/m$. In that case, the density ρ satisfies the *Liouville equation*

$$\frac{\partial \rho}{\partial t} + \boldsymbol{v} \cdot \boldsymbol{\nabla} \rho = 0 \quad .$$

In the presence of collision processes, the Liouville equation (6.7.1) becomes

$$\frac{\partial \rho}{\partial t} + \boldsymbol{v} \cdot \boldsymbol{\nabla} f = \mathcal{C}(f) \,, \tag{6.35}$$

where $\mathcal{C}(f)$ is the term arising from collision processes, for which we will find an explicit form shortly.

The elastic scattering and absorption rates λ_{el} and λ_{abs} are products of the elementary cross-sections, the density of scattering centers n_{239}, and the mean velocity v

$$\lambda_{\mathrm{el}} = v n_{239} \sigma_{\mathrm{el}} \qquad \lambda_{\mathrm{abs}} = v n_{239} \sigma_{\mathrm{abs}} \,. \tag{6.36}$$

The absorption is due to both (n,γ) reactions and to fission

$$\sigma_{\mathrm{abs}} = \sigma_{(\mathrm{n},\gamma)} + \sigma_{\mathrm{fis}} \,. \tag{6.37}$$

The collision term is therefore

$$\mathcal{C}(f(\boldsymbol{p})) = n_{239} \int \mathrm{d}^3p' v(\boldsymbol{p}')\, f(\boldsymbol{r},\boldsymbol{p}',t) \frac{\mathrm{d}\sigma}{\mathrm{d}^3\boldsymbol{p}'}(\boldsymbol{p}' \to \boldsymbol{p}) \tag{6.38}$$

$$- [\lambda_{\mathrm{el}} + \lambda_{\mathrm{abs}}] f(\boldsymbol{r},\boldsymbol{p},t) + S(\boldsymbol{r},\boldsymbol{p}) \,.$$

The first term accounts for neutrons coming from the elements of phase space $\mathrm{d}^3r\,\mathrm{d}^3p'$ which enter the element of phase space $\mathrm{d}^3r\,\mathrm{d}^3p$ by elastic scattering. The second term represents the neutrons which leave the element $d^3r d^3p$ either by elastic scattering or by absorption. The last term $S(r, p)$ is a source term, representing the production of neutrons by fission. We will write its explicit form shortly.

6.7.2 The Lorentz equation

We assume that all neutrons have the same velocity, v, i.e. that the function $f(r, p)$ is strongly peaked near values of momentum satisfying $|p| = m_n v$. This is the case in breeders and in fission explosive devices. The homogeneity assumption is also reasonable in first approximation in these examples.

In that case, the differential elastic scattering cross-section is

$$\frac{d\sigma}{d^3 p'}(p \to p') = p^{-2}\delta(p - p')\frac{d\sigma}{d\Omega} . \tag{6.39}$$

We assume, for simplicity, that the scattering cross section is isotropic

$$\frac{d\sigma_{el}}{d\Omega} = \frac{\sigma_{el}}{4\pi} . \tag{6.40}$$

Using (6.40) we find that the Boltzmann equation (6.35) and (6.38) reduces to the *Lorentz equation*

$$\frac{\partial f}{\partial t} + v \cdot \nabla f = \lambda_{el}(\bar{f} - f) - \lambda_{abs}f + S(r, p) , \tag{6.41}$$

where

$$\bar{f}(r, p, t) = \frac{1}{4\pi} \int f(r, p, t) d\Omega_p , \tag{6.42}$$

is the phase-space density averaged over momentum directions.

It is useful to integrate the Lorentz equation over the direction of the momentum (or the velocity) $d^3\Omega_p$ (we do not integrate on p itself since the modulus of p or v is fixed by assumption). This leads to

$$\frac{\partial n}{\partial t} + \nabla \cdot J = -\lambda_{abs}n + 4\pi S(r) , \tag{6.43}$$

where $4\pi S(r)$ is the momentum integral of $S(p, r)$.

In the or local quasi-equilibrium regime which is of interest here, the mean free path between two collisions is small compared to the size of the medium.[1] We make the usual approximation of Fick's law, where the current is proportional to the density gradient:

$$J = -Dv\nabla n . \tag{6.44}$$

The *diffusion coefficient* D is related to the elastic-scattering rate or to the mean free path by

---

[1] A comparison of the mean free path and the critical radius in Table 6.4 indicates that this is, admittedly, not an excellent approximation.

$$D = \frac{v}{3\lambda_{el}} = \frac{l}{3} \ . \tag{6.45}$$

The value of this coefficient will be justified in Appendix D. In general, the diffusion coefficient depends on the velocity and on the position r, but this difficulty is by-passed within our assumptions.

Inserting (6.45) into (6.44) and (6.43) leads to

$$\frac{\partial n}{\partial t} - Dv\nabla^2 n = -\lambda_{abs}n + 4\pi S(r) \ . \tag{6.46}$$

In the absence of absorption and sources, we recover a Fourier diffusion equation.

6.7.3 Divergence, critical mass

Consider a very simplistic fast-neutron reactor in which we assume that the neutrons have all the same velocity v, and a kinetic energy of $\sim 2\,\mathrm{MeV}$. These neutrons evolve in a homogeneous fissile medium which contains n_{239} $^{239}\mathrm{Pu}$ nuclei per unit volume. The motion of the neutrons is a random walk. At each collision they can be absorbed by the nuclei of the medium, with a cross-section σ_{abs}, or they can be scattered elastically with a cross-section σ_{el}.

The total cross-section is $\sigma_{tot} = \sigma_{abs} + \sigma_{el}$, and the total reaction rate is

$$\lambda_{tot} = n_{239}\sigma_{tot} = v/l \ , \tag{6.47}$$

where l is the mean free path of the neutrons in the medium.

The source term in (6.43) corresponds to the rate of neutron production by fission. If σ_{fis} is the fission cross-section ($\sigma_{fis} < \sigma_{abs}$), and $\bar{\nu}$ is the average number of neutrons produced in a fission, the rate of increase of the density $n(r)$ due to fissions is

$$4\pi S(r) = \bar{\nu}n_{239}\,n(r)\,v\sigma_{fis} \ . \tag{6.48}$$

If we insert the expression (6.44) for J and this source term into (6.43), we obtain the evolution equation for n[2]

$$\nabla^2 n + \frac{(\bar{\nu}\lambda_{fis} - \lambda_{abs})}{vD}n = \frac{1}{vD}\frac{\partial n}{\partial t} \ . \tag{6.49}$$

If we set $k = \bar{\nu}\sigma_{fis}/\sigma_{abs}$, as done previously, we obtain

$$\nabla^2 n + B^2 n = \frac{1}{vD}\frac{\partial n}{\partial t} \ , \tag{6.50}$$

where we have defined

$$B^2 = (k-1)\frac{\lambda_{abs}}{vD} \ . \tag{6.51}$$

---

[2] Traditionally, in neutron transport equations one works with the so-called *neutron scalar flux* $\phi = vn$. In order to avoid introducing too many symbols, we stick to the variable n since this makes no difference in the simple case considered here.

(We assume that $k \geq 1$.)

Since we assume that the medium is finite, spherical, of radius R, vn depends only on the distance r from the center. The conditions we must impose on vn are the following : $vn \geq 0$ for $r \leq R$, and $vn(0,t)$ is finite.

However, equation (6.50) is only valid *inside* the medium. There are no *incoming* neutrons from the outside. In diffusion theory, a simple but accurate empirical way to simulate this condition is to impose that n vanishes at an "extrapolated distance" R_e :

$$n(R_\mathrm{e},t) = 0 \quad \text{with} \quad R_\mathrm{e} = R + 0.71l , \tag{6.52}$$

where l is the mean free path $1/n\sigma_\mathrm{tot}$.

Of particular interest is the *stationary solution* (critical regime) of (6.50), i.e. a solution for which $(\partial n/\partial t = 0)$. We then have to solve $\nabla^2 n + B^2 n = 0$ or, in spherical coordinates,

$$\frac{1}{r}\frac{\mathrm{d}^2}{\mathrm{d}r^2}rn + B^2 n = 0 \quad . \tag{6.53}$$

Setting $u(r) = rn(r)$, this equation is readily solved:

$$u(r) = \alpha \sin Br + \beta \cos Br \quad , \tag{6.54}$$

and, since vn must be regular at the origin,

$$n(r) = \alpha\frac{\sin Br}{r} \quad . \tag{6.55}$$

The limiting condition (6.52) imposes

$$B\,R_\mathrm{e} = \pi \quad . \tag{6.56}$$

In other words, there is only one value R_c of the radius R of the fissile sphere for which a critical regime exists (permanent or stationary regime) :

$$R_\mathrm{c} = \pi/B - 0.71\,\lambda . \tag{6.57}$$

For plutonium

$$D = 1.14\ 10^{-2}\ \mathrm{m} \quad ,$$

$$B = 39\ \mathrm{m}^{-1} \quad .$$

Therefore, there is a *critical radius* R_c and a *critical mass* M_c:

$$R_\mathrm{c} = 5.63\ 10^{-2}\mathrm{m} \quad , \tag{6.58}$$

$$M_\mathrm{c} = = \rho_{239}(4\pi/3)R_\mathrm{c}^3 = 14.7\,\mathrm{kg} \quad . \tag{6.59}$$

For $R \neq R_\mathrm{c}$, a stationary regime cannot occur. One can readily check this by searching for solutions of the type :

$$n(r,t) = e^{\gamma t}f(r) , \tag{6.60}$$

that:

– for $R > R_c$, necessarily $\gamma > 0$, this corresponds to a *supercritical* regime, the system diverges and explodes;
– for $R < R_c$, necessarily $\gamma < 0$, this corresponds to a *sub-critical regime*; the leaks (finite medium) are not compensated and the chain reaction cannot take place. The neutron density decreases exponentially in time.

The calculation leading to the plutonium critical mass is oversimplified. The actual values for critical masses of spheres of pure metals are $M_c = 6 \, \text{kg}$ for ^{239}Pu and $M_c = 50 \, \text{kg}$ for ^{235}U. These values can be reduced if the material is surrounded by a non-fissile medium consisting of heavy nuclei so that neutrons have a high probability of scattering back into the fissile material.

6.8 Nuclear reactors

Fission was discovered in 1939 when Hahn and Strassman discovered the presence of rare-earth elements in uranium after irradiation by neutrons. L. Meitner and O. Frisch then interpreted this production as being due to neutron-induced fission of uranium. This discovery was followed rapidly by applications since, on December 2 1942, Enrico Fermi at the University of Chicago produced a chain reaction in a system consisting in a periodic stack of natural uranium spheres separated by graphite moderators. Fermi thus demonstrated experimentally the notion of criticality of the size of the stack in order to ensure a chain reaction. This was achieved with a very small total power of the system, $\sim 1 \, \text{W}$. Present power reactors attain powers of $\sim 3 \, \text{GW}$. The increase in power does not present by any means the same complication as in fusion, as we shall see in the next chapter. Indeed, in the fission process all phenomena are more or less linear, in (great) contrast with controlled fusion systems.

A fission reactor core consists of the following essential elements

- Fuel elements, generally consisting of bars containing natural uranium, uranium enriched in ^{235}U, or ^{239}Pu. If there is to be a self-sustaining chain reaction, the amount of fuel must be greater than the critical mass defined by geometric losses.
- A heat extraction system, generally a fluid, e.g. water in thermal-neutron reactors or sodium in fast-neutron reactors. Its role is to limit the temperature of the core and, in power reactors, to transfer the core's thermal energy to electric generators.
- (Thermal-neutron reactors only) A moderating system to thermalize the neutrons. This is most simply done by bathing the fuel bars in water. In this case, the moderator also serves as the heat transporter.

In the following subsections, we will briefly describe three basic types of fission reactors, those based on thermal neutrons, fast neutrons, and proposed schemes where reactors are driven by particle accelerators.

6.8.1 Thermal reactors

Pressurized water reactors (PWRs). This is the most widely used category. In PWRs the pressurized water is both the moderator and the coolant. Typical characteristics are given in Table 6.5.

Table 6.5. Structure and operating characteristics of a typical pressurized water reactor.

Volume	$V \sim 27, \mathrm{m}^3$
Initial mass of U	$3 \times 24\,\mathrm{ton} = 72\,\mathrm{ton}$
Initial mass of ^{235}U	$3 \times 0.84\,\mathrm{ton} = 2.52\,\mathrm{ton}$
Initial number of ^{235}U nuclei	$N_{235}(t=0) = 6.4 \times 10^{27}$
Mean neutron flux	$\phi = 5 \times 10^{14}\,\mathrm{cm}^{-2}\mathrm{s}^{-1}$
Mean neutron velocity	$v = \sqrt{3kT/m_\mathrm{n}} = 2.5 \times 10^3\,\mathrm{m\,sec}^{-1}$
Mean neutron density	$n = \phi/v = 2 \times 10^{11}\,\mathrm{m}^{-3}$
Fission rate	$\lambda_\mathrm{fis} = \phi\sigma_\mathrm{fis}N_{235} = 9.4 \times 10^{19}\,\mathrm{s}^{-1}$
Thermal power	$P = \lambda_\mathrm{fis} \times 200\,\mathrm{MeV} = 3000\,\mathrm{MW}$
Final mass of ^{235}U	$3 \times 220\,\mathrm{kg} = 660\,\mathrm{kg}$
Final mass of ^{238}U	$3 \times 23\,\mathrm{ton} = 70\,\mathrm{ton}$
Final mass of ^{239}Pu	$3 \times 145\,\mathrm{kg} = 435\,\mathrm{kg}$
Final mass of fission products	$3 \times 400\,\mathrm{kg} = 1.2\,\mathrm{ton}$

The fuel rods are made with ceramic pellets of UO_2, typically enriched to about 3% in ^{235}U. The pellets are sealed inside tubes made of zirconium, a material chosen for its low neutron absorption cross-section and its material strength. The tubes are arranged in bundles in a steel structure inside which the coolant circulates. The bundles are secured and are arranged vertically in a rigid configuration called the core structure, as shown on Fig. 6.9. The core is inside a massive steel pressure container as shown on Fig. 6.10.

The maximum permissible power is determined by the requirement that the temperature in the fuel rods be less than the melting point of UO_2, 3100 K. Assuming the neutron flux is constant inside the tube, the emitted heat per unit volume, Q, is position independent. The integration of the heat transfer equations gives

$$Qa^2 = 4k(T_\mathrm{max} - T_\mathrm{s}) \tag{6.61}$$

where a is the tube radius, T_max is the maximal temperature inside the tube, T_s is the temperature of surface of the tube and k is the conduction coefficient of the uranium oxide. Using $T_\mathrm{max} = 1800\,\mathrm{C}$, $T_\mathrm{s} = 450\,\mathrm{C}$ and $k = 0.06\,\mathrm{W\,cm}^{-1}\mathrm{K}^{-1}$, we have $Qa^2 = 300\,\mathrm{W\,cm}^{-1}$. For a diameter of 1 cm, we obtain

$$Q\rho = 40\,\mathrm{W\,g}^{-1} . \tag{6.62}$$

Fig. 6.9. Fuel element holder for a PWR reactor. The figure shows the control rods in place. When filled, the structure contains 264 zirconium tubes of length 4 m and diameter 9.5 mm containing pellets of UO_2.

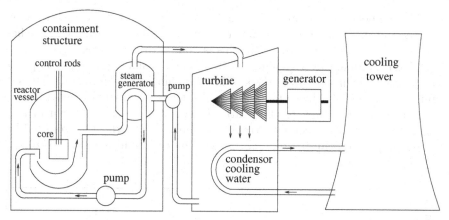

Fig. 6.10. Schematic of a pressurized water reactor.

(The density of the oxide is $\sim \rho = 10\,\mathrm{g\,cm^{-3}}$.) The size of the reactor is therefore fixed by the cooling conditions. For 3000 MW, we need a fuel mass of

$$M_{\mathrm{fuel}} = \frac{3000\,\mathrm{MW}}{40\,\mathrm{W\,g^{-1}}} = 75\,\mathrm{ton} . \tag{6.63}$$

This corresponds to 10^5 m of fuel tubes of 1 cm diameter.

The neutron transport considerations (efficient thermalization, minimal losses out the edges, etc.) imposes a moderation ratio (i.e. the ratio between the volume of water and the total volume of the tubes) of roughly 3. The reactor therefore has a diameter of at least three meters. Such large diameters and heights are necessary in order to limit the neutron leakage.

Water is used both as a moderator and as a coolant fluid. Its maximal temperature in the vicinity of the zircalloy tubes is 300 C. In order to prevent the water from boiling, it is permanently pressurized to 150 atmospheres. It circulates and feeds a heat exchanger which leads to a secondary water circuit which generates steam to drive turbines for the production of electricity. The maximum efficiency with which the thermal energy of the steam can be transformed to work on the turbines and then to electrical energy is limited by Carnot's principle

$$\epsilon < 1 - \frac{T_1}{T_2} , \tag{6.64}$$

where T_1 and T_2 are the final and initial water temperatures. T_1 is fixed by a cold source, generally a river or the ocean.[3]

---

[3] The local temperature increase of water has an obvious ecological impact. In particular it favors the development of various unwanted amebae and other microorganisms.

At the beginning of operation, fissions are initiated with a small neutron source and the fission rate is allowed to increase until the neutron flux reaches $\sim 5 \times 10^{14} cm^{-2} s^{-1}$. Precisely how this is done is discussed below.

After ignition, the original fuel is gradually transformed into fission products and trans-uranium elements. The abundances of selected nuclei as a function of time are shown in Fig. 6.11. The trans-uranium elements are produced by neutron capture. The path of neutron captures is shown in Fig. 6.12. By far the most common trans-uranium is ^{239}Pu which is produced by radiative neutron capture on ^{238}U followed by two rapid β-decays (6.16).

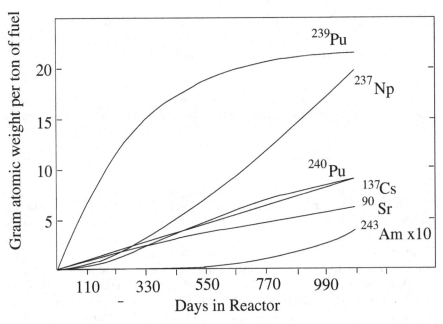

Fig. 6.11. The buildup of selected fission products and trans-uranium elements inside a nuclear reactor [65]. Long-lived fission-products (e.g. ^{137}Cs) generally have abundances that increase linearly with time. The trans-uranium nuclide ^{239}Pu is produced from a single neutron capture on ^{238}U followed by two rapid β-decays and therefore increases linearly at small time and then reaches an equilibrium when its production rate is balanced by its destruction rate from induced fission and neutron capture. The production of ^{237}Np requires two neutron captures (on ^{235}U) and therefore has an abundance that varies quadratically with time.

The fuel can last 3 years before it is too contaminated by fission products. One third of the reactor fuel is renewed each year. Each recharge contains 24 tons of enriched uranium.

When it leaves the reactor, the fuel still contains 23 tons of uranium enriched to 0.95%, i.e. 220 kg of ^{235}U and 145 kg of plutonium. 271 kg of the

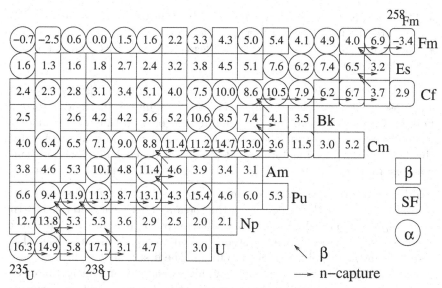

Fig. 6.12. The production of trans-uranium elements by neutron capture on ^{235}U and ^{238}U in a nuclear reactor. The production proceeds by a series of neutron captures and β^--decays. The figures in each box give the logarithm of the half-life in seconds ($7.5 \Rightarrow t_{1/2} = 1\,\mathrm{yr}$) and the shape of the box gives the dominant decay mode. Short-lived nuclei ($\log t_{1/2} < 6$) generally decay while long-lived nuclei capture neutrons. The sequence ends at ^{258}Fm that decays by spontaneous fission with $t_{1/2} = 370\,\mu$ s. Elements beyond ^{258}Fm cannot be fabricated via neutron capture.

^{238}U have been burnt by neutron absorption leading to the production of ^{239}Pu. Altogether, 826 kg of uranium have been consumed.

This balance shows that the efficiency is poor. One consumes 92 tons of natural uranium in order to recover the fission energy of 826 kg of nuclear matter, i.e. 0.9%. This observation explains the interest for breeders, which can increase the energetic potential by a factor of 100.

Heavy water thermal-neutron reactors. Despite its high price, heavy water has been chosen as a moderator in the Canadian system CANDU. In such reactors, one can use natural uranium as a fuel (Table 6.2), which is an obvious advantage and compensates for the price of the moderator, since no enrichment is necessary.

In the CANDU system, the pellets are placed in pressurized tubes where heavy water circulates at a temperature of 200 C (pressure of 90 atm). Heavy water is both the moderator and the coolant.

Graphite-gas thermal-neutron reactors. In these systems, the fuel is natural metallic uranium. The moderator is graphite, and the coolant is carbon dioxide CO_2.

Among the drawbacks, there are :

- The use of metallic uranium whose low melting point limits the core temperature and therefore the thermal efficiency.
- The smaller moderation power which leads to larger sizes than water reactors for the same power.

The RMBK reactors in the former Soviet Union are slightly different. They use weakly enriched uranium (1.8%). Graphite is the moderator and boiling water is the coolant. The major drawback is that k can increase when the temperature increases (see below), and therefore increases when the density decreases due, for example, to loss of the coolant. Therefore, such reactors are not self-regulated, as illustrated in the Chernobyl accident.

Void coefficient. A very important characteristic of the various reactors is the variation of k with the density of the coolant. This is characterized by the *void coefficient*. This determines the behavior of the reactor if, for instance, there is a leak in the coolant system, and the temperature increases. This results from the combination of two effects. The first is that if the amount of moderator diminishes, the neutrons have higher energies and deposit more energy. However, if the neutrons are more energetic, the cross-sections decreases as can be seen on Fig. 6.7, and this reduces the neutron flux. Altogether, in PWR reactors, the void coefficient is negative, and therefore such reactors are self regulated against fluctuations or leak of the coolant.

In RMBK reactors, the net effect of these two opposing characteristics varies with the power level. At the high power level of normal operation, the temperature effect predominates, so the global void coefficient is negative. However, at a lower power output of less than 20% the maximum, the positive void coefficient effect is dominant and the reactor becomes unstable and prone to sudden power surges. This was a major factor in the development of the Chernobyl accident (which was above all the product of a lack of "safety culture").

Reactor control . The number density of neutrons obeys the equation

$$\frac{dn}{dt} = \lambda_{\mathrm{fis}}(k - 1)n \tag{6.65}$$

where k is the number of neutrons produced per fission, taking into account absorption and geometrical losses. The fission rate per neutron is given by

$$\lambda_{\mathrm{fis}} = n_{\mathrm{fuel}}\langle\sigma_{\mathrm{fis}}v\rangle \tag{6.66}$$

where n_{fuel} is the number density of fissionable nuclei and $\langle\sigma v\rangle$ is the mean fission cross-section times velocity. The solution is

$$n = n_0 \exp\frac{(k - 1)t}{\tau} \tag{6.67}$$

where $\tau = 1/\lambda_{\mathrm{fis}}$ is of the order of 0.1 ms in a thermal reactor. If $k > 1$, we observe a rapid exponential increase of the flux and the reactor will become impossible to control. For instance, if k=1.1, in one tenth of a second the power of the reactor is multiplied by 20000.

This would have been a terribly difficult problem to solve technically, if it had not been for the existence of a small number of *delayed neutrons* produced in fission reactions. Most neutrons are emitted immediately in the fission reaction. However, a small number of neutrons (7.5 per thousand for ^{235}U) are emitted by highly-excited daughters of β-unstable fission fragments (Sect. 2.7). An example of a delayed neutron, due to the decay of ^{87}Br, is shown on Fig. 6.13.

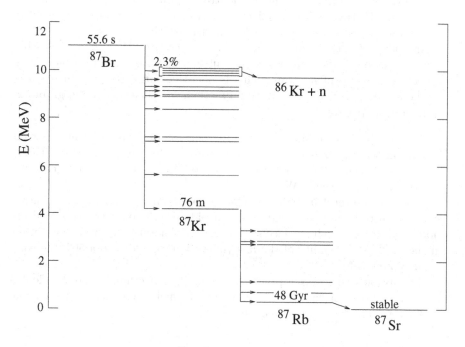

Fig. 6.13. Decay diagram of ^{87}Br. 2.3% of the β-decays are to excited states of ^{87}Kr that have excitation energies greater than the neutron separation energy of ^{87}Kr, $S_n = 5.518$ MeV. These states decay by neutron emission.

Before the reactor is ignited, control rods are in the core of the reactor. The rods are made of steel impregnated with strong neutron absorbers such as boron or cadmium. If the control rods are deep inside the core, the chain reaction cannot occur. The rods are then slowly withdrawn to allow for a steady regime. In case the neutron flux increases too much the rods are lowered in the core. These operations take several seconds to complete. In order for this process to effectively prevent an supercritical reactor from diverging, we must have

$$k < \frac{1}{1-\beta}, \tag{6.68}$$

where β is the fraction of fission neutrons that are delayed neutrons. If this condition is not satisfied, the divergence would be due to prompt neutrons and it would not be possible to control it by mechanically lowering the control rods.

Another characteristic that is important for reactor control is that k should decrease if the temperature increases by error or fluctuation or if the cooling fluid is lost. This is the case for PWRs:

- Temperature increase. In PWRs, k decreases with increasing temperature because of the resonances in the neutron capture of ^{238}U (Fig. 6.7). The increased thermal agitation effectively widens the resonances, resulting in increased neutron absorption before thermalization.
- An unexpected emptying of the coolant in the core. Here, k decreases because the loss of fluid increases the neutron mean-free-path thus increasing geometric losses of neutrons. This is not necessarily the case in reactors where the cooling and moderating functions are separated, as in graphite-moderated reactors.

Present projects to improve reactor performance are concentrating on increasing security. For example the European Pressurized Reactor (EPR) project, a French–German collaboration should design with more secure reactors than the present PWRs.

There are several ingredients. One is to use more sophisticated control systems. In particular enriched boron carbide (B4C) control rods, sealed inside hafnium tubes will replace soluble boron. Another improvement will be in the mechanical quality of the fuel tubes. This can lead to a 50% increase in the power delivered.

Finally, several research programs are made concerning severe accidents, in particular the risk that the "corium," i.e. the melted core which is a magma at 3000 K, can traverse the steel vessel.

6.8.2 Fast neutron reactors

Fast neutron reactors can be *breeders* that produce more nuclear fuel than they consume, by using an intermediate fertile nucleus such as ^{238}U or ^{232}Th. This is possible because more than two neutrons per fission are produced in fast-neutron ^{239}Pu reactors (Table 6.2). One of these neutrons can be used to maintain the chain reaction and the others can create further ^{239}Pu via neutron absorption on ^{238}U. If the probability for this to happen is sufficiently close to unity, the ^{239}Pu destroyed by fission can be replaced by a ^{239}Pu created by radiative capture on ^{238}U. The final result is that ^{238}U is the effective fuel of the reactor.

The following remarks are in order. Consider, for definiteness, a breeder with the fertile nucleus ^{238}U. The fissile nucleus is ^{239}Pu for two reasons. Firstly, it is produced in neutron absorption by ^{238}U, which leads to a closed cycle, secondly it is produced abundantly in nuclear technologies, whereas one can only rely on the natural resources of ^{235}U. In order for a fertile capture of a neutron to produce an appreciable amount of the fissile ^{239}Pu inside the fuel, the probability for this capture must not be too small compared to the probability that the various nuclei in the medium undergo fission. This probability depends both on the amounts of ^{239}Pu, ^{235}U and ^{238}U, and on the physical design of the fuel elements. It can be calculated in terms of the amount of various nuclides and of the capture and fission cross-sections of, respectively, ^{238}U and the pair ^{239}Pu $- \, ^{235}$U. We can read off from table (6.2) that for thermal neutrons one has $\sigma_\gamma(238)/\sigma_f(239) \sim 3.6 \, 10^{-3}$ whereas for fast neutrons, on the contrary, the same ration is of order 1, i.e. 300 times larger. The same feature appears for the ^{232}Th $- \, ^{233}$U pair. This is why fast neutron reactors are used in breeders.

Furthermore, this also explains why in ^{238}U - ^{239}Pu breeders, the design of fuel elements consists in a central core of ^{239}Pu surrounded by a mantle of ^{238}U *depleted* in ^{235}U in order to lower the amount of fission in the external fertile region.

Consequently fast neutron reactors are used as breeders, and two fertile-fissile pairs are *a priori* possible, ^{238}U $-^{239}$ Pu and ^{232}Th $-^{233}$ U. Present nuclear industry is oriented toward the first, because the thermal reactors produce plutonium which is separated in the fuel processing operation.

The ^{232}Th $- \, ^{233}$U couple is under study at present. It has many advantages, among which that it does not lead to appreciable amounts of dangerous trans-uranium elements such as americium and curium. These "minor actinides" are dangerous, because they are produced in appreciable amounts and they have half-lives, and therefore activities, lying in the dangerous region, neither small enough to decay sufficiently rapidly on a human scale nor long enough to be ignored, such as natural uranium and thorium. The half lives of some of these isotopes are 432 years for ^{241}Am, 7400 years for ^{243}Am and 8500 years for ^{247}Cm.

The consequences of using fast neutrons. The use of fast neutrons has several important consequences:

- One must avoid the slowing down of neutrons through the presence of light nuclei. In particular, water cannot be used a coolant.
- Since fission cross-sections are much smaller than with thermal neutrons, so it is more difficult to reach the critical regime.
- Great care must be taken concerning the mechanical damage caused by the fast neutrons to the structures and construction materials.

In order to obtain the divergence of the reactor, one must use a fuel containing a high proportion of ^{239}Pu (of the order of 15 %). The mass of fissile material inside a breeder is therefore larger than for a thermal neutron reactor. In the core of the Superphenix breeder in France, there was the equivalent of 4.8 tons of ^{239}Pu for an electric power of 1200 MW, whereas the core of a usual thermal reactor contains roughly 3.2 tons of ^{235}U for an electric power of 1300 MW. Therefore, the initial investment in fissile material is larger for a fast neutron reactor. This is an economic drawback, but it has the advantage of having a more compact core and a larger neutron flux.

Materials. In the core of a fast-neutron reactor one must minimize the neutron leakage in order to ensure breeding. Therefore the core has a particular composition. There are three concentric shells:

- An internal shell composed of fissile fuel. In the case of Superphenix, this was a mixture of 15 % of plutonium oxide and of 85 % of uranium oxide.
- An intermediate shell of fertile material called the "mantle " in which the ^{238}U is converted into ^{239}Pu.
- An external shell contains steel elements which protect the vessel of the reactor and backscatter neutrons into the core.

A breeder produces more plutonium than it burns. It can therefore self-feed itself in fissile plutonium, provided that it is associated with a processing plant. This plant extracts from the irradiated fuel the useful plutonium necessary for further use.

Under these conditions, everything occurs as if the reactor consumes only ^{238}U. As an example, the yearly consumption of uranium by the reactor Superphenix (1200 MW electric power) was one ton of ^{238}U (an initial 4.8 ton of plutonium was provided). This is to be compared with the consumption of a PWR of 900 MW electric power which uses 92 ton of natural uranium per year.

This explains the interest of breeders in view of the preservation of uranium resources.

The use of sodium. Fast neutron reactors must use a coolant fluid with only heavy nuclei so as to avoid neutron-energy losses. For this reason, water cannot be used, and the general choice is to use liquid sodium.

Melted sodium has very good thermal exchange properties. Furthermore, it melts at 98 C and boils at 882 C. It is used at a maximal temperature of 550 C, and it does not necessitate any pressurizing, which is a favorable feature for the mechanical conception and safety of the installations. Another feature is that it has hydraulic properties similar to water at room temperature, which is a useful coincidence for testing materials.

Another favorable feature is that sodium can be used at higher temperatures than water, which improves the thermal/electric conversion factor.

On the other hand, sodium has the big drawback that it burns spontaneously in the air and in water. This is a major disadvantage for security considerations. Furthermore, sodium becomes radioactive by neutron activation. Therefore, two circuits are necessary. In order to prevent the primary circuit from leaking into the secondary circuit, the latter is maintained at a higher pressure than the former.

6.8.3 Accelerator-coupled sub-critical reactors

We have made much of the fact that for a chain reaction to be self-sustaining, the number of neutrons per fission available to produce further fissions must be greater than unity, $k > 1$. Sub-critical system ($k < 1$) still generate energy (200 MeV per fission) but the nuclear "fire" must be continuously "re-lit" by the injection of neutrons from an external source. Such systems have the advantage that they cannot burn out of control and can be stopped by removing the neutron source.

The most convenient neutron source is a particle accelerator that directs a beam of charged particles on a target. This produces neutrons through (spallation) reactions that breakup nuclei in the target. The target is surrounded by a sub-critical mantle containing fissile material. A schematic example is shown in Fig. 6.14. Such accelerator-coupled detectors are often called *hybrid reactors*.

The spallation target. The interaction of protons with energies $E_p >$ 100 MeV in a target sufficiently thick to stop the beam, gives rise to a copious emission of neutrons. They are produced both in the primary proton-nucleus reaction and by further interactions of secondary particles $(p,n,\pi...)$ inside the thick target. This results in a cascade contained in the target cylinder, whose length is roughly the stopping distance of the incident beam. The diameter of the beam is optimized so that a maximum number of neutrons leave the target and interact with the outside material.

The mean number ν_p of spallation neutrons emitted per incident proton is observed to be

$$\nu_p \sim 30 \times E_p \tag{6.69}$$

where E_p is the incident beam energy in GeV. The high value of this number determines, as we shall see, the energetic feasibility of the system. The energy spectrum of the neutrons emitted goes from a few keV to the beam energy according to their production mechanism (spallation, fission, evaporation). The low energy part (1–2 MeV) is favored by a thick target, compared to the high energy part which corresponds to direct spallation.

The sub-critical system. The thick target is surrounded by a sub-critical system characterized by $\bar{\nu}$ and k, the numbers of neutrons per fission before and after correction for absorption and geometrical losses.

The ν_p neutrons injected into the sub-critical system by the proton are therefore successively multiplied by k. The total number of neutrons per incident proton is equal to:

$$N_t = \nu_p(1 + k + k^2 + ...) = \frac{\nu_p}{(1 - k)} \tag{6.70}$$

Among these N_t neutrons, $N_t - \nu_p$ are produced by fission. In this respect, we notice that the higher the value of $k < 1$, the larger the proportion of fission neutrons in the sub-critical medium. Therefore, the neutron energy

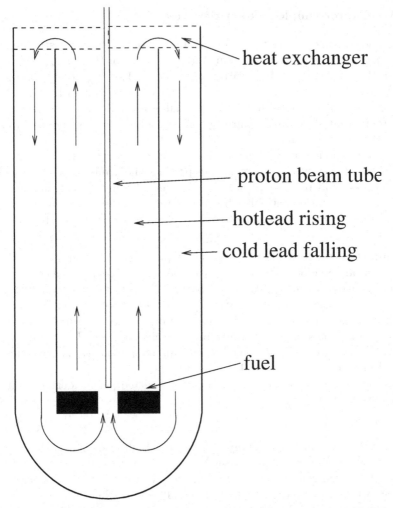

Fig. 6.14. A possible design of a hybrid reactor [66]. The proton beam is directed downward entering a region containing molten lead and surrounded by thorium fuel. The neutrons created in the lead induce fission in the thorium. The heat generated causes the lead to rise toward heat exchangers.

spectrum will be determined more by the sub-critical medium than by the origin of the primary neutrons.

Since each fission produces $\overline{\nu}$ neutrons, the total number of fissions per incident proton is equal to :

$$N_{\text{fis}} = \frac{(N - \nu_{\text{p}})}{\overline{\nu}} = \frac{\nu_{\text{p}}}{\overline{\nu}} \frac{k}{1 - k} \quad . \tag{6.71}$$

The gain of the system, i.e. the energy released in fission divided by the incident proton energy is

$$G = \frac{N_t E_{\text{fis}}}{E_p} = \frac{E_{\text{fis}}}{E_p} \frac{\nu_p}{\overline{\nu}} \frac{k}{1-k} ,$$ (6.72)

where $E_{\text{fis}} \sim 200\,\text{MeV}$ is the energy release per fission. The thermal power is:

$$P(\text{MW}) = E_{\text{fis}}(\text{MeV}) I(\text{A}) \frac{\nu_p}{\overline{\nu}} \frac{k}{1-k}$$ (6.73)

where $I(\text{A})$ is the beam current. We notice that the beam intensity necessary to reach a given power of the reactor decreases as criticality ($k = 1$) is approached.

This system can achieve an energetic self-sufficiency condition provided the power delivered by the reactor is greater than the power which is necessary in order to make the accelerator run (neglecting all other energetic needs). In fact, given the powers under consideration, nearly all the power given to the accelerator is absorbed by the beam. The self-sufficiency condition is expressed by:

$$G\epsilon_1\epsilon_2 > 1$$ (6.74)

where ϵ_1 and ϵ_2 represent respectively the thermal conversion efficiency of the reactor, and the electric efficiency of the accelerator. This condition, which is called the "break-even" provides a definition of a minimum value k_{min} for k, below which there is no longer energetic self-sufficiency :

$$k_{\text{min}} = \frac{1}{1 + \epsilon_1\epsilon_2 \left(E_{\text{fis}}/E_p\right)\left(\nu_p/\overline{\nu}\right)} .$$ (6.75)

The value of k_{min} lies between 0.61 and 0.72, depending on the values of ϵ_1 and ϵ_2. The factor ϵ_1 depends mainly on the temperature of the reactor; it lies between 0.3 and 0.4. The other factor is more difficult to estimate.

The high intensity accelerator. The intensity of a high energy beam for a system whose power is equivalent to a standard reactor of 3000 MW (thermal) lies between a few mA to 360 mA. State-of-the-art accelerators have somewhat less power. The most powerful are either linear accelerators (linacs) working in a continuous mode, (e.g. the Los Alamos 800 MeV, 1 mA linac) or cyclotrons, (e.g., the PSI 600 MeV, 0.8 mA cyclotron).

The technological feasibility of reaching the necessary power requires progress in several areas. Beam losses must be limited to very small values (of the order of 10^{-8}m^{-1} for a linac) in order to avoid activating the structures of the accelerator up to a level where manual interventions become impossible. The efficiency of the RF system (radio frequency) must be increased in order to obtain the best possible efficiency ϵ_2. This efficiency ϵ_2 depends on the losses by Joule effect in the surface of the RF cavities. The power loss in the cavities is of the order of several times that of the beam for a hot linac and it becomes negligible for a superconducting linac.

The proposal shown in Fig. 6.14 is based on the thorium cycle with fast neutrons. The fertile material is ^{232}Th. The fissile material is ^{233}U. The proposal is called an *energy amplifier* because the gain which is aimed at is very large ($k = 0.98$). Molten lead is used both as spallation source and as coolant, using its natural convection. The advantages of such a system are the following :

- The thorium cycle produces little plutonium and heavier elements. It is therefore "cleaner " than the Uranium cycle.
- Thorium is twice as abundant as uranium and does not require isotopic separation.
- If one masters the technology of molten lead, one avoids the drawbacks of liquid sodium.
- The accelerator introduces a better control of the system thanks to the quickness of reactions (of the order of a microsecond) compared to the mechanical handling of control bars (of the order of one second).

6.8.4 Treatment and re-treatment of nuclear fuel

Nuclear fuel requires considerable treatment before burning and careful "management" after burning because of its high radioactivity. The global processing of the nuclear fuel starts with the extraction of the ore. This is followed by the concentration, conversion and enrichment of the uranium, the manufacturing of the fuel elements, and, finally the processing of the used fuel, and the disposal and storage of the waste.

For thermal-neutron reactors, the production of 75 ton of fuel requires 156 ton of natural uranium. The unburned natural uranium contains 1.1 ton of ^{235}U out of which only 775 kg is retained in the enriching procedure.

The most critical step in the preparation of uranium-based fuels is isotopic separation. The possibilities for this are

- Diffusive enrichment using the higher drift speed of gaseous ^{235}UF$_6$ compared to ^{238}UF$_6$ when traversing a capillary system. This is the most commonly used system.
- Centrifugal enrichment using higher centrifugal force on ^{238}UF$_6$ compared to that on ^{235}UF$_6$ in high-speed rotors.
- Mass-spectrometer enrichment using the differing values of Q/M of the two isotopes. This method is only possible for small quantities.
- Laser ionization methods, using the slight isotopic shifts in atomic spectra of the two isotopes. Isotopes are selectively ionized by photon absorption and then collected in an electric field.

The nuclear fuel stays between three and four years in the core of the reactor. Gradually, it loses its fissile matter and is enriched in plutonium and other trans-uranium elements, and fission fragments many of which are highly radioactive.

In 100 kg of initial fuel, i.e. 3 kg of ^{235}U and 97 kg of ^{238}U, there remains, after three years of running:

- 0.9 kg of ^{235}U (2 kg have fissioned)
- 97 kg of ^{238}U (2 kg have been transformed into ^{239}Pu)
- 0.6 kg of ^{239}Pu (in the above 2 kg, 1 kg fissioned)
- 1.7 kg of fission products.

It is necessary to renew the fuel elements periodically. The reactor is stopped each year for 3 or 4 weeks. One third or one quarter of the fuel is renewed.

The fuel re-processing . The re-processing consists of separating, in the irradiated fuel, the uranium and the plutonium (which can be used again) from the fission products (which are essentially useless and are very radioactive, as shown in Fig. 6.15). The separation is basically a chemical process, using various solvents.

The storage of fission products. The volume of liquid containing the fission products generated in one year by a 900 MW reactor is roughly $20 \, m^3$. The liquid is first stored in special vessels refrigerated by water in order to evacuate the residual heat. After a few years, it is possible to evaporate the solvents and to vitrify the waste. The vitrified blocks can be stored underground in ventilated areas.

After several years, the problem of long term storage of the radioactive vitrified waste arises. Several methods are envisaged, in particular deep underground storage in salt or granite.

Accelerators coupled to fissile matter could have an interesting application in that they can be designed to destroy long lifetime radioactive waste. Under neutron irradiation, plutonium, trans-plutonium elements such as Americium and Curium, and other dangerous long-lived fission products such as Technetium can absorb neutrons and either be transformed to short-lived or stable isotopes or fission into lighter elements. (as a general rule, the lighter the elements, the less dangerous they are).

6.9 The Oklo prehistoric nuclear reactor

It is interesting to note that if civilization had taken 2.5 billion years to develop on Earth rather than 4.5 billion years, the use of nuclear reactors would have been much simpler because at that time the abundance of ^{235}U was sufficiently high that enrichment would have been unnecessary. In fact, on at least one occasion, the necessary conditions for stable reactor operation were united apparently without intelligent intervention. These conditions include a sufficiently high concentration of uranium ($> 10\%$ by weight), a sufficiently low concentration of nuclei with high neutron-absorption cross-sections, and sufficient water ($> 50\%$ by weight) to serve as the moderator.

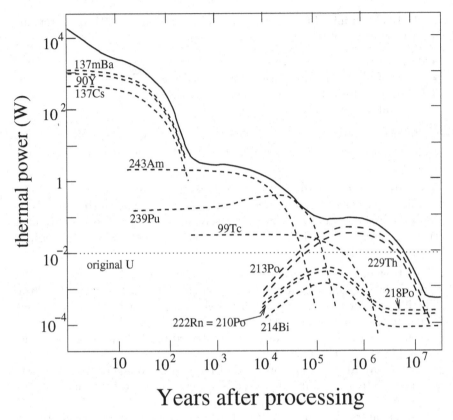

Fig. 6.15. Radioactivity of the fission products and trans-uranium elements in fuel that has produced 100 Mw-yr of electrical energy [65]. It is assumed that 99.5% of the uranium and plutonium was removed for reprocessing. For comparison, also shown is the activity of the original uranium.

These conditions were present in a uranium-ore deposit in Oklo, Gabon [67]. As illustrated in Fig. 6.16, the uranium ore in this deposit is depleted in ^{235}U (as little as 0.42% instead of 0.72%) and enriched in fission products. It is believed that the reactor operated $\sim 1.8 \times 10^9$ years ago for a period of $\sim 10^6$yr.

One interesting result of studies of the Oklo reactor is a very stringent limit on the time variation of the fundamental constants [68], even stronger than those derived from the study of ^{187}Re decay in Sect. 5.5.2. The limit comes from the observation that the nuclide ^{149}Sm has an Oklo abundance that is typical of reactor wastes, i.e. about 40 times less than the natural isotopic abundance of 13.8%. This low abundance comes about because of a resonance for thermal neutron capture (Fig. 6.17) that transforms ^{149}Sm to ^{150}Sm in the high neutron flux.

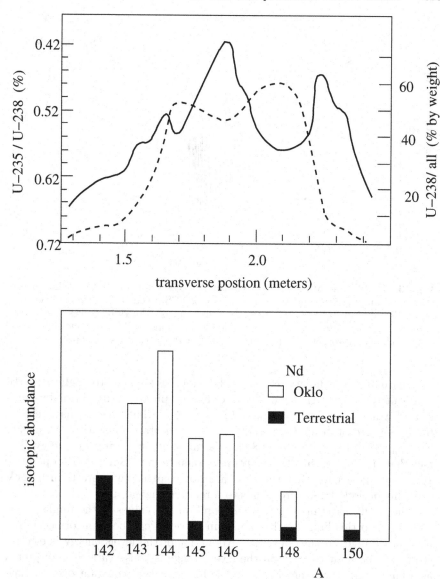

Fig. 6.16. The composition of a uranium deposit in Oklo, Gabon [67]. The deposit contains several layers, each about 1 meter thick, containing very rich ore, about 50% uranium by weight. The top panel shows the uranium abundance profile across the layer (dashed line). The solid line shows that ^{235}U is highly depleted in the layer, as little as 0.42% compared to the normal 0.72%. The bottom panel shows the abundances of Nd isotopes. The fission products ^{143}Nd $-^{150}$Nd all have larger than normal abundances compared to that of ^{142}Nd which cannot be produced by fission (Exercise 6.5).

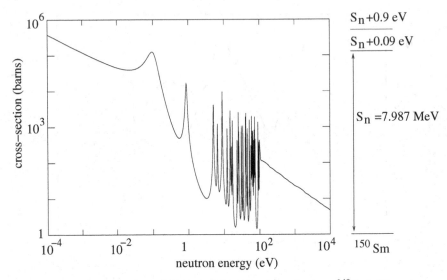

Fig. 6.17. The radiative neutron capture cross-section on ^{149}Sm. The absorption resonances correspond to excited states of ^{150}Sm that are above the neutron-separation energy $S_n = 7.987$ MeV. The diagram on the right shows the ground state and the first two states responsible for the resonances. The first one at $S_n + 0.09$ can resonantly absorb thermal neutrons ($3kT = 0.078$ eV for $T = 300$ K).

Thermal neutron-capture on nuclide (A, Z) are due to highly excited states of $(A + 1, Z)$ that can decay either by photon emission or by emission of a neutron of energy $E_n \sim kT \sim 0.02$ eV. This means that, relative to the ground state of $(A + 1, Z)$, the excited state must have an energy $E \sim S_n + 0.02$ eV where $S_n \sim 8$ MeV is the neutron separation energy. The fact that the thermal-neutron-capture resonance in ^{149}Sm was still operative 10^9 years ago means that the level has not changed by more than 0.02 eV over this period. This corresponds to a relative change $< 10^{-8}$.

The positions of nuclear levels depend on the values of the fundamental constants, so this limit on the change in the level can be transformed into a limit on changes in the constants. About 1% of the levels energy is electrostatic, so the limit of 10^{-8} on the level change is conservatively interpreted as a limit of 10^{-6} on the change in the fine-structure constant over the last 2 billion years [68].

6.10 Bibliography

1. O. Hahn and F. Strassman, *Naturwissenschaften*, **27**, 11 (1939).
2. L. Meitner and O.R. Frisch, *Nature*, **143**, 239 (1939).
3. N. Bohr and J.A. Wheeler, *Phys. Rev.*, **56**, 426, (1939).

4. A.M. Weinberg and E.P. Wigner, *The Physical Theory of Neutron Chain Reactors*, University of Chicago Press, Chicago, Ill., 1958.

5. L. Wilets, *Theories of Nuclear Fission*, Clarendon Press, Oxford, 1964.

6. S. Glasstone and A. Sesonke, *Nuclear Reactor Engineering*, Van Nostrand, New York, 1967.

Exercises

6.1 Consider the typical fission process induced by thermal neutrons

$$\mathrm{n}\,{}^{235}_{92}\mathrm{U} \;\rightarrow\; {}^{140}_{54}\mathrm{Xe} + {}^{93}_{38}\mathrm{Sr} + \nu \text{ neutrons} \quad . \tag{6.76}$$

What are the values of ν and Q_{fis} for this reaction? Calculate the total fission energy of 1 kg of ${}^{235}_{92}\mathrm{U}$ assuming all fissions proceed via this reaction.

6.2 Assume that the average time τ between production and absorption of a neutron in a reactor is 10^{-3} s. Calculate the number of free neutrons present at any time in the core when the reactor is operating at a power level of 1 GW.

6.3 A beam of neutrons of 0.1 eV is incident on 1 cm^3 of natural uranium. The beam flux is 10^{12} neutrons s^{-1}cm^2. The fission cross section of ^{235}U at that energy is 250 b. The amount of ^{235}U is 0.72%. The density of uranium is 19 g cm^{-3}. Each fission produces 165 MeV in the material. What is the nuclear power produced?

6.4 Consider a nuclear plant producing an electric power of 900 MW with thermal neutrons and enriched uranium at 3.32% in ^{235}U. The total yield of nuclear energy into electric energy is $R = 1/3$ (including the thermal yield). The total uranium mass is 70 tons.

1. How many ^{235}U atoms are burnt per second?
2. What mass of ^{235}U is used per day?
3. Assuming the plant works at constant full power, how long can it run before changing the fuel?

6.5 Using the Table in Appendix G, explain why ^{142}Nd would not be expected to be abundantly produced in a nuclear reactor, unlike the other stable Nd isotopes.

6.6 Estimate the amount of uranium needed to create 100 Mw-yr of electrical energy assuming a thermal-to-electricity efficiency of 0.3. This is the amount of uranium considered in Fig. 6.15. In this figure, translate the thermal power to decay rate (in Bq) by assuming $\sim 5\,\mathrm{MeV}$ per decay. Discuss the origin of the nuclides shown in the figure.

7. Fusion

Fusion reactions, taking place in the Sun, have always been the main source of energy on Earth. Even fossil fuels like coal and petroleum were fabricated through photosynthesis and should therefore be considered as stored solar energy. The only exception is fission energy due to the heavy elements uranium and thorium, which were synthesized during supernova explosions.

Since the first explosive occurrence of fusion on Earth in 1952, mankind has had the ambition to tame that form of energy. It is cleaner than fission, creating much less long-lived radioactive waste. Its resources are unlimited over historical time scales. In 300 liters of sea water, there is 1 g of deuterium. The fusion of two ^{2}H nuclei yields about $10 \, \text{MeV} \sim 2 \times 10^{-12} \, \text{J}$. Therefore the water in the oceans could provide sufficient energy for human needs during time scales of several hundred millions of years. It is particularly frustrating to see that, unlike fission which was used industrially a few years after its discovery, fusion is still in a prospective stage more than 50 years after its first terrestrial use.

The apparent difficulty in using fusion reactions comes from the fact that, unlike neutron-induced fission reactions, fusion reaction involve positively charged nuclei that, at normal temperatures and densities, are prevented from reacting by the Coulomb barrier. The challenge of taming fusion is to maintain a (non-explosive) plasma that is sufficiently hot and dense to have a useful rate of fusion.

In this chapter, we will first catalog the possible fusion reactions with the conclusion that the most promising reaction for terrestrial energy reaction is deuterium-tritium fusion

$$d \, t \; \rightarrow \; ^4\text{He} \, n \; + 17.5 \, \text{MeV} \,. \tag{7.1}$$

We will then calculate fusion rates in a hot plasma as a function of temperature and density. In Sect. 7.2 we derive performance criteria for fusion reactors, in particular the *Lawson criterion* for effective energy generation. The basic problem will be to maintain a plasma at a sufficiently high temperature and pressure for a sufficiently long time before it cools down, mostly through photon emission by electrons (bremsstrahlung) and atomic impurities. Sections 7.3 and 7.4 will then discuss the problems of the two most widely discussed method of maintaining the plasma, those using magnetic confinement and those using laser induced inertial implosion.

7.1 Fusion reactions

Because the binding energy per nucleon generally increases with A for $A < 50$, the fusion of two light nuclei is frequently exothermic. Some examples are :

$$\mathrm{d\,d} \rightarrow {}^3\mathrm{He\ n} + 3.25\,\mathrm{MeV} \tag{7.2}$$

$$\mathrm{d\,d} \rightarrow {}^3\mathrm{H\ p} + 4\,\mathrm{MeV} \tag{7.3}$$

$$\mathrm{d\,t} \rightarrow {}^4\mathrm{He\ n} + 17.5\,\mathrm{MeV} \quad . \tag{7.4}$$

Reactions like (7.4) that produce ${}^4\mathrm{He}$ are particularly exothermic owing to the large binding energy of that nucleus. Other examples are reactions that yield ${}^4\mathrm{He}$ from the weakly bound Li-Be-B nuclei:

$$\mathrm{n\,{}^6Li} \rightarrow {}^3\mathrm{H\,{}^4He} + 4.8\,\mathrm{MeV} \tag{7.5}$$

$$\mathrm{d\,{}^6Li} \rightarrow 2{}^4\mathrm{He} + 22.4\,\mathrm{MeV} \tag{7.6}$$

$$\mathrm{p\,{}^{11}B} \rightarrow 3\ {}^4\mathrm{He} + 8.8\,\mathrm{MeV} \ . \tag{7.7}$$

The above reactions are the "terrestrial" reactions with which we are concerned in this chapter. The basic fusion reaction in the Sun is

$$\mathrm{p\,p} \rightarrow {}^2\mathrm{H\ e^+\ \nu_e} + 0.42\,\mathrm{MeV} \quad . \tag{7.8}$$

This reaction transforms primordial hydrogen into deuterium, which subsequently fuses to ${}^4\mathrm{He}$, and then to the heavier elements of which we are made. Unlike the terrestrial fusion reactions, (7.8) is due to the weak interactions because of the necessity for transforming a proton into a neutron. Consequently, it has a tiny cross-section which, forgetting the Coulomb barrier penetration probability, is of order $\sim (G_F/(\hbar c)^3)^2\,(\hbar c)^2\,E^2 \sim 10^{-47}\mathrm{cm}^2$, with $E = 0.42\,\mathrm{MeV}$. This makes it impossible to observe in a laboratory. One would have to observe $\sim 10^{20}$ proton–proton collisions in order to have a chance of seeing one event of the type (7.8).

In the terrestrial reactions (7.2-7.7), the neutrons are already present in the initial nuclei, having been produced in the primordial Universe (Chap. 9). Therefore the reactions can proceed through the strong and electromagnetic interactions.

The interest of fusion lies in two main facts. For civil uses, the main advantage over fission is the large amount of available fuel, i.e. hydrogen isotopes. For military uses, the advantage is the absence of any critical mass constraint. The power of a fission device is limited by the fact that the pre-ignition masses of its components cannot exceed the critical mass. This is not the case for fusion devices since the device can explode if and only if it is brought to sufficiently high temperatures and densities. The amount of material is irrelevant. For the same reason, a controlled fusion installation does not have the same risks of a nuclear accident that a critical fission system presents.

If fusion is still not yet exploited, it is due to the fact that there are tremendous unsolved technical problems in achieving it.

The best fuel for terrestrial fusion is the deuterium-tritium mixture generating energy through reaction (7.4). This is done in association with some amount of ^{6}Li which absorbs neutrons and regenerates the tritium through reaction (7.5).

Tritium itself is β-unstable,

$$^3\text{H} \rightarrow {}^3\text{He e}^- \bar{\nu}_e \quad t_{1/2} = 12.33 \,\text{yr} , \tag{7.9}$$

and is consequently not present in large quantities on Earth. It is produced by neutron irradiation of ^{6}Li, exploiting the exothermic reaction (7.5), or by radiative neutron capture by deuterium. The radioactivity of tritium obviously creates practical problems for its manipulation.

7.1.1 The Coulomb barrier

The physical difficulty in achieving fusion reactions is due to the fact that, unlike neutron absorption in fission reactions, the nuclei which interact are charged and, therefore, have a Coulomb repulsion. In order for the nuclei to have strong interactions, they must approach one another at a distance of the order of the nuclear forces, i.e. the radius of the nuclei. This is once again a situation where one must cross an electrostatic potential barrier. The barrier has a height given by

$$\frac{Z_1 Z_2 e^2}{4\pi\epsilon_0 a} = \frac{Z_1 Z_2 \alpha \hbar c}{a} = 1.4 \,\text{MeV} \times Z_1 Z_2 \frac{1 \,\text{fm}}{a} , \tag{7.10}$$

where a is the distance within which the attractive nuclear forces become larger than the Coulomb force. For energies or temperatures less than $\sim$ 1 MeV the barrier is operative.

If E is the energy of the particle impinging on this barrier, the probability to cross the barrier by quantum tunneling is proportional to the Gamow factor we have seen in Sect. 2.6:

$$P \sim \exp\left[-2 \int_a^b \sqrt{\frac{2m(V(r) - E)}{\hbar^2}} \,\mathrm{d}r\right] \tag{7.11}$$

where m is the reduced mass $m = m_1 m_2/(m_1 + m_2)$ of the two interacting nuclei and b is the classical turning point defined by $V(b) = E$ where $V(r)$ is the repulsive Coulomb potential.

The integral in the right hand side of (7.11) can be calculated easily, since we note that

$$a \ll b = \frac{Z_1 Z_2 e^2}{4\pi\epsilon_0 E} = \frac{Z_1 Z_2 \alpha \hbar c}{E} = 143 \,\text{fm} \times Z_1 Z_2 \frac{10 \,\text{keV}}{E} , \tag{7.12}$$

and therefore the radius a can be taken equal to zero in good approximation. This leads to the following expression due to Gamow (in 1934) for the tunneling probability:

$$P \sim \exp\left(\frac{-2\pi Z_1 Z_2 e^2}{4\pi\epsilon_0 \hbar v}\right) = \exp\left(\sqrt{-E_B/E}\right) . \tag{7.13}$$

where v is the relative velocity and $E = \mu v^2/2$ is the center-of-mass kinetic energy for a reduced mass μ. The barrier is characterized by the parameter

$$E_B = 2\pi^2 Z_1^2 Z_2^2 \alpha^2 \mu c^2 = 1052\,\text{keV} \times Z_1^2 Z_2^2 \frac{\mu c^2}{1\,\text{GeV}} . \tag{7.14}$$

Table 7.1. Some fusion reactions. The first three are used in terrestrial fusion reactors. The last three make up the "PPI" cycle responsible for most of the energy generation in the Sun. Note the tiny $S(E)$ for the weak-reaction pp $\rightarrow$ de$^+$v$_e$. It can only be calculated using weak-interaction theory.

reaction	Q (MeV)	$S(10\,\text{keV})$ (keV b)	E_B (keV)	$E_G(1\,\text{keV})$ (keV)	$E_G(20\,\text{keV})$ (keV)
d d $\rightarrow$ n ^{3}He	3.25	58.3	987.	5.1	37.5
d d $\rightarrow$ p ^{3}H	4.	57.3	987.	5.1	37.5
d t $\rightarrow$ n ^{4}He	17.5	14000.	1185	6.8	50.1
p p $\rightarrow$ d e$^+$ v$_e$	1.442	3.8×10^{-22}	526	5.1	37.5
p d $\rightarrow$ ^{3}He γ	5.493	2.5×10^{-4}	701	5.6	41.2
2 ^{3}He $\rightarrow$ p p ^{4}He	12.859	5×10^3	25200.	18.5	136

We note that the argument of the exponential increases in absolute value with the product of the charges and that it decreases as the inverse of the velocity. The higher the energy of the nuclei is, the greater the probability to tunnel through the barrier. Likewise, the larger the product of charges $Z_1 Z_2$ is, the higher the barrier, therefore at a given energy, it is the lighter nuclei which can undergo fusion reactions. For particles of charge $+1$ (e.g. d + d) we have

$$E = 1\text{keV} \Rightarrow P \sim 10^{-13} \qquad E = 10\,\text{keV} \Rightarrow P \sim 10^{-3} .$$

This suggests that $\sim 10\,\text{keV}$ is the order of magnitude of the kinetic energy that the nuclei must have in order for fusion reactions to take place. (Much below that energy, the cross section vanishes for all practical purposes.) The energy of the nuclei comes from their thermal motion, therefore from the *temperature* of the medium where they are contained. Hence the name *thermonuclear reactions* for fusion reactions.

We must find the factors of proportionality between the tunneling probability and the reaction cross-section. In Sect. 3.6, where we treated absorption reaction inhibited by potential barriers, we argued that the cross-section should be of the form

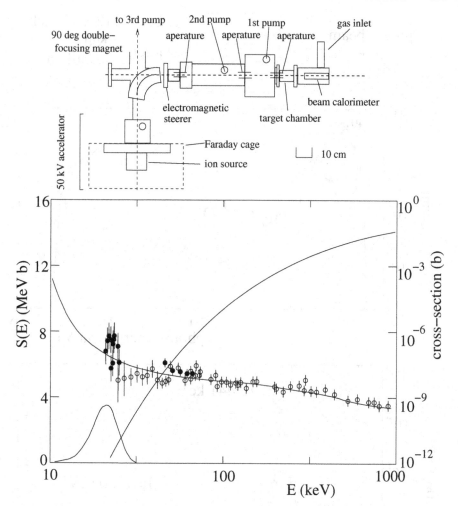

Fig. 7.1. Cross-section and $S(E)$ for ^{3}He ^{3}He $\rightarrow^4$ He p p, as measured by the LUNA underground accelerator facility [69]. The top panel shows the small ($\sim 1\,\text{m}^2$) experiment consisting of a ^{3}He ion source, a 50 kV electrostatic accelerator, an analyzing magnetic spectrometer, a gaseous (^{3}He) target chamber, and a beam calorimeter to measure the beam intensity. The sides of the target chamber are instrumented with silicon ionization counters that measure dE/dx and E of protons produced by ^{3}He $+^3$ He $\rightarrow^4$ He $+$ p p in the chamber. Because of the very small cross-sections to be measured, the experiment is in the deep underground laboratory LNGS, Gran Sasso, Italy, where cosmic-ray background is eliminated. The bottom panel shows the LUNA measurements as well as higher energy measurements [70]. The lowest energy measurements cover the region of the solar Gamow peak for this reaction (Fig. 7.3). Note that while the cross-section varies by more than 10 orders of magnitude between $E = 20\,\text{keV}$ and 1 MeV, the factor $S(E)$ varies only by a factor ~ 2.

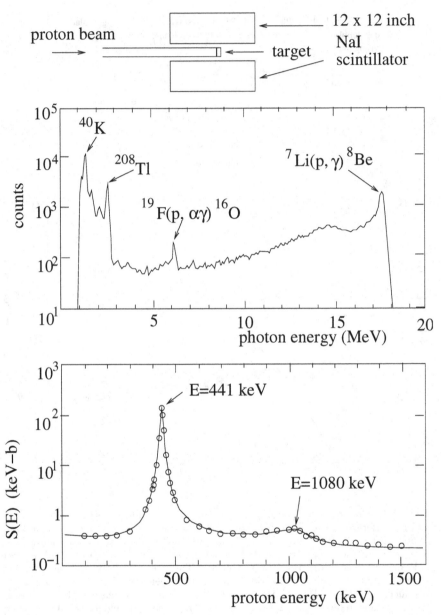

Fig. 7.2. $S(E)$ for $p\,^7\mathrm{Li} \rightarrow\,^8\mathrm{Be}\,\gamma$ as measured by [71]. The top panel shows how a proton beam impinges upon a target consisting of $10\,\mu\mathrm{g\,cm^{-2}}$ of LiF evaporated on a copper backing. The target is inside a large NaI scintillator that detects photons emerging from the target. The middle panel shows a typical photon energy spectrum showing peaks due to $^7\mathrm{Li}(p,\gamma)^8\mathrm{Be}$, in addition to peaks due to $^{19}\mathrm{F}(p,\alpha\gamma)^{16}\mathrm{O}$ and to natural radioactivity in the laboratory walls. The S-factor deduced from the photon counting rate is shown on the bottom panel as a function of proton energy. It shows the presence of two resonances due to excited states of $^8\mathrm{Be}$.

$$\sigma(E) = \frac{S(E)}{E} \exp\left(-\sqrt{E_B/E}\right) . \tag{7.15}$$

where $S(E)$ is a slowly varying function of the center-of-mass energy and E_B is given by (7.14).

The experimental determination of the nuclear factors $S(E)$ is a problem of major interest for all calculations in astrophysical and cosmological nucleosynthesis, as we shall see in the next chapters. Examples are shown in Figs. 7.1 and 7.2. Considerable effort has been made in recent years to measure the cross-sections at energies comparable to stellar temperatures. Without such data it was necessary to extrapolate the $S(E)$.

Table 7.1 lists the $S(E)$ for some important fusion reactions. Note that the tiny cross-section for the stellar reaction pp $\rightarrow$ ^{2}He$^+\nu_e$ makes this reaction unobservable. The $S(E)$ must therefore be calculated using weak-interaction theory.

The Gamow formula (7.15) can also be obtained quite easily in the Born approximation. A cross-section involves, like any transition rate, the square of a matrix element between initial and final states $|\langle f|M|i\rangle|^2$. When forces are short range, the asymptotic states $|i\rangle$ and $|f\rangle$ are monochromatic plane waves.

In the presence of a Coulomb interaction, which is of infinite range, the asymptotic behavior is different. The wave function is exactly calculable (see for instance A. Messiah, *Quantum Mechanics* vol. 1, chap. XI-7) and, asymptotically, the argument of the exponential has additional terms of the form

$$\varphi(r) \sim \exp(i(\boldsymbol{k}r + \gamma \log kr)) \quad \text{with} \quad \gamma = Z_1 Z_2 e^2/4\pi\epsilon_0 \hbar v . \tag{7.16}$$

These asymptotic wave functions are called Coulomb scattering states. Consider, for instance, the reaction d + t $\rightarrow$ ^{4}He + n. In the initial state, we must use Coulomb scattering states (and not the usual asymptotic states). Since the strong interaction is short range, we can, in good approximation, simply multiply the usual matrix element by the value of the Coulomb scattering wave at the origin $\psi_{coul}(0) = \Gamma(1+i\gamma)e^{-\pi\gamma/2}$, i.e. multiply the nuclear cross section by the factor $|\Gamma(1+i\gamma)e^{-\pi\gamma/2}|^2 = \pi\gamma e^{-\pi\gamma}/\sinh\pi\gamma$. Once the kinematic factors are taken into account, one recovers the tunnel effect factor $e^{-2\pi\gamma} \sim e^{-\pi\gamma}/\sinh\pi\gamma$ introduced empirically for $\gamma = Z_1 Z_2 e^2/4\pi\epsilon_0 \hbar v \gg 1$.

7.1.2 Reaction rate in a medium

Consider the specific case of a mixture of deuterium and tritium. We want to evaluate the reaction rate per unit volume, i.e. the number R of reactions per unit volume and unit time, if the number densities of d and t are respectively n_1 and n_2. Knowing this rate, we can readily calculate the power emitted.

Consider a deuterium nucleus, of velocity v with respect to the tritium nucleus. As discussed in Sect. 3.1.4, the probability per unit time λ that a fusion reaction occurs is:

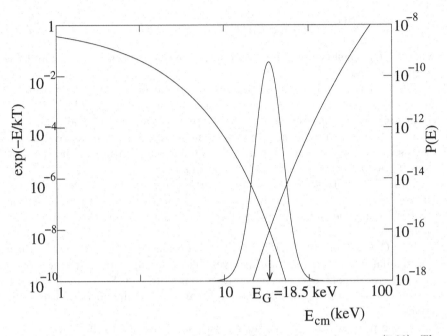

Fig. 7.3. Factors entering the calculation of the pair reaction rate (7.22). The Boltzmann factor $\exp(-E/kT)$ (logarithmic scale on the left) and the barrier penetration probability $P(E) = \exp(-\sqrt{E_B/E})$ (7.13) (logarithmic scale on the right) are calculated for $kT = 1\,\text{keV}$ (corresponding to the center of the Sun) and for the reaction $^3\text{He}\,^3\text{He} \rightarrow {}^4\text{He}\,\text{pp}$. The product is the Gaussian-like curve in the center (shown on a linear scale). It is maximized at $E_G = (\sqrt{E_B}kT/2)^{2/3} \sim 18.5\,keV$ and most reactions occur within $\sim 5\,\text{keV}$ of this value. Note the small values of $\exp(-E_{Gm}/kT) \sim 10^{-8}$ and $P(E_G) \sim 10^{-16}$.

$$\lambda = n_2\,\sigma(v)\,v \tag{7.17}$$

where σ is the fusion cross-section (7.15). The rate per unit volume is then found by multiplying the rate per deuterium nucleus by n_1:

$$R = n_1 n_2 \sigma(v) v \quad . \tag{7.18}$$

We must average this expression over the velocity distribution in the medium at temperature T:

$$R = n_1 n_2 \langle \sigma(v) v \rangle \tag{7.19}$$

where $\langle \sigma(v) v \rangle$ is the average of the product $\sigma(v) v$, the probability of v being determined by the Maxwell distribution at temperature T. Because of the decreasing Coulomb barrier, the product σv increases rapidly with the energy. In the averaging, it is however in competition with the decrease of the Maxwell distribution with increasing velocity. If only one of the two species is in motion, we would have

$$\langle \sigma v \rangle \sim \int d^3 v \, e^{-E/kT} \sigma(v) v \sim \int v^3 e^{-mv^2/2kT} \sigma(v) dv \quad . \tag{7.20}$$

In reality, both species are in motion so the integral is slightly more complicated. For nuclear of masses of m_1 and m_2, we have

$$\langle \sigma v \rangle = \left(\frac{m_1}{2\pi kT} \right)^{3/2} \left(\frac{m_2}{2\pi kT} \right)^{3/2} \int e^{-(m_1 v_1^2 + m_2 v_2^2)/2kT} \sigma(v) v \, d^3 v_1 d^3 v_2 \; ,$$

where $v = |\boldsymbol{v}_1 - \boldsymbol{v}_2|$ is the relative velocity. Turning to center-of-mass vari-
ables, $\mu = m_1 m_2/(m_1 + m_2)$ being the reduced mass, we can integrate over
the total momentum (or the velocity of the center of gravity). This leads to

$$\langle \sigma v \rangle = \sqrt{\frac{8}{\pi \mu (kT)^3}} \int e^{-E/kT} E \sigma(E) dE \quad . \tag{7.21}$$

Using (7.15) this is

$$\langle \sigma v \rangle = \sqrt{\frac{8}{\pi \mu (kT)^3}} \int e^{-\sqrt{E_B/E}} e^{-E/kT} S(E) dE \quad . \tag{7.22}$$

The integrand contains the product of two exponentials shown in Fig. 7.3.
Their product peaks at the *Gamow energy*

$$E_G = E_B^{1/3} (kT/2)^{2/3} \; , \tag{7.23}$$

where E_B is given by (7.14). As long as $S(E)$ has no resonances (e.g. as
in Fig. 7.2) only the narrow region around the Gamow energy (called the
Gamow peak) contributes significantly to $\langle \sigma(v)v \rangle$. Its position determines
the effective energy at which the reaction takes place.

In the absence of resonances, the nuclear factor $S(E)$ varies slowly and
only the value $S(E_G)$ is relevant so it can be taken out of the integral (7.22).
We can also make a Taylor expansion of the argument of the exponential in
the region E_G:

$$\sqrt{E_B/E} + E/kT \sim \frac{3}{2} \left(\frac{E_B}{kT/2} \right)^{1/3} + \frac{1}{2} \frac{(E - E_G)^2}{\Delta_E^2} \; , \tag{7.24}$$

where the width of the Gamow peak is

$$\Delta_E = \frac{2}{\sqrt{3}} E_G \left(\frac{kT}{E_G} \right)^{1/2} = \frac{2}{\sqrt{3}} E_B^{1/6} \left(\frac{kT}{2} \right)^{5/6} \quad . \tag{7.25}$$

[Note that the Gamow peak is relatively narrow: $\Delta_E/E_G \sim (kT/E_B)^{1/6}$.] We
then have

$$\langle \sigma v \rangle = \frac{8\pi}{\sqrt{\mu}} (kT)^{-3/2} S(E_G) \exp \left[-(3/2) \left(\frac{E_B}{kT/2} \right)^{1/3} \right]$$

$$\times \int \exp \left(\frac{(E - E_G)^2}{2\Delta_E^2} \right) dE \quad . \tag{7.26}$$

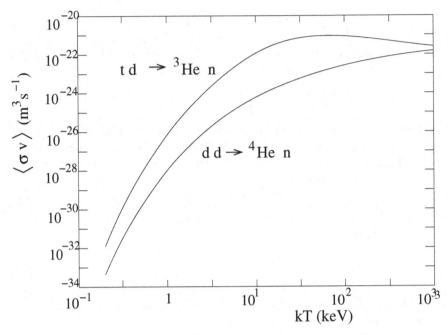

Fig. 7.4. Variation of the pair reaction rate $\langle v\sigma \rangle$ as a function of the temperature for d-d and d-t mixtures.

The Gaussian integral just gives a factor Δ_E so we end up with

$$\langle \sigma v \rangle = \frac{8\pi}{\sqrt{\mu}} (kT)^{-2/3} E_B^{1/6} S(E_G) \exp \left[-(3/2) \left(\frac{E_B}{kT/2} \right)^{1/3} \right] \qquad (7.27)$$

Figure 7.4 shows $\langle \sigma v \rangle$ as a function of temperature for $d\,d \rightarrow n\,{}^3\text{He}$ and for $d\,t \rightarrow n\,{}^4\text{He}$. The rate rises rapidly for $kT < 10\,\text{keV}$ before leveling off. We can say that $kT \sim 10\,keV$, i.e. $T \sim 1.5 \times 10^8$ K, defines an optimal temperature for a fusion reactor.

7.1.3 Resonant reaction rates

If a reaction cross-section exhibits resonances, like those in Fig. 7.2, then the integral in (7.21) may also receive an important contribution from energies near the resonance, in addition to the Gamow peak. Such resonances are due to states of the compound nucleus consisting of the two interacting particles. For example, the resonance in Fig. 7.2 is due to an excited state of ${}^8\text{Be}$ (Exercise 7.3). This state can decay to both the initial particles

$${}^8\text{Be}^* \rightarrow p\,{}^7\text{Li} \qquad \Gamma_p = 6\,\text{keV} , \qquad (7.28)$$

and to the ground state of ${}^8\text{Be}$

$$^8\text{Be}^* \rightarrow \gamma\, ^8\text{Be} \quad \varGamma_\gamma = 12\,\text{eV}\,. \tag{7.29}$$

Near the peak of the resonance, the cross-section for $\text{p}\,^7\text{Li} \rightarrow\, ^8\text{Be}\,\gamma$ is given by (3.183)

$$\sigma_{i\rightarrow f}(E) \;\sim\; 4\pi \frac{(\hbar c)^2}{2\mu E}\,\frac{(\varGamma_\text{p}/2)(\varGamma_\gamma/2)}{(E - E_0)^2 + \varGamma^2/4}\;, \tag{7.30}$$

where we have neglected the spin factors and where $\varGamma = \varGamma_\gamma + \varGamma_\text{p}$. The contribution to the integral in (7.21) coming from the resonance region is then just proportional to the cross-section on resonance, $4\pi(\varGamma_\gamma/\varGamma)(\hbar c)^2/(2\mu E)$ times the width $\varGamma$:

$$\int_\text{res} \text{e}^{-E/kT} E\sigma(E)\text{d}E \;\sim\; \text{e}^{-E_\text{res}/kT}\,\frac{(\hbar c)^2}{\mu c^2}\,\varGamma_\gamma\,. \tag{7.31}$$

Comparing this with the non-resonant rate (7.22) we get the ratio of the contributions of the resonance and the Gamow peak

$$\frac{\langle\sigma v\rangle_\text{res}}{\langle\sigma v\rangle_\text{Gamow}} \;\sim\; \frac{\text{e}^{-E_\text{res}/kT}}{\text{e}^{-E_\text{G}/kT}}\,\frac{1}{\text{e}^{-\sqrt{E_B/E_G}}}\,\frac{(\hbar c)^2/\mu c^2}{S(E_\text{G})}\,\frac{\varGamma_\gamma}{\varDelta_E}\,. \tag{7.32}$$

We have written the ratio as the product of four dimensionless factors. The fourth favors the Gamow contribution, $\varGamma_\gamma/\varDelta_E \sim 10^{-3}$, while the third, $\sim 10^3$ favors the resonance. The relative importance is then determined by the strongly temperature dependent first and second terms. We leave it to Exercise 7.5 to show that in the Sun ($kT \sim 1\,\text{keV}$) the Boltzmann factor suppresses the resonant contribution to a negligible level.

However, for resonances near or below the Gamow peak, the resonance is generally more important. The most important example in astrophysics is the reaction

$$^4\text{He}\,^8\text{Be} \rightarrow\, ^{12}\text{C}\,\gamma\,. \tag{7.33}$$

which proceeds through an excited state of ^{12}C corresponding to a $^4\text{He} - \,^8\text{Be}$ center-of-mass energy of $E_\text{res} = 283\,\text{keV}$. This reaction is responsible for the production of carbon and takes place in stars at $kT \sim 15\,\text{keV}$ corresponding to $E_\text{G} \sim 300\,\text{keV}$. In this case, the Boltzmann factors in (7.32) cancel. The factor $\exp(-\sqrt{E_B/E_G}) \sim 10^{-11}$ then ensures that the resonant contribution dominates.

7.2 Reactor performance criteria

A fusion reactor consists of three essential elements

- A confined plasma containing positively charged ions (generally ^2H and ^3H) and electrons to maintain the neutrality.
- An energy injector to create and, if necessary, maintain a high temperature.

- An energy recovery mechanism that collects the energy of escaping fusion-produced neutrons and thermal photons.

In practice, the confinement of the plasma is achieved by three different mechanisms: gravitational confinement, inertial confinement and magnetic confinement. Gravitational confinement is achieved naturally in stars. The plasma is maintained indefinitely by the self-gravitation of the star. Inertial confinement is used in laser induced fusion. It also occurs in supernovae explosions (and in explosive devices). Magnetic confinement has been the main method investigated for controlled fusion before laser induced inertial confinement was declassified.

Let V be the plasma volume, R the reaction rate as defined by (7.19) and Q the energy released in an elementary fusion reaction. The fusion-generated power P (before any losses are taken into account) is given by

$$P = R V Q \ . \tag{7.34}$$

For a given value of the temperature, in order to increase the reaction rate R, one must increase the densities n_1 and n_2. For a given value of $n = n_1 + n_2$, the best proportion, which maximizes the product $n_1 n_2$ corresponds to $n_1 = n_2 = n/2$, i.e. equal amounts of reagents, at high densities. One therefore seeks *high temperatures* ($\sim 10^7$ K, $kT \sim 1$ keV) and a *strong compression*.

Fusion reactors are judged by how much power they create compared to how much was used in heating the plasma. Three goals, in order of decreasing difficulty, are defined for any fusion reactor:

- Ignition. After heating, the reaction rate is sufficiently high to maintain the temperature without further injection of energy. In reaction (7.4), the neutron escapes so the ^{4}He energy must be used to compensate for cooling by radiation of photons and neutrons.
- Breakeven. Power generated by fusion is equal to the input power that must be continually injected to compensate for energy losses (neutrons and photons).
- One-shot breakeven (Lawson Criterion). The energy generated by fusion is equal to the input energy necessary to heat the plasma. A reactor satisfying only this criterion is similar to the sub-critical fission reactors of Sect. 6.8.

The present goal is to satisfy the Lawson criterion which we now make more precise. The time τ during which the plasma maintains its temperature T and its cohesion, after its creation, is called the confinement time. This time is effectively infinite in stars, in the sense that confinement lasts at least as long as there is some nuclear fuel left.

In order to heat the plasma to the temperature T, one must furnish, per unit volume, the energy $3nkT$ where n is the number density of nuclei. (We assume one free electron per ion).

Let η be the efficiency to transform nuclear energy into electric energy in the reactor, in other words the efficiency to recover the energy produced by the plasma. The reactor will run in *ignition* conditions if the energy spent to heat it is smaller than the electric energy it can produce:

$$3nkT < (n^2/4\eta)\langle v\sigma \rangle Q\tau \tag{7.35}$$

where we assume that there are two species (d-t mixture) of number densities $n_1 = n_2 = n/2$. We must therefore have

$$n\tau > (\frac{1}{\eta}) \frac{12kT}{(Q\langle v\sigma \rangle)} \tag{7.36}$$

where Q is the energy produced in an elementary fusion reaction. This formula is called the Lawson criterion.

From Fig. 7.4 we see that $\langle \sigma v \rangle \propto T^2$ for $kT \sim 10\,\mathrm{keV}$:

$$\langle \sigma v \rangle (\mathrm{dt} \to {}^4\mathrm{He\,n}) \sim 10^{-22}\,\mathrm{m^3 s^{-1}} \left(\frac{kT}{10\,\mathrm{keV}} \right)^2 . \tag{7.37}$$

Substituting this into (7.36) and using $Q = 17.5\,\mathrm{MeV}$ we get the Lawson criterion for the d-t reaction

$$n\tau \frac{kT}{10\,\mathrm{keV}} > \eta^{-1} 1.5 \times 10^{20}\,\mathrm{m^{-3}s} \ , \tag{7.38}$$

or equivalently

$$n\tau kT > \eta^{-1} 1.5 \times 10^{21}\,\mathrm{m^{-3}s\,keV} \ , \tag{7.39}$$

In gravitational confinement, τ is effectively infinite so the Lawson criterion is irrelevant. In magnetic confinement, confinement times achieved so far are $\sim 1\,\mathrm{s}$, limited by excess cooling due to impurities in the plasma and to various plasma instabilities. The density is planned to be of the order of $10^{20}\,\mathrm{m^{-3}}$ in the project ITER (International Thermonuclear Experimental Reactor) and the temperature of the order of 10 to 20 keV, which is the ignition threshold. In laser induced inertial confinement, the confinement time is much shorter $\tau \simeq 10^{-11}$ s. However, the density is much higher $\simeq 10^{31}\,\mathrm{m^{-3}}$. The temperature is comparable, of the order of 10 keV.

Table 7.2. Comparison of the parameters in three fusion confinement schemes.

Type	$n(\mathrm{m^{-3}})$	$\tau(\mathrm{s})$	$T(\mathrm{keV})$
ITER Tokamak	10^{20}	1	10 to 20
Inertial; laser	10^{31}	10^{-11}	10
Sun (grav)	$7.\,10^{30}$	10^{17}	1.3

7.3 Magnetic confinement

The magnetic confinement method consists in using a magnetic field to contain the plasma. Tokamaks are toroidal machines where the plasma is maintained inside a tube by a strong toroidal field superimposed on a poloidal field which maintains it in the horizontal plane.

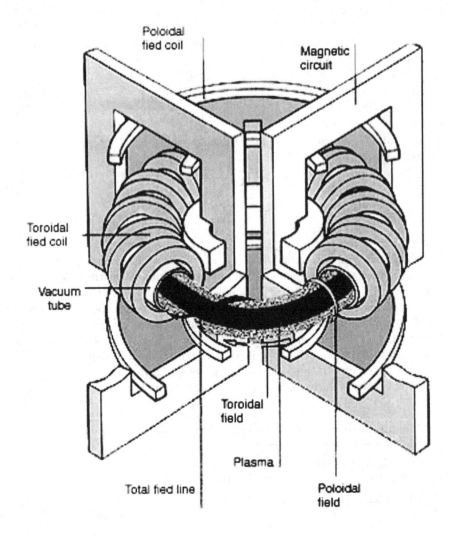

Fig. 7.5. Diagram of a Tokamak.

In the tokamak configuration shown on Fig. 7.5, the toroidal field, created by external currents, is rotationally symmetric around the vertical axis. The poloidal field comes from a high intensity current of several million amperes which runs in the tube. The design is of course quite involved since one does not deal with a beam of particles, as in accelerators, but with an entire large-size plasma.

Creation and heating of the plasma. The magnetic configuration is created inside a toroidal tube. The evacuated tube is filled with the hydrogen–deuterium or tritium–deuterium mixture with a pressure of 300 P, which corresponds to the desired density of 10^{14} particles per cubic centimeter. The gas is then completely ionized in a few milliseconds.

In order to reach and to maintain a temperature of several keV, several heating techniques are used.

- First, there is the *ohmic heating*, where the high intensity current in the plasma transfers heat by Joule effect. This type of heating allows one to reach temperatures of 2 to 3 keV.
- In order to reach the 10 to 20 keV necessary for fusion, *fast neutral atoms* are injected. One first accelerates ions of the same nature as those of the plasma (hydrogen or deuterium), outside the tokamak. These ions are neutralized by capturing electrons in a medium. The neutral atoms which are formed have considerably higher energies than the ions in the plasma. They are unaffected by the magnetic field. They cross the magnetic field, enter the chamber and are then ionized by the plasma. Finally the whole system thermalizes.
- The *microwave heating* comes from the absorption of high-frequency electromagnetic waves. The modeling of this process is in itself a field of magneto-hydrodynamics. Waves are sent in the plasma by antennas inside the chamber, near the walls. One can inject powers of the order of 10 megawatts. The power in the European tokamak JET (Joint European Torus) is 40 megawatts.
- Finally, the α particles produced in the fusion reaction and which carry 20% of the fusion energy release this energy by thermalizing with the ions of the plasma. This should maintain the plasma temperature in the ignition regime.

Loss mechanisms; energy balance. The plasma loses energy by radiation of photons. These radiation losses come mainly from the emission lines of incompletely ionized atoms. Most of these atoms are impurities of oxygen, carbon, iron and nickel. The losses due to this effect represent, at present, 10 to 50% of the energy balance. They can be minimized by a proper conditioning of the walls of the vacuum chamber and by controlling the quality of the plasma near the walls. Both operations are intended to prevent the production of impurities. Maintaining the concentration of impurities at an acceptable level is a necessary condition to have a positive fusion energy balance.

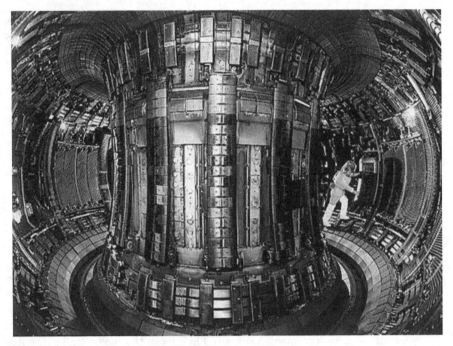

Fig. 7.6. The interior of JET (Joint European Torus).

The energy losses by the *bremsstrahlung* of electrons scattering on deuterium and tritium ions is an unavoidable source of energy loss. At a given density, they require a high enough plasma temperature since the fusion cross-section rises more rapidly with temperature than the bremsstrahlung cross-section. Losses by bremsstrahlung are proportional to $n^2 T^{1/2}$ (Exercise 7.6) whereas $\langle \sigma v \rangle$ for fusion is proportional to T^2 in the $10 - 20$ keV range. The overall result is that the temperature should be larger than 5 keV.

Stability of the plasma. The instability mechanisms of the plasma itself are a field of complex fundamental research. The most dangerous instabilities are magneto-hydrodynamic instabilities because they are highly non-linear and they involve the plasma as a whole.

Perspectives of fusion by magnetic confinement. Since the first results obtained by Soviet physicists in 1968 on the tokamak T-3, about 30 machines of that type have been constructed in the world. Much progress has been made toward satisfying the Lawson criterion.

Three large tokamaks have been installed successively between 1982 and 1985 and have attained their nominal performances : the TFTR (Tokamak Fusion Test Reactor) by the United States at the Princeton laboratory in New Jersey ; The JET (Joint European Torus), by the EC, located in England at Culham, near Oxford ; the JT-60 (Jaeri Tokamak), by Japan, at the Jaeri laboratory in Naka, near Kyoto. The current in the plasma ring increased from 100 kA in the T-3 to 7 MA in the JET. The break-even zone was approached in 1991 by JET and in 1993 by the JT-60.

The performances of various machines are represented in Fig. 7.7.

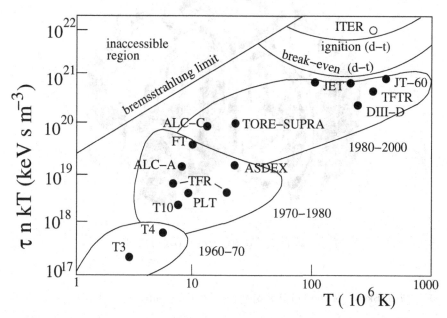

Fig. 7.7. Performance of various Tokamaks. The ITER project is projected to start construction in 2005. Above bremsstrahlung limit, cooling by radiation limits the temperature (Exercise 7.6).

The planned successor of JET is ITER (International Thermonuclear Experimental Reactor), a world-wide collaboration initiated by the USA and the Soviet Union in 1985. ITER is now a collaboration of Europe, Japan, Canada, the USA, and the Russian Federation.

The ambition is not limited to obtaining ignition conditions of the deuterium-tritium mixture (one aims at a nuclear power of 1.5 GW). It is also to maintain the plasma in stationary equilibrium for a period of 1000 seconds.

The project is gigantic. The volume of the plasma should be $2000\,\mathrm{m}^3$ instead of $100\,\mathrm{m}^3$ in JET. ITER is supposed to deliver a power of 1500 MW. The magnetic system will be entirely superconducting.

The construction should start in 2005 and should last ten years. The exploitation phase should last for 20 years.

The next step would be the construction of an experimental plant which would produce electricity, called DEMO. This plant could be exploited start-

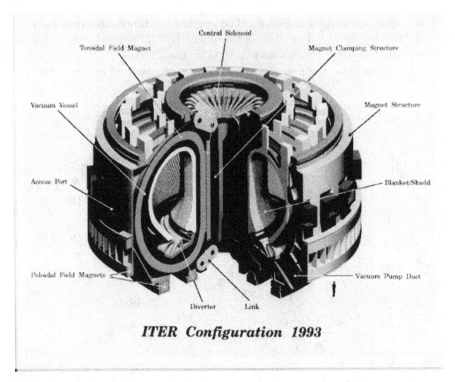

Fig. 7.8. Sketch of the ITER project; notice the comparative size.

ing in 2035, and the construction of the first commercial plant would only start after 2050.

7.4 Inertial confinement by lasers

The principle of inertial confinement by lasers is to adiabatically compress a small ($\sim$ 1mg) sphere containing deuterium and tritium in order to increase its density by $\sim 10^4$, and to obtain temperatures of $\sim$ 10 keV. The core of the sphere ignites during a time of the order of 10^{-10} s and then explodes (see Fig. 7.9).

One calls *ignition* the regime where the temperature and density conditions of the core allow the burning of the d-t mixture.

Initially, this method was classified, because of its military applications. Since 1993 civilian-physicists have worked in the field.

The principle is simpler than in tokamaks. Additionally, research on this method has been funded partly owing to its importance for the understanding of thermonuclear explosions.

Principle of the method. The radiation of a set of laser beams delivering a very large power (TW) for a short time (ns) is directed toward a sphere of the order of a mm^3 of a solid deuterium–tritium mixture. There is an ablation, or sudden vaporization, of the periphery of the sphere and the formation of a corona of plasma.

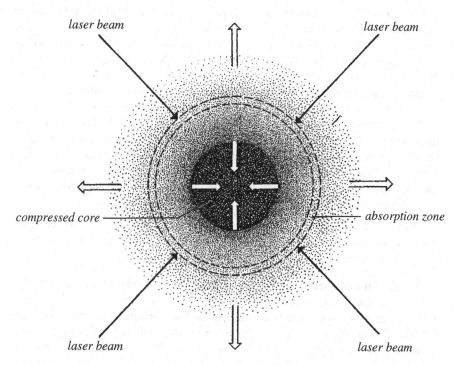

Fig. 7.9. Sketch of laser induced fusion. The d-t sphere interacts with the laser beams and it is vaporized superficially. By reaction, the corona compresses the central core.

The electrons of the medium which oscillate in the laser field transfer energy to the plasma by colliding with the ions. The energy is transfered to the cold regions of the center of the target by thermal conduction, by fast electrons and by UV and X radiation. A shock wave is created which compresses and heats the central region of the deuterium-tritium sphere, called the *core*.

Under that implosion, the core is compressed by a factor of 1000 to 10000, i.e. densities of $\sim 10^{31}$ m^{-3}, and its temperature reaches ~ 10 keV. Under these conditions, the fusion of the d-t nuclei occurs abundantly. The core burns for about 10^{-11}s. Its cohesion is maintained by inertia, it explodes because of the thermonuclear energy release.

The laser energy goes mainly into the compression of the d-t mixture. The energy necessary to heat the plasma comes mainly from the fusion energy release. This results in a reduction of the laser energy which is necessary to make the target burn.

In 1992, in the United States, the Lawrence Livermore National Laboratory (LLNL) declassified the principle of inertial confinement fusion. At Ann Arbor, the KMS laboratory (named after its creator, K.M. Siegel) was the first, around 1973, to achieve the implosion of glass "micro-balloons" containing gaseous deuterium-tritium. The experiment was then performed by other laboratories. The 100 kJ Nova laser of LLNL reached a production of 10^{13} neutrons per laser pulse.

Such experiments led to rapid development of computer-simulated explosions. The experimental inputs to these calculations involved the observation of X-rays and neutrons emitted by the target, the spectroscopy of tracers incorporated in the d-t mixture, such as argon and neon, and pictures of the α particles produced in the fusion reactions. Such measurements were compared with the results of computer simulations, in order to validate the assumptions entering the codes, in particular the fact that neutron emission is of thermonuclear origin.

Technical problems. An important problem is that the irradiation of the target must be as uniform as possible, in order for the implosion to be as spherical as possible. Several methods are used.

- One can use a direct attack by several laser beams symmetrically distributed about the target.
- The indirect attack consists in sending the laser beams inside a cavity made with a heavy element. This results in an emission of thermal X-rays. The X rays irradiate the sphere in the center of the cavity much more uniformly. Using gold, one can reach conversion rates of laser radiation into X-rays of 80%.
- Another approach consists in replacing the laser beams electron with ion beams of ~ 1 MeV. High power beams are easy to produce but sufficiently accurate focusing of the beam is difficult to achieve. The construction of a 10 GeV heavy ion accelerator (for instance uranium) that could deliver 1 MJ in 10 ns is under study.

Projects. Two important projects are underway.

The NIF (National Ignition Facility) project, in the United States, consists of 192 laser beams. It will deliver an energy of 1.8 MJ in each pulse of 1 ns. Its installation is scheduled for 2003; its cost should be 1200 million dollars. Ignition is expected in 2008-2012.

The Megajoule laser project, LMJ, in France, will consist of 240 laser beams. It should deliver 1.8 MJ in 18 ns. Its installation is expected in 2010, for a cost of 1000 million euros. Ignition is expected in 2015.

7.5 Bibliography

1. James Glanz *Turbulence may sink Titanic reactor*, Science, 274, December 1996.
2. M. N. Rosenbluth et A. M. Sessler et T. H. Stix, *Build the International Thermonuclear Experimental Reactor? Yes/No*, Physics Today, June 1996.
3. I. Fodor, B. Coppi and J. Lawyer, *Views ITERated on proposed new reactor, Ignitor, fusion power*, Physics Today, December 1996.
4. M. N. Rosenbluth et A. M. Sessler et T. H. Stix, *ITER debaters reply to pro and cons fusion comments from readers*, Physics Today, January 1997.
5. W. E. Parkins, J. A. Krumhansl et C. Starr, *Insurmountable engineering problems seen as ruling out fusion power to the people in 21st century*, Physics Today, Mach 1997.

Exercises

7.1 Calculate the nuclear energy content in joules in 1 kg of heavy water, assuming that all deuterium atoms undergo fusion into ^{4}He.

7.2 Using the reaction rate in Fig. 7.4, calculate the fusion power in a deuterium plasma at a temperature of $kT = 10\,\text{keV}$ and a density of 10^{14}cm^{-3}.

7.3 Calculate the energy of the excited state of ^{8}Be (relative to the ground state) that leads to the 441 keV resonance shown in Fig. 7.2.

7.4 Consider the experiment to measure the reaction ^{7}Li$(p, \gamma)^8$Be shown in Fig. 7.2. Calculate the photon energy for a proton kinetic energy of 200 keV. Estimate the energy loss of the proton as it traverses the $10\,\mu\text{g cm}^{-2}$ of LiF. How much does the cross-section vary over this energy range? Discuss how this limits the acceptable thickness of the target.

7.5 Evaluate (7.32) for ^{7}Li$(p, \gamma)\,^8$Be for solar conditions, $kT \sim$ keV, to show that the resonant contribution to the rate can be ignored.

7.6 The bremsstrahlung cross-section for radiation of a photon of energy E_γ by non-relativistic electrons scattering on protons is approximately

$$\frac{\mathrm{d}\sigma}{\mathrm{d}E_\gamma} \sim \alpha \left(\frac{c^2}{v^2}\right) \frac{\sigma_\text{T}}{E_\gamma}, \tag{7.40}$$

where $v \ll c$ is the relative electron–proton velocity and σ_T is the Thomson cross-section. The rate of photon radiation (per unit time and per unit volume) in a plasma of electron and proton density n is then

$$\frac{dN_\gamma}{dt\,dV\,dE_\gamma} \sim n^2 v \frac{d\sigma}{dE_\gamma} \sim \frac{c}{v} n^2 \alpha c \frac{\sigma_T}{E_\gamma} \,, \tag{7.41}$$

where v is now the mean velocity. Integrate this expression up to the maximum photon energy ($\sim kT$) to show that the plasma luminosity (energy radiated per unit time and per unit volume) is

$$L = n^2 \alpha c \sigma_T \sqrt{m_e c^2 kT} \,. \tag{7.42}$$

Argue that the characteristic time, τ_{brem}, for the plasma to cool by photon radiation is

$$\tau_{\mathrm{brem}} = \frac{nkT}{L} \sim \frac{1}{n\alpha c \sigma_T} \sqrt{\frac{kT}{m_e c^2}} \,. \tag{7.43}$$

Plot the quantity $nkT\tau_{\mathrm{brem}}$ as a function of kT and compare with the bremsstrahlung limit in Fig. 7.7. Above this line, the confinement time is longer than the cooling time so the plasma cools until the confinement time is shorter.

8. Nuclear Astrophysics

Perhaps the most important accomplishment of nuclear physics is its explanation of stellar energy production and nucleosynthesis. Astrophysicists now believe that they understand in some detail stellar histories from their origins as diffuse clouds through their successive stages of nuclear burning until their "deaths" as white dwarfs or supernovae. Through this process, nuclear physics has allowed us to understand how the initial mix of hydrogen and helium produced in the primordial Universe has been transformed into the interesting mixture of heavy elements that makes terrestrial life possible.

In this brief chapter, we will first give the very minimum of stellar structure theory that is necessary to understand stars. We will see how stars work as "gravitational confinement" fusion reactors where one naturally reaches a self-regulating state with a stable temperature sufficiently high to allow nuclei to penetrate their mutual Coulomb barriers.

In the second section, we will describe in some detail the stages of nuclear burning where the nuclear fuels are successively hydrogen, helium, carbon, oxygen and so forth until elements in the iron group are produced.

In Sect. 8.3 we describe the estimations of the abundances of the elements in the solar system and how the existence of elements beyond the iron-group can be explained by stellar neutron-capture reactions.

Finally, in Sect. 8.4 we describe observations of neutrinos and γ-rays that have empirically confirmed that the processes discussed in this chapter do indeed take place inside stars.

8.1 Stellar Structure

Stars are gravitationally bound collections of nuclei and electrons. They also contain blackbody photons in thermal equilibrium with the particles. Photons cannot be gravitationally bound in stars so they diffuse through the star until they escape. As we will see, the role of nuclear reactions is to replace the energy leaving the stars so as to maintain stable conditions.

In this section, we will first study *classical* stars, by which we mean bodies that are sufficiently dilute that we can neglect the quantum nature of the particles. This is the case for stars like the Sun but fails for objects like white dwarfs and neutron stars where fermionic particles are "degenerate," i.e. the phase-space density approaches the limit allowed by Pauli exclusion principle. Such stars will be treated in Sect. 8.1.2.

8.1.1 Classical stars

We would like to understand the density $\rho(r)$, pressure $P(r)$, and temperature $T(r)$ profiles of stars. Since stars evolved slowly, it is a good first approximation to take them to be time-independent. This is possible if the star is held in quasi-hydrostatic equilibrium where the gravitational force working for collapse is balanced by a pressure gradient. As illustrated in Fig. 8.1, the necessary condition for hydrostatic equilibrium is that the pressure gradient satisfy.

$$\frac{dP}{dr} = -\rho(r)\frac{GM(r)}{r^2} \,, \tag{8.1}$$

where $M(r)$ is the mass contained within a sphere of radius r. Clearly, the pressure must decrease as r increases.

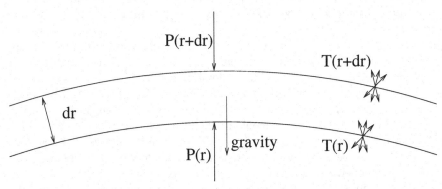

Fig. 8.1. The pressure and temperature gradients of a stable star can be qualitatively understood by considering a thin layer of material at a distance r from the center of a star. The material in the layer experiences downward forces from gravity and from the pressure $P(r + dr)$. The upward force comes only from the pressure $P(r)$. If the pressure gradient satisfies (8.1) the upward and downward forces balance. Additionally, the surfaces at r and at $r + dr$ radiate isotropically blackbody photons. If the temperature gradient satisfies (8.4), the difference between the energy radiated outward from the surface at r and the energy radiated inward from the surface $r + l_\gamma$ is equal to the net luminosity (l_γ=photon mean free path).

Through the equation of state $P = nkT$, the pressure gradient (8.1) generates density and temperature gradients. We then need another equation to determine one of these two gradients. For many stars, the temperature gradient can be determined by considering energy transport by blackbody radiation. The energy liberated by nuclear reactions is ultimately evacuated by radiation of photons. The surface of a black body radiates an energy per surface area of σT^4 where σ is the Stefan–Boltzmann constant. A spherical blackbody of radius R must then have a luminosity given by

$$\frac{L(R)}{4\pi R^2} = \sigma T(R)^4 \,, \tag{8.2}$$

where $T(R)$ is the surface temperature. Inside the star, the situation is more complicated since, as illustrated in Fig. 8.1, a surface at radius r both radiates energy and absorbs energy from layers further from the center. We must require that the difference between these two energies be equal to the luminosity $L(r)$ generated inside the surface. The energy the surface receives from outer layers originates about one photon absorption length (l_γ) higher so we would expect that the net energy flux per unit area to be roughly the difference between the values of σT^4 at values of r that differ by l_γ:

$$\frac{L(r)}{4\pi r^2} \sim \sigma \left[T^4(r) - T^4(r + l_\gamma) \right] \,. \tag{8.3}$$

A careful treatment (e.g. [72]) gives a factor 4/3:

$$\frac{L(r)}{4\pi r^2} = (4/3)\sigma \frac{\mathrm{d}T^4}{\mathrm{d}r} l_\gamma \,. \tag{8.4}$$

The absorption length is usually written as $(\kappa\rho)^{-1}$, where the function κ is called the "Rossland mean opacity." It is a function of the temperature, density and chemical composition.

Equation (8.4) determines the solar temperature gradient over its inner 70%. The outer 30% has adiabatic convective mixing which determines the temperature gradient by standard thermal arguments (Exercise 8.1).

To equations (8.1) and (8.4) we must add the effect of the release of nuclear binding energy in fusion reactions. For a steady state, this energy release is balanced by the photon luminosity so we have

$$\frac{\mathrm{d}L}{\mathrm{d}r} = 4\pi r^2 \epsilon(r) \,, \tag{8.5}$$

where $\epsilon(r)$ is the total rate of energy release per unit volume. For a given reaction, it is the Q-value times the temperature- and density-dependent reaction rates calculated in the previous chapter, (7.19) and (7.27).

The three coupled differential equations, (8.1), (8.4) and (8.5), can be solved numerically to find the internal structure of a star. The calculated density, temperature and pressure profiles of the Sun are shown in Fig. 8.2. The central temperature of the Sun is $T(0) \sim 1.568 \times 10^7 \mathrm{K}$ or $kT \sim 1.35\,\mathrm{keV}$.

To understand the slow evolution of a star, it is useful to take a global approach and consider the total energy of the star. The end result of these considerations will be an expression (8.20) that gives the total energy as a function of the number particles in the star and the radius of the star.

We consider an object of mass M consisting of free massive particles (atoms and free nuclei and electrons) gravitationally bound in a sphere of radius R. The total energy is

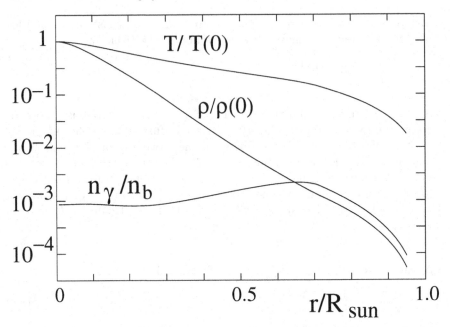

Fig. 8.2. The calculated [73] solar density and temperature profiles normalized to the central values $\rho(0) = 152.4\,\mathrm{g\,cm^{-3}}$ and $T(0) = 1.568 \times 10^7$ K. The density scale height (i.e. the distance over which the density changes by a factor e) is $\sim 0.1 R_\odot$. Also shown is the photon–baryon ratio n_γ/n_b. This ratio is nearly constant at $\sim 10^{-3}$ except for $r > 0.7 R_\odot$ where convection forces the ratio to decrease (Exercise 8.1)

$$E_{\text{tot}} = \sum_{\text{particles}} \left(mc^2 + \frac{p^2}{2m} \right) + E_{\text{grav}} + E_{\text{photons}} . \tag{8.6}$$

We have used the non-relativistic form, $mc^2 + p^2/2m$, for the energy of the particles because, as we will see, only non-relativistic particles can be bound by gravitation. The second term is the negative gravitational energy of the object

$$E_{\text{grav}} = - \int_0^R \frac{GM(r)}{r} \rho(r) 4\pi r^2 \mathrm{d}r . \tag{8.7}$$

For an object of uniform density, $E_{\text{grav}} = -(3/5)GM^2/R$ so it is useful to introduce the effective "gravitational" radius of the star, R_{grav}

$$\frac{1}{R_{\text{grav}}} = (5/3)M^{-2} \int_0^R \frac{M(r)}{r} \rho(r) 4\pi r^2 \mathrm{d}r , \tag{8.8}$$

so that E_{grav} is simply

$$E_{\text{grav}} = -(3/5) \frac{GM^2}{R_{\text{grav}}} . \tag{8.9}$$

R_{grav} is just the radius of the uniform sphere of matter that would have the same gravitational energy as the star in question. For the Sun, a numerical integration of the density profile of Fig. 8.2 gives $R_{\text{grav}} = 0.37 R_\odot$.

We can relate the kinetic and gravitational energies through the "virial theorem." Multiplying (8.1) by $4\pi r^3$ and integrating by parts. we find an expression relating the mean pressure $\bar{P}$ and E_{grav}:

$$3\bar{P}V = -E_{\text{grav}} = (3/5)\frac{GM^2}{R_{\text{grav}}} . \tag{8.10}$$

If the photon pressure is negligible, we can use the ideal gas law, $PV = NkT$, to estimate the mean temperature of the star

$$k\bar{T} = (1/5)\frac{N_{\text{b}}}{N_{\text{part}}}\frac{GN_{\text{b}}m_{\text{p}}^2}{R_{\text{grav}}} \sim (1/10)\frac{GN_{\text{b}}m_{\text{p}}^2}{R_{\text{grav}}} , \tag{8.11}$$

where N_{part} is the total number of free massive particles in the star and $N_{\text{b}} \sim M/m_{\text{p}}$ is the number of baryons (nucleons) in the star. [We neglect the $\sim 1\%$ difference between m_{p} and $m(A, Z)/A$.] In the second form we have assumed totally ionized hydrogen, $N_{\text{part}} = N_{\text{e}} + N_{\text{p}} = 2N_{\text{b}}$. For the Sun, $N_{\text{b}} \sim 10^{57}$ we find $kT \sim 500\,\text{eV}$ in reasonable agreement with the mean temperature in solar models.

Equation (8.11) tells us that a star has a negative specific heat, i.e. as it loses energy and contracts, its temperature increases. This fact will turn out to be crucial in a star's ability to maintain a stable nuclear-burning regime.

The kinetic energy per particle is $(3/2)kT$ so the total kinetic energy in the star is simply related to E_{grav}

$$\sum_{\text{particles}} \left(\frac{p^2}{2m}\right) = -(1/2)E_{\text{grav}} . \tag{8.12}$$

This is just a form of the virial theorem. Another proof that does not appeal to hydrostatic equilibrium is given in Exercise 8.2.

The final term in (8.6) is the energy of the photons that must be present if a quasi-stationary thermal equilibrium is reached:

$$E_{\text{photons}} = \int_0^R \rho_\gamma 4\pi r^2 dr , \tag{8.13}$$

where the photon energy density is given by the Stefan–Boltzmann law

$$\rho_\gamma = \frac{2\pi^2}{30}\frac{(kT)^4}{(\hbar c)^3} . \tag{8.14}$$

It turns out to be more interesting to work with the photon number density:

$$n_\gamma = \frac{2.5}{\pi^2}\left(\frac{kT}{\hbar c}\right)^3 \sim \frac{\rho_\gamma}{3kT} . \tag{8.15}$$

We can use the mean temperature (8.11) to estimate the total number of photons in the sun $N_\gamma \sim n_\gamma(\bar{T})4\pi R_{\text{grav}}^3/3$:

$$N_\gamma \sim \alpha_G^3 N_b^3 \frac{1}{10^3} \frac{4 \times 2.4}{3\pi} \sim 10^{-3} \alpha_G^3 N_b^3 , \tag{8.16}$$

where the "gravitational fine structure constant" is

$$\alpha_G = \frac{Gm_p^2}{\hbar c} = 6.707 \times 10^{-39} . \tag{8.17}$$

The photon–baryon ratio is then

$$\frac{N_\gamma}{N_b} \sim 10^{-3} \alpha_G^3 N_b^2 . \tag{8.18}$$

We see that that the photon–baryon ratio is independent of the radius and proportional to the square of the number of baryons. For the Sun, $N_b = 1.2 \times 10^{57}$ giving $\alpha_G^3 N_b^2 = 0.4$ so the photon–baryon ratio is about 10^{-3}, in agreement with solar models (Fig. 8.2). For a star of $M \sim 30 M_\odot$, the photon–baryon ratio is 30^2 times larger, approaching unity. It turns out that this makes the star unstable and the radiation pressure expels mass from the surface until the mass falls below $30 M_\odot$.

The mean energy of photons in a star is of order kT so we have a total photon energy content of

$$E_{\text{photons}} \sim 10^{-4} \alpha_G^3 N_b^3 \frac{Gm_p^2 N_b}{R_{\text{grav}}} . \tag{8.19}$$

Since, for the Sun, the number of photons is $\sim 10^{-3} N_b$, E_{photons} is of order 10^{-3} of the total kinetic energy of the particles. This also implies that the photon pressure is of order 10^{-3} the thermal pressure due to the massive particles.

If we neglect E_{photons} we can have a compact formula for the energy of a star as a function of its radius. Substituting (8.12) into (8.6), we get

$$E_{\text{tot}} = \sum_{\text{particles}} (mc^2) - \frac{3GM^2}{10R_{\text{grav}}} . \tag{8.20}$$

We are now in a position to understand the effect of the thermal photons diffusing out of a star. These photons have *positive* energy so, as photons leave, E_{tot} must diminish in order to conserve energy. Equation (8.20) suggests that this can be done either by contraction, leading to a decrease in $E_{\text{grav}} \propto -1/R_{\text{grav}}$ or by exothermic nuclear reactions which decrease the mc^2 term.[1] The stellar luminosity, dE_{tot}/dt, must then be

$$L = -\frac{dmc^2}{dt} - \frac{3GM^2}{10R_{\text{grav}}^2} \frac{dR_{\text{grav}}}{dt} . \tag{8.21}$$

In fact, as illustrated in Fig. 8.3, it is generally the case that one or the other of the terms on the left side of (8.21) dominates at a given time. Stars

---

[1] Lowering the mass would also *increase* slightly E_{grav} but, since the mc^2 term dominates, this effect is negligible.

start their lives as diffuse clouds of gas that are too cold to initiate nuclear reactions. Photon radiation luminosity is thus supplied by the second term in (8.21). The decreasing radius increases the internal temperature of the star as required by the virial theorem (8.11). As the temperature rises, the rate of nuclear reactions increases until the photon luminosity can be provided by the nuclear reactions. At this point, the star can reach a stable regime at constant R_{grav}. After the fuel is exhausted, the contraction begins again.

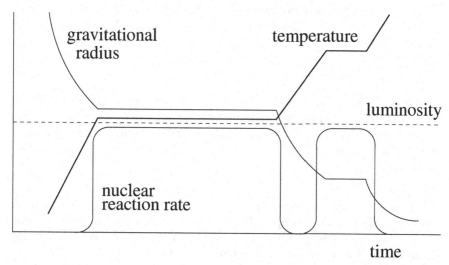

Fig. 8.3. The simplified evolution of a classical star. The star initially contracts and the temperature rises until hydrogen fusion is initiated. The radius and temperature then remain constant until the fuel is exhausted, at which point another contraction phase begins. The temperature rises until another fuel (helium) can be burned. During all this time, the luminosity is constant if the mean photon cross-section remains constant.

The stability of a nuclear burning regime illustrated in Fig. 8.3 is due to a thermostatic effect of the negative specific heat of gravitationally bound structures. If, for a moment, the rate of nuclear energy production increases above the steady state value, the local temperature increases. This causes the pressure to increase and the star to expand. This lowers the temperature and, as a result, the reaction rate. On the other hand, if the nuclear energy generation rate decreases, the star starts to contract, thus increasing the temperature and the reaction rate. A gravitationally-confined fusion reactor is thus self-regulating.

As soon as the nuclear fuel is used up, the reactions cease and the star must resume its contraction. The contraction can stop if a new type of nuclear reaction using a different fuel reaches a rate where it can supply the star's luminosity. Stars can then pass through a series of stable phases where the fuels are first hydrogen, then helium, and then carbon and oxygen until the stars core consists of iron-group nuclei. The order of fuels follows a sequence of increasing Z because the associated higher Coulomb barriers require higher temperatures to supply the necessary reaction rate. This sequence of burning stages will be discussed in more detail in Sect. 8.2. The sequence may be interrupted if, during the contraction, the star approaches its "ground state" where fermion degeneracy prevents further contraction (Sect. 8.1.2). This effect prevents all but the heaviest stars ($M > 10 M_\odot$) from burning their fuel all the way to the iron group. Stars with $M < 0.07 M_\odot$ reach their ground state before any nuclear reactions are ignited.

It is interesting to estimate the luminosity of a star. Very roughly speaking, it is given by

$$L = \frac{E_{\text{photons}}}{\tau_\gamma} \tag{8.22}$$

where τ_γ is the mean time for a photon to diffuse out of the star. Photons leave the sun as soon as their random walk takes them to the "photosphere," beyond which their mean free paths are essentially infinite. The mean time for this is of order

$$\tau \sim \frac{3R^2}{\lambda c} , \tag{8.23}$$

where λ is the typical photon mean free path. There are two limits where the photon cross-section is simple. The first is at sufficiently high temperature so that the medium is completely ionized. In this case photons can only Thomson scatter with a mean free path given by

$$\lambda^{-1} \sim \sigma_T N_b / R_{\text{grav}}^3 . \tag{8.24}$$

This regime occurs, depending on the density, for temperatures greater than $10^5 - 10^6$K. The second simple case occurs when the temperature is so low that no atoms are ionized and few photons have enough energy to ionize them. In this case, only Rayleigh scattering is important and the mean free path is much longer than that given by (8.24). In between these two regimes, absorption of photons by bound electrons (photo-ionization) and on continuum electrons in the field of a positive ion ("free–free scattering" or inverse bremsstrahlung) dominate the mean free path so the effective cross section will be higher than σ_T.

The mean escape time is then estimated as

$$\tau \sim \frac{\langle \sigma \rangle N_b}{Rc} . \tag{8.25}$$

For the Sun, the average photon cross-section $\langle \sigma \rangle$ is of order $\sim 10 \sigma_T$, giving $\tau \sim 10^5$ yr.

The luminosity (8.22) is ratio of the E_{photons} given by (8.19) and the photon escape time (8.25). We see that the factor $1/R$ in E_{photons} (8.19) is canceled by the factor $1/R$ in the mean escape time so the energy luminosity is independent of R but proportional to $\langle \sigma \rangle$

$$L \sim \alpha_G^3 N_b^3 \frac{G m_p^2 \langle \sigma \rangle}{c} . \tag{8.26}$$

For the Sun, this gives $L \sim 10^{26}$ W compared of the observed luminosity $L_\odot = 3.8 \times 10^{26}$ W. Given the many approximations made in this estimation, it is satisfying that we find a number of the correct order of magnitude.

Note that if the effective photon cross-section $\langle \sigma \rangle$ were the same in all stars, the luminosity (8.26) would simply be proportional to N_b^3, i.e. to the third power of the stars mass. This is in fact observed to be a good approximation for hydrogen-burning stars. Note also that if the effective photon cross-section were temperature independent so that it remained unchanged as the star contracts, the star's luminosity would not change as the star evolves through contraction and nuclear-burning stages. This is nearly true for very heavy stars where the medium is mostly ionized with $\langle \sigma \rangle \sim \sigma_T$. This idealized evolution is illustrated in Fig. 8.3 where the luminosity is independent of time.

The total time a star spends in a particular stage of its evolution is given by $T = \bar{L}/\Delta E$ where $\bar{L}$ and ΔE are the mean luminosity and available energy during the phase. The Sun is in its hydrogen burning phase which liberates ~ 6 MeV per proton. Since there is no convective mixing in the inner parts of the Sun, only the inner 10% of the hydrogen will actually be burned. Using the present luminosity, this gives a total hydrogen-burning time of $\sim 10^{10}$ yr. The Sun's present age is $\sim 4.5 \times 10^9$ yr so the Sun is a middle-aged star.

The time the Sun required to contract to its present radius before burning hydrogen can by calculated by using $\Delta E = (3/5) G M_\odot^2 / R_{grav}$. Assuming that the solar luminosity was the same during the contraction phase as in the hydrogen-burning phase, this gives a contraction time of $\sim 10^7$ yr. In the nineteenth century before nuclear energy was discovered, this was the estimated total age of the Sun, in clear conflict with the age of the Earth estimated by geologists.

Coming back to the photon–baryon ratio in a star, the fact that it is greater than unity for $M > 30 M_\odot$ means that the pressure due to photons is greater than that due to massive particles. This is in inherently dangerous situation because it makes the condition (8.10) difficult to maintain because photons dominate the left-hand side whereas only particles contribute to the right-hand side. Detailed calculations indicate that stars greater than about $30 M_\odot$ are unstable and generally evaporate particles until their mass reaches this value.

8.1.2 Degenerate stars

We have seen that a collection of particles must radiate photons and contract, with the contraction pausing whenever nuclear reactions are ignited to provide the photon luminosity. This process must stop when the collection reaches its quantum-mechanical "ground state."

We want to estimate the ground-state energy of a collection of electrons and nucleons bound by gravitation. We can do this in the same way as was done in Chap. 1 where we estimated the ground-state energy of a collection of nucleons bound by the strong force.

The maximum phase-space density allowed for fermions is two particles per $(\Delta p \Delta x)^3 = (2\pi\hbar)^3$. We assume that N fermions are spread uniformly over a spatial volume $V = 4\pi R^3/3$. This means that in the ground state the single-particle orbitals are filled up to those corresponding to a Fermi momentum p_F determined by

$$\frac{N}{V \times (4\pi/3)p_F^3} = \frac{1}{(2\pi\hbar)^3} \tag{8.27}$$

i.e.

$$\frac{4\pi p_F^3}{3} = N\frac{(2\pi\hbar)^3}{2}\frac{1}{4\pi R^3/3} . \tag{8.28}$$

The mean momentum squared of this assembly is

$$\langle p^2 \rangle = \frac{\int_0^{p_F} p^2 p^2 \mathrm{d}p}{\int_0^{p_F} p^2 \mathrm{d}p} = (3/5)p_F^2 . \tag{8.29}$$

Since, the $\langle p^2 \rangle$ is independent of the particle mass, the mean kinetic energy, $p^2/2m$, will be dominated by the light electrons. (This is not the case for a classical star where thermal equilibrium requires that the mean kinetic energies of all particles be equal.) The energy of the star is then

$$E \sim N_e\sqrt{\langle p^2 \rangle c^2 + m_e^2 c^4} - (3/5)\frac{GM^2}{R} .$$

(We do not include the rest-energy of the nucleons in this formula.) Using $M = N_b m_p$ and equations (8.28) and (8.29) this is

$$E \sim N_e\sqrt{(3/5)p_F^2 c^2 + m_e^2 c^4} - \frac{3Gm_p^2 N_b^2}{5N_e^{1/3}}\frac{p_F}{2\pi\hbar}\left(\frac{32\pi^2}{9}\right)^{1/3} , \tag{8.30}$$

where N_e is the number of electrons and N_b is the number of baryons. Dropping the numerical factors for this rough estimate, and taking the derivative of E with respect to the momentum we find that the minimum energy occurs for

$$\frac{p_F}{\sqrt{p_F^2 + m_e^2 c^2}} = \frac{Gm_p^2}{\hbar c}\frac{N_b^2}{N_e^{4/3}} = \left(\frac{N_b}{N_c}\right)^{2/3} , \tag{8.31}$$

where the "critical" number of baryons is

$$N_c = \left(\frac{1}{\alpha_G}\right)^{3/2}\left(\frac{N_e}{N_b}\right)^2 = 1.82 \times 10^{57}\left(\frac{N_e}{N_b}\right)^2 , \tag{8.32}$$

where $\alpha_G = Gm_p^2/\hbar c$. This corresponds to a total mass

$$M_c = m_p N_c = 3.04 \times 10^{30} \text{ kg} \left(\frac{N_e}{N_b}\right)^2 = 1.52 M_\odot \left(\frac{N_e}{N_b}\right)^2 \quad (8.33)$$

Within a factor of order unity, this is just the celebrated *Chandrasekhar mass*, the critical mass estimated by more sophisticated reasoning:

$$M_{Ch} \sim 5 M_\odot \left(\frac{N_e}{N_b}\right)^2 \sim 1.25 M_\odot \quad \text{for } N_e/N_b = 1/2. \quad (8.34)$$

The fate of a star depends on whether its mass is greater than or less than the Chandrasekhar mass. Stars with a number of baryons less than N_c have a well-defined ground state. The radius of the star in the ground state found by substituting (8.31) into (8.28) giving

$$R_{gs} \sim \frac{\hbar}{m_e c} N_e^{1/3} \left(\frac{N_c}{N_b}\right)^{2/3} \left(1 - \frac{N_b}{N_c}\right)^{1/2}. \quad (8.35)$$

For the Sun, $N_b = 1.2 \times 10^{57}$ and $N_e/N_b = 0.7$ which gives $R_{gs} \sim 10^4$ km. This is considerably less than the actual solar radius $R_\odot = 6.96 \times 10^5$ km implying that the Sun is far from its ground state. In fact, 10^4 km is a typical radius for white dwarfs, believed to be degenerate stars (nearly) in their ground states.

For a very light stars, $N_b \to 0$, $R_{gs} \to \infty$ so we can anticipate that such a star will contract to its ground state before becoming hot enough to ignite nuclear reactions. Detailed models suggest that this happens for stars with $M < 0.07 M_\odot$. Planets are examples of such objects.

Considering stars more massive than the critical mass, we note that (8.31) has no solution if the $N_b > N_c$, so such stars apparently have no ground state. This suggests that after burning their nuclear fuel, they must collapse to a blackhole unless they can shed mass until $N_b < N_c$. We note however that during the final collapse, p_F rises until the majority of the electrons have energies above the threshold for electron capture

$$e^-(A, Z) \to (A, Z-1)\nu_e. \quad (8.36)$$

As the protons are transformed to neutrons by this reaction (neutronization), the temperature rises to the point where nuclei are dissociated to create a star made mostly of neutrons. We must then consider the ground state of a gravitationally bound collection of neutrons. To estimate its parameters, we can repeat the previous analysis replacing m_e with m_n and N_e/N_b with unity. This last replacement means that the critical number of baryons for a neutron star is larger than the critical number of a white dwarf meaning that some of the stars that could not be saved from collapse by electron degeneracy will reach a stable state as a neutron star. The radius of such a star, estimated by replacing m_e with m_n in (8.35), is:

$$R_{gs} \sim \frac{\hbar}{m_n c} N_b^{1/3} \qquad N_b \ll N_c. \quad (8.37)$$

For a solar mass this gives $R_{gs} \sim 3\,\mathrm{km}$, comparable with the observed sizes of neutron stars. It corresponds to a density about 10 times greater than that of normal nuclei.

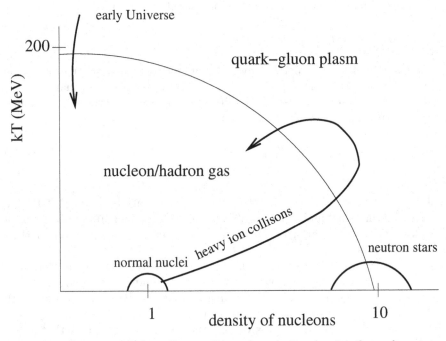

Fig. 8.4. The expected phase diagram for nuclear matter showing the nuclear state as a function of temperature and baryon density (minus the antibaryon density). At high temperature and density, quarks and gluons act as free particles in a *quark–gluon plasma*. At low temperature and density the quarks and gluons combine to form hadrons and nucleons. At vanishing temperature, the transition corresponds to a density about 10 times that of normal nuclei, i.e. nucleons in contact. Neutron stars are believed to have densities near this point. At low density, the transition is at $kT \sim 200\,\mathrm{MeV}$, i.e. $\sim m_\pi c^2$. Such a phase transition is believed to have occurred in the early universe (Chap. 9). High-energy ($> 100\,\mathrm{GeV\,nucleon^{-1}}$) heavy ion collisions are believed to sometimes create quark–gluon plasmas that quickly cool back to the nucleon–hadron phase.

We emphasize that the calculated radius cannot be taken too seriously because at neutron-star densities neighboring nucleons are in contact and it is necessary to take into account the nucleon–nucleon interactions. The situation is further complicated by the possibility that gases of nucleons undergo a phase transition at high densities and temperatures where the constituent quarks and gluons are liberated, forming a *quark–gluon plasma*. The expected phase diagram is shown in Fig. 8.4. At zero temperature, the transition is expected to take place when the nucleons are in contact, as may be the case in neutron stars.

8.2 Nuclear burning stages in stars

In this section we will give some of the details of the various stages of stellar nuclear burning. These stages are summarized in Table 8.1.

Table 8.1. The energy released in the (idealized) stages that transform 56 protons and 56 electrons to one ^{56}Fe nucleus and 26 electrons. Columns 2 and 3 give the available energy. We see that most of the energy comes in the first stage when hydrogen is fused to helium. The fourth column gives f_ν, the fraction of the energy released that goes to neutrinos and is therefore not available for heating the medium. (For hydrogen burning, this fraction depends on the precise reaction chain that dominates and we have taken those in the Sun.) The final two columns give the approximate stellar ignition temperatures. It should be emphasize that the five stages listed here do not represent distinct stages in real stars, in which many different reactions may take place simultaneously but at different depths.

reaction	$Q/56$ (MeV)	Q/m 10^{12} J kg^{-1}	f_ν	T $(10^9$K)	kT (keV)
$14[4^1\text{H} \rightarrow {}^4\text{He}]$	6.683	640	0.02	0.015	1.3
$2[7^4\text{He} \rightarrow {}^{12}\text{C}\,^{16}\text{O}]$	0.775	75	0	0.15	15
$2[^{12}\text{C}^{16}\text{O} \rightarrow {}^{28}\text{Si}]$	0.598	57	0	0.8-2.0	100
$2^{28}\text{Si} \rightarrow {}^{56}\text{Ni}$	0.195	19	0	3.5	300
$^{56}\text{Ni} \rightarrow {}^{56}\text{Co} \rightarrow {}^{56}\text{Fe}$	0.120	12	0.2		
total	8.371	803	.02		

8.2.1 Hydrogen burning

There are four principal ways of converting 4^1H to ^{4}He via exothermic reactions.[2] The first, called "PPI" uses only ^{1}H as the primary ingredient. The second two, "PPII" and "PPIII" use ^{4}He as a catalyst. The fourth, the "CNO" cycle, is catalyzed by ^{12}C and, as such, is only possible in second generation stars. (The first stars created in the Universe contained essentially only the ^{1}H and ^{4}He produced in the primordial Universe.) Since two protons must be converted to neutrons, each chain has two reactions due to weak interactions, producing two ν_e.

---

[2] Only exothermic reactions are allowed because of the relatively low temperatures of hydrogen-burning stars.

The PPI chain proceeds as follows:

$$^1\text{H}\,^1\text{H} \rightarrow\, ^2\text{H}\,e^+\nu_e \qquad\quad ^1\text{H}\,^1\text{H} \rightarrow\, ^2\text{H}\,e^+\nu_e \tag{8.38}$$

$$^1\text{H}\,^2\text{H} \rightarrow\, ^3\text{He}\,\gamma \qquad\quad ^1\text{H}\,^2\text{H} \rightarrow\, ^3\text{He}\,\gamma. \tag{8.39}$$

$$^3\text{He}\,^3\text{He} \rightarrow\, ^4\text{He}\,2^1\text{H} \tag{8.40}$$

Because of the non-existence of ^2He and ^2n, reaction (8.38) is the only reaction involving only ^1H. PPI thus starts with two such reactions, transforming two protons to neutrons.

Once deuterium is formed by (8.38), it is in principle possible to directly produce ^4He via the reaction $^2\text{H}^2\text{H} \rightarrow\, ^4\text{He}\gamma$. Since there is little deuterium initially present in the star, a much more likely possibility is that the ^2H produced by (8.38) quickly captures one of the abundant protons in reaction (8.39) to produce ^3He.

There are no non-weak exothermic reactions between ^1H and ^3He so, unless one uses ^4He as a catalyst, the star has to wait until the quantity of ^3He builds to the point where ^4He can be produced by reaction (8.40). This reaction terminates PPI. After annihilation of the positrons from (8.38) with electrons in the medium, the complete chain is

$$4e^- \; 4^1\text{H} \; [2\,^1\text{H}] \rightarrow 2e^- \; ^4\text{He} \; 2\nu_e \; [2\,^1\text{H}]\,. \tag{8.41}$$

where the two ^1H in brackets are catalysts returned in the final reaction (8.40). In this expression, we ignore the photons which, along with the kinetic energy of the charged particles, are thermalized.

To avoid waiting for the buildup of ^3He, a star can make use of the abundant ^4He as a catalyst. This is done in the PPII and PPIII chains as follows:

$$^1\text{H}\,^1\text{H} \rightarrow\, ^2\text{H}\,e^+\nu_e \tag{8.42}$$

$$^1\text{H}\,^2\text{H} \rightarrow\, ^3\text{He}\,\gamma \tag{8.43}$$

$$^4\text{He}\,^3\text{He} \rightarrow\, ^7\text{Be}\,\gamma \tag{8.44}$$

$$e^-\,^7\text{Be} \rightarrow\, ^7\text{Li}\,\nu_e \;\; (\text{PPII}) \qquad \text{or} \qquad ^1\text{H}\,^7\text{Be} \rightarrow\, ^8\text{B}\,\gamma \;\; (\text{PPIII}) \tag{8.45}$$

$$^1\text{H}\,^7\text{Li} \rightarrow\, ^8\text{Be}\,\gamma \qquad\qquad\qquad\quad ^8\text{B} \rightarrow\, ^8\text{Be}\,e^+\nu_e \tag{8.46}$$

$$^8\text{Be} \rightarrow 2^4\text{He}\,. \tag{8.47}$$

In these two cycles, the ^3He capture (8.40) in PPI is replaced by ^4He capture producing ^7Be (8.44). PPII and PPIII then differ through the fate of the ^7Be. In the PPII chain, the ^7Be captures an electron to produce ^7Li which then captures a proton to produce the α-unstable ^8Be. In the PPIII chain, the ^7Be captures a proton to produce ^8B which then β-decays to ^8Be. In both cases, the final decay $^8\text{Be} \rightarrow\, ^4\text{He}\,^4\text{He}$ returns the catalyzing ^4He introduced in (8.44).

The complete PPII and PPIII chains are the same as PPI (8.41) except that the catalyst is ^4He instead of 2^1H.

The final "CNO" chain uses ^{12}C or ^{14}N as a catalyst.

$$^1\text{H} \ ^{12}\text{C} \ \rightarrow \ ^{13}\text{N} \ \gamma \tag{8.48}$$

$$^{13}\text{N} \ \rightarrow \ ^{13}\text{C} \ e^+ \ \nu_e \tag{8.49}$$

$$^1\text{H} \ ^{13}\text{C} \ \rightarrow \ ^{14}\text{N} \ \gamma \tag{8.50}$$

$$^1\text{H} \ ^{14}\text{N} \ \rightarrow \ ^{15}\text{O} \ \gamma \tag{8.51}$$

$$^{15}\text{O} \ \rightarrow \ ^{15}\text{N} \ e^+ \ \nu_e \tag{8.52}$$

$$^1\text{H} \ ^{15}\text{N} \ \rightarrow \ ^{12}\text{C} \ ^4\text{He} \qquad \text{or} \qquad ^1\text{H} \ ^{15}\text{N} \ \rightarrow \ ^{16}\text{O} \ \gamma \tag{8.53}$$

$$^1\text{H} \ ^{16}\text{O} \ \rightarrow \ ^{17}\text{F} \ \gamma \tag{8.54}$$

$$^{17}\text{F} \ \rightarrow \ ^{17}\text{O} \ e^+ \ \nu_e \tag{8.55}$$

$$^1\text{H} \ ^{17}\text{O} \ \rightarrow \ ^{14}\text{N} \ ^4\text{He} \tag{8.56}$$

The cycle branches at (8.53) when the proton capture can lead to the production of ^{12}C or, through radiative capture, of ^{16}O. The catalyzing ^{12}C or ^{14}N introduced at (8.48) or (8.51) is returned at the terminating reaction (8.53) or (8.56).

The CNO chain avoids the use of the weak two-body reaction (8.38), the protons being transformed to neutrons via the β-decays of ^{13}N and ^{15}O or of ^{15}O and ^{17}F. Since it uses heavy nuclei as catalysts, this reaction chain was not possible for the first generation of stars.

The relative importance of the four hydrogen burning chains depends on the mass and chemical composition of the star. First generation stars with practically no carbon cannot use the CNO cycle. For stars with Solar-type compositions, the CNO chain avoids the initial weak interaction (8.39) but it is still estimated to be unimportant in the Sun because of the high Coulomb barrier for proton capture on carbon. The CNO cycle is important in high-mass hydrogen-burning stars because the burning temperature is an increasing function of the mass. This is because the luminosity is proportional to the third power of the mass and the high reaction rate needed to provide the large luminosity requires a high burning temperature.

In stars burning with the PP chains, the (PPII+PPIII) to PPI ratio is determined by the relative probability for a ^3He to capture another ^3He or a ^4He. Capture of a ^3He has a higher probability for Coulomb barrier penetration because of the higher thermal velocity of ^3He compared to ^4He. Low-mass stars therefore prefer PPI. This is the case in the Sun, where 90% of the energy is produced by PPI. In low-mass stars, PPII is favored by over PPIII because of the lack of a Coulomb barrier for (8.45). In the Sun, 10% of the luminosity is generated by PPII and 0.02% by PPIII. About 1% is generated by the CNO cycle.

8.2.2 Helium burning

After the core of a main sequence star is transformed to ^{4}He, it contracts[3] until the temperature is sufficiently high that the star's luminosity can be provided by helium burning. The mechanism of ^{4}He burning is quite different from that of hydrogen burning. There is no lack of neutrons so weak interactions are not needed. On the other hand, helium burning is strongly inhibited by the fact that there are no exothermic two-body reactions involving only ^{4}He. In particular, the mass of ^{8}Be is 92 keV greater than twice the mass of ^{4}He and therefore decays immediately ($\tau = 2 \times 10^{-16}$s) back to ^{4}He^{4}He:

$$^4\text{He}\,^4\text{He} \leftrightarrow \,^8\text{Be} \qquad\qquad |Q| = 92\,\text{keV} \qquad\qquad (8.57)$$

Unlike the irreversible production of ^{4}He from ^{1}H due to strongly exothermic reactions, the production of ^{8}Be is thus endothermic and reversible. As we showed in Sect. 4.1.5, a thermal equilibrium abundance of ^{8}Be is built up given by:

$$\frac{n_{8\text{Be}}}{n_{4\text{He}}} = \frac{n_{4\text{He}}}{(mkT)^{3/2}/(4\pi^2\hbar^3)}e^{-92\,\text{keV}/kT}\,, \qquad\qquad (8.58)$$

where m is the ^{4}He$-\,^4$He reduced mass. The typical density in a ^{4}He-burning core are $\rho \sim 10^5 \text{g cm}^{-3}$ and $kT \sim 15$ keV, so (8.58) gives only a tiny ^{8}Be abundance of $\sim 10^{-9}$ of the ^{4}He abundance.

Using this small abundances of ^{8}Be, it is possible to produce ^{12}C through the reaction

$$^4\text{He}\,^8\text{Be} \rightarrow \,^{12}\text{C}\,\gamma \qquad\qquad Q = 7.366\,\text{MeV} \qquad\qquad (8.59)$$

Because of the very small quantity of ^{8}Be, this would normally lead to a very small production rate of ^{12}C. However, as we noted in Sect. 7.1.3, the rate can be greatly increased if ^{12}C has an excited state near the Gamow energy for the reaction, $E_G \sim 200$ keV for $kT \sim 15$ keV. This lead Hoyle [75] to predict the existence of such a state and subsequent measurements lead to its discovery (Fig. 8.5). This 0^+ excited state of ^{12}C is 7654 keV above the ^{12}C ground state and 283 keV above ^{4}He $-^8$ Be. It decays mostly via α decay, returning the original ^{8}Be, but also has a $\sim 10^{-3}$ branching ratio to the ground state of ^{12}C:

$$^4\text{He}\,^8\text{Be} \rightarrow \,^{12}\text{C}^* \qquad\qquad Q = -283\,\text{keV}$$
$$^{12}\text{C}^* \rightarrow \,^8\text{Be}\,^4\text{He} \qquad\qquad \Gamma = 8.3\,\text{eV}$$
$$^{12}\text{C}^* \rightarrow \,^{12}\text{C}\,\gamma\gamma \qquad\qquad \Gamma_\gamma = 3 \times 10^{-3}\,\text{eV} \qquad\qquad (8.60)$$

The irreversible production of ^{12}C thus proceeds through

$$3^4\text{He} \rightarrow \,^4\text{He}\,^8\text{Be} \rightarrow \,^{12}\text{C}^* \rightarrow^{12}\text{C}\,\gamma\gamma\,. \qquad\qquad (8.61)$$

This sequence is called the "triple-α" process.

---

[3] Only the core must contract so that the gravitational radius decreases. The envelope of the star expands so that the stars appears as a *red giant*.

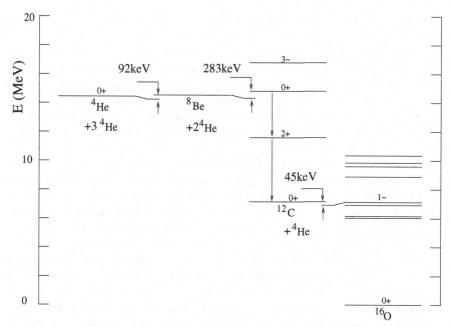

Fig. 8.5. The energy levels of four ^{4}He nuclei.

The energy liberated by the triple-α process can generate the star's luminosity when the central temperature reaches $kT \sim 10\,\mathrm{keV}$, i.e. $T \sim 10^8\mathrm{K}$. As the ^{4}He in the core is depleted, ^{12}C burning is initiated via the non-resonant reaction

$$^4\mathrm{He}\,^{12}\mathrm{C} \rightarrow\, ^{16}\mathrm{O}\,\gamma \qquad Q = 7.162\,\mathrm{MeV} \tag{8.62}$$

This reaction competes favorably with the triple-α process once the ^{4}He is depleted because its rate is linear in the concentration of ^{4}He while the rate of the triple-α process is proportional to the third power of the ^{4}He concentration. The helium-burning stage thus generates a mixture of ^{12}C and ^{16}O.

A peculiar characteristic of the triple-α process is that its end result depends critically on the details of the three remarkable energy alignments of the 0^+ states of ^{8}Be, ^{12}C and the 1^- state of ^{16}O (Fig. 8.5):

$$^4\mathrm{He}\,^4\mathrm{He} \rightarrow\, ^8\mathrm{Be} \qquad\qquad Q_4 = -0.092\,\mathrm{MeV}\,, \tag{8.63}$$

$$^4\mathrm{He}\,^8\mathrm{Be} \rightarrow\, ^{12}\mathrm{C}^* \qquad\qquad Q_8 = -0.283\,\mathrm{MeV}\,, \tag{8.64}$$

$$^4\mathrm{He}\,^{12}\mathrm{C} \rightarrow\, ^{16}\mathrm{O}^*\,\gamma \qquad\quad Q_{12} = +0.045\,\mathrm{MeV}\,. \tag{8.65}$$

The alignment results in the three reactions being respectively slightly endothermic, slightly endothermic, and slightly exothermic. It turns out that this is the only possible arrangement that leads to significant production of ^{12}C.

First, making reactions (8.63) or (8.64) more endothermic by increasing their Q-values would have the effect of increasing the temperature at which the triple-α process takes place. At this higher temperature the Coulomb barrier for the ^{4}He $-^{12}$C reaction would be less effective so the carbon produced by the triple-α process would be quickly burned to ^{16}O, leaving little carbon. According to [76], an increase of 250 keV in the ^{12}C* resonance leads to negligible production of ^{12}C.

Changing the signs of the Q-values (while keeping them small) leads to more interesting scenarios. If (8.63) were exothermic, the hydrogen burning phase would be followed be a helium burning phase producing only ^{8}Be through ^{4}He ^{4}He $\rightarrow$ ^{8}Beγ. This phase would then be followed at a higher temperature with a beryllium burning phase with the production of oxygen via 2^8Be $\rightarrow$ ^{16}Oγ. ^{12}C would be largely bypassed in this scenario.

On the other hand, if reaction (8.64) were exothermic, the triple-α process would not be possible at all since the production of ^{12}C* would not be resonant.

Finally, if reaction (8.65) were slightly endothermic rather than slightly exothermic, ^{4}He absorption by ^{12}C would be resonant so the ^{12}C would be quickly burned to ^{16}O. Once again, little ^{12}C would be produced.

Carbon is unique among low-mass elements as having a chemistry that is sufficiently rich to allow for life "as we know it" on Earth. Its production in stars depends upon a delicate alignment of nuclear levels. This alignment is, in turn, sensitive to the values of the fundamental parameters of physics like the electroweak and strong interaction couplings. In particular, the aforementioned increase by ~ 200 keV in the 0^+ level of ^{12}C would require a change in the nucleon–nucleon potential of order 0.5% or in the fine-structure constant of order 4% [77]. Such estimates should, however, be treated with caution since many correlated changes in physics might occur if the parameters changed.

This sensitivity of stellar nucleosynthesis to nuclear levels is similar to the sensitivity of cosmological nucleosynthesis to the neutron–proton mass difference and the binding energies of the $A = 2$ nuclei. In that case, the physical parameters are such that they prevent hydrogen from being eliminated in the primordial Universe, thus leaving us with a store of available free energy.

As emphasized in the introduction, these facts have inspired speculations concerning the possibility that the physical constants are dynamical variables that can take on different values in different parts of the Universe. We know that physics seems to be the same in other parts of the *visible* Universe, so these variations must take place on scales larger than our "horizon," i.e. the distance to the furthest visible objects. At any rate, in such a picture, there will be some parts of the Universe where the parameters take on values that allow for the production of large quantities of carbon. To the extent that a large quantity of carbon increases the probability for the emergence of life, it would then be natural that we find ourselves in such a region.

8.2.3 Advanced nuclear-burning stages

The later stages of nuclear burning are rather complicated for reasons of both astrophysics and nuclear physics.

The nuclear reaction chains are rather complicated because of the multiple final states for reactions involving two nuclei. For example, in carbon burning, there are three possible exothermic reactions:

$$^{12}\text{C}\,^{12}\text{C} \rightarrow\,^{24}\text{Mn}\,\gamma \qquad Q = 13.93\,\text{MeV}$$

$$\rightarrow\,^{23}\text{Na}\,\text{p} \qquad Q = 2.24\,\text{MeV}$$

$$\rightarrow\,^{20}\text{Ne}\,^4\text{He} \qquad Q = 4.62\,\text{MeV}\,.$$

These three reactions can be considered to be a single reaction consisting of the formation of a "compound nucleus," i.e. an excited state of ^{24}Mn which then decays by photon, proton, or α emission

$$^{12}\text{C}\,^{12}\text{C} \rightarrow\,^{24}\text{Mn}^* \rightarrow \text{x}\,\text{y}\,. \tag{8.66}$$

Proton or α emission have larger probabilities than photon emission. The protons and α-particles produced in $^{24}\text{Mn}^*$ decay is are then absorbed by ^{12}C to produce ^{13}N or ^{16}O.

After carbon, neon and oxygen burning, the temperature becomes sufficiently high that some heavy nuclei are photo-dissociated by the ejection of α's. The abundance of ^4He becomes sufficiently high that α-capture becomes the major reaction leading up to the production of iron-peak nuclei.

The astrophysics of the later stages is complicated by several facts. First, in massive stars, an advanced nuclear stage in the hot core of a star may alternate with "shell burning" in the outer regions of an earlier (hydrogen or helium) stage. Second, a core will be hot enough to initiate the thermal production of neutrinos. Because of their long mean-free paths, neutrino radiation is the dominant source of stellar energy loss and the total (neutrino plus photon) luminosity given by (8.26) increases. This results in a shortening of the duration of these neutrino-radiating stages. Finally, a particular core burning stage may be reached when the electrons are degenerate. As already emphasized, under such circumstances, the nuclear thermostat is not operative and the burning may be explosive.

The best examples of explosive burning are type Ia supernovae. The progenitor of these supernovae are believed to be carbon–oxygen white dwarfs supported by degenerate electrons. Such an object would be an star with a core that has completed it helium-burning and that has lost to a large extent its helium–hydrogen envelope through various processes. An isolated white dwarf will simply cool down but if it has a binary companion, it can accrete matter until it reaches the Chandrasekhar mass (8.33). At this point, the star becomes unstable and will explosively burn its carbon and oxygen to ^{56}Ni. The energy liberated is enough to completely disrupt the gravitationally bound star (Exercise 8.3).

After disruption of the star, the supernova continues to produce energy via the β-decays of ^{56}Ni and ^{56}Co (Fig. 8.6). The lifetimes of these two nuclei determine the luminosity as a function of time of the supernova starting about a week after explosion. An example of this is shown in Fig. 0.1 for Kepler's supernova.

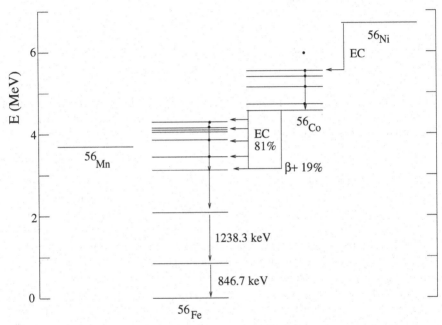

Fig. 8.6. The $A = 56$ system showing the final stage of stellar burning, the β-decays of ^{56}Ni and ^{56}Co.

8.2.4 Core-collapse

Stars with $M > 20 M_\odot$ will burn their cores to ^{56}Fe. At this point, the star profile will look something like the pre-supernova configuration shown in Fig. 8.7. The various core- and shell-burning phases may have left an onion like structure with heavy elements at the center and the original hydrogen–helium mix at the surface. Depending on mass-loss during the lifetime of the star, some of the outer layers may be missing

When the star reaches this configuration there is nothing left to burn in the core. The inert core will then accumulate mass until it reaches the

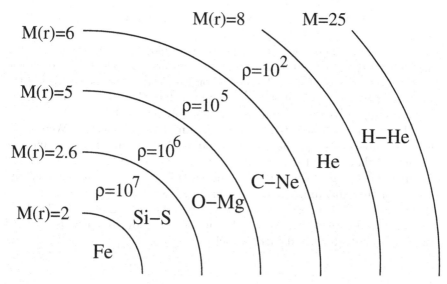

Fig. 8.7. The profile of a $25 M_\odot$ star when its core has burned to ^{56}Fe. For each concentric shell, the characteristic density (in $\mathrm{gm\,cm^{-3}}$) and dominant nuclear species are shown.

Chandrasekhar mass. At this point, it will start to implode since no thermal pressure can balance gravitation. As the temperature rises, the nuclei will start to evaporate to ^{4}He and then to nucleons. At the same time, the Fermi level of the electrons becomes sufficiently high that the electron are sufficiently energetic to be captured on protons to form neutrons

$$e^- p \rightarrow n \nu_e .\tag{8.67}$$

The ν_e will escape from the star (Exercise 8.4). That these three distinct events, collapse, nuclear evaporation, and neutronization, all occur at roughly the same moment is due to the "coincidence" that the electron mass, nuclear binding energies, and the proton–neutron mass difference are all in the MeV range.

Once the protons have been converted to neutrons, the collapse may be halted at the radius corresponding to a degenerate gas of neutrons. The energy change of the $1.4 M_\odot$ core in the process of collapsing from $R \sim 1000\,\mathrm{km}$ to $R \sim 10\,\mathrm{km}$ is

$$\Delta E \sim (3/5)\frac{GM^2}{R} \sim 3 \times 10^{46}\,\mathrm{J} \quad \text{for } M = 1.5 M_\odot .\tag{8.68}$$

Because of their long mean-free paths, the energy is almost entirely evacuated in the form of thermally produced neutrinos and antineutrinos. The $\bar{\nu}_e$ produced in the supernova SN1987a in the neighboring galaxy the Large Magellanic Cloud were detected (Exercise 3.2) confirming the core-collapse mechanism for type II supernova. These observations are described in Sect. 8.4.2.

The outer layers of the of the supernova are blown off into the interstellar medium. These layers are rich in intermediate mass nuclei $A = 4$ to 56 and are a major source of the nuclei present on Earth. It is also believed that the proto-neutron star generates a large neutron flux that creates many heavy nuclei through neutron capture, as discussed in Sect. 8.3.3.

Many of the nuclei ejected from the supernova are radioactive. Were it not for this source of energy, the matter would quickly cool and become invisible. Figure 8.8 shows how the observed luminosity of SN1987a declined with the lifetime of ^{56}Co indicating that this decay (Fig. 8.6) is the primary energy source for the expanding cloud. A confirming observation (Sect. 8.4.3) is that the γ-rays from ^{60}Co decay were observed in the direction of SN1987a. After several ^{60}Co lifetimes, it is believed that the supernova remnant is powered by other, longer-lived, radioactive nuclei.

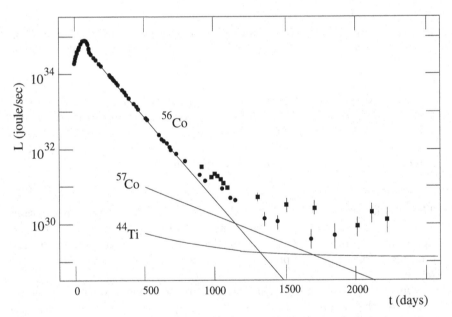

Fig. 8.8. The total luminosity of SN1987a as a function of time [78]. The labeled curves show the calculated contribution to the luminosity from the β-decay of ^{56}Co, ^{57}Co, and ^{44}Ti.

8.3 Stellar nucleosynthesis

One of the great triumphs of nuclear physics has been its ability to provide semiquantitative understanding of the observed abundances of the elements and their isotopes. Most attention has focused on "solar-system abundances" that mostly reflect the initial composition of the pre-solar cloud that condensed 4.5 million years ago to form the Sun, the planets, and meteorites. About 98% of the solar system mass consists of ^{1}H and ^{4}He and most of these two nuclei were produced in the primordial Universe when the cosmological temperature was $kT \sim 60\mathrm{keV}$. The processes leading to the formation of these two elements will be discussed in Chap. 9. The remaining 2% of the solar system mass consists of heavier elements that are believed to have been produced in stars. Prior to the formation of the solar system, these elements were dispersed into the interstellar medium either by continuous mass loss or by supernovae-like events. This pollution of the primordial mixture of helium and hydrogen was essential for the formation of Earth-like planets and the emergence of life.

8.3.1 Solar-system abundances

The estimated solar-system abundances are shown in Fig. 8.9 [79]. The distribution falls with increasing A with peaks at the α nuclei ^{4}He, ^{12}C, ^{16}O, ^{20}Ne, ^{24}Mg, ^{28}Si, ^{32}S, ^{36}Ar and ^{40}Ca that all consist of an integer number of ^{4}He nuclei. A prominent peak is also seen at ^{56}Fe which is believed to be the result of the decay of the α nuclei ^{56}Ni produced in the last stage of stellar nuclear burning.

Beyond the iron peak, the distribution continues to fall with increasing A but shows peaks at $A = 80, 87, 130, 138, 195, 208$. We will see that these peaks are due to the systematics of neutron captures responsible for the production of heavy elements.

The solar-system abundances are derived from a variety of sources. Only the Sun and the giant planets could be expected to have started with an entirely representative sample of material since gravitational attraction was the dominant factor in the formation of these bodies. The small planets and meteorites condensed out via processes that depended on chemistry and caution must therefore be exercised in using elemental abundances derived from these bodies. While chemical separation is important, the formation of small bodies would not have been expected to result in isotopic separation (with a few exceptions). Therefore, isotopic abundance ratios for the Earth or meteorites can generally be taken as representative of the Solar System.

Since most of the mass of the solar system is in the Sun, it would seem best to use "photospheric abundances" from the absorption lines that appear in the Sun's continuous spectrum. Reliable elemental abundances can be derived for most elements. It should be emphasized that photosphere estimates depend on detailed models of the Solar atmosphere since the importance of

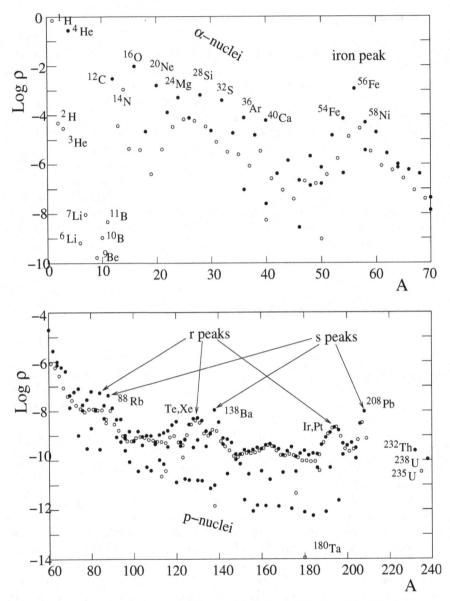

Fig. 8.9. The solar system abundances $\rho(A,Z)/\rho_{tot}$ [79]. The filled circles correspond to even-even nuclei. For $A < 70$, the distribution is visually dominated by cosmogenic ^1H and ^4He and by "iron-peak" elements near $A = 56$ Between these two features, the distribution is dominated by "α-nuclei" comprised of an integer number of ^4He nuclei. For $A > 60$, the distribution has peaks corresponding to neutron magic numbers that are produced by the s-process. The r-process produces peaks shifted to lower A after neutron-rich magic-N nuclei β-decay to the bottom of the valley of stability. Rare elements are produced by the p-process.

a particular line is highly dependent on the photosphere temperature and density which determines the populations of atomic levels. In this regard, it should be noted that one of the most prominent lines in the solar spectrum is due to calcium, a rather rare element. This caused a great deal of confusion before the development of quantum mechanics allowed one to understand the physics behind the creation of absorption lines. A more fundamental problem with the use of photosphere abundances is that they give only *elemental* abundances since *isotopic* splittings of lines are generally narrower than the thermally determined line widths. Exceptions are the isotopic abundances of carbon and oxygen where the vibrational and rotational lines of the CO molecule are directly determined by the atomic weights of the constituent atoms.

The most accurate elemental abundances for most elements come from the analysis of "carbonaceous chondritic meteorites" that are thought to have a representative sample of elements with the important exceptions of hydrogen, carbon, nitrogen, oxygen and the noble gases. While the formation of meteorites was a complicated process involving chemical separation, this type contains three representative phases (silicate, sulfide, and metallic) that give consistent results. With a few exceptions, agreement with photospheric abundances is to within $\pm 10\%$.

The most important elemental abundance that is accurately determined neither by photosphere nor meteor abundances is that of ^{4}He. This nuclide was not retained in the formation of small bodies. In fact, the majority of terrestrial ^{4}He is believed to be due to α decay of heavy elements after the formation of the Earth. In the Sun, helium lines are seen only in the chromosphere where its abundance may not be entirely representative. In fact, the most reliable estimation of the helium abundance appears to be that derived from solar models where the initial helium abundance is a *free parameter* that is adjusted so as to predict that correct solar radius and luminosity [73, 74]. The derived helium abundance is confirmed by measurements of the helioseismological oscillation frequencies [73]. These frequencies depend directly on the sound speed $v_s \sim kT/\mu$ where μ is the mean atomic weight. The temperature profile is well determined in solar models so the sound speed determines μ. Since the Sun is essentially made of hydrogen and helium, this then determines the abundance of helium.

Once the elemental abundances are determined, the nuclear abundances are generally found by multiplying elemental abundances by the terrestrial isotopic abundance. Once again, this does not work for the noble gases. An extreme example is argon where the atomic weight listed in the periodic table is 39.948 reflecting the fact that ^{40}Ar is the dominant terrestrial isotope. In reality, most terrestrial argon comes from ^{40}K decay (Fig. 5.1) while the α nucleus ^{36}Ar dominates in the Sun. The isotopic abundance of this element is therefore best determined from the solar wind.

8.3.2 Production of $A < 60$ nuclei

In previous sections we have seen how a succession of stellar burning stages produces elements from ^{4}He up to the iron-peak. This process favors α-nuclei since their high binding energy involves them in many exothermic reactions. The distribution shown in Fig. 8.9 has peaks at these elements so it is reasonable to suppose that they were produced in stars and then later dispersed into the interstellar medium.

One possible dispersion mechanism would be core-collapse supernovae. As illustrated in Figs. 8.7, the iron core that will collapse to a neutron star is surrounded by concentric shells of the ashes of the different stages of nuclear burning. These shells will be blown off after the core-collapse.

To see whether this mechanism can account for the observed abundances, we note that about $2M_\odot$ of ^{16}O would be dispersed per supernova. The rate for core-collapse supernova explosions in Milky Way-type galaxies is about 2 per century so there have been about 10^8 core-collapse supernova over the $\sim 10^{10}$ yr life of our galaxy, about half of which occurred before the formation of the solar system. This gives about $2 \times 10^8\, M_\odot$ of ^{16}O or about 1% of the total mass of the Milky Way. This nicely matches the observed solar-system abundance of 0.9% ! While this comparison can hardly be considered quantitative, it seems reasonable to say that a significant fraction of the observed intermediate mass nuclei were dispersed in supernovae.

Core-collapse supernovae are not very efficient in dispersing iron-peak nuclei since it is these nuclei that collapsed to form a neutron star. On the other hand, we saw in Sect. 8.2.3 that type Ia supernovae lead to the destruction of a $\sim 1.4M_\odot$ carbon–oxygen white dwarf after its explosive burning to ^{56}Ni. After β-decay to ^{56}Fe there would be about $0.7\,M_\odot$ of ^{56}Fe dispersed into the interstellar medium. The rate for type Ia supernovae is a about 1/4 the core-collapse rate so the total amount of ^{56}Fe produced is not far from the 0.1% observed in the solar system.

8.3.3 $A > 60$: the s-, r- and p-processes

The production of elements with $A > 56$ is paradoxical in the sense that thermal equilibrium requires the dominance of either iron-peak elements (at low temperatures) or of ^{4}He and nucleons (at high temperatures). In spite of this, captures of single protons or neutrons by nuclei near the stability line are always exothermic, so it is possible to create heavy nuclei from a non-equilibrium mixture of nucleons and iron-peak nuclei. Thermal equilibrium could then be approached by later fissioning of the heavy nuclei, though the time scale for spontaneous fission is so long that equilibrium is reached only after time scales that are enormous even for astrophysics.

The modern theory of the nucleosynthesis of elements beyond the iron peak was spelled out in a classic paper by Burbidge, Burbidge, Fowler and Hoyle [80]. These authors realized the importance of neutron captures in the production of heavy elements. While neutrons are present in only tiny numbers in most stars, neutron capture has the advantage not having the Coulomb barrier associated with proton captures. Its importance is immediately suggested by the fact that for values of $A > 56$ with more than one β-stable isobar, the most neutron-rich isobar is the most abundant.

Of course the problem with neutron captures is that normally very few neutrons are present in stars. Important neutron-producing exothermic reactions are (α, n) reactions on the relatively rare nuclei ^{13}C and ^{22}Ne

$$^4\text{He} \, ^{13}\text{C} \rightarrow \text{n} \, ^{16}\text{O} \qquad Q = 3.00 \, \text{MeV} \tag{8.69}$$

$$^4\text{He} \, ^{22}\text{Ne} \rightarrow \text{n} \, ^{25}\text{Mn} \qquad Q = 0.30 \, \text{MeV} \, . \tag{8.70}$$

The slow build-up of heavy nuclei by capture of such neutrons is called the "s-process" for "slow" neutron capture.

Neutrons would also be expected to be present in large numbers in explosive events like supernovae or neutron-star collisions. This type of nucleosynthesis is called the "r-process" for "rapid" neutron capture. In fact, we will see that the existence of both the s- and r- process is necessary to explain the observed abundances.

The workings of the s and r processes are illustrated in Fig. 8.10 which shows how heavy nuclei can be constructed by neutron captures on ^{56}Fe. After first reaching the stable nuclei $^{57}_{26}\text{Fe}_{31}$ and $^{58}_{26}\text{Fe}_{32}$, the β-unstable nucleus $^{59}_{26}\text{Fe}_{33}$ ($t_{1/2} = 44.5$ day) is produced. If the rate of neutron captures is slow compared to the β-decay rate, the $^{59}_{26}\text{Fe}_{33}$ decays "immediately" to $^{59}_{27}\text{Co}_{32}$. The next neutron capture produces $^{60}_{27}\text{Co}_{33}$ which immediately decays to $^{60}_{28}\text{Ni}_{32}$ and so on. This is the *s-process* where neutron capture is *slow* compared to β-decay. The s-process produces nuclei by following a well-defined path along the bottom of the valley of stability. The path sometimes bifurcates (at $^{64}_{29}\text{Cu}_{35}$ in Fig. 8.10) only to quickly merge (at $^{65}_{29}\text{Cu}_{36}$).

On the other hand, if the rate of neutron capture is higher than typical decay rates, it is possible to produce nuclei far from the bottom of the valley of stability. Referring to Fig. 8.10, starting with $^{56}_{26}\text{Fe}_{30}$, the r-process can sequentially produce ^{57}Fe, ^{58}Fe, ^{59}Fe ^{60}Fe and so on until the addition of a further neutron yields an unstable nucleus that decays by neutron emission, $^A\text{Fe} \rightarrow \text{n}^{A-1}\text{Fe}$. In fact, the r-process is expected to take place in at high temperature in the presence of many photons so the limiting reaction is most likely photon dissociation $\gamma^A\text{Fe} \rightarrow \text{n}^{A-1}\text{Fe}$. In Fig. 8.10 we show this happening at $^{66}_{40}\text{Fe}_{26}$, though this is uncertain because nuclei so far from the bottom of the stability valley have not been studied in the laboratory. ^{66}Fe has a very short half-life, 0.4 s, for β-decay producing ^{66}Co. This nucleus can then captures neutrons continuing the r-process.

Events producing the r-process are expected to last a very short time (about 10 sec for a type II supernova) and at the end of the event nuclei would have been produced along the southern slope of the stability valley

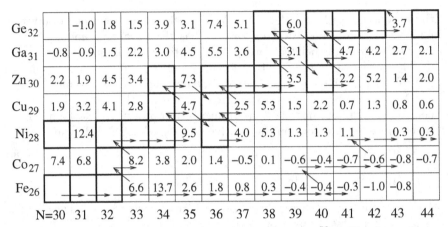

Fig. 8.10. Nucleosynthesis by neutron capture starting at ^{56}Fe. The decimal logarithm of the half-life in seconds is shown for β-unstable nuclei. If the neutron flux is small, β-decay occurs "immediately" after neutron absorption and the path follows the nuclei at the bottom of the stability valley indicated by the arrows. This is the s-process. On the other hand, if the neutron flux is sufficiently high, the r-process is operative where nuclei can absorb many neutrons before β-decaying so the path may ascend the sides of the valley until the nuclei are either photo-dissociated or β decay. Along the Fe-line, this is shown as happening at $^{66}_{40}$Fe$_{26}$. After the neutron flux is turned off, the nuclei on the slopes of the valley β-decay down to the bottom. Note that the neutron capture path in a nuclear reactor (Fig. 6.12) is intermediate between the astrophysical s- and r- processes since the time for neutron absorption is typically a month or so.

as shown in Fig. 8.11. After the neutron flux is turned off, the nuclei then β-decay to the bottom of the valley.

The systematics of the s- and r-process explain the peaks in the abundances of $A > 80$ nuclei shown in Fig. 8.9. The s-process produces an accumulation of nuclei with magic N where subsequent neutron captures are difficult because of the low cross-section. The s-process peaks would occur at $N = 50, 82, 126$ corresponding to $A = 90$, 138, and 208.

On the other hand, the r-process path in Fig. 8.11 follows the lower edge of the valley but takes a northern direction at the neutron magic numbers 50, 82 and 126. This is because the addition of a neutron to a closed shell is inhibited both because of a small cross-section for neutron capture and because, once captured, the extra neutron is weakly bound and easily ejected by a photon. We can therefore expect that at the end of the r-process event there is an accumulation of nuclei along the segments of the path following the magic N lines. After the subsequent β-decays, this will lead to an accumulation of nuclei at $A \sim 80$, 130 and 195. Such r-process peaks are observed in the solar-system abundances (Fig. 8.9).

We note that the details of the r-process peaks depend on the extent to which the shell structure and their magic numbers is maintained for very

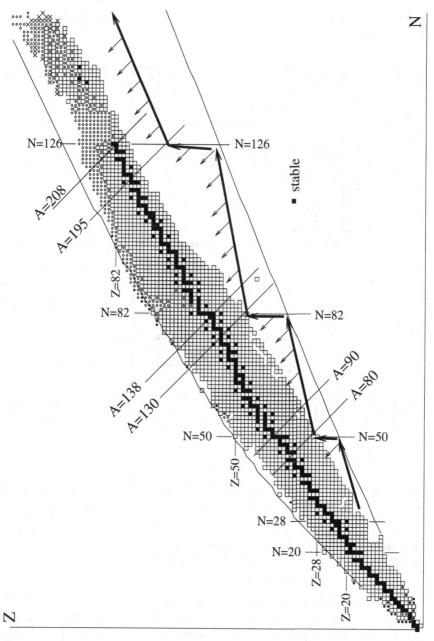

Fig. 8.11. The paths for nucleosynthesis by slow and fast neutron capture. In the s-process, the prompt β-decays insure that all produced nuclei are near the bottom the of the stability valley. In the r-process, rapid neutron absorption during an intense pulse of neutrons leads to nuclei distributed along the thick arrow on the southern slope of the valley. After the pulse is turned off, the nuclei β-decay down to the bottom of the valley, as indicated by the thin arrows.

neutron-rich nuclei. This is an important area of investigation in experimental nuclear physics.

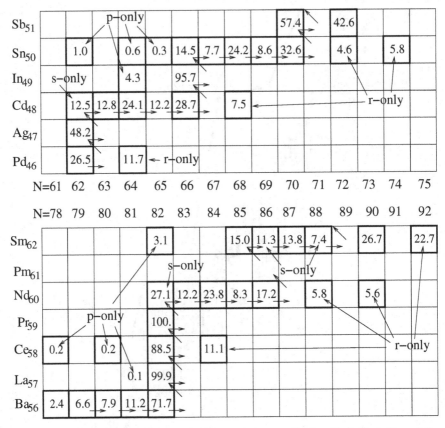

Fig. 8.12. Isotopic abundances and the s-process path for $46 \leq Z \leq 51$ and for $56 \leq Z \leq 62$. Nuclei below the s-process path are labeled "r-only" and can only be produced by rapid neutron capture in, for instance, supernovae. Nuclei that are shielded from the r-process by a r-only nucleus are labeled "s-only." Nuclei above the s-process path have low isotopic abundances. The can only be produced by the "p-process," due, for instance, to (p, γ) or (γ, n) reactions.

Further indication that both s- and r-processes are needed to explain the solar-system abundances comes from the fact that there are nuclei that can only be produced by the s-process and also nuclei that can only be produced by the r-process. The s-process is clearly incapable of producing elements heavier than ^{209}Bi since the β-stable elements for $A = 210, 211$ are the short-lived α-emitters ^{210}Po ($t_{1/2} = 138.376$ day) and ^{211}Bi ($t_{1/2} = 2.14$ m). The existence of natural Uranium and Thorium therefore necessitates the existence of the r-process.

Selected r-only and s-only nuclei are shown in Fig. 8.12. This figure show the solar-system isotopic abundances of the stable nuclei and the s-process path. We can see in Fig. 8.12 that ^{110}Pd, ^{122}Sn, and ^{124}Sn cannot be produced by the s-process. The same is true of ^{148}Nd, ^{150}Nd, ^{152}Sm, and ^{154}Sm. On the other hand, ^{110}Cd can only be produced by the s-process since it is "shielded" by the β-stable nucleus ^{110}Pd from the β-decay of neutron-rich nuclei at the end of an r-process event. In Fig. 8.12, we see that ^{148}Sm and ^{150}Sm are shielded from the r-process by ^{148}Nd and ^{150}Nd.

Comparing the abundances of the s-only Sm isotopes with those of the r-only isotopes, we see that the two processes produce comparable quantities of nuclei. This is surprising in view of the quite different nature of the two processes.

In Fig. 8.12 we also see a number of proton-rich nuclei that can be produced neither by the s- nor the r-process. By definition, these nuclei are created by the "p-process." These nuclei all have very low solar-system abundances indicating that the p-process is less important than the s- and r-processes. Originally, it was thought that these nuclei would be created by proton capture, but more recent work has indicated that (γ, n) reactions (photo-ejection of a neutron) in explosive environments may be the dominant process.

8.4 Nuclear astronomy

While the nuclear processes discussed in this chapter explain much about energy production in stars and nuclear abundances, direct evidence for nuclear reactions in astrophysical conditions is extremely difficult to obtain. Most reactions take place deep inside stars so the only products that escape are neutrinos produced in hydrogen burning and in stellar collapse. The cross-sections for reactions that can be used to to detect these neutrinos are, unfortunately, so small that astronomical neutrinos have only been observed from the Sun and from the nearby (~ 50 kpc) core-collapse supernova SN1987A in the Large Magellanic Cloud.

The other observable reaction product are γ-rays from regions of space that are sufficiently transparent to MeV-photons. One source is long-lived radioactive nuclei that have been dispersed into the interstellar medium by supernovae. Much more abundant are continuous spectra of γ-rays associated with a variety of mechanisms, often related to the acceleration and interactions of cosmic-rays. All these photons have a small probability of penetrating the Earth's atmosphere so are better observed by γ-telescopes in orbit about the Earth.

8.4.1 Solar Neutrinos

Neutrinos are necessarily produced in hydrogen burning stars by the weak-interaction processes that transform protons to neutrons. Most of the 27 MeV liberated in the hydrogen-helium transformation is thermalized in the Sun and eventually escapes from the surface as photons. (On average, only about 500 keV immediately escapes with the neutrinos.) Therefore, the total neutrino production rate is simply determined by the observed solar luminosity, $L_\odot$:

$$\frac{dN_\nu}{dt} \sim \frac{2L_\odot}{27\,\mathrm{MeV}} = 2 \times 10^{38}\,\mathrm{Bq}. \tag{8.71}$$

(This formula assumes that the sun is in a steady state, an assumption that is justified by detailed solar models.) At Earth, this gives a solar neutrino flux

$$\phi_\nu = 6 \times 10^6 \mathrm{m}^{-2}\mathrm{s}^{-1}. \tag{8.72}$$

While the total neutrino flux is easy to calculate, the energy spectrum is not since it depends on the nature of the nuclear reactions that create them. As discussed in Sect. 8.2.1, energy production in the Sun is believed to be dominated by the PPI, PPII, and PPIII cycles, which produce neutrinos through the reactions:

$$^1\mathrm{H}\,^1\mathrm{H} \rightarrow\,^2\mathrm{H}\,e^+\,\nu_e \quad Q = 420\,\mathrm{keV} \qquad \mathrm{PPI, II, III} \tag{8.73}$$

$$e^-\,^7\mathrm{Be} \rightarrow\,^7\mathrm{Li}\,\nu_e \quad Q = 862\,\mathrm{keV} \qquad \mathrm{PPII} \tag{8.74}$$

$$^8\mathrm{B} \rightarrow\,^8\mathrm{Be}\,e^+\,\nu_e \quad Q = 16\,\mathrm{MeV} \quad \mathrm{PPIII} \tag{8.75}$$

The Q's determine the energies of the produced neutrinos. Neutrinos from reaction (8.73) share the Q with the final-state positron, so they have a continuous spectrum with an endpoint of 420 keV. Neutrinos produced in this reaction are often called "ν_{pp}." In PPI, two ν_{pp} are produced.

In PPII, one ν_{pp} is produced along with one "ν_{Be}" in the electron capture (8.74). Being produced in a two body reaction, the ν_{Be} is monochromatic with an energy of 861 keV (branching ratio 90 percent) or 383 keV (branching ratio 10 percent with the production of an excited state of $^7\mathrm{Li}$). In PPIII, one ν_{pp} is produced along with one "ν_B" in the β-decay (8.75). The ν_B have a β-spectrum with an endpoint of 16 MeV.

Additional neutrinos come from the CNO decays

$$^{13}\mathrm{N} \rightarrow\,^{13}\mathrm{C}\,e^+\nu_e \qquad Q = 1.20\,\mathrm{MeV}\,, \tag{8.76}$$

$$^{15}\mathrm{O} \rightarrow\,^{15}\mathrm{N}\,e^+\nu_e \qquad Q = 1.73\,\mathrm{MeV}\,, \tag{8.77}$$

$$^{17}\mathrm{F} \rightarrow\,^{17}\mathrm{O}\,e^+\nu_e \qquad Q = 1.74\,\mathrm{MeV}\,, \tag{8.78}$$

and to the "pep" reaction

$$e^- \, {}^1\text{H} \, {}^1\text{H} \; \rightarrow \; {}^2\text{H} \, \nu_e \qquad Q \; = \; 1.442 \, \text{MeV} \, . \tag{8.79}$$

This reaction sometimes substitutes for (8.73). Finally, the highest energy neutrinos come from the very rare "hep" reaction

$$^1\text{H} \, {}^3\text{He} \; \rightarrow \; {}^4\text{He} \, e^+ \nu_e \qquad Q \; = \; 18.8 \, \text{MeV} \, , \, . \tag{8.80}$$

As explained in Sect. 8.2.1, the Sun is dominated by the PPI cycle so the solar neutrino spectrum is dominated by ν_{pp}. Figure 8.13 shows the calculated flux [73]. The fluxes of the three important neutrinos types are in the ratio $\nu_{pp} : \nu_{Be} : \nu_B = 1 : 0.08 : 10^{-4}$.

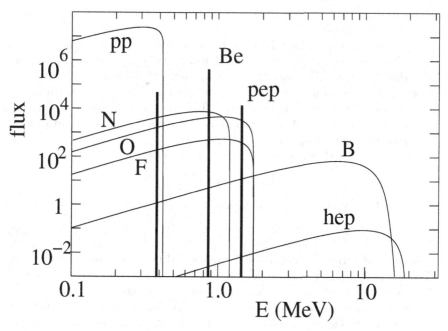

Fig. 8.13. The solar-neutrino flux at Earth predicted by standard solar models [73]. For the continuous components, the flux is in units of $\text{m}^{-2}\text{s}^{-1}\text{MeV}^{-1}$. The flux of the line components is given in $\text{m}^{-2}\text{s}^{-1}$.

All solar neutrinos are produced as ν_e. As explained in Sect. 4.4, this corresponds to a mixture of three massive neutrinos, ν_1, ν_2 and ν_3. Depending on the neutrino masses and mixing angles, these neutrinos should reach us after being affected by vacuum oscillations and/or the MSW effect. The original ν_e spectrum of Fig. 8.13 is transformed into an energy-dependent combination of ν_e, ν_μ, ν_τ. Solar-neutrino detectors based on neutral-current reactions have rates that are independent of the mixture but charged-current detectors are only sensitive to the ν_e-component.

The difficulty of detecting solar neutrinos comes from their tiny interaction cross-sections. For solar neutrinos of energy $\sim 1\text{MeV}$, a first estimate of the cross sections is then $\sigma \sim G_F^2\, 1\mathrm{MeV}^2/(\hbar c)^4$ which is about $5 \times 10^{-48}\mathrm{m}^2$. Taking the full neutrino flux, this gives an event rate of at most a few events per day for a 1 ton detector. This sets the scale for experiments.

It should be emphasized that the rate is much less than the decay rate from natural radioactivity even in the most pure material, say, 10^{-9} uranium.

Water Cherenkov detectors. Because of these difficulties, only the high-energy ν_B can be said to have been satisfactorily investigated. These neutrinos have energies that are sufficiently high that the final-state particles have energies greater than the energies of decay products of natural radioactivity. This fact has allowed them to be detected by two large underground water Cherenkov detectors, Superkamiokande [82] and the Sudbury Neutrino Observatory (SNO) [81].

The most complete results have been obtained by SNO, operating in a nickel mine in Sudbury, Ontario, and using 1000 ton of heavy water as a neutrino detector.[4] A schematic of the apparatus is shown in Fig. 8.14. The heavy water is surrounded by photomultipliers that can measure Cherenkov light emitted by relativistic particles. [5] This makes the detector sensitive to electrons of energy $> 1\,\mathrm{MeV}$ and to photons of energy $> 2\,\mathrm{MeV}$ that produce electrons through Compton scattering.

The reactions that SNO uses to detect neutrinos are

$$\nu_e\,{}^2\mathrm{H} \rightarrow e^-\,\mathrm{pp} \qquad E_\nu > 1.44\,\mathrm{MeV}\,, \tag{8.81}$$

$$\nu\,{}^2\mathrm{H} \rightarrow \nu\mathrm{p\,n} \qquad E_\nu > 2.2\,\mathrm{MeV}\,. \tag{8.82}$$

$$\nu\,e^- \rightarrow \nu e^-\,, \tag{8.83}$$

The charged current reaction dissociation of ${}^2\mathrm{H}$ (8.81) is sensitive only to ν_e. The final-state electron takes all the neutrino energy (minus the 1.44 MeV threshold) so the electron energy spectrum yields directly the neutrino spectrum. The neutral-current dissociation of ${}^2\mathrm{H}$ (8.82) is equally sensitive to all neutrino species so its rate yields the total ν_B flux. The final state neutrons are identified through the $6.25\,\mathrm{MeV}$ photon emitted after the neutron is thermalized and then captured on ${}^2\mathrm{H}$:

$$\mathrm{n}\,{}^2\mathrm{H} \rightarrow {}^3\mathrm{H}\,\gamma \qquad E_\gamma = 6.25\,\mathrm{MeV}\,. \tag{8.84}$$

Finally, neutrino-electron elastic scattering (8.83) is primarily sensitive to ν_e but also to ν_μ and ν_τ with a cross-section about five times smaller. Even for ν_e the cross-section is about ten times smaller than for (8.81) since the first is proportional to $G_F^2 E_{cm}^2 \sim G_F^2 E_\nu m_e c^2$, while the second is proportional to $G_F^2 E_\nu^2$ with $E_\nu \sim 10 m_e c^2$.

---

[4] This large amount of heavy water was provided by the Canadian government who, as explained in Chap. 6, uses heavy water as a neutron moderator in their nuclear reactors.

[5] Cherenkov light is emitted by particles of velocity greater than the phase velocity in the medium, c/n where n is the refraction index.

photomultipliers

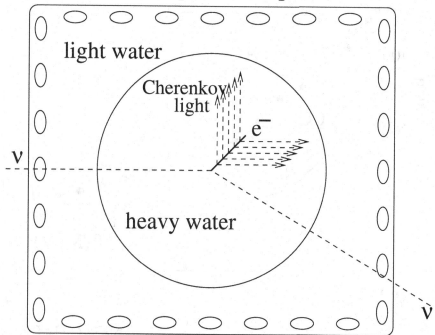

Fig. 8.14. A schematic of the Sudbury Neutrino Observatory (SNO) detector. One kiloton of heavy water is observed by photomultipliers through a light-water shield. [The purpose of the shield is to prevent neutrons from entering the heavy water and simulating the reaction (8.82).] Relativistic particles emit Cherenkov light that is detected by the photomultipliers. The quantity of light is proportional to the energy of the relativistic particles. The Cherenkov light is emitted in the forward direction so direction of the recoil electron can be determined from the pattern of photomultiplier hits.

The energy spectrum of SNO events is shown in Fig. 8.15 along with the expectations for the three types of reactions. The angular distribution with respect to the Sun in the figure shows the peaking in the forward direction due to neutrino-electron elastic scattering. These events are forward peaked because the neutrino energy is much greater than the target mass so the center-of-mass moves forward in the laboratory. The angular distribution for charged-current events show a small expected asymmetry due to the weak-interaction matrix element.

The rates for the three reactions allow one to deduce the flux of ν_B neutrinos. The neutral-current reactions gives a total flux of [81]

$$\phi_{\nu_e} + \phi_{\nu_\mu} + \phi_{\nu_\tau} = (6.42 \pm 1.57 \pm 0.58) \times 10^6 \text{cm}^{-2}\text{s}^{-1} , \qquad (8.85)$$

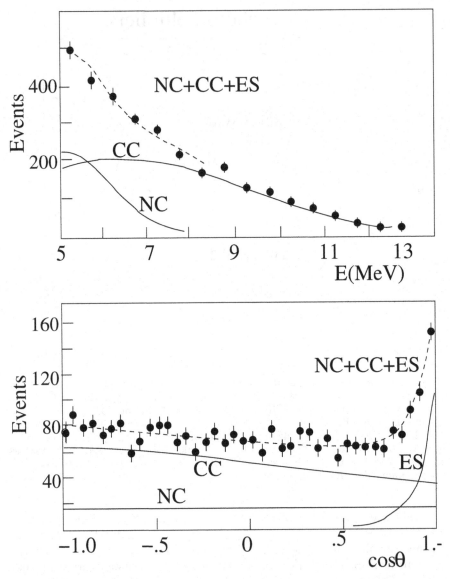

Fig. 8.15. The SNO measurement of the solar-neutrino flux from ^{8}B β-decay [81]. Most events are due to the charged current (CC) reaction (8.81) and to the neutral current (NC) reaction (8.82). The top figure shows the spectrum of event energy as measured by the quantity of Cherenkov light. The two solid curves show the distributions expected for (8.81) and (8.82). The bottom figure shows the direction of recoil electrons with respect to the direction of the Sun. The peak at small angles is due to neutrino-electron elastic scattering (ES), (8.83).

where the first and second errors are respectively statistical and systematic. This is in quite satisfactory agreement with the solar model predictions of the flux of ν_B of $(5.0 \pm 1.0) \times 10^6\,\mathrm{cm}^{-2}\mathrm{s}^{-1}$.

The flux of ν_e as determined from the charged-current reaction is [81]

$$\phi_{\nu_e} = (1.76 \pm 0.06 \pm 0.09) \times 10^6 \mathrm{cm}^{-2}\mathrm{s}^{-1} . \tag{8.86}$$

As expected from the considerations of Chap. 4, the solar ν_e have been partially transformed into neutrinos of other species. The number transformed is consistent with that expected for the presently accepted neutrino masses and mixing angles (4.139).

The SNO results are consistent with those deduced using the data from the Super Kamiokande experiment [82]. This experiment is similar to SNO but uses only light water. As such it is sensitive to ν_B only through reaction (8.83) but has an enormous mass (20 kton) and thus has very small statistical errors.

Radiochemical detectors. Because of their low energy, the ν_{pp} and ν_{Be} cannot be easily identified because electrons produced by such neutrinos cannot be distinguished from the β-decays due to natural radioactivity. These solar neutrinos can, however be detected through neutrino absorption

$$\nu_e\,(A, Z) \rightarrow e^-\,(A, Z+1) ,$$

if the final-state nucleus can be identified. Generally the nucleus is chemically separated from the target. For this reason, these experiments are called *radiochemical experiments*.

The first such experiment was that of R. Davis and collaborators [83] who used the reaction:

$$\nu_e\ ^{37}\mathrm{Cl} \rightarrow e^-\ ^{37}\mathrm{Ar} \quad Q = -814\,\mathrm{keV}$$

The threshold for this reaction corresponding to the energy necessary to produce the ground state of $^{37}\mathrm{Ar}$ is 814 keV. Because of this threshold, the experiment is insensitive to ν_{pp}.

To interpret the Davis results, it is essential to remember that the $^{37}\mathrm{Ar}$ nucleus can be produced in an excited state and it is necessary to know the cross-section for each state. The spectrum of states of $^{37}\mathrm{Ar}$ is shown in Fig. 8.16. The ν_{Be} yield only the ground state of $^{37}\mathrm{Ar}$. The cross-section for this state can be deduced from the lifetime of $^{37}\mathrm{Ar}$ since the nuclear matrix element is the same. The ν_B yield many excited states of $^{37}\mathrm{Ar}$ but luckily the nuclear matrix elements for these transitions can be accurately derived from the rates for the mirror decays $^{37}\mathrm{Ca} \rightarrow\ ^{37}\mathrm{K}^* e^- \nu_e$ [83].

The total absorption rate then depends on ϕ_{Be}, ϕ_B, and on ϕ_{other} (CNO and pep neutrinos). It is traditional to express the rate in "solar neutrino units" or "SNU's" ($1\,\mathrm{SNU} = 10^{-36}$ events per target atom per second). The calculated rate is [73]:

$$\int \phi(E_\nu)\sigma(E_\nu)dE_\nu = [5.76\,\hat\phi_B + 1.15\,\hat\phi_{Be} + 0.7\,\hat\phi_{other}\,]\,\mathrm{SNU}$$

where we have normalized the ν_e fluxes to those calculated by Bahcall [73]: $\hat{\phi}_B \equiv \phi_B/\phi_B(\text{Bahcall})$, etc. Since SNO has measured $\hat{\phi}_B$ to be 0.35 ± 0.04, this gives

$$\int \phi(E_\nu)\sigma(E_\nu)dE_\nu = [2.01 + 1.15\,\hat{\phi}_{Be} + 0.7\,\hat{\phi}_{other}\,]\,\text{SNU}$$

The rate measured with a chlorine detector can then be used to deduce the flux of ν_{Be} and the CNO/pep neutrinos. If these ν_e components are present, the rate should be about 4 SNU.

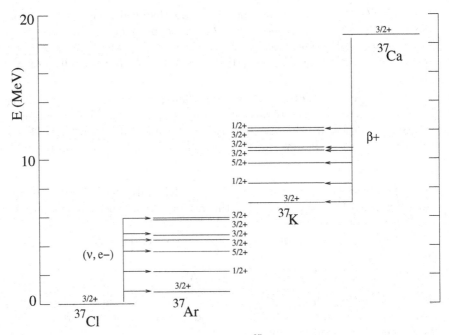

Fig. 8.16. The $A = 37$ nuclei. The states of ^{37}Ar populated by neutrino absorption on ^{37}Cl are shown. The weak matrix elements for these transitions can be determined from the rates for the mirror decays of ^{37}Ca to ^{37}K.

The Homestake experiment in the Homestake gold mine of Lead, North Dakota uses 600 tons of C_2Cl_6 as the neutrino target. It is shown schematically in Fig. 8.17. It is contained in a large tank along with about 1 mg of non-radioactive argon "carrier." After a three month exposure time, the carrier and any ^{37}Ar produced by solar neutrinos is extracted from the C_2Cl_6 by bubbling helium through the tank. The argon is then separated from the helium and trapped in a small proportional counter where the electron-capture decays of the ^{37}Ar ($t_{1/2} = 37$ day) are detected via their Auger electrons.

Over a 25 year period, the three month exposures have yielded, on average about six detected decays of ^{37}Ar. After taking into account the detection

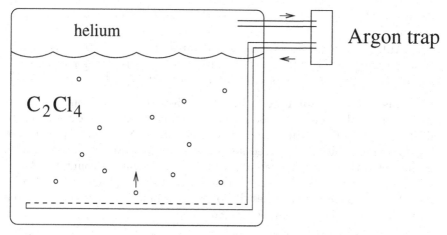

Fig. 8.17. The principle of the Homestake neutrino experiment. ^{37}Ar atoms produced by neutrino absorption are swept out the the C_2Cl_4 by bubbling helium gas through the liquid. The argon is then separated in a cold trap and introduced into a small proportional counter where the electron-capture decays of ^{37}Ar are observed

efficiency, this corresponds to 2.56 ± 0.16 SNU [83]. Compared with the SNO-corrected standard model prediction of 4 SNU, this indicates that the ν_{Be} are also strongly suppressed by neutrino oscillations and/or the MSW effect, as expected for the estimated values of the neutrino masses and mixing angles (4.139).

Finally, two experiments, SAGE [85] and GALLEX [84], have observed solar neutrinos via neutrino absorption on gallium:

$$\nu_e \, ^{71}\text{Ga} \rightarrow e^- \, ^{71}\text{Ge}$$

The threshold for this reaction is 233 keV, making the experiment sensitive to all known sources of solar neutrinos. The ν_{pp} yield only the ground state of ^{71}Ge so the cross section can be calculated from the lifetime of ^{71}Ge. The ν_{Be} are also expected to yield primarily the ground state while the ν_B yield a multitude of excited states. The calculated rate is [73]

$$\int \phi(E_\nu)\sigma(E_\nu)dE_\nu =$$

$$[69.7\,\hat{\phi}_{pp} + 34.2\,\hat{\phi}_{Be} + 12.1\,\hat{\phi}_B + 12\hat{\phi}_{\text{other}}]\,\text{SNU} \qquad (8.87)$$

The coefficient in front of $\hat{\phi}_B$ is accurate only to a factor two but this is of no great importance in view of the small flux measured by SNO, $\hat{\phi}_B = 0.35$.

The gallium experiments follow rather closely the technique of the chlorine experiment. The targets are in the form of a liquid (100 tons of $GaCl_3$ for GALLEX and 60 tons of gallium metal for SAGE), and the volatile carrier ($GeCl_4$) is extracted along with the produced ^{71}Ge. The electron capture decays of ^{71}Ge $(t_{1/2} = 11.43\,\text{day})$ are then observed in small proportional counters.

GALLEX and SAGE have measured rates of $75 \pm 7\,\text{SNU}$. This suggest that the majority of the ν_{pp} are present as ν_e, as expected for the present estimates of the neutrinos masses and mixing angles (4.139).

Conclusions from solar neutrino detection. For many years, the interpretation of solar neutrino experiments was hampered by a lack of independent evidence for neutrino masses and oscillations. Now that the Kamland reactor experiment [48] has seen neutrino oscillations, one has no hesitation to interpret the rates of solar experiments taking into account the MSW effect and neutrino oscillations. This has lead to very precise tests of solar models. One recent paper makes the following comparison between measured and calculated solar neutrino fluxes [86]

$$\phi(\text{pp})_{\text{measured}} = (1.02 \pm 0.02 \pm 0.01)\,\phi(\text{pp})_{\text{theory}} \,, \tag{8.88}$$

$$\phi(^8\text{B})_{\text{measured}} = (0.88 \pm 0.04 \pm 0.23)\,\phi(^8\text{B})_{\text{theory}} \,, \tag{8.89}$$

$$\phi(^7\text{Be})_{\text{measured}} = (0.91\,^{+0.24}_{-0.62} \pm 0.11)\,\phi(^7\text{Be})_{\text{theory}} \,, \tag{8.90}$$

where the first errors are observational and the second are theoretical. If the excellent agreement between observation and theory survives future improved measurement of neutrino oscillation parameters, it will confirm a triumph of experimental and theoretical nuclear astrophysics.

8.4.2 Supernova neutrinos

As explained in Sect. 8.2.3, massive stars may end their lives with a "core-collapse" supernova in which the ^{56}Fe core collapses to a neutron star. Such events are powerful radiators of all species of neutrinos. The first to be created are "neutronization" ν_e from electron captures transforming protons to neutrons

$$e^- (A, Z) \rightarrow \nu_e (A, Z - 1) \,. \tag{8.91}$$

While these neutrinos are essential for the production of a neutron star, the most plentiful neutrinos come from purely thermal process that create neutrinos through reactions like

$$\gamma\gamma \leftrightarrow e^+ e^- \leftrightarrow \bar{\nu}\nu \,. \tag{8.92}$$

These reactions are due to the neutral currents and therefore produce all species of neutrinos. The proto-neutron star is so dense that the neutrinos do not immediately escape, but rather diffuse out just like photons in normal stars, with the escape time given by (8.25)

$$\tau \sim \frac{\langle\sigma\rangle N_b}{Rc} \,, \tag{8.93}$$

but now with σ given by the weak interactions rather than σ_T. Taking $\sigma = 10^{-45}\,\mathrm{m}^2$ and $R \sim 10\,\mathrm{km}$, this gives $\tau \sim 10\,\mathrm{s}$.

Because their cross-section is so much smaller than that of photons implying a correspondingly shorter escape time, the energy evacuated during the collapse to a neutron star is almost entirely taken up by the neutrinos. The total energy was calculated as (8.68)

$$\Delta E \sim (3/5)\frac{GM^2}{R} \sim 3 \times 10^{46}\,\mathrm{J} \qquad M = 1.5\,M_\odot. \qquad (8.94)$$

This energy is expected to be approximately equally divided among the 6 types of neutrinos and antineutrinos.

The neutrinos are expected to have a roughly thermal distribution with a temperature equal to that of the "neutrinosphere," the layer of material where the neutrinos last scattered before escaping. Calculations give different temperatures for the different species:

$$kT_{\nu_e} \sim 3.5\,\mathrm{MeV} \quad kT_{\bar{\nu}_e} \sim 5\,\mathrm{MeV} \quad kT_{\nu \neq \nu_e, \bar{\nu}_e} \sim 8\,\mathrm{MeV} \qquad (8.95)$$

The mean energies are $\sim 3kT$. The non-electron neutrinos and antineutrinos have the highest temperature because they interact only through neutral current interactions. (Their energies are insufficient to create muons or tauons.) They therefore have the smallest cross-sections and therefore escape the proto-neutron star from the deepest, hottest depth. The $\bar{\nu}_e$ and ν_e have also charged current interaction so they continue scatter to a greater distance from the center implying a smaller temperature. The ν_e have a higher cross-section than $\bar{\nu}_e$ because ν_e charged-current scatter on the plentiful neutrons while $\bar{\nu}_e$ charged-current scatter on the rarer protons.

The $\bar{\nu}_e$ from the core-collapse supernovae 1987a in the Large Magellanic Cloud were observed by the Kamiokande [87] and IMB [88] experiments. Both experiments were kiloton-scale experiments similar to SNO (Fig. 8.14) except that they contained only light water. (Kamiokande experiment was an earlier version of Super-Kamiokande.) The supernova $\bar{\nu}_e$ therefore scattered on the free protons in the water through the reaction

$$\bar{\nu}_e\,\mathrm{p} \rightarrow \mathrm{e}^+\,\mathrm{n}. \qquad (8.96)$$

The final state positron creates Cherenkov light as in the SNO experiment. The energy of the $\bar{\nu}_e$ is deduced from the quantity of light.

The 11 $\bar{\nu}_e$ events observed by Kamioka and the 8 events by IMB over a period of 10 s on February 23, 1987 were sufficient to confirm the basic picture of gravitational collapse with the evacuation of energy by neutrinos. Precision tests of the scenario will require taking into account neutrino oscillations and the MSW effect.

Further progress in the field of supernova-neutrino astronomy will be difficult because the total rate of supernova in our galaxy and the nearby Magellanic Clouds is believed to be about 2 per century. Present detectors are not large enough to detect supernovae in more distant galaxies.

It would be very interesting to detect the non-$\bar{\nu}_e$ components of the neutrino spectrum. If SNO is still in operation, it will observe the ν_e component through the reaction (8.81) and the other components through (8.82) and (8.83). Another recently suggested neutral-current detector [89] would use

$$\nu\, p\, \to\, \nu\, p\,. \tag{8.97}$$

The energy spectrum of the recoil protons would give information on the ν_τ and ν_μ energy spectra.

8.4.3 γ-astronomy

γ-astronomy is a vast field that observes high-energy photons from a variety of sources. Many deal with sources related to the production or propagation of cosmic rays. The recently discovered γ-ray bursts are now suspected to be related in some way to gravitational-collapse events.

In the context of nuclear physics, the most interesting sources of γ-rays are radioactive nuclei produced in supernovae and then ejected into interstellar space. Once the expanding cloud of matter becomes sufficiently dilute, it becomes transparent of γ-rays so they become detectable.

Table 8.2 lists the radioactive nuclei that emit γ-rays that have a lifetime sufficiently long to be observed. Most are expected to come from supernovae but some may be ejected in less violent "novae" that are thought to be superficial explosions on white dwarfs acreting matter from a binary companion. Such novae favor the production of relatively low mass nuclei. Some radioactive nuclei are expected to be ejected through continuous mass loss by Wolf-Rayet (WR) and Asymptotic Giant Branch (AGB) stars.

The radioactive decays of the first three groups, $A = 56$, 57 and 44 are believed to be the major heating source in the expanding clouds around supernovae. In fact, the luminosities of some supernovae are observed to decrease with the lifetimes of ^{56}Co, then ^{57}Co and the ^{44}Ti. This is illustrated in Fig. 8.8. As long as the γ-rays are involved in heating the cloud, it cannot be entirely transparent, so the expected flux is complicated to calculate.

As shown in Fig. 8.18, the ^{56}Co γ-rays were observed from the remnant of SN1987a and ^{44}Ti γ-rays from the supernova remnant Cas A. We see that the fluxes are just at the limit of detectability for these two nearby remnants. They nevertheless give strong evidence that radioactive nuclei are, indeed, ejected by supernovae. Another γ-ray that has been detected is that from ^{26}Al $(t_{1/2} = 7.17 \times 10^5\,\mathrm{yr})$. Its long half-life means that its flux represents the combined effects of many supernovae. Indeed, a diffuse flux of ^{26}Al photons has been detected throughout the galactic plane and gives constraints on the rate of nucleosynthesis over the last 10^6 yr.

Gamma-ray observatories now in orbit, especially the INTEGRAL mission, should greatly increase the amount of significant data available.

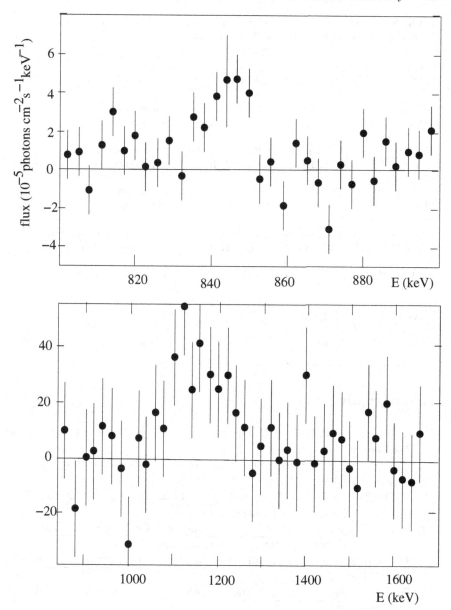

Fig. 8.18. γ-ray spectra from supernova remnants. The top panel shows the 847 keV line (^{58}Co) from SN1987A [90]. The bottom panel shows the 1157 keV line (^{44}Ti) from the region Cas A [91]. This region is believed to be the remnant of a supernova that exploded about 300 years ago.

Table 8.2. Radioactive nuclei giving rise to observable γ-rays. β-emitters with 511 keV annihilation radiation are labeled with e^+.

Decay	$t_{1/2}$	E_γ (keV)	sites
^{56}Ni → ^{56}Co	6.077 day	158, 269, 480	supernovae
^{56}Co → ^{56}Fe	77.27 day	847, 1238, e^+	supernovae
^{57}Ni → ^{57}Co	35.60 hr	127, 1377, 1919	supernovae
^{57}Co → ^{57}Fe	271.79 day	14, 122, 136	supernovae
^{44}Ti → ^{44}Sc	63 yr	68, 78, e^+	supernovae
^{44}Sc → ^{44}Ca	3.927 hr	1157, e^+	supernovae
^{60}Fe → ^{60}Co	1.5×10^6 yr		supernovae, WR, AGB
^{60}Co → ^{60}Ni	5.2714 yr	1173, 1332	supernovae, WR, AGB
^{26}Al → ^{26}Mg	7.17×10^5 yr	1160	supernovae, novae, WR,AGB
^{7}Be → ^{7}Li	53.12 day	478	novae
^{22}Na → ^{22}Ne	2.6019 yr	e^+, 1274	novae

Bibliography

1. H. Karttunen, P. Krüger, H. Oja, M. Poutanen, K. Donner, Fundamental Astronomy, Springer, Berlin, 2003.
2. Hansen, Carl J., Kawaler, Steven D., Trimble, Virginia, Stellar Interiors, Springer, Berlin, 2004.
3. D. D. Clayton, Principles of Stellar Evolution and Nucleosynthesis, University of Chicago Press, Chicago, 1983.
4. B. E. J. Pagel, Nucleosynthesis and Chemical Evolution of Galaxies, Cambridge University Press, Cambridge, 1997.
5. J. Audouze and S. Vauclair, An introduction to nuclear Astrophysics, Reidel, 1980.
6. D. Arnett, Supernovae and Nucleosynthesis, Princeton University Press, Princeton, 1996.

Exercises

8.1 Find a relation between the stellar temperature and density gradients assuming that they are due to adiabatic convection of material between lower and higher regions. Show that the photon–baryon density ($\propto T^3/\rho$) is proportional to ρ, consistent with Fig. 8.2 in the outer convective zone, $r > 0.7 R_\odot$.

8.2 Poincaré gave the following proof of the virial theorem, in the absence of radiation pressure. Consider a self-gravitating system of N particles whose potential energy is

$$\phi = -G \sum_{i=1}^{N} \sum_{j=i+1}^{N} \frac{m_i m_j}{|r_i - r_j|}$$

where G is Newton's constant. We denote by E_k the sum of kinetic energies of the N particles.

The moment of inertia of this system with respect to its center of gravity, which we take as the origin of coordinates is $I = \sum(m_i r_i^2)$. Assuming that the second derivative of I is zero, which is the case for quiet stars such as the sun, prove that one has

$$2E_k + \phi = 0 \quad \text{hence} \quad E = -E_k = \frac{\phi}{2} .$$

8.3 Calculate the energy liberated in the transformation of carbon and oxygen to ^{56}Ni in a white dwarf of mass $10^{57} m_p$. Compare this to the gravitational binding energy of the white dwarf.

8.4 Estimate the time of a neutrino of energy $10\,\text{MeV}$ to escape from a neutron star.

8.5 The fundamental reaction in the sun can be written globally as

$$4p + 2e^- \to\, ^4\text{He} + 2\bar{\nu}_e .\tag{8.98}$$

The binding energy of ^{4}He is $-28.30\,\text{MeV}$. What is the energy release in this reaction?

The total luminosity of the sun is $L_0 = 4\,10^{23}\,\text{kW}$. How many protons are consumed per second by the reaction (1)?

In its present phase, since it was formed, the sun burns quietly with a constant luminosity. Its mass is $M = 2\,10^{30}$ kg, 75% of which was initially hydrogen and 25% helium. Only 15% of the hydrogen can actually be burnt in the reaction (1) in the solar core. For what length of time T can the sun burn before it becomes a red giant?

8.6 The binding energy of ^{3}He is 7.72 MeV. Is the reaction $pd \to\, ^3\text{He} + \gamma$ exothermic? Is the inverse reaction possible in the core of a star like the sun? (The mean photon blackbody energy is $\sim 2.7kT$ with $kT \sim 1\,\text{keV}$.)

8.7 Consider a completely degenerate fermion gas. What is the equation of state $P(n,T)$ of the gas in, respectively, the non-relativistic and the relativistic regimes. We call kT_F the maximum energy of these fermions (T_F is the Fermi temperature), and p_F their maximum momentum. What does the limit $T \ll T_F$ correspond to? What is the corresponding pressure? Compare this pressure with the pressure of an ideal gas for a density $n \sim 10^{30} \mathrm{cm}^{-3}$ and a temperature $T \sim 10^7 \, \mathrm{K}$ (for the ideal gas).

We recall the virial theorem which relates the total internal energy U of a star to its gravitational energy $U_G = -\alpha GM^2/R$ with $\alpha \sim 1$. A star is formed by gravitational contraction of a mass M of gas. How does the temperature of an ideal gas evolve during that collapse? What happens when $T \geq 10^6 \mathrm{K}$? (We recall that the mean temperature of the sun given by the virial theorem is $\simeq 2 \times 10^6 \, \mathrm{K}$.)

Show that there exists a maximum temperature T_{max} in the contraction if one takes into account the possibility that the electron gas becomes degenerate. Show that the electron gas becomes degenerate long before the ion gas, so that the latter can be considered as an ideal gas.

If $T_{max} \leq 10^6 \mathrm{K}$, the star is called a brown dwarf. What is the peculiarity of such stars? What is the minimum mass of a star that does not become a brown dwarf?

How is the gravitational stability of a planet ensured? What is the difference between a planet and a brown dwarf?

8.8 In a massive star at the end of its life, the fusion $^{28}\mathrm{Si} \rightarrow {}^{56}\mathrm{Fe}$ takes place in a shell around an iron core. The core is degenerate. What happens when the mass of the core increases because of the Si fusion?

What is the order of magnitude of the Fermi energy of electrons at the beginning of the collapse? In what follows, we shall use

$$\epsilon_F \simeq 3\,\mathrm{MeV}\,(\rho/10^9 \mathrm{g\,cm}^{-3}) \tag{8.99}$$

and $\rho \sim 10^9 \mathrm{g\,cm}^{-3}$) initially. If the electron capture reactions $\mathrm{e}^- + (\mathrm{A,Z}) \rightarrow (\mathrm{A,Z}-1) + \nu_e$ are endothermic with $Q \sim$ a few MeV, why do these reactions take place when $\rho \geq 10^{11} \mathrm{g\,cm}^{-3}$? Why don't the inverse reactions take place?

Neutrinos interact with heavy nuclei in the core with a cross-section of $\sigma \sim 10^{-42} (E_\nu/1\,\mathrm{MeV})^2 \mathrm{cm}^2$.
The radius of the core evolves as $R \sim 100\,\mathrm{km}\,(\rho/10^{12} \mathrm{g\,cm}^{-3})^{-1/3} (M/M_\odot)^{1/3}$. Show that the neutrinos produced are trapped when $\rho \geq 10^{12} \mathrm{g\,cm}^{-3}$.

The diffusion time of neutrinos up to the "neutrinosphere ," i.e. the region where neutrinos cease to be coupled to matter is of the order of $1 - 10$ sec. The temperature of the emitted neutrinos is $T \sim 5$ MeV. Calculate the total energy E_ν^{tot} carried away by the neutrinos, assuming neutronization is complete.

Assuming that the collapse ceases at a density of $\rho \sim 10^{14} \mathrm{g\,cm}^{-3}$, calculate the gravitational energy which is liberated, and compare it with E_ν^{tot}.

9. Nuclear Cosmology

As discussed in the previous chapter, natural nuclear reactions are now mostly confined to stellar interiors. In the distant past, nuclear reactions are believed to have occurred throughout the Universe during the first few minutes after the "Big Bang." At this time the Universe was sufficiently hot that the Coulomb barrier could not prevent the fusion of nuclei.

We believe that the Universe went through such a hot epoch because the Universe is now expanding. By this we mean that the distances between galaxies are observed to be increasing with time. Specifically, galaxies are observed to recede from us with a recession velocity, v, proportional to their distance R:

$$v = H_0 R \qquad H_0 = (70 \pm 4)(\text{km s}^{-1}) \text{Mpc}^{-1}, \qquad (9.1)$$

where we use the conventional astronomic distance unit, $1\,\text{pc} = 3.085 \times 10^{16}$ m. If the galactic velocities were constant in time, this would imply that all galaxies were superimposed and the cosmological density infinite at a time in the past given by the Hubble time

$$t_\text{H} \equiv H_0^{-1} = 1.4 \times 10^{10} \,\text{yr} . \qquad (9.2)$$

Gravitational forces would be expected to make the galactic velocities time-dependent but the Hubble time nevertheless gives the order of magnitude of what one calculates for the age of the Universe if the known laws of physics are used to extrapolate into the past. We do not know if such an extrapolation to infinite density is justified but we can say with confidence that the expansion has been proceeding at least since an epoch when the density was 45 orders of magnitude greater than at present, with a temperature of $kT > 1\,\text{GeV}$. Going backward in time, we would see stars and galaxies melt into a uniform plasma of elementary particles.

Under such conditions, reactions between elementary particles and nuclei took place until the Universal expansion caused the temperature to drop to a level where the reactions cease because of the decreasing density and increasing efficiency of the Coulomb barrier. It is believed the period of reactions lasted about 3 minutes, ending when the Universe had a temperature of $kT \sim 40\,\text{keV}$ and leaving a mixture of 75% ^{1}H and 25% ^{4}He (by mass). This mixture provided the initial conditions for stellar nucleosynthesis that started once stars were formed some millions of years later.

The job of primordial cosmology is to understand how this process proceeded from the earliest possible moments. While we cannot say what existed at "the beginning" (if there was a beginning), we can extrapolate backward in time to temperatures of order 1 GeV when the Universe was believed to consist of a thermal plasma of relativistic quarks, antiquarks, gluons, neutrinos and photons. When the temperature dropped below a transition temperature estimated to be $kT \sim 200$ MeV, the quarks and antiquarks combined to form bound hadrons (mostly pions) which then, for the most part, annihilated leaving nothing but photons and neutrinos. If there had been equal numbers of quarks and antiquarks this would have been pretty much the end of the story. However, the small excess of order 10^{-9} of quarks over antiquarks meant that a small number nucleons remained at $T \sim 100$ MeV, of order 10^{-9} with respect to the photons and neutrinos. This was the initial condition for cosmological nucleosynthesis that came about when some nucleons later combined ($kT \sim 40$ keV) to form nuclei.

The process of cosmological nucleosynthesis differs from that of stellar nucleosynthesis in several important respects. Among them are

- The presence of neutrons. Because of the lack of neutrons in stars, stellar nucleosynthesis must start with the weak reaction $2 {}^1\text{H} \to {}^2\text{He}^+\nu_e$. Cosmological nucleosynthesis starts with an abundant supply of neutrons which need only combine with protons to form nuclei, starting with the reaction $\text{n}\,\text{p} \to {}^2\text{H}\gamma$.

- A low baryon–photon ratio. Whereas the baryon to photon ratio in stars is greater than unity ($\sim 10^3$ for the sun, Fig. 8.2), it is of order 5×10^{-9} in the primordial Universe. This has the important consequence of delaying nucleosynthesis since the abundant photons quickly dissociate any nuclei that are produced until the temperature drops to ~ 100 keV at which point the probability of a thermal photon having sufficient energy to break a nucleus becomes small enough.

- A low baryon–neutrino ratio, nearly equal to the baryon–photon ratio. This was important during the time when most nucleons were free since weak reactions like $\nu_e\text{n} \leftrightarrow \text{e}^-\text{p}$ can change neutrons into protons and vice versa. In fact, for $kT > 800$ keV it turns out that these reactions are sufficiently rapid that they can maintain a "chemical" equilibrium between the neutrons and protons so that the neutron to proton ration takes the thermal value of $\exp(-(m_n - m_p)c^2/kT)$. The weak interaction rate drops to negligible values at a temperature of around 800 keV when the neutron to proton ratio is about 0.2. Free neutron decay then lowers the ratio to about 0.1 when nucleosynthesis starts at $kT \sim 100$ keV. The proton excess results in a large quantity of post-nucleosynthesis ${}^1\text{H}$.

- A limited amount of time. The elapsed time from the quark–gluon phase transition to the end of cosmological nucleosynthesis is about three minutes. Below temperatures of ~ 60 keV, the Coulomb barrier prevents further reactions, leaving the primordial mixture of 75% hydrogen and 25% ^{4}He.

In this chapter we provide a basic introduction to this process. Section 9.1 summarizes what is believed to be the present state of the observable Universe and sec. 9.2 discusses how this state evolves as the Universe expands. Section 9.3 shows how the universal expansion is governed by gravity. Section 9.5 discusses the basic physics that governs the particle and nuclear reactions in the primordial Universe and the following sections apply this physics to electrons, neutrons, nuclei and some more speculative elementary particles that may have played an important role in cosmology.

9.1 The Universe today

When averaged over large volumes containing many galaxies, the observable Universe is believed to have the following characteristics:

- A tiny density, $\rho \sim 10^{-26}\,\mathrm{kg\,m^{-3}}$, and starlight output, $J \sim 10^{-39}\,\mathrm{W\,m^{-3}}$;
- A curious "chemical" composition (Table 9.1) with most particles being cold photons or neutrinos and most mass being in the form of ordinary "baryonic" matter (protons and nuclei plus electrons), and not-so-ordinary "cold dark matter" (CDM). Most of the energy is in the form of an effective vacuum energy or, equivalently, a cosmological constant.
- A non-equilibrium thermal state characterized by a deficit of highly-bound heavy nuclei.
- A hierarchy of gravitationally bound structures ranging from planets and stars to galaxies and clusters of galaxies.

In this section, we will review the major components of the Universe listed in Table 9.1. It will be convenient to express the mean densities in units of the "critical density"

$$\rho_c = \frac{3H_0^2}{8\pi G} \tag{9.3}$$

where H_0 is the "Hubble constant"

$$H_0 = (70 \pm 4)\mathrm{km\ sec^{-1}\ Mpc^{-1}} \tag{9.4}$$

The physical significance of the critical density and the Hubble constant will be discussed in the next section but here we note only the very low value of ρ_c:

$$\rho_c = 0.92\,h_{70}^2 \times 10^{-26}\,\mathrm{kg\ m^{-3}} \tag{9.5}$$

$$= 1.4\,h_{70}^2 \times 10^{11} M_\odot\ \mathrm{Mpc^{-3}} = 0.51\,h_{70}^2 \times 10^{10}\,\mathrm{eV\ m^{-3}}\ . \tag{9.6}$$

where the reduced Hubble constant is $h_{70} = H_0/70\,\mathrm{km\ sec^{-1}\ Mpc^{-1}}$.

Table 9.1. The known and suspected occupants of the Universe. For each species, i, the table gives estimated number density of particles, n_i, and the estimated mass or energy density, $\Omega_i = \rho_i/\rho_c$, normalized to the "critical density," $\rho_c = (0.92 \pm 0.08) \times 10^{-26}\,\mathrm{kg\,m^{-3}}$. Other than the photon density taken from the COBE data [92], the numbers are taken from the global fit of cosmological parameters by the WMAP collaboration [94]. The lower limit on the neutrino density comes from the oscillation data of [49].

species	$n_i\ (\mathrm{m^{-3}})$	$\Omega_i = \rho_i/\rho_c$
γ (CBR)	$n_\gamma = (4.104 \pm 0.009) \times 10^8$ $T_\gamma = (2.725 \pm 0.002)\,\mathrm{K}$	$\Omega_\gamma = (5.06 \pm 0.4) \times 10^{-5}$
ν_e, ν_μ, ν_τ	$n_\nu = (3/11)n_\gamma$ (per species)	$0.0006 < \Omega_\nu < 0.015$
baryons	$n_b \sim 0.25 \pm 0.01$	$\Omega_b = (0.044 \pm 0.004)$
cold dark matter	$n_\chi = \rho_{CDM}/m_\chi$ (m_χ unknown)	$\Omega_{CDM} = 0.226 \pm 0.04$
"vacuum"	0	$\Omega_\Lambda \sim 0.73 \pm 0.04$
total		$\Omega_T \sim 1.02 \pm 0.02$

9.1.1 The visible Universe

The "building blocks" of the visible Universe are galaxies which are themselves composed of stars, interstellar gas, and unidentified dark matter. Galaxies have a wide variety of shapes (spirals, ellipticals and irregulars), masses, and luminosities. Nevertheless, most of the light in the Universe is produced in galaxies containing 10^{10} to 10^{11} stars that generate a typical galactic luminosity of

$$\langle L_{\mathrm{gal}}\rangle \sim 2 \times 10^{10} L_\odot\,, \tag{9.7}$$

where $L_\odot$ is the solar luminosity, $L_\odot = 2.4 \times 10^{45}\,\mathrm{eV\,s^{-1}}$. The mass of the visible matter in a galaxy is typically

$$\langle M_{\mathrm{gal}}\rangle \sim 4 \times 10^{10} M_\odot\,, \tag{9.8}$$

Galaxies are not uniformly distributed in space, a fact that is not surprising in view of their mutual gravitational attraction. Galaxies are often grouped in bound clusters, the largest of which contain thousands of galaxies. In spite of this "small-scale" inhomogeneity, at large scales $> 100\,\mathrm{Mpc}$ the Universe appears to be uniform with a number density of visible galaxies of

$$n_{\mathrm{gal}} \sim 0.005\,\mathrm{Mpc^{-3}} \tag{9.9}$$

corresponding to a typical intergalactic distance of ~ 6 Mpc. The total mass density ρ_{vis} associated with the "visible" parts of galaxies is

$$\rho_{\text{vis}} = \langle M_{\text{gal}} \rangle n_{\text{gal}} \sim 3 \times 10^8 \, M_\odot \, \text{Mpc}^{-3} \,. \tag{9.10}$$

The density of visible matter (9.10) divided by ρ_c gives

$$\Omega_{\text{vis}} \equiv \frac{\rho_{\text{vis}}}{\rho_c} \sim 0.002 \tag{9.11}$$

9.1.2 Baryons

The total density of ordinary "baryonic matter" (protons, nuclei and electrons) is estimated to be an order of magnitude greater than that of visible baryons (9.11):

$$\Omega_b = 0.044 \pm 0.004 \,. \tag{9.12}$$

This estimate came originally from the theory of the nucleosynthesis of the light elements which correctly predicts the *relative* abundances of the light elements only if Ω_b is near this value.

Since $\Omega_b > \Omega_{\text{vis}}$ one can wonder where the missing "dark" baryons are. Most of them are thought to be in the intergalactic medium in the form of an ionized gas [98]. Some fraction of them may be in dark compact objects such as dead stars (neutron stars or white dwarfs) or stars too light to burn hydrogen (brown dwarfs). It has also been suggested [99] that a significant fraction of the baryons are contained in cold molecular clouds.

Finally, we mention that there are apparently very few antibaryons in the visible Universe [100]. Any antimatter consisting of antibaryons and positrons would quickly annihilate in collisions with ordinary matter. Even if the antimatter were somehow separated from the matter, annihilations in intergalactic space at the boundaries between matter and antimatter domains would lead to a flux of high-energy annihilation photons. It thus seems probable that the density of antimatter is extremely small within the accessible part of the Universe.

9.1.3 Cold dark matter

About 90% of the matter in galaxies and clusters of galaxies is not visible. Structure formation theories suggest that the invisible "dark matter" is "cold dark matter," i.e. non-relativistic matter that has only weak interactions with baryonic matter and photons. Often, it is assumed to be in the form of nonbaryonic weakly interacting massive particles called generically "wimps." The present density of CDM is estimated to be an order of magnitude greater than that of baryons but still less than critical

$$\Omega_{\text{CDM}} = 0.226 \pm 0.04 \,. \tag{9.13}$$

This dark matter is believed to make up most of the mass of galactic halos and galaxy clusters.

Unfortunately, there are no wimps in the zoo of known elementary particles and their existence is a bold prediction of cosmology. Some extensions of the standard model of particle physics predict the existence of wimps that are sufficiently heavy that they would not yet have been produced at accelerators. For example, a class of models that predict the existence of wimps are "supersymmetric" models. In these models, each of the known fermions (bosons) is paired with a heavy supersymmetric partner that is a boson (fermion). The lightest of the supersymmetric partners (LSP) is expected to be stable and to have only weak interactions, making it an ideal wimp candidate. In Sect. 9.8 we will see that the parameters of the supersymmetric model can be chosen so that the wimp has the required present-day density (9.13). The mass would be expected to be between 10 GeV and 10 TeV. Efforts are underway to produce supersymmetric particles at accelerators and to detect them in the Milky Way (Exercises 9.7).

9.1.4 Photons

The most abundant particles in the Universe are the photons of the "cosmic (microwave) background radiation" (CBR) (also referred to in the literature as "CMB" for "Cosmic Microwave Background"). These photons have a nearly perfect thermal spectrum as shown in Fig. 9.1. The photon temperature is

$$T_\gamma = 2.725\,\mathrm{K} \quad \Rightarrow \quad kT = 2.35 \times 10^{-4}\,\mathrm{eV} \tag{9.14}$$

corresponding to a number density of

$$n_\gamma = 410.4\,\mathrm{cm}^{-3} . \tag{9.15}$$

This is considerably greater than the number of photons that have been generated by stars (Exercise 9.1). Despite their great abundance, the low temperature of the CBR results in a small photon energy density, $\rho_\gamma \sim n_\gamma kT$:

$$\Omega_\gamma = (5.06 \pm 0.4) \times 10^{-5} . \tag{9.16}$$

The present-day Universe is nearly transparent to photons (Exercise 9.2) in the sense that the photon mean free path is greater than the distance to the most distant visible objects. The CBR photons were thermalized in the early Universe when the temperature was $> 0.26\,\mathrm{eV}$ and baryonic matter was completely ionized. At $T \sim 0.26\,\mathrm{eV}$, baryonic matter "recombined" to form atoms and the resulting decrease in the photon–matter cross-section made the Universe transparent.

The CBR temperature is not completely isotropic but is observed to vary by factors of order 10^{-5} according to the direction of observation. These small variations are believed to be due to the density inhomogeneities present at the moment of recombination. As such, the temperature anisotropies provide

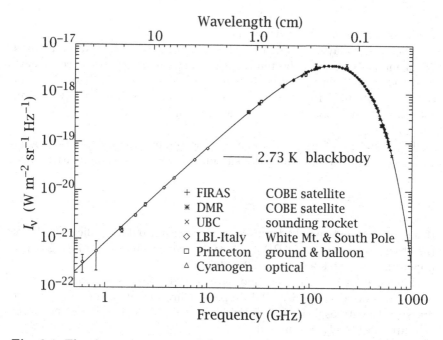

Fig. 9.1. The observed spectrum of the cosmic (microwave) background radiation (CBR) [1]. The points at wavelengths < 1 cm come from ground-based experiments. At shorter wavelengths the Earth's atmosphere is opaque and measurements must be made from balloons, rockets or satellites. The high precision points around the peak of the spectrum were made by the FIRAS instrument of the COBE satellite which observed from 1989 to 1995 [92]. Compilation courtesy of the Particle Data Group.

information about the "initial conditions" for structure formation. The spectrum of anisotropies is the primary source on the values of the cosmological densities in Table 9.1. In particular, it constrains Ω_T to be very near unity.

9.1.5 Neutrinos

In addition to thermal photons, it is believed that the Universe is filled with neutrinos, ν_e, ν_μ and ν_τ and the corresponding antineutrinos.

Neutrinos interact even less than the CBR photons but they had a sufficiently high interaction rate at $T > 1\,\mathrm{MeV}$ to have been thermalized. In Sect. 9.6 we will see that by the time the temperature dropped below $kT \sim m_e c^2$, relativistic neutrinos had a temperature slightly less than the photon temperature:

$$T_\nu = (4/11)^{1/3}\, T_\gamma \,. \tag{9.17}$$

This corresponds to a neutrino (+ antineutrino) number density of

$$n_\nu = (3/11)n_\gamma \qquad \text{per species.} \tag{9.18}$$

corresponding to a present day value of 112 cm^{-3}.

The present-day energy density of a neutrino species depends on its mass. For a massless species, the temperature relation (9.17) would continue to hold today and consequently the energy density would be a bit less than that for photons (9.16). This would be approximately true for any neutrino species with mass much less than the photon temperature, $m_\nu c^2 \ll kT_\gamma \sim 2 \times 10^{-4}$eV.

For a species of mass greater than the temperature, the neutrinos are currently non-relativistic and the summed neutrino and antineutrino mass density is

$$\Omega_\nu = \frac{m_\nu n_\nu}{\rho_c} = 0.2\, h_{70}^{-2}\, \frac{m_\nu}{10\,\text{eV}} \qquad \text{if } m_\nu \gg 10^{-4}\,\text{eV} . \tag{9.19}$$

If one of the neutrinos had a mass > 0.7 eV, it would significantly distort the observed spectrum of CBR anisotropies. Requiring $m_\nu < 0.7$ eV gives the upper limit in Table 9.1, $\Omega_\nu < 0.015$ [94].

Evidence for non-zero neutrino masses comes from searches for neutrino oscillations discussed in Sect. 4.4. In particular, oscillations of neutrinos produced in the atmospheric interactions of cosmic rays [49] have given results that are most easily interpreted as

$$(m_3^2 - m_2^2)c^4 = (3 \pm 1) \times 10^{-3}\,\text{eV}^2 \quad \Rightarrow\ m_3 > 0.04\,\text{eV} , \tag{9.20}$$

where m_3 is the heaviest of the three neutrinos. This implies:

$$\Omega_\nu \sim \frac{n_\nu m_3}{\rho_c} > 0.0006 . \tag{9.21}$$

It is often supposed that neutrinos have a "hierarchical" mass pattern like that of the charged leptons, i.e. $m_3 \gg m_2 \gg m_1$. If this is the case, the above inequality becomes an approximate equality, $m_3 \sim 0.04$ eV. It is not possible, for the moment, to directly verify this hypothesis.

Finally, we note that, because of their extremely weak interactions, there is little hope of directly detecting the cosmic neutrino background [101].

9.1.6 The vacuum

Perhaps the most surprising recent discovery is that the Universe appears to be dominated by an apparent "vacuum energy" or "cosmological constant" Λ:

$$\Omega_\Lambda \sim \frac{\Lambda}{3H_0^2} \sim 0.7 . \tag{9.22}$$

The observational evidence for the existence of vacuum energy involves the apparent luminosity of high redshift objects which can provide information on whether the universal expansion is accelerating or decelerating (as would be expected from normal gravitation). The observations indicate that the expansion is accelerating and, as we will see in Sec. 9.3, this can be explained by a positive vacuum energy density.

Vacuum energy is, by definition, energy that is not associated with particles and is therefore not diluted by the expansion of the Universe. Unless the present vacuum is only metastable, this implies the the vacuum energy density is independent of time. The value implied by $\Omega_\Lambda = 0.7$ is

$$\rho_\Lambda(t) \sim 3\, h_{70}^2 \times 10^9\, \text{eV}\, \text{m}^{-3} . \tag{9.23}$$

Fundamental physics cannot currently be used to calculate the value of the vacuum energy even though it is a concept used throughout modern gauge theories of particle physics. It is expected to be a temperature-dependent quantity which changes in a calculable manner during phase transitions, e.g. the electroweak transition at $T \sim 300\, \text{GeV}$ when the intermediate vector bosons, $W^\pm$ and Z^0, became massive. While the vacuum energy does not change in particle collisions, so its existence can usually be ignored, it does lead to certain observable effects like the Casimir force between uncharged conductors. Unfortunately, all calculable quantities involving vacuum energy concern differences in energy densities and there are no good ideas on how to calculate the absolute value.

Despite the lack of ideas, the existence of a vacuum energy density of the magnitude given by (9.23) is especially surprising. A vacuum energy density can be associated with a "mass scale" M by writing

$$\rho_\Lambda \sim \frac{(Mc^2)^4}{(\hbar c)^3} . \tag{9.24}$$

Particle physicists are tempted to choose the Planck mass

$$m_{\text{pl}} c^2 = (\hbar c^5/G)^{1/2} \sim 10^{19}\, \text{GeV} \tag{9.25}$$

as the most fundamental scale since it is the only mass that can be formed from the fundamental constants. This gives $\rho_\Lambda \sim 3 \times 10^{132}\, \text{eV}\, \text{m}^{-3}$, i.e about 123 orders of magnitude too large making it perhaps the worst guess in the history of physics. In fact, the density (9.23) implies a scale of $M \sim 10^{-3}\, \text{eV}$ which is not obviously associated with any other fundamental scale in particle physics, though it is near the estimated masses of the neutrinos.

A second problem with an energy density (9.22) is that it is comparable to the matter density $\Omega_M \sim 0.3$. Since the matter density changes with the expansion of the Universe while the vacuum energy does not, it appears that we live in a special epoch when the two energies are comparable. This coincidence merits an explanation.

9.2 The expansion of the Universe

Modern cosmology started with Hubble's discovery that galaxies are receding from us with a velocity dR/dt proportional to their distance R (Fig. 9.2) :

Hybrid Cluster Sample

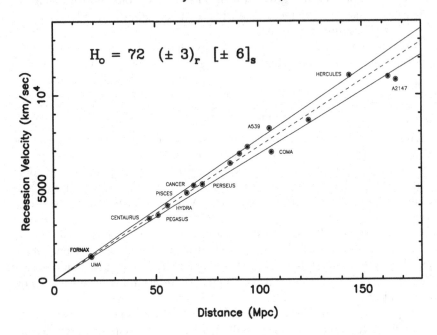

Fig. 9.2. The "Hubble diagram," of galactic recession velocities versus distance for a set of galaxy clusters as determined by the Hubble Key Project [97]. The slope of the line is the Hubble constant H_0. The velocities are determined by the galaxy redshifts and the distances are determined by a variety of methods. For example, if a supernova of known luminosity L is observed in a galaxy or galaxy cluster, the distance can be derived from the observed photon flux $\phi = L/4\pi R^2$.

$$\frac{dR}{dt} = H_0 R . \tag{9.26}$$

The factor of proportionality, H_0, is called the *Hubble constant*.[1] In establishing Hubble's law, the velocity dR/dt is derived from the "redshifts" of a galaxy's photon spectrum:

$$1 + z \equiv \frac{\lambda_0}{\lambda_1} \sim 1 + v/c , \tag{9.27}$$

where λ_1 is the wavelength of photons emitted by the galaxy and λ_0 is the wavelength measured later by us. The equivalence of the redshift z with $v/c = (dR/dt)/c$ is the non-relativistic Doppler effect. As such, the formula is valid only for $z << 1$, which, according to Hubble's law, corresponds to $R << c/H_0 \sim 4300 \mathrm{Mpc}$. A more general interpretation is given below (9.36).

---

[1] In writing (9.26), we ignore the quasi-random "peculiar" velocities v_p that are typically of order $v_p \sim 10^{-3}c \sim 300 \mathrm{km\,s}^{-1}$. These are smaller than the Hubble velocity $H_0 R$ for $R > 300 \mathrm{km\,s}^{-1}/H_0 \sim 5 \mathrm{Mpc}$.

Because photon wavelengths appear[2] to increase with time, photon energies ($E_\gamma = hc/\lambda_\gamma$) decrease with time. The energy loss between emission in a galaxy of redshift $z = H_0 R/c$ and reception by us is

$$E_1 - E_0 = hc(\lambda_1^{-1} - \lambda_0^{-1}) = E_0[(\lambda_0/\lambda_1) - 1] \sim E_0 z .\qquad(9.28)$$

Since the time to travel a distance R is $R/c = z/H_0$, we have

$$\frac{dE_\gamma}{dt} = -H_0 E_\gamma .\qquad(9.29)$$

CBR photons are not different from photons emitted by galaxies, so these photons must lose energy at the same rate. This implies that the cosmological photon temperature decreases with time at a rate that is currently given by

$$\frac{dT_\gamma}{dt} = -H_0 T_\gamma .\qquad(9.30)$$

9.2.1 The scale factor $a(t)$

It is useful to parameterize the expansion of the Universe by a time-dependent function that is proportional to the distances between galaxies. This function is called the "scale factor," $a(t)$:

$$a(t) \propto \langle\text{intergalactic distances}\rangle .\qquad(9.31)$$

The normalization of $a(t)$ can be taken to be arbitrary but we call its value today

$$a(t_0) \equiv a_0 \qquad t_0 \equiv \text{today} .\qquad(9.32)$$

Hubble's law (9.26), $\dot{R}/R = H_0$ informs us that the logarithmic derivative of $a(t)$ is currently equal to H_0:

$$\left[\frac{\dot{a}}{a}\right]_{t_0} = H_0 .\qquad(9.33)$$

Substituting this into (9.29) or (9.30) and generalizing to an arbitrary time we get

$$\frac{dE_\gamma}{da} = -\frac{E_\gamma}{a} \qquad \frac{dT_\gamma}{da} = -\frac{T_\gamma}{a} .\qquad(9.34)$$

This then says that the energies of photons or their temperature fall simply with the scale parameters:

$$E_\gamma \propto a^{-1} \qquad T_\gamma \propto a^{-1} .\qquad(9.35)$$

The photon temperature can therefore be used to define the scale parameter.

---

[2] We say "appear" because the photon changes Galilean reference frames as it travels from galaxy to galaxy. The observer that measures the original high energy is in a different reference frame from the observer who later measures the redshifted energy.

Relation (9.35) applied to photons emitted by galaxies implies a more general interpretation of the galactic redshifts:

$$1 + z \equiv \frac{\lambda_0}{\lambda_1} = \frac{a(t_0)}{a(t_1)}, \tag{9.36}$$

where t_1 is the time of emission. The redshift of photons coming from a galaxy of redshift z then gives the factor of expansion of the Universe between emission and detection.

The scale parameter also determines the time dependences of number densities of objects whose numbers are conserved. For example, the number density of galaxies n_{gal} is proportional to $\langle \text{distances} \rangle^{-3}$ so, if the number of galaxies does not change with time, we have

$$n_{\mathrm{gal}}(t) = n_{\mathrm{gal}}(t_0) \left(\frac{a_0}{a(t)} \right)^3. \tag{9.37}$$

While the number density of galaxies is, in fact, not conserved (there were none in the early Universe), baryon number is conserved to good approximation so we have

$$n_{\mathrm{b}}(a) = n_{\mathrm{b}}(a_0) \left(\frac{a_0}{a(t)} \right)^3. \tag{9.38}$$

This dependence on a is also respected by any species of particles whose number does not vary with time. Hence, we expect the number density of cold dark matter particles and photons of cosmological origin to fall like a^{-3}.

The energy densities in the Universe as a function of time can also be simply expressed in terms of the scale factor. The energy density of non-relativistic particles like baryons and CDM particles is proportional to the number density of particles times their mass. It follows that

$$\rho_{\mathrm{CDM}}(t) + \rho_{\mathrm{b}}(t) \equiv \rho_{\mathrm{M}}(t) = \rho_{\mathrm{M}}(t_0) \left(\frac{a_0}{a(t)} \right)^3. \tag{9.39}$$

On the other hand, the energy per photon falls like a^{-1} which implies an extra factor of a^{-1} in addition to the three factors associated with the falling number density:

$$\rho_\gamma(a) = \rho_\gamma(a_0) \left(\frac{a_0}{a(t)} \right)^4. \tag{9.40}$$

Finally, we need to know how the vacuum energy density, ρ_Λ, evolves with the expansion of the Universe. The evolution of the matter density is due to the changing density of particles as the Universe expands. Since there are, by definition, no particles associated with the vacuum energy, it seems plausible that the vacuum energy density will remain unchanged as the Universe evolves:

$$\rho_\Lambda(a) = \rho_\Lambda(a_0). \tag{9.41}$$

This behavior is confirmed by an analysis using general relativity.

Since the density of relativistic matter (9.40) is proportional to a^{-4} while that of non-relativistic matter (9.39) is proportional to a^{-3}, relativistic matter must come to dominate for $a \to 0$. Equating the two energy densities we find the moment when the photon and matter densities are equal occurred when the scale parameter took the value

$$\frac{a_{\mathrm{eq}}}{a_0} = \frac{\Omega_\gamma}{\Omega_{\mathrm{M}}} \sim 2 \times 10^{-4}, \tag{9.42}$$

corresponding to a temperature of

$$kT_{\mathrm{eq}} \sim 1\,\mathrm{eV}. \tag{9.43}$$

For temperatures higher than this, radiation dominates the energy of the Universe.

Since the matter density falls like a^{-3}, the future Universe will be increasingly dominated by the constant vacuum energy (if it is truly constant).

We see that the Universe passes through a succession of epochs when the energy density is dominated by radiation, by non-relativistic matter, and then by vacuum energy. The energy densities as a function of temperature are shown in Fig. 9.3.

In Table 9.2, we list some formative events in the history of the Universe according to this scenario. Non-controversial physics allows us to follow with confidence the succession of events starting at, say, $T \sim 1\,\mathrm{GeV}$ when the Universe was a nearly homogeneous soup of quarks, gluons, and leptons. With time, the Universe cooled and a succession of bound states were formed, hadrons, nuclei, atoms, and finally the gravitationally bound stars and galaxies. The moments of the formation of bound states are called "recombinations." The recombination that resulted in the formation of atoms caused the Universe to become effectively transparent to photons.

We should add that the nature of the radiation changed with temperature. Today, the radiation consists of photons and, perhaps, relativistic neutrinos. At temperatures $T > m_{\mathrm{e}}$, electron–positron pairs could be produced and we will show in Sec. 9.6 that these pairs were in thermal equilibrium with the photons and formed a blackbody spectrum similar to that of the photons and neutrinos. Going back in time, each time the temperature rose above a particle-antiparticle threshold, a new blackbody component was created. During this period, the numbers of particles and antiparticles were nearly equal. The small number of electrons and baryons present today resulted from the small excess ($\sim 10^{-9}$) of particles over antiparticles present when $T \gg m_{\mathrm{e}}$.

Finally, we note that the two earliest epochs in Table 9.2, those of baryogenesis and inflation, are speculative and involve physics that is not well-understood. The existence of these epochs is postulated to solve certain mysteries in the standard scenario, e.g. the existence of the small particle-antiparticle asymmetry and the origin of the density fluctuations leading to structure formation.

9.3 Gravitation and the Friedmann equation

While the temperature of the Universe $T \propto 1/a$ is simply determined by the scale parameter $a(t)$, in order to calculate the end product of cosmological particle or nuclear reactions, we will need to know how long the Universe spends near a given temperature, i.e. we need to know the time dependence of $a(t)$.

In the absence of gravitation, the recession velocities of galaxies would be constant, implying $\ddot{a} = 0$. In the presence of the attractive effects of gravitation, we might expect that the expansion would be decelerated, $\ddot{a} < 0$.

The correct equation for $\ddot{a}$ can only be found within the framework of a relativistic theory of gravitation, i.e. general relativity. The standard Newtonian formalism is inadequate for two reasons. First, the Newtonian integrals

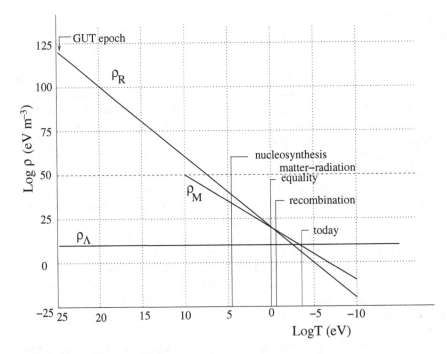

Fig. 9.3. The energy density of matter, radiation and vacuum (assumed constant) as a function of temperature. The temperature scale starts at the expected "grand unification" scale of $\sim 10^{16}$ GeV. We suppose that the CDM particles have masses greater than ~ 10 GeV so the line of ρ_M starts at 10 GeV.

Table 9.2. Some formative events in the past. The values of t_0, t_{rec}, and t_{eq} depend on $(h_{70}, \Omega_M, \Omega_\Lambda)$ and we have used $(1, 0.3, 0.7)$.

t	kT_γ (eV)	event
$t_0 \sim 1.5 \times 10^{10}$ yr	2.349×10^{-4}	today
$\sim 10^9$ yr	$\sim 10^{-3}$	formation of the first structures,
$t_{rec} \sim 5 \times 10^5$ yr	0.26	"recombination" (formation of atoms), Universe becomes transparent
$t_{eq} \sim 5 \times 10^4$ yr	0.8	matter-radiation equality
3 min	6×10^4	nucleosynthesis (formation of light nuclei, $A = 2, 3, 4, 6, 7$)
1 s	10^6	$e^+ e^- \to \gamma\gamma$
4×10^{-6} s	$\sim 4 \times 10^8$	QCD phase transition (formation of hadrons from quarks and gluons)
$< 4 \times 10^{-6}$ s	$> 10^9$	baryogenesis (?) (generation of baryon–antibaryon asymmetry)
		inflation (?) (generation of density fluctuations)

used to calculate the force on a particle or the gravitational potential at a particular point are not defined in an infinite medium of uniform density. Second, the source of Newtonian gravitation is the mass density. We can expect that a relativistic theory will have a more general source, just as in electromagnetism both charge densities and charge currents can lead to fields.

General relativity treats gravitation geometrically. One way of approaching the problem is is illustrated in Fig. 9.4. The figure shows a small region of a two-dimensional homogeneous Universe with a coordinate grid attached to free test particles (galaxies). The test particles are initially equally spaced, as determined by light propagation times between galaxies. With this choice, over a small region, the distance dS between test particles separated by coordinate differences (dx, dy) is just

$$dS^2 = dx^2 + dy^2 .\tag{9.44}$$

As the Universe expands, the test particles withdraw from each other. Compared to the original situation, all intergalactic distances are multiplied by a scale factor $a(t)$. It is convenient to keep the coordinates of the galaxies constant during the expansion, as shown in Fig. 9.4. Such coordinates are called "comoving" coordinates. Since the coordinate separations do not change with time, at a later time t the relation between dS and the coordinate differences is given by

$$dS^2 = = a(t)^2 \, (dx^2 + dy^2) \, . \tag{9.45}$$

For this homogeneous Universe, all the effects of gravity should be stored in the geometric scale factor $a(t)$. The equations of general relativity will tell us how $a(t)$ evolves with time.

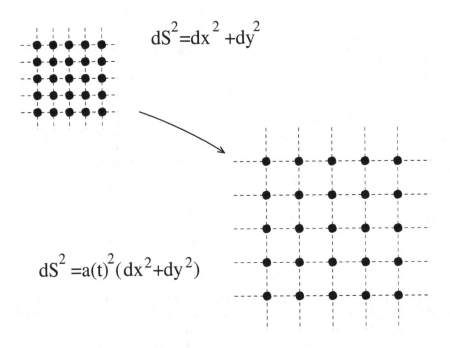

Fig. 9.4. A small region of a homogeneous two-dimensional Universe. A coordinate grid has lines of constant x and y that intersect at equally spaced test particles (galaxies). As the Universe expands, the distances dS between galaxies increases [by a factor $a(t)$], but we choose to keep the coordinates of the galaxies fixed (comoving coordinates).

We note that the geometrization of gravitation is possible only because of the *Principle of Equivalence* which insures that gravity affects all test particles in the same way, independent of their mass and composition. This means that if, in Fig 9.4 we place two test particles at the same point with vanishing relative velocity, then in the future they will stay at the same point. If this were not the case, then the expansion of the coordinate grid would not be unique.

We also note that the equations of general relativity for a homogeneous Universe will be simple because the coordinate grid can be described by one function $a(t)$. In an inhomogeneous Universe, the situation is much more complicated, as illustrated in Fig. 9.5. In the figure we have supposed that there is an excess of mass in the center of the figure. The gravitational attraction of matter to this excess will cause the test particles to expand in a non-uniform manner. Even if the coordinate system is simple at the beginning, it will evolve into a complicated system with distances between test particles defined by three metric functions

$$dS^2 = = g_{xx}(t, x, y)dx^2 + g_{xy}(t, x, y)dxdy + g_{yy}(t, x, y)dy^2 . \qquad (9.46)$$

We now have 3 functions of position and time to describe the coordinate system. The three functions g_{xx}, g_{xy} and g_{yy} define the (spatial) metric of the coordinate system. The equations that govern the metric can be quite complicated.

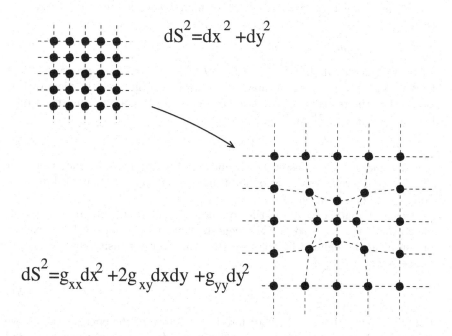

Fig. 9.5. Same as Fig. 9.4 but with a mass excess near the center of the region. The excess gravity causes the test particles to separate less rapidly near the center than in the homogeneous case. The resulting formula for dS is more complicated.

Fortunately for us, the metric of a homogeneous Universe is described by only one function $a(t)$ and we can expect the Einstein equation to be simple. It turns out to be

$$\frac{\ddot{a}}{a} = \frac{-4\pi G}{3c^2}(\rho + 3p), \tag{9.47}$$

where ρ is the energy density and p is the "pressure." As expected, a positive energy density $\rho > 0$ works to decelerate the Universe. In fact, the part of the r.h.s. that is proportional to ρ can be guessed through a simple Newtonian argument, as shown in Exercise 9.4. On the other hand, the appearance of a correction proportional to the pressure p is surprising, especially since a positive pressure also works to decelerate the Universe, out of line with our intuition of thermal pressure encouraging the expansion of a gas. This is however not the role of the pressure in the gravitational context where it acts as a source of gravitation that is normally ignored in Newtonian problems. To see why it can often be ignored, we note that for a collection of particles (number density n), the pressure can be defined to be proportional to the mean value of $|\boldsymbol{p}|^2 c^2 / 3E$:

$$p \equiv n\langle |\boldsymbol{p}|^2 c^2 / 3E \rangle . \tag{9.48}$$

For instance, for a thermal collection of nonrelativistic particles of mass m, the energy density is $\rho = mc^2 n$ and the pressure is given by

$$p = n\langle (|\boldsymbol{p}|c)^2 / 3E \rangle = nkT , \tag{9.49}$$

where we have used $\langle |\boldsymbol{p}|^2 / 2m \rangle = (3/2)kT$. This value of the pressure is just that of the ideal gas law. Because $kT \ll mc^2$ for a non-relativistic gas, this shows that the relativistic correction is small. On the other hand, for a relativistic gas ($|\boldsymbol{p}|c \sim E$) we have

$$p = n\langle (|\boldsymbol{p}|c)^2 / 3E \rangle = \rho/3 . \tag{9.50}$$

An example of such a gas is the cosmic (photon) background radiation. For such a gas, the relativistic correction proportional to the pressure is large and doubles the deceleration rate.

In an adiabatic expansion, the volume and energy of a gas change according to the law $dE = -pdV$. This suggests that in a cosmological context the pressure can be defined by the way that the energy density changes as the Universe expands:

$$p = -\frac{d\rho a^3}{da^3} . \tag{9.51}$$

Using the laws (9.39) and (9.40) for the evolution of the energy density, we see that (9.51) implies that, as expected, $p = 0$ for a non-relativistic gas and $p = \rho/3$ for a relativistic gas. On the other hand, the fact that the vacuum energy density is constant (9.41) implies that for vacuum energy we have

$$p = -\rho \qquad \text{vacuum} . \tag{9.52}$$

A negative pressure is perhaps counter intuitive but we should not forget that our intuition for pressure is based on the pressure of collections of particles whereas the vacuum is the absence of particles.

The solution $\dot{a}(t)$ of (9.47) with the pressure (9.51) is

$$\dot{a}^2 = \left[\frac{8\pi G\rho}{3c^2}\right] a^2 + \text{const}.$$

To evaluate the constant we use the present values $\dot{a}(t_0) = H_0 a_0$ and $8\pi G\rho(t_0)/3c^2 = H_0^2 \Omega_T$ to find

$$\dot{a}^2 = \frac{8\pi G\rho a^2}{3c^2} + H_0^2 a_0^2 (1 - \Omega_T). \tag{9.53}$$

Dividing (9.53) by a^2 we find the "Friedmann equation":

$$\left(\frac{\dot{a}}{a}\right)^2 = \frac{8\pi G\rho}{3c^2} + H_0^2 (1 - \Omega_T) \left(\frac{a_0}{a}\right)^2. \tag{9.54}$$

We note that the measurement previously discussed of the Cosmic Background Radiation anisotropies indicate $\Omega_T \sim 1$ so that the second term on the r.h.s. nearly vanishes. At any rate, it can be ignored in the primordial Universe since it diverges as a^{-2} whereas the radiation density diverges as a^{-4}.

We will be mostly interested in the expansion rate in the primordial Universe when the energy density and pressure were dominated by relativistic radiation given by the Stefan–Boltzmann law:

$$\rho(T) = g(T)\frac{\pi^2}{30}\frac{(kT)^4}{(\hbar c)^3}, \tag{9.55}$$

where $g(T)$ is the effective number of relativistic spin degrees of freedom in thermal equilibrium at the temperature T ($g = 2$ for a blackbody photon gas). At $T = 1\,\text{MeV}$, photons, three neutrino species and electron–positron pairs are in equilibrium corresponding to $g(kT = 1\,\text{MeV}) \sim 10.75$. (Note that a fermion degree of freedom counts for only $\Delta g = 7/8$ because the Pauli principles restricts the number of fermions.) When nucleosynthesis occurs at $kT \sim 50\,\text{keV}$, only photons and neutrinos [with $T_\nu \sim (4/11)^{1/3}T_\gamma$] are present giving $g(kT = 50\,\text{keV}) \sim 3.36$.

We then have

$$\frac{\dot{a}}{a} = \left(\frac{8\pi G}{3}g(T)\frac{\pi^2}{30}\frac{(kT)^4}{(\hbar c)^3}\right)^{1/2}$$

$$\sim 0.65\,\text{s}^{-1}\left(\frac{kT}{1\,\text{MeV}}\right)^2\left(\frac{g(T)}{10}\right)^{1/2}. \tag{9.56}$$

The Hubble time, i.e. the time for significant changes in the temperature is the inverse of $\dot{a}/a$. It is about 1 sec for $T \sim 1\,\text{MeV}$ and about 100 sec for $kT \sim 100\,\text{keV}$ when nucleosynthesis begins.

9.4 High-redshift supernovae and the vacuum energy

The general relativistic equation (9.47) combined with the surprising nega-tive pressure for a positive vacuum energy (9.52) means that the expansion *accelerates* in a Universe dominated by vacuum energy:

$$\frac{\ddot{a}}{a} = +\frac{8\pi G}{3c^2} \rho_V \qquad \text{if } \rho_{tot} \sim \rho_V . \tag{9.57}$$

The observational evidence for acceleration involves the observed photon flux from high redshift supernovae.

Understanding the relation between the measured flux of a supernova and its redshift requires the use of general relativity. However, we can understand the relationship qualitatively through the following argument. The flux is determined by the present distance R to the supernova

$$\phi \propto \frac{1}{R^2} \sim \frac{1}{c^2(t_0 - t_1)^2} , \tag{9.58}$$

where in the second form we replace the distance by the flight time $c(t_0 - t_1)$ where t_1 and t_0 are the explosion and detection times.

On the other hand, the redshift, z, is defined by the emitted (λ_0) and observed (λ_1) wavelengths of photons

$$1 + z \equiv \frac{\lambda_1}{\lambda_0} . \tag{9.59}$$

For nearby objects, this redshift is most simply interpreted as a Doppler shift (9.27). However, equation (9.35) tells us that more generally the redshift is the universal expansion factor between the time of the explosion and the time of the observation

$$1 + z = \frac{a(t_0)}{a(t_1)} . \tag{9.60}$$

If the expansion is accelerating (decelerating), the expansion rate in the past was relatively slower (faster) implying a longer (shorter) time between the explosion and observation. Equation (9.58) then implies that in an ac-celerating (decelerating) Universe, a fixed redshift therefore corresponds to a relatively small (large) photon flux.

Two teams [95, 96] have observed that the fluxes of supernova at $z \sim 0.5$ are about 40% smaller than that expected for a decelerated Universe of mass density equal to the critical density. The effect can be explained by a vacuum energy $\Omega_\Lambda \sim 0.7$.

9.5 Reaction rates in the early Universe

These days, not much goes on in intergalactic space. There are no nuclear reactions occurring and photons and neutrinos very rarely scatter on matter.

More quantitatively, the number of reactions per particle per unit time, λ, is much less than one reaction per Hubble time. For example the rate of Compton scattering per photon is

$$\lambda_{\gamma e \to \gamma e} = n_e \sigma_T c \sim 1.4 \times 10^{-3} H_0 , \qquad (9.61)$$

where the free electron density is set equal to the baryon density from Table 9.1 and σ_T is the Thomson cross-section. The numerical value in the second form means that only one photon out of 700 will scatter in the next Hubble time. As long as the Universe continues to expand, reactions will become rarer and rarer as the density decreases. In fact, we will see that the typical photon will *never* scatter again.

Things were quite different in the early Universe. Just before electrons and nuclei recombined to form atoms, the temperature was 1000 times its present value so the density of electrons was 10^9 times the present density. The expansion rate, given by the Friedmann equation (9.54), was only $\sqrt{\Omega_M} 1000^{3/2} \sim 2 \times 10^4$ times the present rate:

$$\lambda_{\gamma e \to \gamma e}(t_{\rm rec}) \sim n_e(t_{\rm rec}) \sigma_T c \sim 80 H(t_{\rm rec}) . \qquad (9.62)$$

At this epoch, a typical photon suffered 80 collisions per Hubble time.

The thermal spectrum of photons resulted from the high reaction rate in the early Universe. Elastic scattering, e.g.

$$\gamma e^- \leftrightarrow \gamma e^- , \qquad (9.63)$$

caused energy exchanges between particles and generated "kinetic" equilibrium, i.e. a thermal momentum distribution. Inelastic collisions changed the number of particles and generated "chemical" equilibrium where the particle densities have thermal values. For example, bremsstrahlung and photon absorption

$$e^- p \leftrightarrow e^- p \gamma , \qquad (9.64)$$

creates and destroys photons, generating a thermal number density

$$n_\gamma = \frac{2.4}{\pi^2} \frac{(kT)^3}{(\hbar c)^3} \qquad (9.65)$$

The elementary reactions

$$\gamma\gamma \leftrightarrow e^+ e^- \leftrightarrow \nu\bar{\nu} \qquad (9.66)$$

generated thermal (blackbody) densities of electron–positron pairs and neutrinos:

$$n_{e^+} = n_{e^-} = n_{\nu_i} + n_{\bar{\nu}_i} = (3/4) n_\gamma \qquad kT \gg m_e c^2 . \qquad (9.67)$$

The factor $(3/4)$ comes from the fact that fermions must respect the Pauli principle and, as such, have smaller numbers in thermal equilibrium. Also important are the neutron–proton transitions:

$$\nu_e n \leftrightarrow e^- p \qquad\qquad \bar{\nu}_e p \leftrightarrow e^+ n . \qquad (9.68)$$

These reactions established chemical equilibrium between protons and neutrons (4.35):

$$\frac{n_\mathrm{n}}{n_\mathrm{p}} = \exp\left(\frac{-(m_\mathrm{n} - m_\mathrm{p})c^2}{kT}\right) \qquad kT > 1\,\mathrm{MeV}\,. \tag{9.69}$$

This determines the number of neutrons available for nucleosynthesis.

The minimal requirement for the establishment of thermal equilibrium is that the reaction rate per particle be greater than the expansion rate:

$$\lambda \gg \frac{\dot{a}}{a} \qquad \Rightarrow \qquad \text{thermal equilibrium}\,. \tag{9.70}$$

The expansion rate is the relevant parameter because its inverse, the Hubble time t_H, gives the characteristic time for temperature and density changes due to the universal expansion. Collisions can therefore perform the necessary readjustments of momentum and chemical distributions only if each particle reacts at least once per Hubble time.

Because of the expansion, the collision epoch was bound to end when the reaction rate became less than the expansion rate, $\lambda \ll \dot{a}/a$. What happens to the thermal distributions once the collisions cease depends on the type of equilibrium. For purely kinetic equilibrium, i.e. the momentum spectrum of particles, the thermal character of the spectrum may be maintained by the expansion. This is the case for the CBR photons.

On the other hand, chemical thermal equilibrium is maintained only by reactions and the equilibrium is lost once the reactions cease. At the present low temperature, chemical equilibrium would imply that nucleons would tend to be in their most bound states near ^{56}Fe. This is not the case because the nuclear reactions necessary to reach this state ceased when the temperature was much higher, $kT \sim 30\,\mathrm{keV}$. Most nucleons were thus "stranded" in hydrogen and helium. We say that the nuclear reactions "froze" at a temperature $T_\mathrm{f} \sim 30,\mathrm{keV}$. The "freeze-out" left the Universe with a "relic" density of hydrogen and helium nuclei that is far from the equilibrium density.

The nuclear freeze-out left the Universe with a reserve of free energy, i.e. energy that can now be degraded by entropy producing (exothermic) nuclear fusion reactions. In particular, hydrogen can be converted to helium and helium to heavier elements once matter is gravitationally confined in stars. Fusion reactions in stellar interiors transform mass into kinetic energy of the reaction products which is then degraded to thermal energy including a multitude of thermal photons. It is this increase in the number of photons which is primarily responsible for the entropy increase.[3]

After the photons escape from a star, entropy production can continue if the photons are intercepted by a cold planetary surface. On Earth, solar photons ($T \sim 6000\,\mathrm{K}$) are multiplied into ~ 20 thermal photons ($T \sim 300\,\mathrm{K}$). The accompanying entropy increase more than compensates for the entropy decrease associated with the organization of life induced by photosynthesis.

---

[3] Entropy is always approximately proportional to the number of particles.

Without the thermal gradient between the Sun and Earth photosynthesis would not be possible because of the second law of thermodynamics. We see that the loss of thermal nuclear equilibrium in the early Universe provides the free energy necessary for life on Earth. Without this energy source, life would depend on the photons produced during the contraction phases of stars (Fig. 8.3).

The explanation for the current thermal disequilibrium is one of the greatest triumphs of modern cosmology. Nineteenth century physicists were puzzled by the disequilibrium because they knew that all isolated systems tend toward thermal equilibrium. They also worried about the future "heat death" of the Universe when equilibrium will be reached, terminating all intelligent activity. Modern cosmology appears to have inverted the sequence of events since the state of thermal equilibrium occurred in the past rather than the future.

It should be noted, however, that the early Universe can be said to have been in thermal equilibrium only if we ignore the possibility of gravitational collapse of inhomogeneities. Gravitational collapse results in the radiation of photons (Sect. 8.1.1) and this process generates entropy. The initial homogeneous (and therefore low entropy) conditions of the Universe are therefore "special" and require an explanation as to how they were established.

In order to establish the formalism for the study of reactions in an expanding Universe, we consider a general two-body reaction

$$i\,j\ \to\ k\,l \tag{9.71}$$

In Sect. 3.1.4 we saw that the reaction rate per particle i is proportional to the number of j particles present and to the cross-section times velocity:

$$\lambda_{ij\to kl}\ \equiv\ n_j\,\langle\sigma_{ij\to kl}\,v\rangle\,. \tag{9.72}$$

We note that $\lambda_{ij\to kl}\neq\lambda_{ji\to kl}$. This is simply because the reaction rate per particle i is proportional to the number of particles j and vice versa.

The Boltzmann equation governing the time dependence of n_i was derived in Chap. 3 for a time-independent volume (3.32). For an expanding Universe, the equation is modified to include the effect of the expansion:

$$\frac{dn_i}{dt}\ =\ -3\,\frac{\dot a}{a}\,n_i\ -\ n_i\,\lambda_{ij\to kl}\ +\ n_k\,\lambda_{kl\to ij}\,. \tag{9.73}$$

The three terms in this equation describe the three effects that change n_i: the expansion of the Universe, destruction of i particles, and creation of i particles.

While there is no analytic solution to the Boltzmann equation, there are two very simple limits with approximate solutions. The first is if the expansion rate is much greater than the destruction rate, $\dot a/a\gg\lambda_{ij\to kl}$, and production rate $\dot a/a\gg(n_k/n_i)\lambda_{kl\to ij}$. In this case the first term dominates

$$\frac{dn_i}{dt}\ =\ -3\,\frac{\dot a}{a}\,n_i\qquad\Rightarrow n_i\propto a^{-3}\,. \tag{9.74}$$

As expected, the number density of i particles just dilutes with the expansion of the Universe.

The second case occurs if the reaction rates are much greater than the expansion rates, the Boltzmann equation is simply that for a gas of particles in a fixed volume

$$\frac{dn_i}{dt} = -n_i \lambda_{ij \to kl} + n_k \lambda_{kl \to ij} \qquad \Rightarrow n_i \sim n_i(T) . \qquad (9.75)$$

The solution in this case is the thermal equilibrium solution (principle of detailed balance). This comes about since if n_i is greater than (less than) the equilibrium value, the first term (second term) dominates and the time derivative of n_i is negative (positive). The equation then pushes n_i to the equilibrium value where the two terms cancel. The slow time dependence due to the temperature decrease with the expansion is then enforced by the first term of (9.73).

We can therefore get a qualitative understanding of the solution of the Boltzmann equation by simply comparing expansion and reaction rates. Depending on which is greater, we will either have free dilution or thermal equilibrium. This will be our strategy in the following sections.

9.6 Electrons, positrons and neutrinos

As a first application of the Boltzmann equation, we will treat the case of electrons and positrons. These particles are created and destroyed principally by the reaction

$$e^+ e^- \leftrightarrow \gamma\gamma . \qquad (9.76)$$

The diagram is shown in Fig. 9.6. For high center-of-mass energies, $E_{cm} \gg m_e$, the annihilation cross-section is

$$\sigma_{e^+ e^- \to \gamma\gamma} = (\hbar c)^2 \frac{2\pi\alpha^2}{E_{cm}^2} [\, 2 \ln(E_{cm}/m_e) - 1\,] \qquad E_{cm} \gg m_e . \quad (9.77)$$

For low energy, $v \ll c$, the cross-section is proportional to $1/v$ as is expected for barrier-free exothermic reactions:

$$\sigma_{e^+ e^- \to \gamma\gamma} \left(\frac{v}{c}\right) = (\hbar c)^2 \frac{\pi\alpha^2}{m_e^2 c^4} \qquad v \ll c . \qquad (9.78)$$

To determine if the electrons and positrons are in chemical equilibrium with the photons, we only need to calculate the thermal equilibrium value of the annihilation rate and compare it with the expansion rate given by the Friedmann equation (9.56). The expansion and annihilation rates are shown in Fig. 9.7. For the annihilation rate, there are two simple limits, $kT \gg m_e c^2$ and $kT \ll m_e c^2$.

For $kT \gg m_e c^2$, we simply replace E_{cm} in (9.77) with its mean value $\sim kT$ so $\langle \sigma v \rangle \propto T^{-2}$ (ignoring the logarithmic factor). The densities of electrons

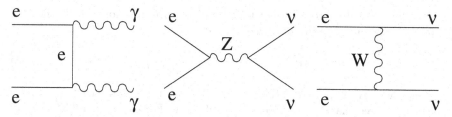

Fig. 9.6. The diagrams for the reactions $e^+e^- \leftrightarrow \gamma\gamma$ and $\nu\bar{\nu} \leftrightarrow e^+e^-$.

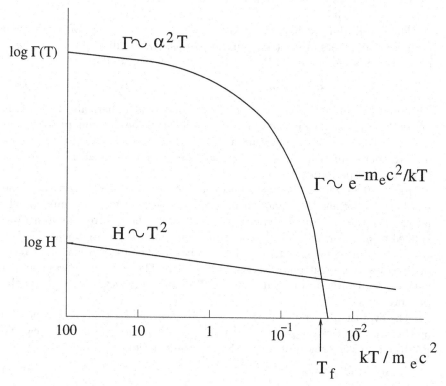

Fig. 9.7. The annihilation rate $\lambda(e^+e^- \to \gamma\gamma)$ and the expansion rate $H = \dot{a}/a$ as a function of temperature under conditions of thermal equilibrium and with $n_{e^-} = n_{e^+} = n_e$. For $T > T_f$, $\lambda > \dot{a}/a$ and n_e will take its equilibrium value. For $T < T_f$, $\lambda < \dot{a}/a$ and the reactions are "frozen." After the freeze-out, the number of electrons and positrons is constant so n_e decreases as $1/a^3$.

and positrons are proportional to T^3 and the annihilation rate is therefore proportional to the temperature:

$$\lambda_{e^+e^-\to\gamma\gamma} \sim \hbar^{-1}\alpha^2 \, kT \sim 10^{18}\,\text{s}^{-1} \left(\frac{kT}{1\,\text{MeV}}\right) \qquad kT \gg m_e c^2 \,, \quad (9.79)$$

where we have suppressed the numerical prefactors. Comparing (9.56) and (9.79), we see that $\lambda > \dot{a}/a$ for $kT < 10^{14}\,\text{GeV}$. For example, at $kT \sim m_e c^2$, $\lambda \sim 10^{18}\dot{a}/a$, i.e. 10^{18} reactions per Hubble time. We can conclude that the electrons and positrons were in chemical equilibrium with the photons for $m_e < T < 10^{14}\,\text{GeV}$.

The equilibrium is inevitably lost for $T \ll m_e$ because, for a Universe with equal numbers of electrons and positrons, the equilibrium density falls rapidly with an exponential Boltzmann factor (Exercise 9.6):

$$n_{e^-} = n_{e^+} \propto \exp(-m_e c^2/kT) \qquad m_e c^2 \ll kT \,. \qquad (9.80)$$

The annihilation rate drops accordingly and one finds numerically that it falls below the expansion rate when $kT \sim m_e c^2/40$ when the electron–photon ratio is $\sim 2 \times 10^{-16}$. The temperature at which this occurs is called the "freeze-out" temperature T_f because after this temperature is reached the number of electrons and positrons is frozen.

Figure 9.8a shows the ratio between the number of electrons and the number of photons as a function of temperature calculated by numerically integrating the Boltzmann equation. For $T \ll T_f$ the electron–photon ratio is fixed at about $\sim 2 \times 10^{-16}$, confirming our qualitative argument.

The evolution of n_{e^-} and n_{e^+} for a universe like our own with an excess of electrons over positrons is shown in Fig. 9.8b. The electron excess coupled with charge conservation leads to a small number of electrons surviving the primordial epoch.

The evolution of the neutrino density is governed by the same principles as that of the electron–positron density. The three neutrino species can be produced and destroyed at $kT \sim \text{MeV}$ by the reaction (Fig. 9.6)

$$\nu \bar{\nu} \leftrightarrow e^+ e^- \,. \qquad (9.81)$$

Since this reaction is due to the weak interactions, the cross-section for all species is of order

$$\sigma \sim \frac{G_F^2 E_\nu^2}{(\hbar c)^4} \qquad m_e c^2 \ll E_\nu \ll m_W c^2 \,. \qquad (9.82)$$

The annihilation rate is therefore

$$\lambda_{\nu\bar{\nu}\to e^+ e^-} = n_\nu \langle \sigma v \rangle \sim \frac{G_F^2 \, (kT)^5}{\hbar(\hbar c)^6}$$

$$\sim 1\,\text{s}^{-1} \left(\frac{kT}{1\,\text{MeV}}\right)^5 \qquad m_e c^2 \ll kT \,, \qquad (9.83)$$

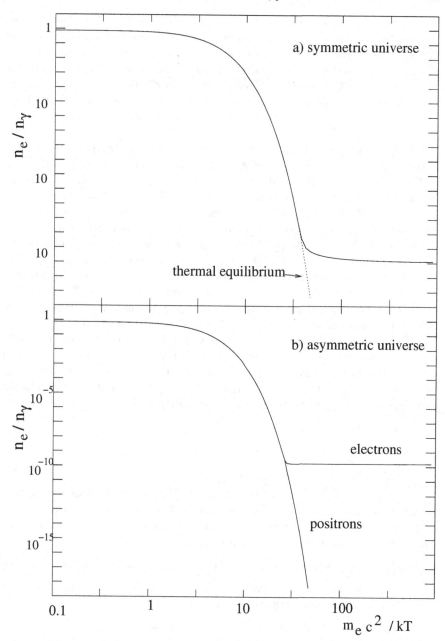

Fig. 9.8. $\log(n_e/n_\gamma)$ versus temperature for a symmetric universe, $n_e = n_{e^-} = n_{e^+}$ (top panel) and for a asymmetric universe, $(n_{e^-} - n_{e^+})/n_\gamma = 3 \times 10^{-10}$ (bottom panel). For the symmetric universe, the dotted line shows $\log(n_e/n_\gamma)$ in the case of thermal equilibrium.

where thee factors of kT come for n_ν and two factors from the cross-section. Since the annihilation rate is proportional to T^5 while the expansion rate (9.56) is proportional to T^2, the reaction rate must win at high temperature. Numerically, one finds that the annihilation rate is greater than the expansion rate for $kT > 1\,\text{MeV}$ so we can conclude that for temperatures greater than 1 MeV, neutrinos were in chemical equilibrium. After the annihilation reactions freeze-out at $T_f \sim 1\,\text{MeV}$, the neutrinos decouple and their number density falls like $n_\nu \propto a^{-3}$.

We are now in a position to understand why the temperature for relativistic neutrinos ends up being lower than the photon temperature. For $kT > 1\,\text{MeV}$, photons, neutrinos, and electron–positrons were in thermal equilibrium with a unique temperature, $T_e = T_\gamma = T_\nu$. Apart from factors due to the Pauli principle and the small electron–positron asymmetry the numbers of electrons, positrons, neutrinos and photons were equal.

When kT_e and kT_γ dropped below $m_e c^2$ the $e^+ e^-$ pairs were transformed into photons (by $e^+ e^- \to \gamma\gamma$) but not into the decoupled neutrinos. After these annihilations, the number of photons was therefore *greater* than the number of neutrinos. Since the distributions are thermal with $n \propto T^3$, it follows that $T_\gamma > T_\nu$ for $kT < m_e c^2$. One says that the photons were "reheated" be electron–positron annihilation.

The ratio between the post-annihilation photon and neutrino temperatures can be calculated by using the fact that the electron–positron–photon system remains in thermal equilibrium until there are very few remaining electrons and positrons which implies that the post- and pre-annihilation entropies of this system are equal. The calculation gives

$$T_\nu = (4/11)^{1/3} T_\gamma. \tag{9.84}$$

This corresponds to a neutrino number density of

$$n_\nu + n_{\bar\nu} = (3/11) n_\gamma \qquad \text{per species}. \tag{9.85}$$

The temperature ratio (9.84) is maintained as long as the neutrinos remain relativistic since in this case both photon and neutrino temperatures fall as a^{-1}. If the neutrinos have masses that are sufficiently small $m_\nu c^2 \ll kT_\gamma(t_0)$, they are still relativistic and have a present temperature of

$$T_\nu(a_0) = (4/11)^{1/3} T_\gamma(a_0) \sim 2\,\text{K} \qquad (\text{if } m_\nu c^2 \ll 10^{-4}\,\text{eV}). \tag{9.86}$$

On the other hand, the density ratio (9.85) is maintained whether or not the neutrinos remain relativistic, so today we expect

$$n_\nu + n_{\bar\nu} = 1.12 \times 10^8 \, m^{-3} \qquad \text{per species}. \tag{9.87}$$

9.7 Cosmological nucleosynthesis

At sufficiently high temperatures and densities, nuclear reactions can take place and it is important for cosmologists to understand the mix of elements that is produced in the early Universe. The calculations were originally performed by Gamow and collaborators in the 1940s with the hope that the relative abundances of all elements could be explained. We now know that nuclear reactions froze at $kT \sim 30\,\mathrm{keV}$ leaving most nuclei in the form of hydrogen and helium. Nucleosynthesis started up again once stars were formed providing "gravitational confinement" for astronomical "fusion reactors."

In this section, we will present a very brief introduction to the theory and confirming observations [106]. The essential theoretical result will be predictions for the relative abundances of the light elements ($A \leq 7$, Table 9.3). Observationally, it is very difficult to determine the primordial abundances because of "pollution" by stellar nucleosynthesis. The best observational estimates are given in Table 9.3. Apart from $^1\mathrm{H}$ and $^4\mathrm{He}$, small quantities of $^2\mathrm{H}$, $^3\mathrm{He}$, $^7\mathrm{Li}$ were produced in the early Universe.

Table 9.3. The important nuclei for nucleosynthesis with their binding energies per nucleon, B/A, their observed primordial abundances, their half-lives and decay modes. (The half-life of $^7\mathrm{Be}$ by electron capture is given for atomic beryllium.) We note the high binding energy of $^4\mathrm{He}$ in comparison with the other light nuclei, which implies that this species will be the primary product of primordial nucleosynthesis. The absence of stable nuclei at $A = 5$ or $A = 8$ prevents the production of heavy elements by two-body reactions between $^1\mathrm{H}$ and $^4\mathrm{He}$. Primordial nucleosynthesis therefore stops at $A = 7$. The production of heavy elements occurs in stars where the triple-α reaction $3\,^4\mathrm{He} \rightarrow\,^{12}\mathrm{C}$ takes place.

nucleus	B/A (MeV)	n_x/n_H primordial (observed)	half-life	decay mode
p	0	1	$> 10^{32}\,\mathrm{yr}$	
n	0	0	$10.24\,\mathrm{min}$	$\mathrm{n} \rightarrow \mathrm{pe}^-\bar{\nu}_\mathrm{e}$
$^2\mathrm{H}$	1.11	$\sim 5 \times 10^{-5}$		
$^3\mathrm{H}$	2.83	0	$12.3\,\mathrm{yr}$	$^3\mathrm{H} \rightarrow\,^3\mathrm{He}\;\mathrm{e}^-\bar{\nu}_\mathrm{e}$
$^3\mathrm{He}$	2.57	?		
$^4\mathrm{He}$	7.07	0.08		
$^5\mathrm{Li}$	5.27	0	$3 \times 10^{-22}\,\mathrm{s}$	$^5\mathrm{Li} \rightarrow \mathrm{p}\,^4\mathrm{He}$
$^6\mathrm{Li}$	5.33	$< 10^{-10}$		
$^7\mathrm{Li}$	5.61	$\sim 3 \times 10^{-10}$		
$^7\mathrm{Be}$	5.37	0	$53.3\,\mathrm{d}$	$\mathrm{e}^-\,^7\mathrm{Be} \rightarrow \nu_\mathrm{e}\,^7\mathrm{Li}$
$^8\mathrm{Be}$	7.06	0	$6.7 \times 10^{-17}\,\mathrm{s}$	$^8\mathrm{Be} \rightarrow\,^4\mathrm{He}\,^4\mathrm{He}$

We will see that the abundances are predicted as a function of the baryon to photon ratio η. The calculated values agree with the best observations for

$$\eta = \frac{n_b}{n_\gamma} \sim 5 \times 10^{-10} \quad \Rightarrow \quad \Omega_b = \frac{\eta n_\gamma(t_0) m_p}{\rho_c} \sim 0.04 . \quad (9.88)$$

This is an extremely important result since it is significantly less than the estimates of the total matter density $\Omega_M \sim 0.3$ implying the existence of nonbaryonic dark matter.

Calculation of the primordial abundances are performed by numerically integrating the appropriate Boltzmann equations for each nuclear species. While this obviously requires a complicated computer code, we can understand things quantitatively because at high temperature most of the nucleons are free and remain so to a surprisingly low temperature, $kT \sim 70\,\mathrm{keV}$. Most of the results can be understood by considering three epochs illustrated in Fig. 9.9:

- $kT > 800\,\mathrm{keV}$. Neutrons and protons are free and in chemical equilibrium implying

$$\frac{n_n}{n_p} \sim \exp\left(-\frac{(m_n - m_p)c^2}{kT}\right) , \quad (9.89)$$

where $(m_n - m_p)c^2 = 1.29\,\mathrm{MeV}$. The chemical equilibrium is possible because of reactions transforming neutrons into protons and vice versa:

$$\nu_e n \leftrightarrow e^- p \qquad \bar{\nu}_e p \leftrightarrow e^+ n . \quad (9.90)$$

The cross-sections for these weak reactions are of the same order of magnitude as that for $\nu\bar{\nu} \leftrightarrow e^+ e^-$ considered in the previous section so the reaction rate per baryon is approximately given by (9.83). The equilibrium is lost when the reaction rates fall below the expansion rate (9.56). The freeze-out temperature turns out to be $kT \sim 800\,\mathrm{keV}$, similar to that for $\nu\nu \leftrightarrow e^+ e^-$. From (9.89) it follows that the neutron–proton ratio at the end of this epoch is of order

$$\frac{n_n}{n_p}(T_f) \sim 0.2 \qquad kT_f \sim 800\,\mathrm{keV} . \quad (9.91)$$

- $800\,\mathrm{keV} > kT > 60\,\mathrm{keV}$. The neutrons decay freely. The duration of this period is $\Delta t = t(60\,\mathrm{keV}) - t(800\,\mathrm{keV}) \sim 3\,\mathrm{min}$ (Exercise 9.5) so about half the neutrons decay leaving a neutron–proton ratio of

$$\frac{n_n}{n_p}(60\,\mathrm{keV}) \sim 0.2\exp(-\Delta t/\tau_n) \sim 0.1 . \quad (9.92)$$

- $kT \sim 60\,\mathrm{keV}$: nucleosynthesis. The remaining neutrons are rapidly incorporated into nuclei via a series of reactions, the most important being:

$$n\,p \to {}^2\mathrm{H}\,\gamma \quad (9.93)$$

$$ {}^2\mathrm{H}\,{}^2\mathrm{H} \to {}^3\mathrm{He}\,n \qquad {}^2\mathrm{H}\,{}^2\mathrm{H} \to {}^3\mathrm{H}\,p$$

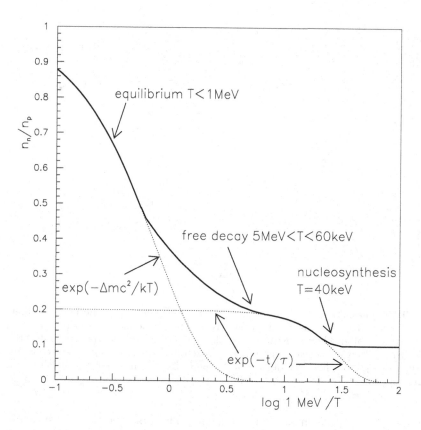

Fig. 9.9. The neutron–proton ratio as a function of temperature, as explained in the text.

$$^{3}\mathrm{He}\,^{2}\mathrm{H} \to {}^{4}\mathrm{He}\,\mathrm{p} \qquad {}^{3}\mathrm{H}\,^{2}\mathrm{H} \to {}^{4}\mathrm{He}\,\mathrm{n} \qquad {}^{2}\mathrm{H}\,^{2}\mathrm{H} \to {}^{4}\mathrm{He}\,\gamma\,.$$

The nuclear abundances versus time are shown in Fig. 9.10. For $\eta \sim 5 \times 10^{-10}$, practically all the neutrons are incorporated into $^{4}\mathrm{He}$, the most bound light nucleus.[4] The number of available neutrons (9.92) therefore determines the quantity of helium:

$$\frac{\rho_{\mathrm{He}}}{\rho_{\mathrm{H}}} = \frac{2n_{\mathrm{n}}/n_{\mathrm{p}}}{1 - n_{\mathrm{n}}/n_{\mathrm{p}}} \sim 0.25 \qquad (\eta \sim 5 \times 10^{-10})\,. \qquad (9.94)$$

There are two obvious questions that we can ask about this scenario: why does nucleosynthesis start so late ($kT \sim 70\,\mathrm{keV}$) and why does it stop so soon

---

[4] The most bound nucleus is abundant not because it is easy to produce but rather because it is very difficult to destroy at $kT \sim 60\,\mathrm{keV}$.

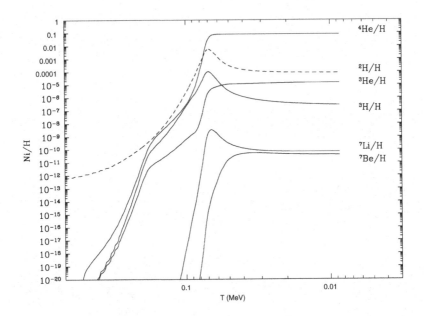

Fig. 9.10. The calculated abundances of the light elements as a function of temperature for $\eta = 3 \times 10^{-10}$ as calculated in [104]. The abundances are negligible until $kT \sim 70\,\mathrm{keV}$, after which most of the available neutrons are incorporated into $^4\mathrm{He}$. After $kT \sim 30\,\mathrm{keV}$, nuclear reactions are frozen and the abundances are constant in time except for the later decays of $^3\mathrm{H}$ and $^7\mathrm{Be}$. Figure courtesy of Elisabeth Vangioni-Flam.

without the production of heavy elements. The first question is especially interesting because the nuclear binding energies are all in the MeV range so it might be expected that nuclei would be produced when $T \sim \mathrm{MeV}$. The reason for the late start is the tiny baryon–photon ratio $\eta \sim 5 \times 10^{-10}$ and its effect on the first step of nucleosynthesis, the formation of $^2\mathrm{H}$ via reaction (9.93). For $\eta \sim 5 \times 10^{-10}$ and for $T \sim \mathrm{MeV}$ the rate per neutron of this reaction is greater than the expansion rate (Exercise 9.3), from which it follows that there is approximate chemical equilibrium between n , p and $^2\mathrm{H}$. Under these conditions, the formation rate of $^2\mathrm{H}$ is equal to the destruction rate by photo-dissociation:

$$n_{\mathrm{p}}\, n_{\mathrm{n}} \langle \sigma_{\mathrm{np}} v \rangle_T \;=\; n_2\, n_\gamma \langle \sigma_{2\gamma} v \rangle_T \;, \tag{9.95}$$

where n_2 is the number density of $^2\mathrm{H}$. Since $n_\gamma \gg n_{\mathrm{p}}, n_{\mathrm{n}}$, (9.95) can be satisfied only if $n_2 \ll n_{\mathrm{p}}, n_{\mathrm{n}}$ to compensate for the large number of photons. This situation persists until the temperature is sufficiently low that $\langle \sigma_{2\gamma} v \rangle_T$ becomes small because very few photons have energies above the threshold for photo-dissociation (2.2 MeV). Using the Saha equation, it can be shown (Exercise 9.3) that $n_2 \ll n_{\mathrm{p}}, n_{\mathrm{n}}$ for $kT > 70\,\mathrm{keV}$. Since heavier nuclei cannot be formed until ^{2}H is formed, it follows that nucleosynthesis cannot start until $kT \sim 70\,\text{keV}$.

The end of nucleosynthesis so soon after its start is due to two effects. The first is the absence of stable or metastable elements with $A = 5$ or $A = 8$ which makes it impossible to form anything from the two primary species, ^{4}He and ^{1}H. The second is the increasing efficiency of the Coulomb barrier between charged nuclei which strongly suppresses the cross-sections for $kT < 60\,\text{keV}$. Nuclear reactions therefore freeze-out at a temperature of $kT \sim 30\,\text{keV}$ with a non-equilibrium relic abundance of nuclei characterized by the complete lack of heavy nuclei.

As we have already mentioned, one of the great interests of the theory of primordial nucleosynthesis is that a comparison with observations permits us to estimate η. The predicted abundances as a function of η are shown in Fig. 9.11. The abundance of ^{4}He, the primary product of primordial nucleosynthesis, is an increasing function of η. The abundances of the loosely bound intermediate nuclei ^{2}H and ^{3}He are decreasing functions of η. The reason for this behavior is quite simple. Nucleosynthesis can proceed only if the reaction rates between nuclei are greater than the expansion rate. The nuclear reaction rates are proportional to densities of initial state nuclei, which are themselves proportional to the total baryon density. For the first reaction (9.93), it is easy to show (Exercise 9.3) that its rate per neutron is smaller than the expansion rate for $\eta < 10^{-13}$ implying that there is essentially no nucleosynthesis for $\eta < 10^{-13}$. For $\eta > 10^{-13}$, nucleosynthesis proceeds with an efficiency that increases with η. More precisely, the temperature of the nuclear freeze-out is a decreasing function of η. The later the freeze-out, the more efficient the destruction of the intermediate nuclei and the production of ^{4}He.

We see in Fig. 9.11 that for $\eta \sim 5 \times 10^{-10}$ the abundance of ^{4}He is rather insensitive to η. This is simply because the great majority of the available neutrons are incorporated into this nucleus. On the other hand, the abundance of ^{2}H is very dependent on η, so a reliable measurement of the primordial abundance of ^{2}H would yield a robust measurement of η.

It is in principle simple to measure the quantity of ^{2}H in intergalactic clouds by measuring the relative absorption by deuterium and hydrogen of photons from background objects. The brightest background objects are quasars, believed to be massive black holes radiating photons as matter falls into them. Figure 9.12 shows the spectrum of a high-redshift quasar. The spectrum exhibits a prominent "Lyman-α" emission line corresponding to the $n = 2 \to n = 1$ states of atomic hydrogen, caused by fluorescence of material surrounding the black hole. The line is at $\lambda = 121\,\text{nm}$ but in the spectrum it is redshifted to $580\,\text{nm}$. Blueward of the quasar's Lyman-α emission, we see the forest of lines corresponding to Ly-α absorption in individual clouds along the line-of-sight. The formation of these absorption lines is illustrated in Fig. 9.13.

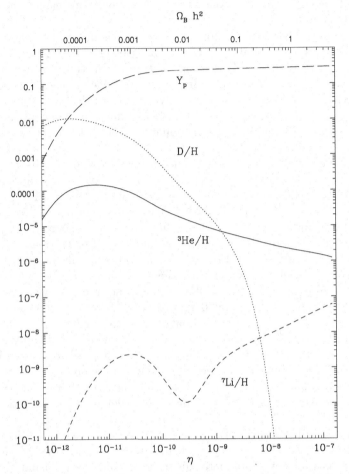

Fig. 9.11. The abundances of the light elements as a function of the assumed baryon–photon ratio η (Bottom Horizontal Axis) or of the assumed value of $\Omega_b h^2$ (Top Horizontal Axis), as calculated in [104]. For ^{4}He, the abundance is given as the fraction $Y_p = \rho_{\mathrm{He}}/\rho_b$ of the total baryonic mass that is in the form of ^{4}He, while the other elements are reported as number densities normalized to ^{1}H. The abundance by mass of ^{4}He is a slowly increasing function of η. The abundances of the loosely bound intermediate nuclei ^{2}H and ^{3}He are decreasing functions of η. The form of the curve for ^{7}Li is due to the fact that the production is mostly direct for $\eta < 3 \times 10^{-10}$ and mostly indirect via production and subsequent decay of ^{7}Be for $\eta > 3 \times 10^{-10}$. Observations [105] indicate that ^{2}H/^{1}H $\sim 3.4 \times 10^{-5}$ in high-redshift intergalactic clouds. If this figure reflects the primordial abundance, we can conclude that $\eta \sim 5 \times 10^{-10}$ corresponding to $\Omega_b \sim 0.04$ ($\Omega_b \sim 0.04$). Figure courtesy of Elisabeth Vangioni-Flam.

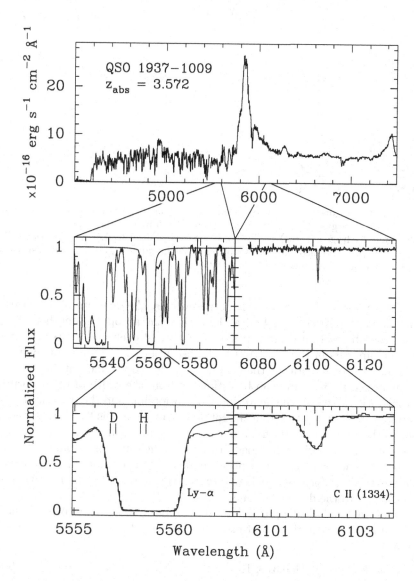

Fig. 9.12. A quasar spectrum showing Ly-α emission at 580 nm and, blueward of this line, the "forest" of Ly-α absorption lines by intervening gas clouds [105]. The zoom on the left shows Ly-α hydrogen and deuterium absorption by one cloud. The ^{2}H line is shifted with respect to the hydrogen line because the atomic energy levels are proportional to the reduced electron–nucleus mass. The ratio between the hydrogen and deuterium absorption can be used to determine the two abundances within the cloud. Courtesy of D. Tytler.

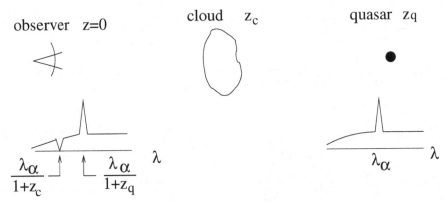

Fig. 9.13. The formation of absorption lines in quasar spectra by intervening clouds of intergalactic gas. The emission at λ_α corresponding to the $n = 2 \to n = 1$ line of atomic hydrogen (right) is observed to be redshifted to $\lambda_\alpha/(1 + z_q)$ (left). The observed spectrum also shows absorption in the cloud at $\lambda_\alpha/(1 + z_c)$ corresponding to the inverse transition, $n = 1 \to n = 2$.

The amount of absorption is determined by the quantity of hydrogen in the cloud. The trick is to find a cloud that has an optical depth that permits the observation of absorption by both hydrogen and deuterium. Such a cloud appears in the spectrum with absorption at 555.8 nm for hydrogen. The absorption is total at the center of the hydrogen line but the quantity of hydrogen in the cloud can be estimated from the width of the absorption profile. Also visible is the deuterium absorption line and the quantity of deuterium can be estimated from the total absorption at this line. The spectrum shows that only clouds within a narrow range of optical depth can be used to measure the deuterium–hydrogen ratio. The measurement would be impossible for clouds with less absorption (making the deuterium line too weak to be observed) or for clouds with more absorption (causing the hydrogen line to widen into the deuterium line).

Only a handful of appropriate absorption systems have been found in quasar spectra. The three best examples give results that are consistent with [105]:

$$n_2/n_1 = (3.4 \pm 0.3) \times 10^{-5} . \tag{9.96}$$

The quantity of heavy elements in the three high-redshift clouds is very small ($\sim 10^{-3}$ solar abundance) which suggests that the ^{2}H abundance might be unperturbed by stellar nucleosynthesis. If we suppose that (9.96) reflects the primordial abundance, the theory of nucleosynthesis gives a precise value of η and Ω_b:

$$\eta \sim (5.1 \pm 0.3) \times 10^{-10} \quad \Rightarrow \quad \Omega_b = 0.039 \pm 0.002 . \tag{9.97}$$

As is often the case in astrophysics, the cited error is purely formal because the real uncertainty comes from the hypotheses necessary to interpret the data. In this case, it is necessary to suppose that the two lines in Fig. 9.12 near 555.8 nm are correctly identified and to suppose that the measured abundances are primordial. If either hypothesis is false, the measurement must be reinterpreted. For instance, the "deuterium" line could be a hydrogen line of a second cloud of a slightly different redshift. This would cause the observers to overestimate the deuterium and therefore underestimate Ω_b. On the other hand, if the measured deuterium is not primordial, the primordial deuterium is underestimated since stellar processes generally destroy deuterium. This would cause an overestimation of Ω_b.

It is clear that the value of η derived from the ^{2}H abundance requires confirmation from independent measurements. The measured abundances of ^{4}He and ^{7}Li (Table 9.3) give qualitative confirmation though some controversy continues [106]. The total amount of gas in the Ly-α forest gives a lower limit on Ω_b that is consistent with the nucleosynthesis value [107]. Finally, the spectra of CBR anisotropies [94] favors a similar value.

We end this section with some comments on how the results of cosmological nucleosynthesis depends on the fundamental constants. The situation is similar to that in astrophysical nucleosynthesis where the production of elements with $A > 8$ depended strongly on the alignment of states of ^{8}Be, ^{12}C and ^{16}O (Fig. 8.5). In the cosmological case, we have already mentioned two important facts

- $m_n > m_p$ by 0.1%. If this mass-ordering were reversed, the now stable neutron would be more abundant than protons when the weak-interactions froze their ratio. Not suffering from the Coulomb barrier, these neutrons would attach themselves to ^{2}H to form ^{3}H which could fuse to form ^{4}He by the reaction ^{3}H ^{3}H $\rightarrow$ ^{4}He nn. The end products of cosmological nucleosynthesis could then be a mixture of stable neutrons and ^{4}He. Later, in stars, nuclear burning would start with n n $\rightarrow$ ^{2}H $e^- \bar{\nu}_e$. Once heavy elements are formed, the free neutrons would be rapidly absorbed by radiative capture.
- The proton–proton system is unbound by $\sim$ 50 keV. If it were bound, diprotons would be formed that would then β-decay to ^{2}H. The end product of cosmological nucleosynthesis would be a mixture of ^{4}He, ^{3}He and any ^{2}H that failed to fuse to form helium. Later, in stars, nuclear burning would start with ^{2}H ^{2}H $\rightarrow$ ^{4}He γ since no weak interactions are required.

In the $m_n > m_p$ scenario, very little hydrogen would be available for the development of life based on organic chemistry. The pp-stable scenario is a bit less hopeless since organic chemistry could still be possible with the small amount of surviving ^{2}H. Note that the fact that stellar nuclear burning would not have to start with a weak interaction does not mean that stars would burn faster. Indeed, we saw in Chap. 8 that the luminosity of a star does not depend on the cross-section for the nuclear reactions producing the luminosity. The higher d-d cross-section compared to p-p cross-section would just mean that hydrogen-burning stars would burn at a lower temperature and be stable at a larger radius.

The results of cosmological nucleosynthesis depend on another relation between the weak interactions and gravity. The neutron–proton ratio was frozen when the expansion rate, $\sim (G(kT)^4/(\hbar c)^3)^{1/2}$, was equal to the weak-interaction rate, $\sim G_{\mathrm{F}}^2 (kT)^5/(\hbar^7 c^6)$. The fact that the freeze-out, $\sim 0.8\,\mathrm{MeV}$ is very near the neutron–proton mass difference $\sim 1.29\,\mathrm{MeV}$ is due to the "coincidence"

$$\frac{G_{\mathrm{F}}^2 (\Delta mc^2)^3}{\sqrt{G}\hbar^{11/2} c^{9/2}} \sim 1 \,. \tag{9.98}$$

If the left side were much greater than unity, the weak interactions would have maintained chemical equilibrium longer, resulting in a smaller neutron–proton ratio. Only hydrogen would have been produced. On the other hand, if the left side were much smaller than unity, chemical equilibrium would have been broken sooner, leading to equal numbers of neutrons and protons. Only ^{4}He would have been produced. In fact, nature finds itself just at the frontier between these two scenarios resulting in the production of an "interesting" mixture of hydrogen and helium.

9.8 Wimps

The three known neutrino species were relativistic when they decoupled ($kT_{\mathrm{f}} \gg m_{\mathrm{v}} c^2$). The consequence of this is that today the number density of neutrinos is of the same order of magnitude as that of photons $n_{\mathrm{v}} = (3/11)n_{\gamma}$. If one of these neutrinos is sufficiently massive to be non-relativistic today, its present mass density would be $\rho_{\mathrm{v}} = m_{\mathrm{v}} n_{\mathrm{v}}$. This gives $\Omega_{\mathrm{M}} \sim 0.3$ for a neutrino mass of order $10\,\mathrm{eV}/c^2$.

Any hypothetical stable particle with $mc^2 > 10\,\mathrm{eV}$ that decoupled when it was relativistic would create problems for cosmology because the calculated mass density would be overcritical. A heavy weakly interacting particle can give an appropriate cosmological density only if it has an annihilation cross-section sufficiently large to keep it in equilibrium until the particle was non-relativistic, $kT_{\mathrm{f}} \ll mc^2$. In this case its number density would be suppressed by the Boltzmann factor and the present density might not be too large. In fact, if the cross-section is chosen correctly, the particle can give a relic density near critical and constitute the desired nonbaryonic dark matter.

Such a compensation between relic density and mass would seem a priori improbable, but stranger things have happened in cosmology. In fact, as it turns out, particles with weak interaction and masses in the GeV range naturally give relic densities within an order of magnitude or so of critical. Such particles are called "wimps" for "weakly interacting massive particle."

A stable wimp is generally predicted by supersymmetric extensions of the standard model of particle physics. The particle is called the "LSP" (lightest supersymmetric particle) and is denoted by χ. Supersymmetric wimps are usually "Majorana" particles, i.e. they are their own antiparticle. Supersymmetric theories have many parameters that are not (yet) fixed by experiment so one can generally choose parameters that give a cross-section yielding the required relic density. The fact that they have not been seen at accelerators means that probably $m_\chi c^2 > 30\,\text{GeV}$ [102].

As with electrons and positrons, an approximate solution of wimp Boltzmann equation is

$$n_\chi \sim n_\chi(T) \qquad\qquad T \gg T_f \qquad\qquad\qquad (9.99)$$

$$n_\chi a^3 \sim \text{constant} \qquad\qquad T \ll T_f\,, \qquad\qquad\qquad (9.100)$$

where T_f is the freeze-out temperature corresponding to the moment when the annihilation rate was equal to the expansion rate:

$$n_\chi(T_f)\,\langle \sigma v\rangle \;=\; H(T_f) \;\sim\; \left(\frac{GT_f^4}{(\hbar c)^3}\right)^{1/2}. \qquad\qquad (9.101)$$

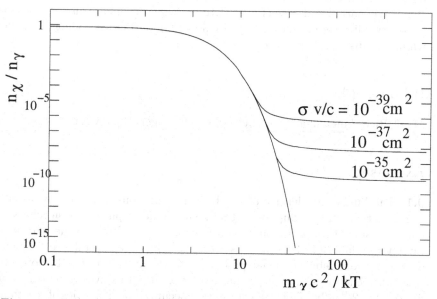

Fig. 9.14. $\log(n_\chi/n_\gamma)$ versus temperature for three values of the annihilation cross-section $\sigma v/c$. The wimp mass was taken to be $m_\chi c^2 = 50\,\text{GeV}$. The dotted line shows $\log(n_\chi/n_\gamma)$ in thermal equilibrium. Freeze-out occurs around $T_f \sim m_\chi/20$. We see that to good approximation the relic density is inversely proportional to the cross-section.

The numerical solution is shown for three values of the cross-section in Fig. 9.14. We see that, because of the exponential dependence of $n_\chi(T)$ for

$T < m_\chi$, the freeze-out temperature is relatively insensitive to the cross-section, $T_f \sim m_\chi/20$. We can therefore easily estimate the χ relic density by equating the annihilation rate and the expansion rate. This gives:

$$\frac{n_\chi(a_f)}{n_\gamma(a_f)} \sim \frac{n_\chi(a_f)}{T_f^3} \propto \frac{1}{\langle \sigma v \rangle m_\chi} . \tag{9.102}$$

We see that the ratio is inversely proportional to the wimp mass and annihilation cross section. This ratio stays roughly constant after the freeze so today's wimp-proton ratio is of the same order of magnitude. Using the present-day photon density, this gives a present-day wimp mass density of

$$\rho_\chi = n_\chi m_\chi \propto \frac{1}{\langle \sigma v \rangle} . \tag{9.103}$$

It turns out that the a cross-section of $\langle \sigma v/c \rangle \sim 10^{-41}\,\mathrm{m}^2$ gives the observed density of cold dark matter. Supersymmetric models that give this cross-section can be constructed.

If it turns out that such supersymmetric models describe nature, the dark-matter mystery would be solved. Efforts are underway to detect supersymmetric particles at accelerators (LHC) . Experiments are also attempting to observe directly Galactic wimps by detecting particles recoiling from rare wimp-nucleus scattering events (Exercise 9.7).

Bibliography

1. J. Rich, Fundamentals of Cosmology, Springer, Berlin, 2001.

Exercises

9.1 The luminosity density of the Universe (photons produced by stars)is $\sim 1.2 \times 10^8 L_\odot \,\mathrm{Mpc}^{-3}$. Supposing that stellar light output has been relatively constant since the formation of the first stars one Hubble time ago, estimate the number of photons ($E \sim 2$ eV) that have been produced by stars. Compare the number of stellar photons with the number of CBR photons.

Stellar energy is mostly produced by the fusion of hydrogen to helium $4p \to {}^4\mathrm{He} + 2e^+ + 2\nu_e$. This transformation occurs through a series of reactions in stellar cores that liberate a total of ~ 25 MeV. After thermalization, the energy emerges from stellar surfaces in the form of starlight. Estimate the number of protons (per Mpc^3) that have been transformed into helium over the last Hubble time. Compare this number with the number of protons available $n_b \sim \Omega_b \rho_c/m_p$.

9.2 Estimate the contribution to the universal photon mean free path of the following processes:

- Thomson scattering of photons on free electrons of number density $n_e \sim n_b$.
- Absorption by stars of number density $n_{\text{stars}} \sim \Omega_{\text{stars}} \rho_c / M_\odot$, $\Omega_{stars} \sim 0.003$ and cross-section $\sim \pi R_\odot^2$.
- Absorption by dust in galaxies with $n_{\text{gal}} \sim 0.005 \, \text{Mpc}^{-3}$ and cross-section $\sim \epsilon \pi R_{\text{gal}}^2$ where $R_{\text{gal}} \sim 10 \, \text{kpc}$ and the fraction of visible light absorbed when passing through a galaxy is $\epsilon \sim 0.1$.

Compare these distances with the "Hubble distance," $d_H \equiv c/H_0$ ($\sim$ the distance of the most distant visible objects). Is the Universe "transparent"?

9.3 The cross-section times velocity for the reaction $np \to {}^2\text{H}\gamma$ is

$$\sigma v \sim 7.4 \times 10^{-20} \, \text{cm}^3 \, \text{s}^{-1} \qquad (v \ll c) . \qquad (9.104)$$

(a) Show that the rate per neutron of this reaction is smaller than the expansion rate at $kT \sim 60 \, \text{keV}$ if $\eta < 4 \times 10^{-12}$. It follows that there is no nucleosynthesis if η is less than this value.

If $\eta > 4 \times 10^{-12}$, ${}^2\text{H}$ is in thermal equilibrium with neutrons and protons. The abundances of ${}^2\text{H}$, protons and neutrons are governed by the Saha equation:

$$\frac{n_2}{n_p n_n} = (2\pi\hbar c)^3 \left(\frac{2\pi}{m_n c^2 kT} \frac{2\pi}{m_p c^2 kT} \frac{m_2 c^2 kT}{2\pi} \right)^{3/2} e^{B/T}$$

where $B = 2.2 \, \text{MeV}$ is the ${}^2\text{H}$ binding energy.
(b) Show that for $\eta \sim 5 \times 10^{-10}$ the great majority of neutrons are free until $kT \sim 60 \, \text{keV}$. (Since the majority of baryons are protons, you can approximate $n_p \sim \eta n_\gamma$.)

9.4

To derive (9.47) in the case of a Universe dominated by non-relativistic matter $(p \sim 0)$, we can use a simple Newtonian argument. Referring to Fig. 9.15, we place a galaxy of mass m at a distance $R = \chi a(t)$ from the "center" of a universe of uniform density ρ. Since the mass distribution is spherically symmetric, Gauss's theorem "suggests" that the galaxy is subject to a gravitational force directed toward the origin that is proportional to the total mass at a distance $< \chi a(t)$ from the origin:

$$|\boldsymbol{F}| = \frac{GM(\chi)m}{\chi^2 a^2} , \qquad (9.105)$$

where

$$M(\chi) = 4\pi(\chi a)^3 \rho/3 . \qquad (9.106)$$

(We ignore the question of whether Gauss's theorem applies in an infinite medium.) Using $|\boldsymbol{F}| = m\ddot{R} = m\chi\ddot{a}$, find the equation (9.47) for the deceleration of the Universe.

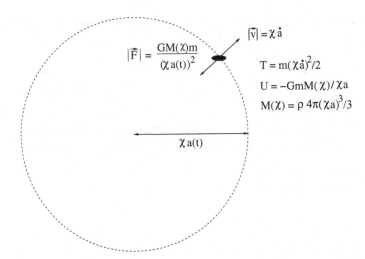

Fig. 9.15. A Newtonian treatment of the universal expansion. A galaxy of mass m placed in a universe of uniform density ρ at a distance $\chi a(t)$ from the "center of the Universe." The spherical symmetry suggests that the Newtonian force on the galaxy will be directed toward the origin with a magnitude $F = GM(\chi)m/(\chi a(t))^2$, where $M(\chi)$ is the total mass at a distance $< \chi a(t)$ from the origin. For a uniform mass density ρ, $M(\chi) = 4\pi(\chi a)^3\rho/3$.

9.5 Estimate the time available for neutron decay, corresponding to the temperature range $800\,\text{keV} > kT > 50\,\text{keV}$, by integrating (9.56) between these limits and using the approximation $a \propto 1/T$. Take $g(T) = 4$ is an appropriate mean value over the interval.

9.6 Consider a universe with equal numbers of electrons and positrons. If the universe is in thermal equilibrium at $kT \ll m_e c^2$, use the principle of detailed balance (Sect. 4.1.5) applied to the reaction $e^+ e^- \leftrightarrow \gamma\gamma$ to show that the number density of electrons is proportional to a Boltzmann factor $\exp(-m_e c^2/kT)$. For $kT \ll m_e c^2$, explain why it is a good approximation to ignore the effects of stimulated emission.

9.7 The dark matter in Milky Way is believed to be made up of weakly-interacting massive particles (wimps). The mass density of galactic wimps at the position of the Earth is estimated to be $\rho c^2 \sim 0.3\,\text{GeV}\,\text{cm}^{-2}$. The wimps move around the galaxy with the same mean speed as stars, $v \sim 10^{-3}c$. Suppose that the wimps have a mass of $50\,\text{GeV}/c^2$. What is their typical kinetic energy? In a collision with a nucleus, would you expect a wimp to be able to excite or break the nucleus?

Suppose that the elastic wimp cross section is $10^{-35}\,\mathrm{cm}^2$ on moderate mass nuclei. What is the mean free path of wimps in the Earth?

Experiments have attempted to detect the wimps with germanium detectors via the elastic scattering

$$\chi\,\mathrm{Ge} \;\rightarrow\; \chi\,\mathrm{Ge}\,. \tag{9.107}$$

The recoiling nucleus creates a signal in the diode. What is the scattering rate for the assumed cross-section? Compare this rate the decay rate of cosmogenic $^{68}\mathrm{Ge}$, $\sim 0.3\,\mathrm{mBq\,kg^{-1}}$ (Sect. 5.2.2).

Current experiments [103] have not observed a signal that must be ascribed to wimp scattering. This negative result places an upper limit on the wimp scattering cross-section.

A. Relativistic kinematics

In this appendix, we briefly review some facts from the special theory of relativity that are useful in nuclear physics. Relativity is used in nuclear physics primarily through the relativistic expressions for the energy and momentum of a free particle of (rest) mass m and velocity $\boldsymbol{v}$:

$$E = \frac{mc^2}{\sqrt{1 - v^2/c^2}} \qquad \boldsymbol{p} = \frac{m\boldsymbol{v}}{\sqrt{1 - v^2/c^2}} . \tag{A.1}$$

The energy and momentum defined in this way are conserved quantities. They satisfy

$$E^2 = m^2 c^4 + p^2 c^2 \qquad \frac{v}{c} = \frac{pc}{E} . \tag{A.2}$$

In nuclear physics, the non-relativistic limit $v \ll c$ ($\Rightarrow pc \ll E$) usually applies for nuclei, in which case we have

$$E \sim mc^2 + \frac{p^2}{2m} \qquad v = \frac{p}{m} . \tag{A.3}$$

For neutrinos and photons, the limit $mc \ll p$ generally applies:

$$E \sim pc + \frac{m^2 c^2}{2p^2} . \qquad v = c \left(1 - \frac{m^2 c^2}{2p^2} \right) . \tag{A.4}$$

It is customary to group energy and momentum in a single object called the energy-momentum *4-vector*

$$P \equiv (E, \boldsymbol{p}) . \tag{A.5}$$

In a particle's rest-frame, it takes the value $(mc^2, 0, 0, 0)$. The squared magnitude of the 4-vector is defined as

$$P^2 \equiv P \cdot P \equiv E^2 - \boldsymbol{p} \cdot \boldsymbol{p} = m^2 c^4 , \tag{A.6}$$

where the last form follows from (A.1). The magnitude is clearly independent of the energy of the particle, i.e. it is invariant with respect to changes of reference frame.

Consider the energy-momentum of a particle, P, viewed in an inertial reference frame. Consider another inertial reference frame moving with velocity v in, say, the z direction with respect to the first. The energy-momentum

4-vector in the second is related to that in the first by a *Lorentz transformation*:

$$
\begin{pmatrix} E' \\ p'_x c \\ p'_y c \\ p'_z c \end{pmatrix} = \begin{pmatrix} \gamma & 0 & 0 & -\beta\gamma \\ 0 & 1 & 0 & 0 \\ 0 & 0 & 1 & 0 \\ -\beta\gamma & 0 & 0 & \gamma \end{pmatrix} \begin{pmatrix} E \\ p_x \\ p_y \\ p_z \end{pmatrix} ,
\tag{A.7}
$$

where

$$
\beta = v/c \qquad \gamma = \frac{1}{\sqrt{1 - \beta^2}} .
\tag{A.8}
$$

The generalizations to other directions are obvious. This relation follows trivially from (A.1) if one of the frames is the rest-frame. It is less obvious in the general case but note that the transformation has the virtue of maintaining the magnitude (A.6), as it must.

Energy momentum conservation can be economically expressed by using 4-vectors. Consider the decay

$$
A \rightarrow BC .
\tag{A.9}
$$

Energy-momentum conservation is

$$
P_A = P_B + P_C ,
\tag{A.10}
$$

which is entirely equivalent to

$$
E_A = E_B + E_C \qquad \boldsymbol{p}_A = \boldsymbol{p}_B + \boldsymbol{p}_C .
\tag{A.11}
$$

One is often called upon to calculate the momentum of the decay products in the rest frame of the decaying particle ($\boldsymbol{p}_A = 0$). Momentum conservation gives $\boldsymbol{p}_B = -\boldsymbol{p}_C$ so energy conservation gives

$$
m_A c^2 = \sqrt{p^2 c^2 + m_B^2 c^4} + \sqrt{p^2 c^2 + m_C^2 c^4} ,
\tag{A.12}
$$

where p is the common momentum we would like to find. This equation is not especially easy to solve. It is much easier to write the 4-vector equation

$$
P_C = P_A - P_B .
\tag{A.13}
$$

We now take the squared magnitude of both sides of this equation:

$$
m_C^2 c^4 = (P_A - P_B)^2 = P_A^2 + P_B^2 - 2P_A \cdot P_B .
\tag{A.14}
$$

The first two terms on the right give $m_A^2 c^4 + m_B^2 c^4$. Since the scalar product $P_A \cdot P_B$ is Lorentz invariant, we can evaluate it in the rest frame of A:

$$
P_A \cdot P_B \equiv E_A E_B - \boldsymbol{p}_A \cdot \boldsymbol{p}_B = m_A c^2 \sqrt{m_B^2 c^4 + p^2 c^2} .
\tag{A.15}
$$

We thus deduce

$$
\sqrt{p^2 c^2 + m_B^2 c^4} = \frac{m_A^2 c^4 + m_B^2 c^4 - m_C^2 c^4}{2 m_A c^2} ,
\tag{A.16}
$$

$$p^2 c^2 = \left(\frac{m_A^2 c^4 + m_B^2 c^4 - m_C^2 c^4}{2 m_A c^2} \right)^2 - m_B^2 c^4 \, . \tag{A.17}$$

We note that in nuclear physics we can often use directly energy conservation (A.12) because all the particles are either ultra-relativistic or non-relativistic so we can eliminate the square roots. For example, in radiative decay of an excited nucleus

$$(A, Z)^* \rightarrow (A, Z)\, \gamma \, , \tag{A.18}$$

the two nuclei are non-relativistic so energy conservation is

$$m_* c^2 = m c^2 + \frac{p^2}{2m} + p c \, , \tag{A.19}$$

The nuclear kinetic energy is $p v / 2 \ll p c$ so we have immediately

$$p c \sim (m_* - m) c^2 \, . \tag{A.20}$$

This also follows from (A.17) in the limit $m_A^2 - m_C^2 \sim 2 m_A (m_A - m_C)$ and $m_B = 0$.

B. Accelerators

The scattering experiments discussed in Chap. 3 generally required the use of beams of charged particles produced by accelerators. A notable exception is the original Rutherford-scattering experiments that used α-particles from natural radioactive decays. Neutron-scattering experiments use neutrons produced at fission reactors or secondary neutrons produced by the scattering of accelerated charged particles.

Particle accelerators require a source of charged particles and an electric field to accelerate them. They can be classified as *DC* machines using static electric fields and *AC* machines using oscillating fields. The second category can be divided into *linear accelerators* where particles are accelerated in straight line and *magnetic accelerators*, i.e. cyclotrons and synchrotrons, where particles move in circular orbits.

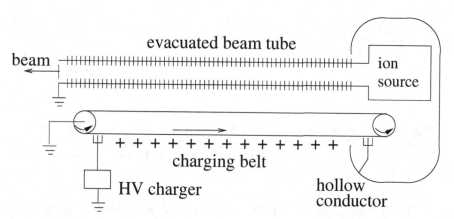

Fig. B.1. A schematic of a simple Van de Graff accelerator. Positive charges are transfered from ground potential to a hollow terminal. The ion source is placed inside the terminal and particles are accelerated through the electrostatic field to ground potential.

In simple electrostatic systems, an ion source is placed at high voltage and extracted ions are accelerated through the electric field. Potentials of $1-2\,\mathrm{MV}$ can be produced with normal rectifier circuits and potentials up to 10 MV can be produced in a *Van de Graff* accelerator, illustrated in Fig. B.1. In this system, charge is transformed to the positive terminal by an insulating belt. Ions are accelerated through an evacuated tube constructed from alternating insulators and electrodes so as to maintain a constant gradient. The maximum potential is limited by breakdown in the surrounding gas. Currents in the mA range can be achieved.

Tandem Van de Graff Accelerators (Fig. B.2) modify the basic design to provide higher energy and an ion source that is at ground potential, making it more accessible. In this case, the source provides singly-charged negative ions, e.g. O^- containing an extra electron. These are accelerating to the positive terminal where a "stripper" consisting of a thin foil or gas-containing tubes removes electrons. The resulting positive ions are then accelerated to ground potential where an analyzing magnet selects a particular value of q/m. Obtainable currents are in the μA range, smaller than simple Van de Graffs because of the difficulty in obtaining negative ions.

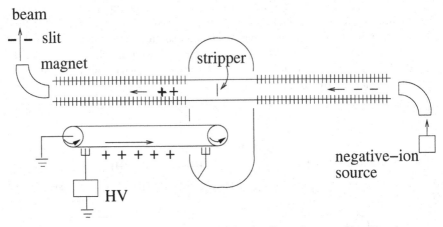

Fig. B.2. A schematic of a tandem Van de Graff accelerator. Negative ions are accelerated to the positive potential where a "stripper" removes electrons. The resulting positive ions are then accelerating to ground potential where a definite charge state is selected by a magnetic field and slit.

The 10 MV limitation of DC machines can be avoided by using radio-frequency (RF) electric fields. The frequency is typically ∼ 30 MHz. The simplest configuration is the linear accelerator, or *linac*, illustrated in Fig. B.3. The RF voltage is applied to alternating conducting "drift tubes" so that charged particles are accelerating between tubes if they arrive at the gaps at appropriate times. The tube lengths must thus decrease in length as the particle velocity increases down the accelerator. Linacs produce a "bunched" beam consisting of pulses of particles. The bunch structure is persists during the acceleration because of the "phase stability" illustrated in Fig. B.4.

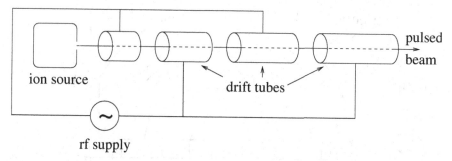

Fig. B.3. A schematic of a drift-tube linear accelerator. Ions are accelerated in the alternating electric field between drift-tubes.

Linear accelerators are most commonly used to accelerate electrons. The largest is the 2-mile long accelerating SLAC at Stanford, California, that produces 20 GeV electrons.

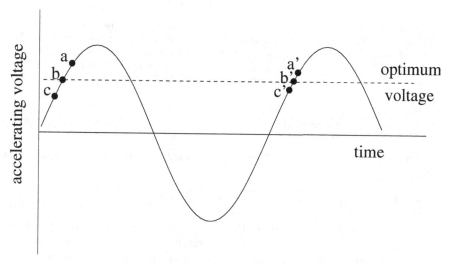

Fig. B.4. The principle of phase stability in a linear accelerator. Particles arriving in a gap at point b are accelerated such that they arrive at the text gap at the point b' with the same phase with respect to the alternating field. Particles arriving at point a (c) receive more (less) acceleration and therefore arrive relatively earlier (later) in the next gap, point a' (c'). Particles in the range a-c are thus "focused" in phase-space toward the point b.

Cyclotrons are a common class of accelerators illustrated in Fig. B.5. Ion orbit in a dipole magnetic field where they are accelerated twice per orbit in a RF field. As they are accelerated, the ions spiral out with the radius of curvature given by

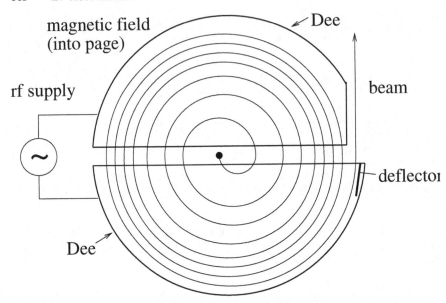

Fig. B.5. Schematic of a cyclotron. Particles are injected near the center of a dipole magnetic field and then spiral outward as they gain energy each time they pass through the alternating electric field between two electrodes called "Dee's." The radio-frequency is tuned to the particle's cyclotron frequency, $\omega_{\mathrm{c}} = qB/m$. Near the maximum radius, the particles are deflected out of the cyclotron.

$$R = \frac{mv}{qB\sqrt{1 - v^2/c^2}} \, , \tag{B.1}$$

for a particle of velocity v, mass m and charge q. The orbital frequency is then

$$f_{\mathrm{c}} = \frac{v}{2\pi R} = \frac{qB}{2\pi m}\sqrt{1 - v^2/c^2} \, , \tag{B.2}$$

and the RF must be tuned to this frequency to accelerate the particles. As long as the particle remains non-relativistic, $v \ll c$, the frequency is a constant, proportional to the *cyclotron frequency*, qB/m, equal to $9.578 \times 10^7 \mathrm{rad\,s^{-1}\,T^{-1}} \times B$. The energy at radius R is, for $v \ll c$

$$\frac{1}{2}mv^2 = \frac{1}{2}m\left(\frac{qB}{m}\right)^2 R^2 \sim 10\,\mathrm{MeV}\left(\frac{B}{1\,\mathrm{T}}\right)^2\left(\frac{R}{0.5\,\mathrm{m}}\right)^2 \, , \tag{B.3}$$

where the numerical example is for a proton. A modest-sized cyclotron can therefore produce particles of energies interesting for nuclear-physics experiments. Currents in the mA range can be produced.

The simple design of a constant-field cyclotron must be modified in practical designs for a number of reasons. Most important is the necessity to prevent the particles from spiraling in the vertical direction. This can be prevented by introducing a small radial variation of the field, as illustrated in

Fig. B.6. This introduces vertical components to the force on particles that are not in the median plane so that particles are focused in the vertical direction. Unfortunately, this simple scheme introduces other problems, among them being that the required RF frequency now depends on position. More popular focusing schemes use magnetic fields that vary azimuthally to obtain the desired effect.

For energies > 1 GeV, cyclotrons become impractical because of the large required radius. It then becomes more practical to use *synchrotron's* where ring of dipole magnets replace the one large dipole. The accelerating force is provided by RF cavities distributed about the ring in spaces between magnets. Vertical and horizontal focusing is provided by quadrupole magnets. Sychrotrons are the most common accelerators in the field of high-energy particle physics.

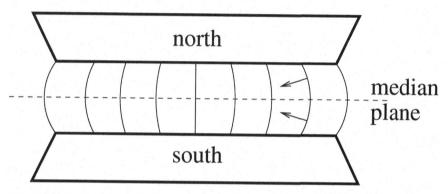

Fig. B.6. Vertical focusing in a radially-decreasing dipole magnetic field. Particles in the median plane experience a horizontal force. The force on particles above or below the median plane has a vertical component that pushes the particle back toward the median.

C. Time-dependent perturbation theory

Perturbation theory is the basis for most of the calculations performed in Chaps 3 and 4. Here we derive the basic equations.

C.0.1 Transition rates between two states

Consider a system described by a Hamiltonian H that is the sum of a "non-perturbed" Hamiltonian H_0 and a small perturbation H_1 which can induce transitions between various eigenstates of H_0. It is useful to express the state of the system as a superposition of eigenstates of H_0:

$$|\psi(t)\rangle = \sum_i \gamma_i(t) e^{-iE_i t/\hbar} |i\rangle , \tag{C.1}$$

where

$$H_0|i\rangle = E_i|i\rangle . \tag{C.2}$$

We suppose that the system is initially in the state $|i\rangle$:

$$\gamma_i(t = 0) = 1 \qquad \gamma_{j \neq i}(0) = 0 . \tag{C.3}$$

At a later time, it has an amplitude $\gamma_f(t)$ to be in some other state $|f\rangle$. This amplitude can be calculated using the Schrödinger equation

$$i\hbar \frac{d}{dt}|\psi(t)\rangle = (H_0 + H_1)|\psi(t)\rangle . \tag{C.4}$$

Substituting (C.1) into this equation, multiplying on the left by $\langle f|$ and using $\langle i|j\rangle = \delta_{ij}$, we find a differential equation for $\gamma_f(t)$:

$$i\hbar \frac{d\gamma_f(t)}{dt} = \sum_k \gamma_k(t)\langle f|H_1|k\rangle e^{i(E_f - E_k)t} . \tag{C.5}$$

This equation is exact but can only be solved numerically. A perturbative solution for small t is found by using (C.3) to use the first approximation $\gamma_i(t) = 1$:

$$i\hbar \frac{d\gamma_f(t)}{dt} = \langle f|H_1|i\rangle e^{i(E_f - E_i)t} . \tag{C.6}$$

This equation can be directly integrated to give the first order time-dependent perturbation theory estimate of γ_f:

$$\gamma_f^1(t) = \frac{-2i}{\pi} e^{i(E_f - E_i)t/2\hbar} \langle f|H_1|i\rangle \, \Delta_t(E_f - E_i) \,. \tag{C.7}$$

In this expression we have introduced a limiting form of the Dirac distribution

$$\Delta_t(E_f - E_i) = \frac{1}{\pi} \frac{\sin(E_f - E_i)t/2\hbar}{E_f - E_i} \tag{C.8}$$

which we discuss below.

Squaring this amplitude, we find the probability that the system is in the state f at time t

$$P_{if}(t) = \frac{2\pi t}{\hbar} |\langle f|H_1|i\rangle|^2 \delta_t(E_f - E_i) \,, \tag{C.9}$$

where $\delta_t(E)$, which we will discuss below, is a function that is peaked at $E = 0$ with a width $\Delta E \sim \hbar/t$:

$$\delta_t(E) = \frac{1}{\pi} \frac{\sin^2(Et/2\hbar)}{E^2 t/2\hbar} \,. \tag{C.10}$$

In the limit $t \to \infty$ δ_t approaches the Dirac delta function:

$$\int_{-\infty}^{\infty} \delta_t(E)dE = 1 \,. \tag{C.11}$$

This means that at large time $[t(E_f - E_i))/\hbar \gg 1]$ the only states that are populated are those that conserve energy to within the Heisenberg condition $\Delta E \, t > \hbar$.

If for some reason the first-order probability vanishes, second-order perturbation theory gives

$$P_{if}(t) = \frac{2\pi t}{\hbar} \left| \sum_{j \neq i,f} \frac{\langle f|H_1|j\rangle\langle j|H_1|i\rangle}{E_j - E_i} \right|^2 \delta_t(E_f - E_i) \,. \tag{C.12}$$

The *transition rate* is found by simply dividing the probability by the time t:

$$\lambda_{if} = \frac{P_{if}(t)}{t} \tag{C.13}$$

Total transition rates are found by summing (C.13) over all final states f.

$$\lambda = \sum_f \frac{P_{if}(t)}{t} = \frac{2\pi}{\hbar} \sum_f |\langle f|H_1|i\rangle|^2 \delta_t(E_f - E_i) \,. \tag{C.14}$$

If the states f form a continuum with $\rho_f(E)dE$ states within the energy interval dE and if all these states have the same matrix element, we can simply replace the sum by an integral and find the *Fermi golden rule*:

$$\lambda = \frac{2\pi}{\hbar} |\langle f|H_1|i\rangle|^2 \rho_f(E_i) \,. \tag{C.15}$$

C.0.2 Limiting forms of the delta function

In the above expressions, it has been useful to introduce the functions :

$$\Delta_T(E) = \frac{1}{\pi} \frac{\sin(ET/2\hbar)}{E} \tag{C.16}$$

and

$$\delta_T(E) = \frac{1}{\pi} \frac{\sin^2(ET/2\hbar)}{E^2 T/2\hbar} \ . \tag{C.17}$$

We note that

$$\int_{-\infty}^{\infty} \Delta_T(E) = 1 \ , \tag{C.18}$$

and

$$\int_{-\infty}^{\infty} \delta_t(E) = 1 \ . \tag{C.19}$$

In the limit $T \to \infty$, these two functions tend, in the sense of distributions, to the Dirac distribution

$$\lim_{T\to\infty} \Delta_T(E) = \lim_{T\to\infty} \delta_T(E) = \delta(E) \ . \tag{C.20}$$

They are related by :

$$(\Delta_T(E))^2 = \frac{T}{2\pi\hbar} \delta_T(E) \ , \quad \forall T \ . \tag{C.21}$$

The generalization to three variables is straightforward:

$$\Delta_L^3(\boldsymbol{p}) = \prod_{i=1}^{3} \Delta_L(p_i) \ , \quad \delta_L^3(\boldsymbol{p}) = \prod_{i=1}^{3} \delta_L(p_i) \ , \tag{C.22}$$

with $\boldsymbol{p} = (p_1, p_2, p_3)$. We have quite obviously

$$\lim_{L\to\infty} \Delta_L^3(\boldsymbol{p}) = \lim_{L\to\infty} \delta_L^3(\boldsymbol{p}) = \delta^3(\boldsymbol{p}) , \tag{C.23}$$

and

$$(\Delta_L^3(\boldsymbol{p}))^2 = \frac{L^3}{(2\pi\hbar)^3} \delta_L^3(\boldsymbol{p}) \quad \forall L \ . \tag{C.24}$$

D. Neutron transport

In this appendix, we give a few more details about neutron transport in matter and the Boltzmann equation used in Sect. 6.7. We refer to the literature[1] for more complete details.

D.0.3 The Boltzmann transport equation

The Boltzmann transport equation governs the behavior of neutrons in matter. We shall write it under the following assumptions:

- The medium is static (neglecting small thermal motions); it is, spherical, homogeneous, and consists of ^{239}Pu nuclei.
- Neutron–neutron scattering is negligible (since the density of neutrons is much smaller than the density of the medium) ;
- Neutron decay is negligible, i.e. the neutron lifetime is very large compared to the typical time differences between two interactions.

The neutrons are described by their density in phase space

$$\frac{\mathrm{d}N}{\mathrm{d}^3p\mathrm{d}^3r} = f(\boldsymbol{r}, \boldsymbol{p}, t) \quad , \tag{D.1}$$

where $\mathrm{d}N$ is the number of neutrons in the phase space element $\mathrm{d}^3p\mathrm{d}^3r$. The space density of neutrons and the current describing the spatial flow of neutrons are the integrals over the momentum

$$n(\boldsymbol{r}, t) = \int f(\boldsymbol{r}, \boldsymbol{p}, t)\mathrm{d}^3p \,,$$

$$\boldsymbol{J}(\boldsymbol{r}, t) = \int \boldsymbol{v} f(\boldsymbol{r}, \boldsymbol{p}, t)\mathrm{d}^3p \,.$$

In the absence of collisions, neutron momenta are time-independent and the flow of particles in phase space is generated by the motion of particle at velocities $\boldsymbol{v} = \boldsymbol{p}/m$. The density f satisfies an equation of the form

---

[1] See for instance, E. M. Lifshitz and L. P. Pitaevskii *Physical Kinetics*, Pergamon Press, 1981; C. Cercignani, *Theory and application of the Boltzmann Equation*, Scottish Academic Press, 1975.

$$\frac{\partial f}{\partial t} + \boldsymbol{v} \cdot \boldsymbol{\nabla} f = \mathcal{C}(f) , \tag{D.2}$$

where $\mathcal{C}(f)$ is the term arising from collision processes, for which we will find an explicit form shortly. For $\mathcal{C}(f) = 0$, (D.2) is called the *Liouville equation*.

The elastic scattering and absorption rates λ_{el} and λ_{abs} are products of the elementary cross-sections, the density of scattering centers n_{239}, and the mean velocity v

$$\lambda_{el} = vn_{239}\sigma_{el} \qquad \lambda_{abs} = vn_{239}\sigma_{abs} \tag{D.3}$$

The absorption is due to both (n, γ) reactions and to fission

$$\sigma_{abs} = \sigma_{(n,\gamma)} + \sigma_{fis} . \tag{D.4}$$

The collision term is then written as

$$\mathcal{C}(f(\boldsymbol{p})) = n_{239} \int d^3 p' v(p') f(\boldsymbol{r}, \boldsymbol{p}', t) \frac{d\sigma}{d^3 \boldsymbol{p}'}(\boldsymbol{p}' \to \boldsymbol{p}) \tag{D.5}$$

$$- [\lambda_{el} + \lambda_{abs}] f(\boldsymbol{r}, \boldsymbol{p}, t) + S(\boldsymbol{r}, \boldsymbol{p}) .$$

The first term accounts for neutrons coming from the elements of phase space $d^3 r d^3 p'$ which enter the element of phase space $d^3 r d^3 p$ by elastic scattering. The second term represents the neutrons which leave the element $d^3 r d^3 p$ either by elastic scattering or by absorption. The last term $S(\boldsymbol{r}, \boldsymbol{p})$ is a source term, representing the production of neutrons by fission.

D.0.4 The Lorentz equation

We recall that we assume the neutrons all have the same time-independent energy, and that the medium is homogeneous.

In that case, the differential elastic scattering cross-section is

$$\frac{d\sigma}{d^3 \boldsymbol{p}'}(\boldsymbol{p} \to \boldsymbol{p}') = p^{-2}\delta(p - p')\frac{d\sigma}{d\Omega} . \tag{D.6}$$

We also assume, for simplicity, that the scattering cross section is isotropic

$$\frac{d\sigma_{el}}{d\Omega} = \frac{\sigma_{el}}{4\pi} . \tag{D.7}$$

Later on, we will also make the assumption that all neutrons have the same velocity, v, i.e. that the function $f(\boldsymbol{r}, \boldsymbol{p})$ is strongly peaked near values of momentum satisfying $|\boldsymbol{p}| = m_n v$.

Using (D.7) we find that the Boltzmann equation (D.2) and (D.5) reduces to the *Lorentz equation*

$$\frac{\partial f}{\partial t} + \boldsymbol{v} \cdot \boldsymbol{\nabla} f = \lambda_{el}(\bar{f} - f) - \lambda_{abs} f + S(\boldsymbol{r}, \boldsymbol{p}) , \tag{D.8}$$

where

$$\bar{f}(\boldsymbol{r}, \boldsymbol{p}, t) = \frac{1}{4\pi} \int f(\boldsymbol{r}, \boldsymbol{p}, t) \mathrm{d}\Omega_p \quad, \tag{D.9}$$

is the phase-space density averaged over momentum directions.

The Lorentz equation has a large range of applications. It applies to electric conduction, to thermalization of electrons in solids, to the transfer of radiation in stars or in the atmosphere, and to the diffusion of heat, as well as to neutron transport.

It is useful to integrate the Lorentz equation over $\mathrm{d}^3\boldsymbol{p}$, yielding

$$\frac{\partial n}{\partial t} + \boldsymbol{\nabla} \cdot \boldsymbol{J} = -\lambda_{\mathrm{abs}} n + 4\pi S(\boldsymbol{r}) \,, \tag{D.10}$$

where $4\pi S(\boldsymbol{r})$ is the momentum integral of $S(\boldsymbol{p}, \boldsymbol{r})$. Furthermore, multiplying the Lorentz equation by $\boldsymbol{v}$ and integrating over $\mathrm{d}^3\boldsymbol{p}$, we obtain :

$$\frac{\partial \boldsymbol{J}}{\partial t} + \int \boldsymbol{v}(\boldsymbol{v} \cdot \boldsymbol{\nabla} f(\boldsymbol{r}, \boldsymbol{p}, t)) \, \mathrm{d}^3\boldsymbol{p} = -(\lambda_{\mathrm{el}} + \lambda_{\mathrm{abs}})\boldsymbol{J} \,, \tag{D.11}$$

where we have assumed that the source term $S(\boldsymbol{p}, \boldsymbol{r})$ is independent of the direction of $\boldsymbol{p}$.

Equations (D.10) and (D.11) are the basic equations that we want to solve. The integral

$$I = \int \boldsymbol{v}(\boldsymbol{v} \cdot \boldsymbol{\nabla} f(\boldsymbol{r}, \boldsymbol{p}, t)) \mathrm{d}^3\boldsymbol{p} \,, \tag{D.12}$$

in the left hand side of (D.11) contains all the physical difficulties of the problem. There are two extreme situations.

1. The first is the *ballistic* regime, where the mean free path is much larger than the size of the medium. Collisions have a weak effect and the drift time $\propto 1/(\boldsymbol{v} \cdot \boldsymbol{\nabla} f(\boldsymbol{r}, \boldsymbol{p}, t))$ controls the evolution. This is the case of electron movement in the base of a transistor.

2. Conversely, in the *diffusive regime* or local quasi-equilibrium regime which is of interest here, the mean free path between two collisions is small compared to the size of the medium. In first approximation, $f(\boldsymbol{r}, \boldsymbol{p}, t)$ is independent of the direction of $\boldsymbol{p}$ so $f(\boldsymbol{r}, \boldsymbol{p}) \sim \bar{f}(\boldsymbol{r}, p)$ and we can write this distribution function in the form

$$f(\boldsymbol{r}, \boldsymbol{p}, t) = \bar{f}(\boldsymbol{r}, \boldsymbol{p}, t) + f_1(\boldsymbol{r}, \boldsymbol{p}, t) \tag{D.13}$$

where $f_1 \ll \bar{f}$ contains all the anisotropy, and $\int f_1(\boldsymbol{r}, \boldsymbol{p}, t)\mathrm{d}^3\boldsymbol{p} = 0$, i.e. f_1 does not contribute to the density n but only to the current $\boldsymbol{J}$.

3. There exist mixed situations, where the medium has large density variations in the vicinity of which none of these approximations holds. This is the case for neutrino transport in the core of supernovae during the rebound of nuclear matter. Such situations require sophisticated numerical techniques.[2]

---

[2] See for instance J-L. Basdevant, Ph.Mellor and J.-P. Chièze, "Neutrinos in Supernovae, An exact treatment of transport," Astronomy and Astrophysics, vol.197, p 123 (1988)

We place ourselves in the case (D.13). (This assumption amounts to expanding the distribution function in Legendre polynomials, or spherical harmonics, and in retaining only the first two terms of the expansion.) We neglect the anisotropic part f_1 in the integral (D.12). Since $\bar{f}$ is independent of the direction of p, the integral over angles is simple

$$I = \frac{v^2}{3}\nabla n(r, t) \tag{D.14}$$

and (D.11) becomes

$$\frac{\partial J}{\partial t} + \frac{v^2}{3}\nabla n = -(\lambda_{\text{el}} + \lambda_{\text{abs}})J . \tag{D.15}$$

Equations (D.10) and (D.15) are now the basic equations to be solved.

Pure Diffusion. We first consider a situation where there is no absorption and no source term, i.e. the case of pure diffusion where where there is only elastic scattering with the nuclei of the medium. The two equations (D.10) and D.15) reduce to

$$\frac{\partial n}{\partial t} + \nabla J = 0 , \tag{D.16}$$

$$\frac{\partial J}{\partial t} + \frac{v^2}{3}\nabla n = -\lambda_{\text{el}}J . \tag{D.17}$$

The first relation expresses the conservation of the number of particles (one can write energy conservation in the same manner). The second expresses the current density in terms of the gradient of the density of particles

$$J = -D\left[v\nabla n + \frac{3}{v}\frac{\partial J}{\partial t}\right] , \tag{D.18}$$

where the *diffusion coefficient* D depends on the velocity and the elastic-scattering rate

$$D = \frac{v}{3\lambda_{\text{el}}} = \frac{l}{3} , \tag{D.19}$$

where in the second form we use the fact that $\sigma_{\text{tot}} = \sigma_{\text{el}}$ implying that the mean free path is $l = v/\lambda_{\text{el}}$. Under the conditions (which occur frequently) where $(3/v)\partial J/\partial t$ can be neglected, this boils down to Fick's law, where the current is proportional to the density gradient:

$$J = -Dv\nabla n . \tag{D.20}$$

Inserting (D.18) into (D.16) leads to

$$\frac{\partial n}{\partial t} + \frac{3D}{v}\frac{\partial^2 n}{\partial t^2} - Dv\nabla^2 n = 0 , \tag{D.21}$$

which is called the *telegraphy equation* .

This equation has the general form of a wave equation where the wavefront propagates with the velocity $v/\sqrt{3}$ but the wave decreases exponentially with the distance. If the mean free path $1/3D$ is small compared to the dimension R of the system under consideration, the propagation time $\tau = v/R\sqrt{3}$ of the wave in the system is very short compared to the time of migration of a neutron by a random walk on the same distance. One can therefore neglect the propagation term $(3D/v)\partial^2 n/\partial t^2$ which amounts to considering the propagation velocity as infinite in the telegraphy equation.[3]

In this approximation, one ends up with the *Fourier diffusion equation*

$$\frac{\partial n}{\partial t} = Dv\nabla^2 n ,$$
(D.22)

which has a large range of applications and which can be solved by taking the Fourier transformation. We set

$$n(\boldsymbol{r},t) = \int e^{i\boldsymbol{k}\cdot\boldsymbol{r}} g(\boldsymbol{k},t)\mathrm{d}^3\boldsymbol{k} ,$$
(D.23)

and, by inserting this into (D.22), we obtain

$$\frac{\partial g}{\partial t} = -k^2 Dv g ,$$
(D.24)

i.e.

$$g(\boldsymbol{k},t) = f(\boldsymbol{k})e^{-k^2 Dvt} ,$$
(D.25)

where $f(\boldsymbol{k})$ is determined by the initial conditions using the inverse Fourier transform

$$n(\boldsymbol{r},t=0) = \int e^{ikr} f(\boldsymbol{k})\mathrm{d}^3\boldsymbol{k} ,$$
(D.26)

i.e.

$$f(\boldsymbol{k}) = (2\pi)^3 \int e^{-i\boldsymbol{k}\cdot\boldsymbol{r}} n(\boldsymbol{r},t=0)\mathrm{d}^3\boldsymbol{r} .$$
(D.27)

If at time $t = 0$ the density n is concentrated at the origin, $n(\boldsymbol{r},t=0) = n_0\delta(\boldsymbol{r})$, $f(\boldsymbol{k})$ is then a constant f, and $n(\boldsymbol{r},t)$ is the Fourier transform of a Gaussian:

$$n(\boldsymbol{r},t) \propto e^{-r^2/4Dvt} .$$
(D.28)

The diffusion time T in a sphere of radius R is of the order of $T \sim (R^2/Dv) = (R/\lambda)^2(\lambda/v)$ where λ/v is the mean time between two elementary collisions.

The telegraphy equation (D.21) can also be treated by Fourier transform. One can directly check under which conditions the propagation term $(3D/v)\partial^2 n/\partial t^2$ can be neglected.

---

[3] We remark that the Fourier equation is a bona fide wave equation with exponential damping at infinity. The wavefronts have a finite velocity $v/\sqrt{3}$, however the propagation effects are completely negligible in the diffusion regime.

E. Solutions and Hints for Selected Exercises

Chapter 1

1.9 One has

$$\langle E \rangle = A \langle \frac{p^2}{2m} \rangle - \frac{A(A-1)}{2} g^2 \langle \frac{1}{r} \rangle \quad .$$

Therefore, owing to the Heisenberg + Pauli inequality $\langle p^2 \rangle \geq A^{2/3} \hbar^2 (\langle 1/r \rangle)^2$ we obtain

$$\langle E \rangle \geq A^{5/3} \hbar^2 \frac{1}{2m} \langle \frac{1}{r} \rangle^2 - \frac{A(A-1)}{2} g^2 \langle \frac{1}{r} \rangle \quad .$$

Minimizing with respect to $\langle 1/r \rangle$, we obtain

$$\langle E \rangle / A \sim -mg^4 A^{4/3} / 8 \hbar^2 \quad \text{and} \langle 1/r \rangle \sim 2 \hbar^2 A^{-1/3} / mg^2 \quad .$$

1.12 The ratio of the quadrupole and magnetic hyperfine splittings for a very elongated nucleus is of order

$$\frac{Z^2 R^2 / a_0^2}{\alpha^2 m_e / m_p} \sim 0.3 \, Z^2$$

where we use $R \sim 5$ fm and $a_0 = \hbar c / \alpha m_e c^2$. For a slightly deformed nucleus, R^2 is replaced by $Q \propto R^2 \Delta R / R$ (Q is the mass quadrupole moment). This lowers the quadrupole splitting by a factor of more then 10. The sign of this splitting is opposite for prolate and oblate nuclei whereas the magnetic splitting is shape independent.

1.13 The energy splitting is $\Delta E = 2 \times 2.79 \mu_N = 1.76 \times 10^{-7}$eV. For $kT = 0.025$ eV this gives a difference in population of

$$\frac{e^{\Delta E / 2kT} - e^{-\Delta E / 2kT}}{e^{\Delta E / 2kT} + e^{-\Delta E / 2kT}} \sim \Delta E / 2kT \sim 3.5 \times 10^{-6} \quad .$$

The absorption frequency is $\Delta E / 2 \pi \hbar = 4.2 \times 10^7$ Hz.

Medical applications of MRI are confined to hydrogen since ^{1}H is the only common nuclide with spin.

The magnetic field due to neighboring spins is of order $(\mu_0 / 4\pi) \mu_N / a_0^3 \sim 5 \times 10^{-3}$T.

1.16 The data indicates that the value of m/Z for ^{48}Mn is about midway between the values for ^{46}Cr and ^{50}Fe. Using a ruler, one can find that $m/Z(\text{Mn}) \sim \text{f} \times m/Z(\text{Fe}) + (1 - \text{f}) \times m/Z(\text{Cr})$ with $f \sim 0.54 \pm 0.01$. The values of B/A for ^{46}Cr and ^{50}Fe imply $m/Z(\text{Cr}) = 1783.624\,\text{MeV}$ and $m/Z(\text{Fe}) = 1789.497$ so $m/Z(\text{Mn}) = 1786.74 \pm 0.06$. This gives $B/A(^{48}\text{Mn}) = (8.26 \pm 0.03)\,\text{MeV}$. The experimenters (not obliged to use a ruler) give an uncertainty of $0.002\,\text{MeV}$.

1.17 The protons initially have kinetic energy $E_p = 11\,\text{MeV}$ corresponding to a momentum $p_p c = \sqrt{2E_p m_p c^2} = 143\,\text{MeV}$. For protons recoiling from Ni nuclei in the $1.35\,\text{MeV}$ excited state, to first approximation, the proton energy is reduced by this amount, i.e. $E'_p = 11 - 1.35 = 9.65\,\text{MeV}$. This corresponds to a proton momentum $p'_p c = 134\,\text{MeV}$. Momentum conservation then allows us to deduce the momentum components of the recoiling ^{64}Ni nucleus if the proton scatters at an angle θ:

$$p_t c = (134 \sin \theta)\,\text{MeV} \qquad p_l c = (143 - 134 \cos \theta)\,\text{MeV} ,$$

for the directions perpendicular to and along the beam direction. For $\theta = 60\,\text{deg}$, this gives a Ni momentum of $pc = 139\,\text{MeV}$ and a kinetic energy of $0.16\,\text{MeV}$. We can then re-estimate the energy of protons recoiling at 60 deg to be $9.65 - 0.16 = 9.49\,\text{MeV}$.

Chapter 2

2.4 The simplest way to demonstrate the equivalence is to write down the 3-d wavefunctions in terms of products of 1-d harmonic oscillatory wavefunctions and show that they are proportional to the appropriate spherical harmonics: $Y_{10} \propto \cos \theta$ and $Y_{1\pm 1} \propto \sin \theta e^{\pm i\phi}$.

2.6 ^{41}Ca has one neutron outside closed shells containing 20 protons and 20 neutrons. The orbital above 20 particles is 1f7/2 so $J = 7/2$ and $l = 3$ implying that the parity is negative (-1^l). So spin$^{\text{parity}} = 7/2^-$ in agreement with observation.

2.7 ^{83}Kr has an odd neutron orbiting closed shells while ^{93}Nb has an odd proton. The odd proton contributes to the magnetic moment through both its spin and orbital angular momentum while the neutron contributes only its spin. For $J = 9/2$, the orbital moment must dominate so we expect ^{93}Nb to have the greater moment. For ^{93}Nb, the Schmidt formulas give (for $l = 4$ or $l = 5$):

$$g = (9/2 - 1/2) + 2.79 = 6.79 \quad \text{or} \quad (9/11)[6 - 2.79] = 2.62 .$$

The shell model suggests $l = 4 \Rightarrow g = 6.79$ to be compared with the experimental value 6.167.

For ^{86}Kr the Schmidt formulas give:

$$g = -1.91 \quad \text{or} \quad (9/11)1.91 = 1.56 .$$

The shell model suggests $l = 4 \Rightarrow g = -1.91$ to be compared with the experimental value -0.97.

In both cases, the experimental values are between the two Schmidt values and somewhat closer to the value predicted by the shell model.

Chapter 3

3.2 The neutrino flux (integrated over the duration of the pulse ~ 15 s) was $F = N/(4\pi R^2) = 10^{57}/(3 \; 10^{43})$, the number of protons in the target was $N_c = (4/3)10^{32}$. The number of events detected is $F \; N_c \; \sigma \simeq 10$. One can meditate on the many elements of the observers good luck. (The Kamiokande detector had been built 2 years before to observe a completely different phenomenon, the as yet unobserved proton decay).

3.6 To first approximation, the scattered electron keeps all of its energy so its momentum components perpendicular and parallel to the beam directions are $p_t c \sim 500 \, \mathrm{MeV} \times \sin\theta$ and $p_l c \sim 500 \, \mathrm{MeV} \times \cos\theta$. Momentum conservation then gives the momentum of the recoiling target particle

$$p_t c \sim 500 \, \mathrm{MeV} \times \sin\theta \qquad p_l c \sim 500 \, \mathrm{MeV} \times (1 - \cos\theta) \; .$$

For $\theta = 45 \deg$ this gives a recoil energy of $78 \, \mathrm{MeV}$ for a nucleon and $39 \, \mathrm{MeV}$ for a deuteron. Subtracting this from the electron energy gives a peak at $422 \, \mathrm{MeV}$ for recoil from a proton and $461 \, \mathrm{MeV}$ for recoil from a deuteron. The proton peak energy should be further reduced by the $2.2 \, \mathrm{MeV}$ necessary to break the deuteron.

3.7 The Rutherford cross-section is

$$\frac{\mathrm{d}\sigma}{\mathrm{d}\Omega} = \frac{\alpha^2 (\hbar c)^2}{16 E^2 \sin^4(\theta/4)} \; .$$

Equating this with the strong-interaction cross-section, $\sim (10 \, \mathrm{fm})^2 \, \mathrm{sr}^{-1}$, gives $\sin \theta/2 \sim 0.13$, i.e. $\theta \sim 16 \deg$. At smaller angles Rutherford scattering dominates while at higher angles strong-interaction scattering dominates. It should be kept in mind that the *amplitudes* for the two interactions must be summed. This permits one to determine their relative phases.

Chapter 4

4.4 The maximum energy photons have about $15 \, \mathrm{keV}$ excess energy out of $1065 \, \mathrm{keV}$ so the decaying nuclei initially have $v/c \sim 1.4 \times 10^{-2}$ corresponding to an energy $\sim 7 \, \mathrm{MeV}$. The Bethe–Bloch formula gives an energy loss of $\sim 2 \times 10^6 \, \mathrm{MeV} (\mathrm{g \, cm^{-2}})^{-1}$ or about $2 \times 10^7 \, \mathrm{MeV \, cm^{-1}}$ in nickel. The Br ions then would stop after $\sim 3 \times 10^{-7} \, \mathrm{cm}$ in a time of about $10^{-15} \, \mathrm{s}$. Since it appears that about half the nuclei decay before stopping, this would mean that the lifetime is of order $10^{-15} \mathrm{s}$. In fact, because at very low velocities the ion attaches electrons reducing its effective charge, the Bethe–Bloch formula overestimates by about a factor ~ 100 the energy loss for Br ions at $v/c \sim 10^{-2}$ (see L.C. Northcliffe and R.F. Schilling Nuclear Data Tables, **A7** (1970) 233; F.S. Goulding and B.G. Harvey, Ann. Rev. Nucl. Sci **25** (1975) 167.). The stopping time is thus a factor of ~ 100 greater and the lifetime is $\sim 0.5 \times 10^{-12}$s.

4.5 The decay of ^{60}Co to the ground and first excited states of ^{60}Ni are forbidden ($\Delta J > 1$) so the decay is primarily by the allowed transition to the 4^+ state. The 4^+ state decay to the ground state is M4 so the decay is primarily through the cascade of two E2 transitions. For $E_\gamma \sim 1\,\mathrm{MeV}$ such transitions have mean lives of $\sim 10^{-12}$ s.

4.9 ^{152m}Eu has an allowed Gamow-Teller decay to the 1^- state of ^{152}Sm while the decays of the ^{152}Eu to the shown states are forbidden.

The kinetic energy of the recoiling Sm is $p^2c^2/2mc^2 = (840\,\mathrm{keV})^2/(2 \times 145\,\mathrm{GeV}) = 2.4\,\mathrm{eV}$ corresponding to a velocity of $v/c \sim 6 \times 10^{-6}$. The 961 keV photons emitted in the direction of the Sm velocity are thus blue shifted to an energy of $961(1 + 6 \times 10^{-6})\,\mathrm{keV}$. This gives them enough energy to excite a second Sm nuclei (taking into account the recoil of the second Sm).

4.10 To good approximation the neutrino conserves its energy $\sim 5\,\mathrm{MeV}$ so a neutron recoiling from a back-scattered neutrinos has an energy $p^2c^2/2m_nc^2 \sim (5\,\mathrm{MeV})^2/2\,\mathrm{GeV} \sim 12\,\mathrm{keV}$. The cross-section for such neutrons on hydrogen nuclei is $\sim 20\,\mathrm{b}$ corresponding to a mean free path of $\sim 1\,\mathrm{cm}$ in CH. This neglects the carbon, which has a smaller cross-section, $\sim 5\,\mathrm{b}$. Since a neutron loses on average half its kinetic energy in a collision with a proton (isotropic scattering at low energy), about 17 collisions are necessary to reduce the energy by five orders of magnitude to a reasonably thermal energy, $0.1\,\mathrm{eV}$.

The absorption cross-section is about 0.1 b for thermal neutrons and they have $v \sim 4 \times 10^5\,\mathrm{cm\,s^{-1}}$. This gives a mean absorption time of

$$\left[10^{-25}\,\mathrm{cm}^2 \times 4 \times 10^5\,\mathrm{cm\,s^{-1}} \times 6 \times 10^{23}/13 \right]^{-1} \sim 0.5\,\mathrm{ms} \ .$$

Chapter 5

5.4

$$t \sim 8200\,\mathrm{yr} \times \ln\left(\frac{0.233}{19.6 \times 10^{-4}} \right) \sim 3.9 \times 10^4\,\mathrm{yr} \ .$$

5.5 Assuming equal initial amounts of ^{235}U and ^{238}U, the elapsed time since creation is

$$\frac{7 \times 10^8\,\mathrm{yr}/\ln 2}{1 - 0.7/4.5} \ln\left(\frac{99.27}{0.72} \right) \sim 5.9 \times 10^9\,\mathrm{yr} \ .$$

Assuming the same for ^{234}U and ^{238}U, one finds 3.5×10^6 yr. The discrepancy is due to the fact that most of the original ^{234}U has decayed so the ^{234}U now present comes from the decay chain initiated by ^{238}U. In this case, one expects $^{234}\mathrm{U}/^{238}\mathrm{U} = t_{1/2}(234)/t_{1/2}(238)$ in agreement with the measured values.

5.6 The α-particle originally has $\beta^2 \sim 2 \times 10^{-3}$ so the initial energy loss is $\sim 1000 \, \text{MeV cm}^{-1}$ for $\rho = 1.8 \, \text{g cm}^{-3}$. The probability that it produces a neutron before losing $1 \, \text{MeV}$ is

$$P = (1/1000) \, \text{cm} \times 0.4 \times 10^{-24} \, \text{cm}^2 \times 1.8 \, \text{g cm}^{-3}$$
$$\times (6 \times 10^{23}/9) \, \text{g}^{-1} \sim 5 \times 10^{-5},$$

so to give $1 \, \text{Bq}$ neutron activity we need $2 \times 10^4 \, \text{Bq}$ of α activity.

Chapter 6

6.3 The mean free path for neutrons is dominated by fission of ^{235}U:

$$l^{-1} = 250 \times 10^{-24} \text{cm}^2 \times \frac{6 \times 10^{23}}{238} \times 0.0072 \times 19 \, \text{g cm}^{-3} \sim 11 \, \text{cm}.$$

If the uranium is in the shape of a cube, the probability of a fission is $P = 1 \, \text{cm}/11 \, \text{cm}$ and the fission rate is

$$(1/11) \times 10^{12} \, \text{cm}^{-2}\text{s}^{-1} \sim 10^{12} \, \text{s}^{-1}$$

corresponding to $\sim 30 \, \text{W}$. The rate is lower if the uranium is deformed so that the dimension in the direction of the beam is comparable to or greater than the mean free path.

6.5 Neutron-rich fission products with $A = 142$ will β^--decay to ^{142}Ce which is a long-live 2β emitter.

6.6 The nuclides with $A \sim 100$ are fission products. The transuraniums ^{243}Am and ^{239}Pu are produced by neutron captures (followed by β-decays) on ^{238}U. The nuclides with $210 < A < 235$ come from the decay chains initiated by the transuraniums.

Chapter 7

7.4 The photon energy is $17.49 \, \text{MeV}$ (as above). Using the Bethe–Bloch formula, the energy loss of the proton in the LiF is

$$\Delta E \sim 10^{-5} \text{g cm}^{-2} \times \frac{1 \, \text{MeV} \, (\text{g cm}^{-2})^{-1}}{\beta^2} \sim 20 \, \text{keV}$$

where we use $\beta^2 = 4 \times 10^{-4}$ for the proton. The actual energy loss is ~ 5 times less since the Bethe–Bloch formula overestimates the energy loss at this velocity.

The cross-section is proportional to $\propto \exp(-\sqrt{E_B/E})$ where $E_B \sim 7.75 \, \text{MeV}$ for this reaction. This gives the variation of the cross-section as the incident proton loses energy in the LiF:

$$\frac{d\sigma}{\sigma} = (1/2)\sqrt{E_B/E}\frac{\Delta E}{E} \sim 3\frac{\Delta E}{E} \sim 0.05,$$

so the cross-section is relatively constant over the thickness of the target. If the target were much thicker, the variation would be substantial and the event rate would not be easily interpretable.

7.5 The parameters entering the calculations are $E_B = 7.75\,\text{MeV}$, $E_G = 12.45\,\text{keV}$, $\Delta E_G = 0.3\,\text{keV}$, $S(E_G) = -.5\,\text{keV b}$, and $\Gamma_\gamma = 12\,\text{eV}$. The factor that deviates most from unity is the Boltzmann factor giving the probability to have a proton with enough energy to excite the resonance: $\exp(-441\,\text{keV}/kT)$. This makes the resonance contribution completely negligible at $kT = 1\,\text{keV}$.

7.6 For $T = 10^6\,\text{K}$, $kT = 0.086\,\text{keV}$ we have $nkT\tau_{\text{brem}} \sim 3 \times 10^{19}\,\text{keV m}^{-3}\,\text{s}$ in agreement with the figure. It is proportional to $T^{3/2}$, also in agreement with the figure.

Chapter 8

8.4 The mean free path of a 10 MeV neutrino in a neutron star is of order

$$l = \left[\frac{10^{57}}{(4\pi/3)(10^4\,\text{m})^3} 10^{-41}\,\text{cm}^2 \right]^{-1} \sim 4\,\text{m}\,,$$

which is much less than the neutron star radius, $R \sim 10^4$ m. The neutrinos therefore diffuse out of the star with a time of order $R^2/cl \sim 0.1\,\text{s}$. In fact, the neutrino pulse from the collapse of stellar core to a neutron star lasts somewhat longer, about 10 s.

8.7 All degenerate gases have a phase space density of order $\hbar^{-3}$. The phase space density of such a gas is the momentum space density ($\sim p_F^{-3}$) times the real space density (n) so the Fermi momentum is $p_F^2 \sim n^{2/3}\hbar^2$. For $p_F \ll mc$ and $p_F \gg mc$ this gives a total energy for N fermions

$$E \sim \frac{Np_F^2}{2m} \sim \frac{Nn^{2/3}\hbar^2}{m} \qquad E \sim Np_F c \sim Nn^{1/3}\hbar c\,.$$

The pressure is the derivative of the energy with respect to the volume. We find for the two limits

$$P \sim n^{5/3} \frac{(\hbar c)^2}{mc^2} \qquad P \sim n^{4/3}\hbar c\,,$$

where n is the number density. (The numerical factor in the first case is $(3\pi^2)^{2/3}/5$).

For a number density of electrons $n \sim 10^{30}$ cm^{-3}, and a temperature $T \sim 10^7$ K the degenerate quantum pressure $\propto n^{5/3}$ is much larger than a classical ideal gas pressure $\propto n$. Between the density where the star can be treated as an ideal gas and that where it becomes a Fermi gas, there is a transition regime. Above a critical density, and for temperatures smaller than the Fermi temperature, the electron gas becomes degenerate. Notice that owing to the mass effect, the gas of nuclei is still an ideal gas.

For a non-relativistic degenerate electron gas, the strong quantum pressure is temperature independent and it resists futher collapse, since the gravitational inward pressure behaves as $n^{4/3}$. There is no further contraction, no further nuclear reactions, the star cools endlessly. This situation corresponds to a white dwarf.

The order of magnitude of the temperature at which the contracting gas reaches this regime can be estimated from the virial theorem, $\langle 3PV \rangle \sim GM^2/R$. We approximate the pressure by the sum of the classical and quantum pressures. This gives

$$NkT \sim \frac{GM^2}{R} - \frac{N^{5/3}\hbar^2/m}{R^2} . \tag{E.1}$$

Minimizing with respect to R we get

$$R \sim \frac{N^{5/3}\hbar^2}{GM^2} \qquad kT_{\max} = \frac{G^2M^2m}{4N^{5/3}\hbar^2} ,$$

where M is the mass of the star and m and N are the mass and number of the degenerate particle.

If $T_{max} \leq 10^6$ K, the star is called a brown dwarf because the temperature has not reached the value where nuclear reactions can take place. The difference between a brown dwarf and a planet is that, in a planet, the individual atoms and molecules have not completely dissociated in a plasma of electrons and nuclei, at least in the crust. The temperature is much lower, the overall cumulative gravitational forces give the object a global spherical shape, but rocks and other non-spherical objects, whose shapes are due to electromagnetic forces, can still exists on the surface.

8.8 The mass of the iron core of a star can increase only up to the Chandrasekhar mass at which point it will collapse. During the collapse, the Fermi energy of the electrons increases until most electrons have sufficient energy to by captured endothermically. The neutrinos produced in the captures do not induce the reverse reaction because they escape from the star after a period of diffusion (Exercise 8.4).

The energy radiated by a neutrino species of temperature T is given by Stefan's law (after a minor modification taking into account the fact that neutrinos are fermions). Taking $kT = 1\,\text{MeV}$, a neutrinosphere radius of $R = 10^4\,\text{m}$, and a pulse duration of $10\,\text{s}$, one finds that the total energy radiated by three neutrino species is

$$5.67 \times 10^{-8}\,\text{W}\,\text{m}^{-2}\,\text{K}^{-4} \times 10\,\text{s} \times (kT)^4\,4\pi R^2 \sim 3 \times 10^{46}\,\text{J} .$$

This agrees with the total energy liberated, $(3/5)GM^2/R$. (The agreement is not fortuitous since the temperature and radius of the neutrino sphere are constrained by this requirement.) Note that the number of neutrinos radiated, $(3 \times 10^{46}\,\text{J})/2kT \sim 2 \times 10^{58}$, is greater than the number of ν_e produced by neutron capture $\sim 10^{57}$. Most of the neutrinos are thermally produced, $\gamma\gamma \leftrightarrow e^+e^- \leftrightarrow \nu\bar{\nu}$.

Chapter 9

9.3 The density of protons when $kT = 60\,\text{keV}$ can be scaled up from the present density by the third power of the temperature:

$$n_{\text{p}} \sim n_{\text{p}}(t_0)(60\,\text{keV}/kT(t_0))^3 = \eta n_\gamma(t_0) \times \left(\frac{60\,\text{keV}}{2 \times 10^{-4}\,\text{eV}}\right)^3 .$$

The reaction rate per neutron is this density multiplied by σv which gives $\sim 3 \times 10^{-3}\text{s}^{-1}$ for $\eta \sim 4 \times 10^{-12}$. This nearly the expansion rate $\sim 0.65\text{s}^{-1}(60\text{keV}/1\,\text{MeV})^2 \sim 2 \times 10^{-3}\text{s}^{-1}$.

The deuteron-neutron ratio at this temperature is

$$\frac{n_2}{n_{\text{n}}} = \eta n_\gamma \left(m_{\text{p}}c^2 kT\right)^{-3/2} (2\pi\hbar c)^3 e^{-B/kT} \sim \eta \times 10^{-18} ,$$

and rises very quickly above unity as the temperature falls.

9.5 We use

$$\dot{a}/a \sim \dot{T}/T \sim 0.65(kT/1\,\text{MeV})^2\,\text{s}^{-1} .$$

Integrating, we get

$$\Delta t \sim \int_{kT\sim 1\,\text{MeV}}^{kT=60\,\text{keV}} \frac{dT}{\dot{T}} \sim 166\,\text{s} .$$

9.7 The wimps have kinetic energies of order $(1/2)mc^2\beta^2 \sim 50\,\text{keV}$. For most nuclear targets, this is much less than the excitation energy of the first excited state so we expect only elastic scattering to be possible.

The mean free path in the Earth is of order

$$l^{-1} \sim 10^{-35}\,\text{cm}^2\,\text{nucleus}^{-1}$$

$$\times(6 \times 10^{23}\text{nucleon}\,\text{g}^{-1}/50\,\text{nucleon/nucleus})\,5\,\text{g}\,\text{cm}^{-3} \sim \frac{1}{10^7\,\text{km}} ,$$

i.e, much greater than the radius of the Earth. The interaction rate in one kg of germanium is

$$\lambda \sim \frac{0.3\,\text{GeV}\,\text{cm}^{-3}}{50\,\text{GeV/wimp}} \times 3 \times 10^7 \text{cm}\,\text{s}^{-1} \times 10^{-35}\text{cm}^2$$

$$\times \frac{6 \times 10^{26}\text{nucleon}\,\text{kg}^{-1}}{72\text{nucleon/nucleus}} \sim 10^{-5}\text{s}^{-1} ,$$

which is less than the rate of ^{68}Ge decay.

F. Tables of numerical values

Table F.1. Selected physical and astronomical constants, adapted from [1].

quantity	symbol	value
speed of light in vacuum	c	$2.99\,792\,458 \times 10^8$ m s^{-1}
Planck constant	$\hbar$	$1.054\,571\,596(82) \times 10^{-34}$ J s
conversion constant	$\hbar c$	$197.326\,960\,2(77)$ MeV fm
conversion constant	$(\hbar c)^2$	$389\,379\,292\,(30)$ MeV^2 b
		$(1\,\mathrm{b} \equiv 10^{-28}\,\mathrm{m}^2)$
e$^-$ charge magnitude	e	$1.602\,176\,462(63) \times 10^{19}$ C
		$\Rightarrow 1\,\mathrm{eV} = 1.602 \times 10^{-19}$ J
Fine structure constant	$\alpha = e^2/4\pi\epsilon_0\hbar c$	$[137.035\,999\,76(50)]^{-1}$
Bohr radius	a_∞	$0.529\,177\,208\,3(39) \times 10^{-10}$ m
Rydberg energy	$\alpha^2 m_e c^2/2$	$13.605\,691\,72(53)$ eV
Thomson cross-section	σ_T	$0.665\,245\,854(15) \times 10^{-28}$ m^2
Gravitational constant	$G_N\ (= G)$	$6.673(10) \times 10^{-11}$ m^{3}kg^{-1}s^{-2}
Planck mass	$m_{pl} = \sqrt{\hbar c/G}$	$1.221\,0(9) \times 10^{19}$ GeV$/c^2$
Fermi coupling constant	$G_F/(\hbar c)^3$	$1.166\,39(1) \times 10^{-5}$GeV^{-2}
electron mass	m_e	$0.510\,998\,902(21)$ MeV$/c^2$
proton mass	m_p	$938.271\,998(38)$ MeV$/c^2$
		$1.672\,621\,58(13)$ kg
		$1836.152\,667\,5(39)m_e$
neutron–proton Δm	$m_n - m_p$	$1.293\,318(9)$ MeV$/c^2$
Avogadro constant	N_A	$6.022\,141\,99(47) \times 10^{23}$ mol^{-1}
nuclear magneton	$\mu_N = e\hbar/2m_p$	$3.152\,451\,238(24)\,10^{-14}$MeVT^{-1}
Boltzmann constant	k	$1.380\,650\,3(24) \times 10^{-23}$ J K^{-1}
		$8.617\,342(15) \times 10^{-5}$ eV K^{-1}
parsec	pc	$3.085\,677\,580\,7(4) \times 10^{16}$ m
		$=3.262...$ ly
solar mass	$M_\odot$	$1.988\,9(30) \times 10^{30}$ kg
		$= 1.189 \times 10^{57}m_p$
solar luminosity	$L_\odot$	$3.846(8) \times 10^{26}$ W s^{-1}
solar equatorial radius	$R_\odot$	6.961×10^8 m

G. Table of Nuclei

The following table lists known nuclei sorted by their mass number A. Binding energies are taken from [2] while decay modes, lifetimes (in seconds), and terrestrial abundances (for long-lived isotopes) are generally taken from [3]. For a given A, the binding energies shown in column 2 are the parabolic functions of Z illustrated in Fig. 2.6. Because of the nucleon-pairing energy, their is only one parabola for odd-A and two parabolas for even-A (one for even-even and one for odd-odd). Nuclei on the neutron-rich side of the parabola are generally β^--unstable while those on the proton-rich side are unstable to electron-capture ($Q_{ec} < 2m_e$) or to both electron-capture and β^+ decay ($Q_{ec} > 2m_e$). A few very weakly bound nuclei can also decay by nucleon emission, e.g. $A = 16$, $Z = 5, 9, 10$.

Because of the single or double parabolic structure, there is only one β-stable nucleus for odd-A and two or three β-stable nuclei for even-A. For even-A, only one nucleus is also stable against double-β decay, but the lifetime for 2β decay is generally greater than 10^{20} yr so nuclei that are only 2β unstable are still present on Earth.

Nuclei with $A > 150$ ($A > 100$) are also usually unstable to α-decay (spontaneous fission). The lifetimes are generally greater than 10^{20} yr for $A < 208$.

Decay and reaction Q's can be calculated from the binding energies in this table. For example

$$Q_{\beta-}[(A, Z) \to (A, Z+1)] = B(Z+1) - B(Z) + (m_n - m_p - m_e)c^2$$

$$= B(Z+1) - B(Z) + 0.782 \, \text{MeV} \,,$$

$$Q_{\beta+}[(A, Z) \to (A, Z-1)] = B(Z-1) - B(Z) - (m_n + m_e - m_p)c^2$$

$$= B(Z-1) - B(Z) - 1.804 \, \text{MeV} \,,$$

$$Q_\alpha[(A, Z) \to (A-4, Z-2)] = B(A-4, Z-2) - B(A, Z) + B(4, 2) \,,$$

$$= B(A-4, Z-2) - B(A, Z) + 28.295 \,.$$

A X Z	B/A (MeV)	→	log $t_{1/2}$ or %	A X Z	B/A (MeV)	→	log $t_{1/2}$ or %
1 n 0	0.0000	β⁻	2.79	11 Li 3	4.1499	β⁻	-2.07
1 H 1	0.0000		99.99%	11 Be 4	5.9528	β⁻	1.14
				11 B 5	6.9277		80.10%
2 H 1	1.1123		0.01%	11 C 6	6.6764	β⁺	3.09
				11 n 7	5.3043	p	-21.05
3 H 1	2.8273	β⁻	8.59				
3 He 2	2.5727		0.00%	12 Be 4	5.7208	β⁻	-1.63
				12 B 5	6.6313	β⁻	-1.69
4 H 1	1.3753	n		12 C 6	7.6801		98.90%
4 He 2	7.0739		100.00%	12 N 7	6.1701	β⁺	-1.96
4 Li 3	1.1545	p		12 O 8	4.8778	p	-20.78
5 H 1	0.2164	n		13 Be 4	5.1261	n	-21.14
5 He 2	5.4811	n	-20.96	13 B 5	6.4964	β⁻	-1.76
5 Li 3	5.2661	p	-21.36	13 C 6	7.4699		1.10%
				13 N 7	7.2389	β⁺	2.78
6 H 1	0.9636	n		13 O 8	5.8121	β⁺	-2.07
6 He 2	4.8782	β⁻	-0.09				
6 Li 3	5.3324		7.50%	14 Be 4	4.9991	β⁻	-2.36
6 Be 4	4.4873	p	-20.15	14 B 5	6.1016	β⁻	-1.86
				14 C 6	7.5203	β⁻	11.26
7 He 2	4.1178	n	-20.39	14 N 7	7.4756		99.63%
7 Li 3	5.6064		92.50%	14 O 8	7.0524	β⁺	1.85
7 Be 4	5.3715	EC	6.66	14 F 9	5.1678	p	
7 B 5	3.5314	p	-21.33				
				15 B 5	5.8794	β⁻	-1.98
8 He 2	3.9260	β⁻	-0.92	15 C 6	7.1002	β⁻	0.39
8 Li 3	5.1598	β⁻	-0.08	15 N 7	7.6995		0.37%
8 Be 4	7.0624	α	-16.01	15 O 8	7.4637	β⁺	2.09
8 B 5	4.7172	β⁺	-0.11	15 F 9	6.4834	p	-21.18
8 C 6	3.0978	p	-20.54	15 Ne 10	4.7907	?	
9 He 2	3.3621	n	-20.66	16 B 5	5.5057	n	-9.70
9 Li 3	5.0379	β⁻	-0.75	16 C 6	6.9221	β⁻	-0.13
9 Be 4	6.4628		100.00%	16 N 7	7.3739	β⁻	0.85
9 B 5	6.2571	α	-17.91	16 O 8	7.9762		99.76%
9 C 6	4.3371	β⁺	-0.90	16 F 9	6.9637	p	-19.78
				16 Ne 10	6.0831	p	-20.27
10 Li 3	4.4922	n	-21.26				
10 Be 4	6.4977	β⁻	13.68	17 B 5	5.2697	β⁻	-2.29
10 B 5	6.4751		19.90%	17 C 6	6.5578	β⁻	-0.71
10 C 6	6.0320	β⁺	1.29	17 N 7	7.2862	β⁻	0.62
10 N 7	3.5538	?		17 O 8	7.7507		0.04%
				17 F 9	7.5423	β⁺	1.81

A X Z	B/A (MeV)	→	$\log t_{1/2}$ or %	A X Z	B/A (MeV)	→	$\log t_{1/2}$ or %
17 Ne 10	6.6414	β^+	-0.96	22 Al 13	6.7825	β^+	-1.15
17 Na 11	5.4961	?		22 Si 14	6.1115	β^+	-2.22
18 B 5	4.9472	?		23 N 7	6.1926	?	
18 C 6	6.4259	β^-	-1.02	23 O 8	7.1637	β^-	-1.09
18 N 7	7.0383	β^-	-0.20	23 F 9	7.6204	β^-	0.35
18 O 8	7.7671		0.20%	23 Ne 10	7.9552	β^-	1.57
18 F 9	7.6316	β^+	3.82	23 Na 11	8.1115		100.00%
18 Ne 10	7.3412	β^+	0.22	23 Mg 12	7.9011	β^+	1.05
18 Na 11	6.1867	?		23 Al 13	7.3349	β^+	-0.33
				23 Si 14	6.5616	?	
19 B 5	4.7410	?					
19 C 6	6.0962	β^-	-1.34	24 N 7	5.8831	?	
19 N 7	6.9483	β^-	-0.52	24 O 8	7.0199	β^-	-1.21
19 O 8	7.5665	β^-	1.43	24 F 9	7.4636	β^-	-0.47
19 F 9	7.7790		100.00%	24 Ne 10	7.9932	β^-	2.31
19 Ne 10	7.5674	β^+	1.24	24 Na 11	8.0635	β^-	4.73
19 Na 11	6.9379	p		24 Mg 12	8.2607		78.99%
19 Mg 12	5.8956	?		24 Al 13	7.6498	β^+	0.31
				24 Si 14	7.1668	β^+	-0.99
20 C 6	5.9586	β^-	-1.85	24 P 15	6.2492	?	
20 N 7	6.7092	β^-	-1.00				
20 O 8	7.5685	β^-	1.13	25 O 8	6.7352	?	
20 F 9	7.7201	β^-	1.04	25 F 9	7.3390	β^-	-1.23
20 Ne 10	8.0322		90.48%	25 Ne 10	7.8407	β^-	-0.22
20 Na 11	7.2988	β^+	-0.35	25 Na 11	8.1014	β^-	1.77
20 Mg 12	6.7234	β^+	-1.02	25 Mg 12	8.2235		10.00%
				25 Al 13	8.0211	β^+	0.86
21 C 6	5.6592	?		25 Si 14	7.4802	β^+	-0.66
21 N 7	6.6090	β^-	-1.07	25 P 15	6.8470	?	
21 O 8	7.3894	β^-	0.53				
21 F 9	7.7383	β^-	0.62	26 O 8	6.4782	?	
21 Ne 10	7.9717		0.27%	26 F 9	7.0971	?	
21 Na 11	7.7655	β^+	1.35	26 Ne 10	7.7539	β^-	-0.71
21 Mg 12	7.1047	β^+	-0.91	26 Na 11	8.0058	β^-	0.03
21 Al 13	6.3432	?		26 Mg 12	8.3339		11.01%
				26 Al 13	8.1498	β^+	13.35
22 C 6	5.4678	?		26 Si 14	7.9248	β^+	0.35
22 N 7	6.3642	β^-	-1.62	26 P 15	7.1979	β^+	-1.70
22 O 8	7.3648	β^-	0.35	26 S 16	6.5910	?	
22 F 9	7.6243	β^-	0.63				
22 Ne 10	8.0805		9.25%	27 F 9	6.8828	?	
22 Na 11	7.9157	β^+	7.91	27 Ne 10	7.5188	β^-	-1.49
22 Mg 12	7.6626	β^+	0.59	27 Na 11	7.9593	β^-	-0.52

A X Z	B/A (MeV)	→	$\log t_{1/2}$ or %	A X Z	B/A (MeV)	→	$\log t_{1/2}$ or %
27 Mg 12	8.2639	β⁻	2.75	31 Ar 18	7.2527	β⁺	-1.82
27 Al 13	8.3316		100.00%				
27 Si 14	8.1244	β⁺	0.62	32 Ne 10	6.6651	?	
27 P 15	7.6646	β⁺	-0.59	32 Na 11	7.2304	β⁻	-1.88
27 S 16	6.9593	β⁺	-1.68	32 Mg 12	7.8028	β⁻	-0.92
				32 Al 13	8.0992	β⁻	-1.48
28 F 9	6.6332	?		32 Si 14	8.4816	β⁻	9.67
28 Ne 10	7.3891	β⁻	-1.77	32 P 15	8.4641	β⁻	6.09
28 Na 11	7.8009	β⁻	-1.52	32 S 16	8.4931		95.02%
28 Mg 12	8.2724	β⁻	4.88	32 Cl 17	8.0723	β⁺	-0.53
28 Al 13	8.3099	β⁻	2.13	32 Ar 18	7.6993	β⁺	-1.01
28 Si 14	8.4477		92.23%	32 K 19	6.9687	?	
28 P 15	7.9080	β⁺	-0.57				
28 S 16	7.4788	β⁺	-0.90	33 Na 11	7.0375	β⁻	-2.09
28 Cl 17	6.6479	?		33 Mg 12	7.6291	β⁻	-1.05
				33 Al 13	8.0208	?	
29 F 9	6.4390	?		33 Si 14	8.3604	β⁻	0.79
29 Ne 10	7.1801	β⁻	-0.70	33 P 15	8.5138	β⁻	6.34
29 Na 11	7.6843	β⁻	-1.35	33 S 16	8.4976		0.75%
29 Mg 12	8.1152	β⁻	0.11	33 Cl 17	8.3048	β⁺	0.40
29 Al 13	8.3487	β⁻	2.60	33 Ar 18	7.9289	β⁺	-0.76
29 Si 14	8.4486		4.67%	33 K 19	7.4159	?	
29 P 15	8.2512	β⁺	0.62				
29 S 16	7.7486	β⁺	-0.73	34 Na 11	6.8621	β⁻	-2.26
29 Cl 17	7.1595	?		34 Mg 12	7.5466	β⁻	-1.70
				34 Al 13	7.8564	β⁻	-1.22
30 Ne 10	7.0693	?		34 Si 14	8.3361	β⁻	0.44
30 Na 11	7.4980	β⁻	-1.32	34 P 15	8.4484	β⁻	1.09
30 Mg 12	8.0545	β⁻	-0.47	34 S 16	8.5835		4.21%
30 Al 13	8.2614	β⁻	0.56	34 Cl 17	8.3990	β⁺	0.18
30 Si 14	8.5207		3.10%	34 Ar 18	8.1977	β⁺	-0.07
30 P 15	8.3535	β⁺	2.18	34 K 19	7.6777	?	
30 S 16	8.1228	β⁺	0.07	34 Ca 20	7.2243	?	
30 Cl 17	7.4799	?					
30 Ar 18	6.9325	p	-7.70	35 Na 11	6.6496	β⁻	-2.82
				35 Mg 12	7.3062	?	
31 Ne 10	6.8241	?		35 Al 13	7.7824	β⁻	-0.82
31 Na 11	7.3852	β⁻	-1.77	35 Si 14	8.1687	β⁻	-0.11
31 Mg 12	7.8722	β⁻	-0.64	35 P 15	8.4462	β⁻	1.67
31 Al 13	8.2256	β⁻	-0.19	35 S 16	8.5379	β⁻	6.88
31 Si 14	8.4583	β⁻	3.97	35 Cl 17	8.5203		75.77%
31 P 15	8.4812		100.00%	35 Ar 18	8.3275	β⁺	0.25
31 S 16	8.2819	β⁺	0.41	35 K 19	7.9657	β⁺	-0.72
31 Cl 17	7.8702	β⁺	-0.82	35 Ca 20	7.4975	β⁺	-1.30

A X Z	B/A (MeV)	→	$\log t_{1/2}$ or %	A X Z	B/A (MeV)	→	$\log t_{1/2}$ or %
				40 P 15	7.9864	β⁻	-0.59
36 Mg 12	7.2296	?		40 S 16	8.3296	β⁻	0.94
36 Al 13	7.6245	?		40 Cl 17	8.4278	β⁻	1.91
36 Si 14	8.1115	β⁻	-0.35	40 Ar 18	8.5953		99.60%
36 P 15	8.3079	β⁻	0.75	40 K 19	8.5381	β⁻	0.01%
36 S 16	8.5754		0.02%	40 Ca 20	8.5513	ββ	96.94%
36 Cl 17	8.5219	β⁻	12.98	40 Sc 21	8.1737	β⁺	-0.74
36 Ar 18	8.5199	ββ	0.34%	40 Ti 22	7.8623	β⁺	-1.30
36 K 19	8.1424	β⁺	-0.47	40 V 23	7.3632	?	
36 Ca 20	7.8155	β⁺	-0.99				
36 Sc 21	7.2289	?		41 Si 14	7.5156	?	
				41 P 15	7.9032	β⁻	-0.92
37 Al 13	7.5369	?		41 S 16	8.2197	?	
37 Si 14	7.9516	?		41 Cl 17	8.4137	β⁻	1.58
37 P 15	8.2675	β⁻	0.36	41 Ar 18	8.5344	β⁻	3.82
37 S 16	8.4599	β⁻	2.48	41 K 19	8.5761		6.73%
37 Cl 17	8.5703		24.23%	41 Ca 20	8.5467	EC	12.51
37 Ar 18	8.5272	EC	6.48	41 Sc 21	8.3692	β⁺	-0.22
37 K 19	8.3398	β⁺	0.09	41 Ti 22	8.0348	β⁺	-1.10
37 Ca 20	8.0041	β⁺	-0.74	41 V 23	7.6383	?	
37 Sc 21	7.5505	?					
				42 P 15	7.7899	β⁻	-0.96
38 Al 13	7.3894	?		42 S 16	8.1838	β⁻	-0.25
38 Si 14	7.8816	?		42 Cl 17	8.3496	β⁻	0.83
38 P 15	8.1432	β⁻	-0.19	42 Ar 18	8.5556	β⁻	9.02
38 S 16	8.4488	β⁻	4.01	42 K 19	8.5512	β⁻	4.65
38 Cl 17	8.5055	β⁻	3.35	42 Ca 20	8.6166		0.65%
38 Ar 18	8.6143		0.06%	42 Sc 21	8.4449	β⁺	-0.17
38 K 19	8.4381	β⁺	2.66	42 Ti 22	8.2596	β⁺	-0.70
38 Ca 20	8.2401	β⁺	-0.36	42 V 23	7.8374	?	
38 Sc 21	7.7689	?		42 Cr 24	7.4817	?	
38 Ti 22	7.3789	?					
				43 P 15	7.7267	β⁻	-1.48
39 Si 14	7.7355	?		43 S 16	8.0705	β⁻	-0.66
39 P 15	8.0948	β⁻	-0.80	43 Cl 17	8.3208	β⁻	0.52
39 S 16	8.3442	β⁻	1.06	43 Ar 18	8.4875	β⁻	2.51
39 Cl 17	8.4944	β⁻	3.52	43 K 19	8.5766	β⁻	4.90
39 Ar 18	8.5626	β⁻	9.93	43 Ca 20	8.6007		0.14%
39 K 19	8.5570		93.26%	43 Sc 21	8.5308	β⁺	4.15
39 Ca 20	8.3695	β⁺	-0.07	43 Ti 22	8.3529	β⁺	-0.29
39 Sc 21	8.0133	?		43 V 23	8.0720	β⁺	-0.10
39 Ti 22	7.5984	β⁺	-1.59	43 Cr 24	7.6843	β⁺	-1.68
40 Si 14	7.6624	?		44 S 16	8.0341	β⁻	-0.91

A X Z	B/A (MeV)	$\rightarrow$	$\log t_{1/2}$ or %	A X Z	B/A (MeV)	$\rightarrow$	$\log t_{1/2}$ or %
44 Cl 17	8.2234	β^-	-0.36	48 Ar 18	8.2617	?	
44 Ar 18	8.4845	β^-	2.85	48 K 19	8.4309	β^-	0.83
44 K 19	8.5474	β^-	3.12	48 Ca 20	8.6665	$\beta\beta$	0.19%
44 Ca 20	8.6582		2.09%	48 Sc 21	8.6560	β^-	5.20
44 Sc 21	8.5574	β^+	4.15	48 Ti 22	8.7229		73.80%
44 Ti 22	8.5335	EC	9.30	48 V 23	8.6230	β^+	6.14
44 V 23	8.2043	β^+	-1.05	48 Cr 24	8.5721	β^+	4.89
44 Cr 24	7.9522	β^+	-1.28	48 Mn 25	8.2740	β^+	-0.80
44 Mn 25	7.4814	?		48 Fe 26	8.0248	β^+	-1.36
				48 Co 27	7.5938	?	
45 S 16	7.9004	β^-	-1.09				
45 Cl 17	8.1960	β^-	-0.40	49 K 19	8.3867	β^-	0.10
45 Ar 18	8.4188	β^-	1.33	49 Ca 20	8.5947	β^-	2.72
45 K 19	8.5545	β^-	3.02	49 Sc 21	8.6861	β^-	3.54
45 Ca 20	8.6306	β^-	7.15	49 Ti 22	8.7110		5.50%
45 Sc 21	8.6189		100.00%	49 V 23	8.6828	EC	7.45
45 Ti 22	8.5557	β^+	4.05	49 Cr 24	8.6131	β^+	3.40
45 V 23	8.3798	β^+	-0.26	49 Mn 25	8.4397	β^+	-0.42
45 Cr 24	8.0854	β^+	-1.30	49 Fe 26	8.1579	β^+	-1.15
45 Mn 25	7.7503	?		49 Co 27	7.8419	?	
45 Fe 26	7.3179	?					
				50 K 19	8.2811	β^-	-0.33
46 Cl 17	8.1038	β^-	-0.65	50 Ca 20	8.5498	β^-	1.14
46 Ar 18	8.4113	β^-	0.92	50 Sc 21	8.6335	β^-	2.01
46 K 19	8.5182	β^-	2.02	50 Ti 22	8.7556		5.40%
46 Ca 20	8.6689	$\beta\beta$	0.00%	50 V 23	8.6958		0.25%
46 Sc 21	8.6220	β^-	6.86	50 Cr 24	8.7009	$\beta\beta$	4.34%
46 Ti 22	8.6564		8.00%	50 Mn 25	8.5326	β^+	-0.55
46 V 23	8.4861	β^+	-0.37	50 Fe 26	8.3539	β^+	-0.82
46 Cr 24	8.3038	β^+	-0.59	50 Co 27	7.9989	β^+	-1.36
46 Mn 25	7.9150	β^+	-1.39	50 Ni 28	7.7090	?	
46 Fe 26	7.6127	β^+	-1.70				
				51 Ca 20	8.4685	β^-	1.00
47 Cl 17	8.0272	β^-		51 Sc 21	8.5966	β^-	1.09
47 Ar 18	8.3229	β^-	-0.15	51 Ti 22	8.7089	β^-	2.54
47 K 19	8.5146	β^-	1.24	51 V 23	8.7420		99.75%
47 Ca 20	8.6393	β^-	5.59	51 Cr 24	8.7119	EC	6.38
47 Sc 21	8.6650	β^-	5.46	51 Mn 25	8.6336	β^+	3.44
47 Ti 22	8.6611		7.30%	51 Fe 26	8.4611	β^+	-0.52
47 V 23	8.5822	β^+	3.29	51 Co 27	8.1958	?	
47 Cr 24	8.4070	β^+	-0.30	51 Ni 28	7.8661	?	
47 Mn 25	8.1289	β^+	-1.00				
47 Fe 26	7.7794	β^+	-1.57	52 Ca 20	8.3956	β^-	0.66
				52 Sc 21	8.5334	β^-	0.91

A X Z	B/A (MeV)	$\rightarrow$	$\log t_{1/2}$ or %	A X Z	B/A (MeV)	$\rightarrow$	$\log t_{1/2}$ or %
52 Ti 22	8.6916	β^-	2.01	56 Cr 24	8.7233	β^-	2.55
52 V 23	8.7145	β^-	2.35	56 Mn 25	8.7382	β^-	3.97
52 Cr 24	8.7759		83.79%	56 Fe 26	8.7903		91.72%
52 Mn 25	8.6702	β^+	5.68	56 Co 27	8.6947	β^+	6.82
52 Fe 26	8.6096	β^+	4.47	56 Ni 28	8.6426	β^+	5.72
52 Co 27	8.3250	β^+	-1.74	56 Cu 29	8.3555	?	
52 Ni 28	8.0857	β^+	-1.42	56 Zn 30	8.1116	?	
52 Cu 29	7.6855	?		56 Ga 31	7.7229	?	
53 Ca 20	8.3025	β^-	-1.05	57 Ti 22	8.3528	β^-	-0.74
53 Sc 21	8.4928	?		57 V 23	8.5324	β^-	-0.49
53 Ti 22	8.6301	β^-	1.51	57 Cr 24	8.6611	β^-	1.32
53 V 23	8.7100	β^-	1.98	57 Mn 25	8.7367	β^-	1.93
53 Cr 24	8.7601		9.50%	57 Fe 26	8.7702		2.20%
53 Mn 25	8.7341	EC	14.07	57 Co 27	8.7418	EC	7.37
53 Fe 26	8.6487	β^+	2.71	57 Ni 28	8.6708	β^+	5.11
53 Co 27	8.4773	β^+	-0.62	57 Cu 29	8.5032	β^+	-0.70
53 Ni 28	8.2123	β^+	-1.35	57 Zn 30	8.2330	β^+	-1.40
53 Cu 29	7.8972	?		57 Ga 31	7.9338	?	
54 Sc 21	8.3967	?		58 V 23	8.4562	β^-	-0.70
54 Ti 22	8.5972	?		58 Cr 24	8.6423	β^-	0.85
54 V 23	8.6619	β^-	1.70	58 Mn 25	8.6979	β^-	0.48
54 Cr 24	8.7778		2.37%	58 Fe 26	8.7922		0.28%
54 Mn 25	8.7379	β^+	7.43	58 Co 27	8.7389	β^+	6.79
54 Fe 26	8.7363	$\beta\beta$	5.80%	58 Ni 28	8.7320	$\beta\beta$	68.08%
54 Co 27	8.5691	β^+	-0.71	58 Cu 29	8.5708	β^+	0.51
54 Ni 28	8.3917	β^+		58 Zn 30	8.3959	β^+	-1.19
54 Cu 29	8.0529	?		58 Ga 31	8.0667	?	
54 Zn 30	7.7583	?		58 Ge 32	7.7841	?	
55 Sc 21	8.2909	?		59 V 23	8.4089	β^-	-0.89
55 Ti 22	8.5167	β^-	-0.49	59 Cr 24	8.5628	β^-	-0.13
55 V 23	8.6378	β^-	0.82	59 Mn 25	8.6800	β^-	0.66
55 Cr 24	8.7318	β^-	2.32	59 Fe 26	8.7547	β^-	6.59
55 Mn 25	8.7649		100.00%	59 Co 27	8.7679		100.00%
55 Fe 26	8.7465	EC	7.94	59 Ni 28	8.7365	β^+	12.38
55 Co 27	8.6695	β^+	4.80	59 Cu 29	8.6419	β^+	1.91
55 Ni 28	8.4972	β^+	-0.67	59 Zn 30	8.4745	β^+	-0.74
55 Cu 29	8.2428	β^+		59 Ga 31	8.2386	?	
55 Zn 30	7.9159	?		59 Ge 32	7.9351	?	
56 Ti 22	8.4628	β^-	-0.80	60 V 23	8.3226	β^-	-0.70
56 V 23	8.5742	β^-	-0.64	60 Cr 24	8.5388	β^-	-0.24

A X Z	B/A (MeV)	→	log $t_{1/2}$ or %	A X Z	B/A (MeV)	→	log $t_{1/2}$ or %
60 Mn 25	8.6249	β^-	1.71	64 Co 27	8.6755	β^-	-0.52
60 Fe 26	8.7558	β^-	13.67	64 Ni 28	8.7774		0.93%
60 Co 27	8.7467	β^-	8.22	64 Cu 29	8.7390	β^+	4.66
60 Ni 28	8.7807		26.22%	64 Zn 30	8.7358	$\beta\beta$	48.60%
60 Cu 29	8.6655	β^+	3.15	64 Ga 31	8.6117	β^+	2.20
60 Zn 30	8.5832	β^+	2.16	64 Ge 32	8.5305	β^+	1.80
60 Ga 31	8.3338	?		64 As 33	8.2875	?	
60 Ge 32	8.1169	?					
60 As 33	7.7477	?		65 Mn 25	8.3995	β^-	-0.96
				65 Fe 26	8.5474	β^-	-0.40
61 Cr 24	8.4646	β^-	-0.57	65 Co 27	8.6566	β^-	0.08
61 Mn 25	8.5961	β^-	-0.15	65 Ni 28	8.7362	β^-	3.96
61 Fe 26	8.7037	β^-	2.56	65 Cu 29	8.7570		30.83%
61 Co 27	8.7561	β^-	3.77	65 Zn 30	8.7242	β^+	7.32
61 Ni 28	8.7649		1.14%	65 Ga 31	8.6621	β^+	2.96
61 Cu 29	8.7155	β^+	4.08	65 Ge 32	8.5540	β^+	1.49
61 Zn 30	8.6102	β^+	1.95	65 As 33	8.3981	β^+	-0.72
61 Ga 31	8.4499	β^+	-0.82	65 Se 34	8.1685	β^+	
61 Ge 32	8.2139	β^+	-1.40				
61 As 33	7.9440	?		66 Fe 26	8.5255	β^-	-0.36
				66 Co 27	8.6005	β^-	-0.63
62 Cr 24	8.4325	β^-	-0.72	66 Ni 28	8.7399	β^-	5.29
62 Mn 25	8.5376	β^-	-0.06	66 Cu 29	8.7314	β^-	2.49
62 Fe 26	8.6932	β^-	1.83	66 Zn 30	8.7596		27.90%
62 Co 27	8.7214	β^-	1.95	66 Ga 31	8.6693	β^+	4.53
62 Ni 28	8.7945		3.63%	66 Ge 32	8.6257	β^+	3.91
62 Cu 29	8.7182	β^+	2.77	66 As 33	8.4691	β^+	-1.02
62 Zn 30	8.6793	β^+	4.52	66 Se 34	8.3004	β^+	
62 Ga 31	8.5188	β^+	-0.94				
62 Ge 32	8.3489	β^+		67 Fe 26	8.4629	β^-	-0.33
62 As 33	8.0575	?		67 Co 27	8.5817	β^-	-0.38
				67 Ni 28	8.6958	β^-	1.32
63 Mn 25	8.5030	β^-	-0.60	67 Cu 29	8.7372	β^-	5.35
63 Fe 26	8.6296	β^-	0.79	67 Zn 30	8.7341		4.10%
63 Co 27	8.7176	β^-	1.44	67 Ga 31	8.7075	EC	5.45
63 Ni 28	8.7634	β^-	9.50	67 Ge 32	8.6328	β^+	3.05
63 Cu 29	8.7521		69.17%	67 As 33	8.5314	β^+	1.63
63 Zn 30	8.6862	β^+	3.36	67 Se 34	8.3682	β^+	-1.22
63 Ga 31	8.5862	β^+	1.51				
63 Ge 32	8.4185	β^+	-1.02	68 Fe 26	8.4227	β^-	-1.00
63 As 33	8.1984	?		68 Co 27	8.5229	β^-	-0.74
				68 Ni 28	8.6828	β^-	1.28
64 Mn 25	8.4392	β^-	-0.85	68 Cu 29	8.7015	β^-	1.49
64 Fe 26	8.6113	β^-	0.30	68 Zn 30	8.7556		18.80%

A X Z	B/A (MeV)	→	log $t_{1/2}$ or %	A X Z	B/A (MeV)	→	log $t_{1/2}$ or %
68 Ga 31	8.7012	β^+	3.61	72 Se 34	8.6449	EC	5.86
68 Ge 32	8.6881	EC	7.37	72 Br 35	8.5130	β^+	1.90
68 As 33	8.5575	β^+	2.18	72 Kr 36	8.4321	β^+	1.24
68 Se 34	8.4764	β^+	1.55	72 Rb 37	8.1987	?	
68 Br 35	8.2406	p	-5.82				
				73 Ni 28	8.4607	β^-	-0.15
69 Co 27	8.5050	β^-	-0.57	73 Cu 29	8.5709	β^-	0.59
69 Ni 28	8.6289	β^-	1.06	73 Zn 30	8.6458	β^-	1.37
69 Cu 29	8.6953	β^-	2.23	73 Ga 31	8.6939	β^-	4.24
69 Zn 30	8.7227	β^-	3.53	73 Ge 32	8.7050		7.73%
69 Ga 31	8.7245		60.11%	73 As 33	8.6897	EC	6.84
69 Ge 32	8.6809	β^+	5.15	73 Se 34	8.6414	β^+	4.41
69 As 33	8.6114	β^+	2.96	73 Br 35	8.5669	β^+	2.31
69 Se 34	8.5017	β^+	1.44	73 Kr 36	8.4648	β^+	1.43
69 Br 35	8.3510	?		73 Rb 37	8.3085	?	
70 Co 27	8.4374	β^-	-0.82	74 Ni 28	8.4338	β^-	-0.27
70 Ni 28	8.6082	?		74 Cu 29	8.5191	β^-	0.20
70 Cu 29	8.6466	β^-	0.65	74 Zn 30	8.6421	β^-	1.98
70 Zn 30	8.7297	$\beta\beta$	0.60%	74 Ga 31	8.6632	β^-	2.69
70 Ga 31	8.7092	β^-	3.10	74 Ge 32	8.7252		35.94%
70 Ge 32	8.7217		21.23%	74 As 33	8.6800	β^-	6.19
70 As 33	8.6217	β^+	3.50	74 Se 34	8.6877	$\beta\beta$	0.89%
70 Se 34	8.5762	β^+	3.39	74 Br 35	8.5838	β^+	3.18
70 Br 35	8.4226	β^+	-1.10	74 Kr 36	8.5308	β^+	2.84
70 Kr 36	8.2543	?		74 Rb 37	8.3791	β^+	-1.19
71 Co 27	8.4071	β^-	-0.68	75 Ni 28	8.3681	β^-	-0.22
71 Ni 28	8.5500	β^-	0.27	75 Cu 29	8.4965	β^-	0.09
71 Cu 29	8.6358	β^-	1.29	75 Zn 30	8.5913	β^-	1.01
71 Zn 30	8.6889	β^-	2.17	75 Ga 31	8.6608	β^-	2.10
71 Ga 31	8.7175		39.89%	75 Ge 32	8.6956	β^-	3.70
71 Ge 32	8.7033	EC	5.99	75 As 33	8.7009		100.00%
71 As 33	8.6639	β^+	5.37	75 Se 34	8.6789	EC	7.01
71 Se 34	8.5905	β^+	2.45	75 Br 35	8.6281	β^+	3.76
71 Br 35	8.4827	β^+	1.33	75 Kr 36	8.5523	β^+	2.41
71 Kr 36	8.3239	β^+	-1.19	75 Rb 37	8.4483	β^+	1.28
				75 Sr 38	8.2969	β^+	-1.15
72 Ni 28	8.5265	β^-	0.32				
72 Cu 29	8.5882	β^-	0.82	76 Ni 28	8.3381	β^-	-0.62
72 Zn 30	8.6915	β^-	5.22	76 Cu 29	8.4404	β^-	-0.19
72 Ga 31	8.6870	β^-	4.71	76 Zn 30	8.5789	β^-	0.76
72 Ge 32	8.7317		27.66%	76 Ga 31	8.6233	β^-	1.51
72 As 33	8.6604	β^+	4.97	76 Ge 32	8.7052	$\beta\beta$	7.44%

A X Z	B/A (MeV)	→	log $t_{1/2}$ or %	A X Z	B/A (MeV)	→	log $t_{1/2}$ or %
76 As 33	8.6828	β⁻	4.97				
76 Se 34	8.7115		9.36%	80 Zn 30	8.4252	β⁻	-0.26
76 Br 35	8.6359	β⁺	4.77	80 Ga 31	8.5065	β⁻	0.23
76 Kr 36	8.6083	β⁺	4.73	80 Ge 32	8.6265	β⁻	1.47
76 Rb 37	8.4862	β⁺	1.56	80 As 33	8.6501	β⁻	1.18
76 Sr 38	8.3958	β⁺	0.95	80 Se 34	8.7108	ββ	49.61%
				80 Br 35	8.6777	β⁻	3.03
77 Ni 28	8.2700	?		80 Kr 36	8.6929		2.25%
77 Cu 29	8.4144	β⁻	-0.33	80 Rb 37	8.6116	β⁺	1.53
77 Zn 30	8.5276	β⁻	0.32	80 Sr 38	8.5785	β⁺	3.80
77 Ga 31	8.6119	β⁻	1.12	80 Y 39	8.4819	β⁺	1.54
77 Ge 32	8.6710	β⁻	4.61	80 Zr 40	8.3719	?	
77 As 33	8.6960	β⁻	5.15				
77 Se 34	8.6947		7.63%	81 Zn 30	8.3510	β⁻	-0.54
77 Br 35	8.6668	β⁺	5.31	81 Ga 31	8.4877	β⁻	0.09
77 Kr 36	8.6168	β⁺	3.65	81 Ge 32	8.5808	β⁻	0.88
77 Rb 37	8.5373	β⁺	2.35	81 As 33	8.6480	β⁻	1.52
77 Sr 38	8.4381	β⁺	0.95	81 Se 34	8.6860	β⁻	3.05
77 Y 39	8.2845	β⁺		81 Br 35	8.6959		49.31%
				81 Kr 36	8.6828	EC	12.86
78 Ni 28	8.2359	β⁻		81 Rb 37	8.6455	β⁺	4.22
78 Cu 29	8.3556	β⁻	-0.47	81 Sr 38	8.5873	β⁺	3.13
78 Zn 30	8.5040	β⁻	0.17	81 Y 39	8.5096	β⁺	1.85
78 Ga 31	8.5766	β⁻	0.71	81 Zr 40	8.4116	β⁺	1.18
78 Ge 32	8.6717	β⁻	3.72				
78 As 33	8.6739	β⁻	3.74	82 Zn 30	8.2981	β⁻	
78 Se 34	8.7178		23.78%	82 Ga 31	8.4212	β⁻	-0.22
78 Br 35	8.6620	β⁺	2.59	82 Ge 32	8.5653	β⁻	0.66
78 Kr 36	8.6610	ββ	0.35%	82 As 33	8.6130	β⁻	1.28
78 Rb 37	8.5583	β⁺	3.03	82 Se 34	8.6932	ββ	8.73%
78 Sr 38	8.5001	β⁺	2.18	82 Br 35	8.6825	β⁻	5.10
78 Y 39	8.3549	?		82 Kr 36	8.7106		11.60%
				82 Rb 37	8.6474	β⁺	1.88
79 Cu 29	8.3247	β⁻	-0.73	82 Sr 38	8.6357	EC	6.34
79 Zn 30	8.4570	β⁻	0.00	82 Y 39	8.5308	β⁺	0.98
79 Ga 31	8.5553	β⁻	0.45	82 Zr 40	8.4725	β⁺	1.51
79 Ge 32	8.6340	β⁻	1.28	82 Nb 41	8.3262	?	
79 As 33	8.6766	β⁻	2.73				
79 Se 34	8.6956	β⁻	13.55	83 Ga 31	8.3754	β⁻	-0.51
79 Br 35	8.6876		50.69%	83 Ge 32	8.5047	β⁻	0.27
79 Kr 36	8.6571	β⁺	5.10	83 As 33	8.6022	β⁻	1.13
79 Rb 37	8.6010	β⁺	3.14	83 Se 34	8.6586	β⁻	3.13
79 Sr 38	8.5238	β⁺	2.13	83 Br 35	8.6933	β⁻	3.94
79 Y 39	8.4238	β⁺	1.17	83 Kr 36	8.6956		11.50%

A X Z	B/A (MeV)	$\rightarrow$	$\log t_{1/2}$ or %	A X Z	B/A (MeV)	$\rightarrow$	$\log t_{1/2}$ or %
83 Rb 37	8.6752	EC	6.87	87 As 33	8.4215	β^-	-0.32
83 Sr 38	8.6384	β^+	5.07	87 Se 34	8.5308	β^-	0.72
83 Y 39	8.5751	β^+	2.63	87 Br 35	8.6055	β^-	1.75
83 Zr 40	8.4950	β^+	1.64	87 Kr 36	8.6752	β^-	3.66
83 Nb 41	8.3952	β^+	0.61	87 Rb 37	8.7109	β^-	27.83%
				87 Sr 38	8.7052		7.00%
84 Ga 31	8.3111	β^-	-1.07	87 Y 39	8.6748	β^+	5.46
84 Ge 32	8.4685	β^-	-0.02	87 Zr 40	8.6237	β^+	3.78
84 As 33	8.5506	β^-	0.65	87 Nb 41	8.5553	β^+	2.19
84 Se 34	8.6588	β^-	2.27	87 Mo 42	8.4717	β^+	1.13
84 Br 35	8.6712	β^-	3.28	87 Tc 43	8.3642	?	
84 Kr 36	8.7173		57.00%				
84 Rb 37	8.6761	β^-	6.45	88 As 33	8.3648	?	
84 Sr 38	8.6774	$\beta\beta$	0.56%	88 Se 34	8.4949	β^-	0.18
84 Y 39	8.5918	β^+	0.66	88 Br 35	8.5639	β^-	1.21
84 Zr 40	8.5499	β^+	3.19	88 Kr 36	8.6568	β^-	4.01
84 Nb 41	8.4261	β^+	1.08	88 Rb 37	8.6810	β^-	3.03
84 Mo 42	8.3445	β^+		88 Sr 38	8.7326		82.58%
				88 Y 39	8.6825	β^+	6.96
85 Ge 32	8.4048	β^-	-0.27	88 Zr 40	8.6660	EC	6.86
85 As 33	8.5149	β^-	0.31	88 Nb 41	8.5713	β^+	2.94
85 Se 34	8.6105	β^-	1.50	88 Mo 42	8.5241	β^+	2.68
85 Br 35	8.6740	β^-	2.24	88 Tc 43	8.4000	β^+	0.81
85 Kr 36	8.6985	β^-	8.53				
85 Rb 37	8.6974		72.17%	89 Se 34	8.4421	β^-	-0.39
85 Sr 38	8.6757	β^+	6.75	89 Br 35	8.5340	β^-	0.64
85 Y 39	8.6282	β^+	3.98	89 Kr 36	8.6169	β^-	2.28
85 Zr 40	8.5638	β^+	2.67	89 Rb 37	8.6641	β^-	2.96
85 Nb 41	8.4840	β^+	1.32	89 Sr 38	8.7059	β^-	6.64
85 Mo 42	8.3796	?		89 Y 39	8.7139		100.00%
				89 Zr 40	8.6733	β^+	5.45
86 Ge 32	8.3622	?		89 Nb 41	8.6163	β^+	3.84
86 As 33	8.4618	β^-	-0.02	89 Mo 42	8.5449	β^+	2.09
86 Se 34	8.5822	β^-	1.18	89 Tc 43	8.4517	β^+	1.11
86 Br 35	8.6324	β^-	1.74	89 Ru 44	8.3532	?	
86 Kr 36	8.7120	$\beta\beta$	17.30%				
86 Rb 37	8.6969	β^-	6.21	90 Se 34	8.4028	?	
86 Sr 38	8.7084		9.86%	90 Br 35	8.4850	β^-	0.28
86 Y 39	8.6384	β^+	4.73	90 Kr 36	8.5913	β^-	1.51
86 Zr 40	8.6122	β^+	4.77	90 Rb 37	8.6314	β^-	2.20
86 Nb 41	8.5103	β^+	1.94	90 Sr 38	8.6959	β^-	8.96
86 Mo 42	8.4453	β^+	1.29	90 Y 39	8.6933	β^-	5.36
86 Tc 43	8.2980	?		90 Zr 40	8.7099		51.45%
				90 Nb 41	8.6333	β^+	4.72

A X Z	B/A (MeV)	→	$\log t_{1/2}$ or %	A X Z	B/A (MeV)	→	$\log t_{1/2}$ or %
90 Mo 42	8.5970	β^+	4.30	94 Zr 40	8.6668	$\beta\beta$	17.38%
90 Tc 43	8.4896	β^+	1.69	94 Nb 41	8.6489	β^-	11.81
90 Ru 44	8.4156	β^+	1.04	94 Mo 42	8.6623		9.25%
				94 Tc 43	8.6087	β^+	4.25
91 Se 34	8.3382	β^-	-0.57	94 Ru 44	8.5834	β^+	3.49
91 Br 35	8.4468	β^-	-0.27	94 Rh 45	8.4727	β^+	1.85
91 Kr 36	8.5459	β^-	0.93	94 Pd 46	8.3943	β^+	0.95
91 Rb 37	8.6080	β^-	1.77				
91 Sr 38	8.6638	β^-	4.54	95 Kr 36	8.3658	β^-	-0.11
91 Y 39	8.6849	β^-	6.70	95 Rb 37	8.4599	β^-	-0.42
91 Zr 40	8.6933		11.22%	95 Sr 38	8.5495	β^-	1.38
91 Nb 41	8.6709	β^+	10.33	95 Y 39	8.6053	β^-	2.79
91 Mo 42	8.6136	β^+	2.97	95 Zr 40	8.6436	β^-	6.74
91 Tc 43	8.5366	β^+	2.27	95 Nb 41	8.6472	β^-	6.48
91 Ru 44	8.4467	β^+	0.95	95 Mo 42	8.6487		15.92%
				95 Tc 43	8.6227	β^+	4.86
92 Br 35	8.3892	β^-	-0.46	95 Ru 44	8.5873	β^+	3.77
92 Kr 36	8.5133	β^-	0.26	95 Rh 45	8.5253	β^+	2.48
92 Rb 37	8.5699	β^-	0.65	95 Pd 46	8.4309	β^+	
92 Sr 38	8.6495	β^-	3.99				
92 Y 39	8.6617	β^-	4.10	96 Kr 36	8.3328	?	
92 Zr 40	8.6926		17.15%	96 Rb 37	8.4076	β^-	-0.70
92 Nb 41	8.6623	β^+	15.04	96 Sr 38	8.5219	β^-	0.03
92 Mo 42	8.6577	$\beta\beta$	14.84%	96 Y 39	8.5697	β^-	0.73
92 Tc 43	8.5637	β^+	2.40	96 Zr 40	8.6354	$\beta\beta$	2.80%
92 Ru 44	8.5059	β^+	2.34	96 Nb 41	8.6289	β^-	4.92
92 Rh 45	8.3773	?		96 Mo 42	8.6540		16.68%
				96 Tc 43	8.6148	β^+	5.57
93 Br 35	8.3468	β^-	-0.99	96 Ru 44	8.6093	$\beta\beta$	5.52%
93 Kr 36	8.4577	β^-	0.11	96 Rh 45	8.5340	β^+	2.77
93 Rb 37	8.5418	β^-	0.77	96 Pd 46	8.4899	β^+	2.09
93 Sr 38	8.6136	β^-	2.65	96 Ag 47	8.3609	β^+	0.71
93 Y 39	8.6491	β^-	4.56				
93 Zr 40	8.6716	β^-	13.68	97 Rb 37	8.3747	β^-	-0.77
93 Nb 41	8.6642		100.00%	97 Sr 38	8.4741	β^-	-0.37
93 Mo 42	8.6514	EC	11.10	97 Y 39	8.5430	β^-	0.57
93 Tc 43	8.6086	β^+	4.00	97 Zr 40	8.6039	β^-	4.78
93 Ru 44	8.5320	β^+	1.78	97 Nb 41	8.6232	β^-	3.64
93 Rh 45	8.4366	?		97 Mo 42	8.6351		9.55%
				97 Tc 43	8.6237	EC	13.91
94 Kr 36	8.4230	β^-	-0.70	97 Ru 44	8.6041	β^+	5.40
94 Rb 37	8.4924	β^-	0.43	97 Rh 45	8.5598	β^+	3.26
94 Sr 38	8.5937	β^-	1.88	97 Pd 46	8.5023	β^+	2.27
94 Y 39	8.6228	β^-	3.05	97 Ag 47	8.4221	β^+	1.28

A X Z	B/A (MeV)	$\rightarrow$	$\log t_{1/2}$ or %	A X Z	B/A (MeV)	$\rightarrow$	$\log t_{1/2}$ or %
				101 Sr 38	8.3256	β^-	-0.93
98 Rb 37	8.3297	β^-	-0.94	101 Y 39	8.4119	β^-	-0.35
98 Sr 38	8.4477	β^-	-0.19	101 Zr 40	8.4888	β^-	0.36
98 Y 39	8.4991	β^-	-0.26	101 Nb 41	8.5354	β^-	0.85
98 Zr 40	8.5812	β^-	1.49	101 Mo 42	8.5728	β^-	2.94
98 Nb 41	8.5963	β^-	0.46	101 Tc 43	8.5931	β^-	2.93
98 Mo 42	8.6351	$\beta\beta$	24.13%	101 Ru 44	8.6013		17.00%
98 Tc 43	8.6100	β^-	14.12	101 Rh 45	8.5882	EC	8.02
98 Ru 44	8.6203		1.88%	101 Pd 46	8.5608	β^+	4.48
98 Rh 45	8.5607	β^+	2.72	101 Ag 47	8.5115	β^+	2.82
98 Pd 46	8.5336	β^+	3.03	101 Cd 48	8.4495	β^+	1.91
98 Ag 47	8.4397	β^+	1.67	101 In 49	8.3691	β^+	1.18
98 Cd 48	8.3764	β^+	0.96	101 Sn 50	8.2737	β^+	0.48
99 Rb 37	8.2933	β^-	-1.30	102 Sr 38	8.3002	β^-	-1.16
99 Sr 38	8.3990	β^-	-0.57	102 Y 39	8.3790	β^-	-0.44
99 Y 39	8.4723	β^-	0.17	102 Zr 40	8.4679	β^-	0.46
99 Zr 40	8.5408	β^-	0.32	102 Nb 41	8.5054	β^-	0.11
99 Nb 41	8.5789	β^-	1.18	102 Mo 42	8.5684	β^-	2.83
99 Mo 42	8.6078	β^-	5.37	102 Tc 43	8.5706	β^-	0.72
99 Tc 43	8.6136	β^-	12.82	102 Ru 44	8.6074		31.60%
99 Ru 44	8.6086		12.70%	102 Rh 45	8.5769	β^-	7.25
99 Rh 45	8.5795	β^+	6.14	102 Pd 46	8.5805	$\beta\beta$	1.02%
99 Pd 46	8.5376	β^+	3.11	102 Ag 47	8.5148	β^+	2.89
99 Ag 47	8.4748	β^+	2.09	102 Cd 48	8.4817	β^+	2.52
99 Cd 48	8.3976	β^+	1.20	102 In 49	8.3868	β^+	1.34
99 In 49	8.2994	?		102 Sn 50	8.3226	β^+	0.65
100 Rb 37	8.2488	β^-	-1.29	103 Y 39	8.3439	β^-	-0.64
100 Sr 38	8.3762	β^-	-0.69	103 Zr 40	8.4313	β^-	0.11
100 Y 39	8.4392	β^-	-0.13	103 Nb 41	8.4912	β^-	0.18
100 Zr 40	8.5244	β^-	0.85	103 Mo 42	8.5373	β^-	1.83
100 Nb 41	8.5500	β^-	0.18	103 Tc 43	8.5661	β^-	1.73
100 Mo 42	8.6046	$\beta\beta$	9.63%	103 Ru 44	8.5843	β^-	6.53
100 Tc 43	8.5951	β^-	1.20	103 Rh 45	8.5841		100.00%
100 Ru 44	8.6193		12.60%	103 Pd 46	8.5712	EC	6.17
100 Rh 45	8.5752	β^+	4.87	103 Ag 47	8.5375	β^+	3.60
100 Pd 46	8.5637	EC	5.50	103 Cd 48	8.4897	β^+	2.64
100 Ag 47	8.4851	β^+	2.08	103 In 49	8.4234	β^+	1.81
100 Cd 48	8.4384	β^+	1.69	103 Sn 50	8.3415	β^+	0.85
100 In 49	8.3253	β^+	0.85				
100 Sn 50	8.2448	β^+	-0.03	104 Y 39	8.3059	?	
				104 Zr 40	8.4083	β^-	0.08
101 Rb 37	8.2164	β^-	-1.49	104 Nb 41	8.4574	β^-	0.68

A X Z	B/A (MeV)	→	log $t_{1/2}$ or %	A X Z	B/A (MeV)	→	log $t_{1/2}$ or %
104 Mo 42	8.5278	β⁻	1.78	107 Ag 47	8.5539		51.84%
104 Tc 43	8.5410	β⁻	3.04	107 Cd 48	8.5333	β⁺	4.37
104 Ru 44	8.5874	ββ	18.70%	107 In 49	8.4940	β⁺	3.29
104 Rh 45	8.5689	β⁻	1.63	107 Sn 50	8.4400	β⁺	2.24
104 Pd 46	8.5848		11.14%	107 Sb 51	8.3587	?	
104 Ag 47	8.5362	β⁺	3.62	107 Te 52	8.2567	α	-2.51
104 Cd 48	8.5177	β⁺	3.54				
104 In 49	8.4341	β⁺	2.03	108 Nb 41	8.3391	β⁻	-0.71
104 Sn 50	8.3832	β⁺	1.32	108 Mo 42	8.4226	β⁻	0.04
104 Sb 51	8.2552	β⁺	-0.36	108 Tc 43	8.4629	β⁻	0.71
				108 Ru 44	8.5272	β⁻	2.44
105 Zr 40	8.3672	β⁻	-0.22	108 Rh 45	8.5325	β⁻	1.23
105 Nb 41	8.4406	β⁻	0.47	108 Pd 46	8.5670		26.46%
105 Mo 42	8.4950	β⁻	1.55	108 Ag 47	8.5420	β⁻	2.15
105 Tc 43	8.5347	β⁻	2.66	108 Cd 48	8.5500	ββ	0.89%
105 Ru 44	8.5619	β⁻	4.20	108 In 49	8.4951	β⁺	3.54
105 Rh 45	8.5727	β⁻	5.10	108 Sn 50	8.4685	β⁺	2.79
105 Pd 46	8.5706		22.33%	108 Sb 51	8.3732	β⁺	0.87
105 Ag 47	8.5504	β⁺	6.55	108 Te 52	8.3028	β⁺	0.32
105 Cd 48	8.5168	β⁺	3.52	108 I 53	8.1741	α	-1.44
105 In 49	8.4632	β⁺	2.48				
105 Sn 50	8.3962	β⁺	1.49	109 Mo 42	8.3878	β⁻	-0.28
105 Sb 51	8.3000	β⁺	0.05	109 Tc 43	8.4465	β⁻	-0.06
				109 Ru 44	8.4973	β⁻	1.54
106 Zr 40	8.3439	?		109 Rh 45	8.5283	β⁻	1.90
106 Nb 41	8.4006	β⁻	0.01	109 Pd 46	8.5449	β⁻	4.69
106 Mo 42	8.4807	β⁻	0.92	109 Ag 47	8.5479		48.16%
106 Tc 43	8.5066	β⁻	1.55	109 Cd 48	8.5388	EC	7.60
106 Ru 44	8.5610	β⁻	7.51	109 In 49	8.5131	β⁺	4.18
106 Rh 45	8.5539	β⁻	1.47	109 Sn 50	8.4706	β⁺	3.03
106 Pd 46	8.5800		27.33%	109 Sb 51	8.4049	β⁺	1.23
106 Ag 47	8.5446	β⁺	3.16	109 Te 52	8.3181	β⁺	0.66
106 Cd 48	8.5391	ββ	1.25%	109 I 53	8.2191	p	-4.00
106 In 49	8.4702	β⁺	2.57				
106 Sn 50	8.4327	β⁺	2.06	110 Mo 42	8.3696	β⁻	-0.52
106 Sb 51	8.3260	?		110 Tc 43	8.4142	β⁻	-0.04
106 Te 52	8.2347	α	-4.22	110 Ru 44	8.4869	β⁻	1.16
				110 Rh 45	8.5054	β⁻	0.51
107 Nb 41	8.3794	β⁻	-0.48	110 Pd 46	8.5473	ββ	11.72%
107 Mo 42	8.4459	β⁻	0.54	110 Ag 47	8.5321	β⁻	1.39
107 Tc 43	8.4962	β⁻	1.33	110 Cd 48	8.5513		12.49%
107 Ru 44	8.5339	β⁻	2.35	110 In 49	8.5089	β⁺	4.25
107 Rh 45	8.5541	β⁻	3.11	110 Sn 50	8.4960	EC	4.17
107 Pd 46	8.5609	β⁻	14.31	110 Sb 51	8.4069	β⁺	1.36

A X Z	B/A (MeV)	$\rightarrow$	$\log t_{1/2}$ or %	A X Z	B/A (MeV)	$\rightarrow$	$\log t_{1/2}$ or %
110 Te 52	8.3586	β^+	1.27				
110 I 53	8.2479	β^+	-0.19	114 Ru 44	8.3912	β^-	-0.28
110 Xe 54	8.1572	β^+	-6.22	114 Rh 45	8.4379	β^-	0.27
				114 Pd 46	8.4880	β^-	2.16
111 Tc 43	8.3972	β^-	-0.52	114 Ag 47	8.4939	β^-	0.66
111 Ru 44	8.4530	β^-	0.33	114 Cd 48	8.5316	$\beta\beta$	28.73%
111 Rh 45	8.4955	β^-	1.04	114 In 49	8.5120	β^-	1.86
111 Pd 46	8.5221	β^-	3.15	114 Sn 50	8.5225		0.65%
111 Ag 47	8.5348	β^-	5.81	114 Sb 51	8.4641	β^+	2.32
111 Cd 48	8.5371		12.80%	114 Te 52	8.4296	β^+	2.96
111 In 49	8.5222	EC	5.38	114 I 53	8.3462	β^+	0.32
111 Sn 50	8.4932	β^+	3.33	114 Xe 54	8.2879	β^+	1.00
111 Sb 51	8.4459	β^+	1.88	114 Cs 55	8.1772	β^+	-0.24
111 Te 52	8.3667	β^+	1.29				
111 I 53	8.2829	β^+	0.40	115 Ru 44	8.3527	β^-	-0.40
111 Xe 54	8.1806	β^+	-0.13	115 Rh 45	8.4122	β^-	0.00
				115 Pd 46	8.4575	β^-	1.40
112 Tc 43	8.3595	β^-	-0.55	115 Ag 47	8.4906	β^-	3.08
112 Ru 44	8.4391	β^-	0.24	115 Cd 48	8.5108	β^-	5.28
112 Rh 45	8.4725	β^-	0.32	115 In 49	8.5165	β^-	95.70%
112 Pd 46	8.5209	β^-	4.88	115 Sn 50	8.5141		0.34%
112 Ag 47	8.5164	β^-	4.05	115 Sb 51	8.4809	β^+	3.29
112 Cd 48	8.5448		24.13%	115 Te 52	8.4338	β^+	2.54
112 In 49	8.5147	β^-	2.95	115 I 53	8.3688	β^+	1.89
112 Sn 50	8.5136	$\beta\beta$	0.97%	115 Xe 54	8.2955	β^+	1.26
112 Sb 51	8.4437	β^+	1.71	115 Cs 55	8.2161	β^+	0.15
112 Te 52	8.3979	β^+	2.08	115 Ba 56	8.1139	β^+	-0.40
112 I 53	8.3002	β^+	0.53				
112 Xe 54	8.2293	β^+	0.43	116 Ru 44	8.3363	?	
112 Cs 55	8.1004	p	-3.30	116 Rh 45	8.3881	β^-	-0.17
				116 Pd 46	8.4503	β^-	1.07
113 Tc 43	8.3397	β^-	-0.89	116 Ag 47	8.4661	β^-	2.21
113 Ru 44	8.4052	β^-	-0.10	116 Cd 48	8.5124	$\beta\beta$	7.49%
113 Rh 45	8.4570	β^-	0.45	116 In 49	8.5016	β^-	1.15
113 Pd 46	8.4935	β^-	1.97	116 Sn 50	8.5231		14.53%
113 Ag 47	8.5161	β^-	4.29	116 Sb 51	8.4758	β^+	2.98
113 Cd 48	8.5270	β^-	12.22%	116 Te 52	8.4561	β^+	3.95
113 In 49	8.5229		4.30%	116 I 53	8.3826	β^+	0.46
113 Sn 50	8.5068	β^+	7.00	116 Xe 54	8.3384	β^+	1.77
113 Sb 51	8.4653	β^+	2.60	116 Cs 55	8.2386	β^+	0.58
113 Te 52	8.4083	β^+	2.01	116 Ba 56	8.1619	β^+	-0.52
113 I 53	8.3338	β^+	0.82				
113 Xe 54	8.2467	β^+	0.44	117 Rh 45	8.3647	β^-	-0.36
113 Cs 55	8.1479	p	-4.77	117 Pd 46	8.4178	β^-	0.63

A X Z	B/A (MeV)	$\rightarrow$	$\log t_{1/2}$ or %	A X Z	B/A (MeV)	$\rightarrow$	$\log t_{1/2}$ or %
117 Ag 47	8.4600	β^-	1.86	120 Sb 51	8.4757	β^+	2.98
117 Cd 48	8.4890	β^-	3.95	120 Te 52	8.4773	$\beta\beta$	0.10%
117 In 49	8.5039	β^-	3.41	120 I 53	8.4240	β^+	3.69
117 Sn 50	8.5096		7.68%	120 Xe 54	8.4011	β^+	3.38
117 Sb 51	8.4879	β^+	4.00	120 Cs 55	8.3286	β^+	1.81
117 Te 52	8.4510	β^+	3.57	120 Ba 56	8.2804	β^+	1.51
117 I 53	8.4045	β^+	2.12	120 La 57	8.1804	β^+	0.45
117 Xe 54	8.3428	β^+	1.79				
117 Cs 55	8.2718	β^+	0.92	121 Pd 46	8.3268	?	
117 Ba 56	8.1929	β^+	0.24	121 Ag 47	8.3835	β^-	-0.11
				121 Cd 48	8.4300	β^-	1.13
118 Rh 45	8.3301	?		121 In 49	8.4639	β^-	1.36
118 Pd 46	8.4065	β^-	0.28	121 Sn 50	8.4852	β^-	4.99
118 Ag 47	8.4347	β^-	0.58	121 Sb 51	8.4820		57.36%
118 Cd 48	8.4879	β^-	3.48	121 Te 52	8.4669	β^+	6.16
118 In 49	8.4857	β^-	0.70	121 I 53	8.4417	β^+	3.88
118 Sn 50	8.5165		24.23%	121 Xe 54	8.4044	β^+	3.38
118 Sb 51	8.4789	β^+	2.33	121 Cs 55	8.3533	β^+	2.19
118 Te 52	8.4699	EC	5.71	121 Ba 56	8.2905	β^+	1.47
118 I 53	8.4036	β^+	2.91	121 La 57	8.2185	β^+	0.72
118 Xe 54	8.3720	β^+	2.36	121 Ce 58	8.1300	?	
118 Cs 55	8.2866	β^+	1.15				
118 Ba 56	8.2255	β^+	0.74	122 Ag 47	8.3554	β^-	-0.32
118 La 57	8.1158	?		122 Cd 48	8.4240	β^-	0.72
				122 In 49	8.4421	β^-	0.18
119 Rh 45	8.3128	?		122 Sn 50	8.4879	$\beta\beta$	4.63%
119 Pd 46	8.3741	β^-	-0.04	122 Sb 51	8.4682	β^-	5.37
119 Ag 47	8.4225	β^-	0.32	122 Te 52	8.4780		2.60%
119 Cd 48	8.4608	β^-	2.21	122 I 53	8.4369	β^+	2.34
119 In 49	8.4862	β^-	2.16	122 Xe 54	8.4232	EC	4.86
119 Sn 50	8.4995		8.59%	122 Cs 55	8.3589	β^+	1.32
119 Sb 51	8.4879	EC	5.14	122 Ba 56	8.3210	β^+	2.07
119 Te 52	8.4620	β^+	4.76	122 La 57	8.2348	β^+	0.94
119 I 53	8.4259	β^+	3.06	122 Ce 58	8.1727	?	
119 Xe 54	8.3773	β^+	2.54				
119 Cs 55	8.3176	β^+	1.63	123 Ag 47	8.3411	β^-	-0.51
119 Ba 56	8.2430	β^+	0.73	123 Cd 48	8.3946	β^-	0.32
119 La 57	8.1572	?		123 In 49	8.4379	β^-	0.78
				123 Sn 50	8.4673	β^-	7.05
120 Pd 46	8.3611	β^-	-0.30	123 Sb 51	8.4723		42.64%
120 Ag 47	8.3963	β^-	0.09	123 Te 52	8.4655	EC	0.91%
120 Cd 48	8.4582	β^-	1.71	123 I 53	8.4491	β^+	4.68
120 In 49	8.4663	β^-	0.49	123 Xe 54	8.4210	β^+	3.87
120 Sn 50	8.5045		32.59%	123 Cs 55	8.3805	β^+	2.55

A X Z	B/A (MeV)	$\rightarrow$	$\log t_{1/2}$ or %	A X Z	B/A (MeV)	$\rightarrow$	$\log t_{1/2}$ or %
123 Ba 56	8.3297	β^+	2.21	127 Cd 48	8.3152	β^-	-0.43
123 La 57	8.2674	β^+	1.23	127 In 49	8.3757	β^-	0.04
123 Ce 58	8.1908	β^+	0.51	127 Sn 50	8.4209	β^-	3.88
				127 Sb 51	8.4399	β^-	5.52
124 Ag 47	8.3117	β^-	-0.76	127 Te 52	8.4462	β^-	4.53
124 Cd 48	8.3871	β^-	0.10	127 I 53	8.4455		100.00%
124 In 49	8.4144	β^-	0.49	127 Xe 54	8.4341	EC	6.50
124 Sn 50	8.4674	$\beta\beta$	5.79%	127 Cs 55	8.4116	β^+	4.35
124 Sb 51	8.4561	β^-	6.72	127 Ba 56	8.3783	β^+	2.88
124 Te 52	8.4733		4.82%	127 La 57	8.3351	β^+	2.49
124 I 53	8.4415	β^+	5.56	127 Ce 58	8.2806	β^+	1.49
124 Xe 54	8.4375	$\beta\beta$	0.10%	127 Pr 59	8.2152	β^+	0.62
124 Cs 55	8.3835	β^+	1.49	127 Nd 60	8.1381	β^+	0.26
124 Ba 56	8.3559	β^+	2.82				
124 La 57	8.2786	β^+	1.46	128 Cd 48	8.3036	β^-	-0.47
124 Ce 58	8.2273	β^+	0.78	128 In 49	8.3528	β^-	-0.08
124 Pr 59	8.1267	β^+	0.08	128 Sn 50	8.4168	β^-	3.55
				128 Sb 51	8.4206	β^-	4.51
125 Cd 48	8.3575	β^-	-0.19	128 Te 52	8.4488	$\beta\beta$	31.69%
125 In 49	8.4085	β^-	0.37	128 I 53	8.4329	β^-	3.18
125 Sn 50	8.4456	β^-	5.92	128 Xe 54	8.4433		1.91%
125 Sb 51	8.4582	β^-	7.94	128 Cs 55	8.4065	β^+	2.34
125 Te 52	8.4581		7.14%	128 Ba 56	8.3963	EC	5.32
125 I 53	8.4503	EC	6.71	128 La 57	8.3382	β^+	2.48
125 Xe 54	8.4309	β^+	4.78	128 Ce 58	8.3072	β^+	0.61
125 Cs 55	8.3999	β^+	3.43	128 Pr 59	8.2289	β^+	0.49
125 Ba 56	8.3571	β^+	2.32	128 Nd 60	8.1748	β^+	0.60
125 La 57	8.3057	β^+	1.88				
125 Ce 58	8.2408	β^+	0.95	129 In 49	8.3398	β^-	-0.21
125 Pr 59	8.1645	β^+	0.52	129 Sn 50	8.3931	β^-	2.13
				129 Sb 51	8.4180	β^-	4.20
126 Cd 48	8.3473	β^-	-0.30	129 Te 52	8.4304	β^-	3.62
126 In 49	8.3846	β^-	0.20	129 I 53	8.4360	β^-	14.69
126 Sn 50	8.4436	β^-	12.50	129 Xe 54	8.4314		26.40%
126 Sb 51	8.4404	β^-	6.03	129 Cs 55	8.4161	β^+	5.06
126 Te 52	8.4633		18.95%	129 Ba 56	8.3911	β^+	3.90
126 I 53	8.4400	β^+	6.05	129 La 57	8.3562	β^+	2.84
126 Xe 54	8.4438	$\beta\beta$	0.09%	129 Ce 58	8.3068	β^+	2.32
126 Cs 55	8.3992	β^+	1.99	129 Pr 59	8.2561	β^+	1.48
126 Ba 56	8.3798	β^+	3.78	129 Nd 60	8.1894	β^+	0.85
126 La 57	8.3135	β^+	1.73				
126 Ce 58	8.2723	β^+	1.70	130 In 49	8.3148	β^-	-0.49
126 Pr 59	8.1832	β^+	0.50	130 Sn 50	8.3877	β^-	2.35
				130 Sb 51	8.3982	β^-	3.37

A X Z	B/A (MeV)	$\rightarrow$	$\log t_{1/2}$ or %	A X Z	B/A (MeV)	$\rightarrow$	$\log t_{1/2}$ or %
130 Te 52	8.4303	ββ	33.80%	133 Cs 55	8.4100		100.00%
130 I 53	8.4211	β⁻	4.65	133 Ba 56	8.4002	EC	8.52
130 Xe 54	8.4377		4.10%	133 La 57	8.3776	β⁺	4.15
130 Cs 55	8.4088	β⁺	3.24	133 Ce 58	8.3496	β⁺	3.76
130 Ba 56	8.4056	ββ	0.11%	133 Pr 59	8.3112	β⁺	2.59
130 La 57	8.3565	β⁺	2.72	133 Nd 60	8.2632	β⁺	1.85
130 Ce 58	8.3335	β⁺	3.18	133 Pm 61	8.2047	β⁺	
130 Pr 59	8.2653	β⁺	1.60	133 Sm 62	8.1357	β⁺	0.57
130 Nd 60	8.2206	β⁺	1.45				
130 Pm 61	8.1309	β⁺	0.34	134 Sn 50	8.2811	β⁻	0.05
				134 Sb 51	8.3256	β⁻	-0.11
131 In 49	8.2993	β⁻	-0.55	134 Te 52	8.3826	β⁻	3.40
131 Sn 50	8.3634	β⁻	1.75	134 I 53	8.3884	β⁻	3.50
131 Sb 51	8.3929	β⁻	3.14	134 Xe 54	8.4137	ββ	10.40%
131 Te 52	8.4112	β⁻	3.18	134 Cs 55	8.3987	β⁻	7.81
131 I 53	8.4223	β⁻	5.84	134 Ba 56	8.4082		2.42%
131 Xe 54	8.4237		21.20%	134 La 57	8.3747	β⁺	2.59
131 Cs 55	8.4151	EC	5.92	134 Ce 58	8.3651	EC	5.44
131 Ba 56	8.3987	β⁺	6.00	134 Pr 59	8.3129	β⁺	3.01
131 La 57	8.3701	β⁺	3.55	134 Nd 60	8.2864	β⁺	2.71
131 Ce 58	8.3334	β⁺	2.79	134 Pm 61	8.2143	β⁺	0.70
131 Pr 59	8.2874	β⁺	1.96	134 Sm 62	8.1680	β⁺	1.00
131 Nd 60	8.2313	β⁺	1.43				
131 Pm 61	8.1635	?		135 Sb 51	8.2921	β⁻	0.23
				135 Te 52	8.3465	β⁻	1.28
132 In 49	8.2583	β⁻	-0.70	135 I 53	8.3848	β⁻	4.37
132 Sn 50	8.3554	β⁻	1.60	135 Xe 54	8.3986	β⁻	4.52
132 Sb 51	8.3745	β⁻	2.22	135 Cs 55	8.4014	β⁻	13.86
132 Te 52	8.4086	β⁻	5.44	135 Ba 56	8.3976		6.59%
132 I 53	8.4065	β⁻	3.92	135 La 57	8.3829	β⁺	4.85
132 Xe 54	8.4276		26.90%	135 Ce 58	8.3621	β⁺	4.80
132 Cs 55	8.4056	β⁺	5.75	135 Pr 59	8.3287	β⁺	3.16
132 Ba 56	8.4094	ββ	0.10%	135 Nd 60	8.2877	β⁺	2.87
132 La 57	8.3678	β⁺	4.24	135 Pm 61	8.2374	β⁺	1.65
132 Ce 58	8.3522	β⁺	4.10	135 Sm 62	8.1788	β⁺	1.01
132 Pr 59	8.2924	β⁺	1.98	135 Eu 63	8.1084	β⁺	0.18
132 Nd 60	8.2582	β⁺	2.02				
132 Pm 61	8.1773	β⁺	0.80	136 Sb 51	8.2565	β⁻	-0.09
				136 Te 52	8.3194	β⁻	1.24
133 Sn 50	8.3120	β⁻	0.16	136 I 53	8.3510	β⁻	1.92
133 Sb 51	8.3650	β⁻	2.18	136 Xe 54	8.3962	ββ	8.90%
133 Te 52	8.3892	β⁻	2.88	136 Cs 55	8.3898	β⁻	6.06
133 I 53	8.4053	β⁻	4.87	136 Ba 56	8.4028		7.85%
133 Xe 54	8.4127	β⁻	5.66	136 La 57	8.3759	β⁺	2.77

A X Z	B/A (MeV)	→	log $t_{1/2}$ or %	A X Z	B/A (MeV)	→	log $t_{1/2}$ or %
136 Ce 58	8.3737	ββ	0.19%	139 Sm 62	8.2408	β⁺	2.19
136 Pr 59	8.3302	β⁺	2.90	139 Eu 63	8.1871	β⁺	1.25
136 Nd 60	8.3082	β⁺	3.48	139 Gd 64	8.1261	β⁺	0.69
136 Pm 61	8.2447	β⁺	1.67	139 Tb 65	8.0537	?	
136 Sm 62	8.2058	β⁺	1.67				
136 Eu 63	8.1233	β⁺	0.52	140 I 53	8.2340	β⁻	-0.07
				140 Xe 54	8.2910	β⁻	1.13
137 Te 52	8.2821	β⁻	0.40	140 Cs 55	8.3144	β⁻	1.80
137 I 53	8.3271	β⁻	1.39	140 Ba 56	8.3532	β⁻	6.04
137 Xe 54	8.3643	β⁻	2.36	140 La 57	8.3551	β⁻	5.16
137 Cs 55	8.3890	β⁻	8.98	140 Ce 58	8.3764		88.48%
137 Ba 56	8.3919		11.23%	140 Pr 59	8.3466	β⁺	2.31
137 La 57	8.3818	EC	12.28	140 Nd 60	8.3394	EC	5.46
137 Ce 58	8.3671	β⁺	4.51	140 Pm 61	8.2904	β⁺	0.96
137 Pr 59	8.3417	β⁺	3.66	140 Sm 62	8.2605	β⁺	2.95
137 Nd 60	8.3091	β⁺	3.36	140 Eu 63	8.1949	β⁺	0.18
137 Pm 61	8.2626	β⁺	2.16	140 Gd 64	8.1550	β⁺	1.20
137 Sm 62	8.2127	β⁺	1.65	140 Tb 65	8.0687	β⁺	0.38
137 Eu 63	8.1521	β⁺	1.04				
137 Gd 64	8.0822	β⁺	0.85	141 I 53	8.2062	β⁻	-0.37
				141 Xe 54	8.2562	β⁻	0.24
138 Te 52	8.2543	β⁻	0.15	141 Cs 55	8.2943	β⁻	1.40
138 I 53	8.2948	β⁻	0.81	141 Ba 56	8.3260	β⁻	3.04
138 Xe 54	8.3458	β⁻	2.93	141 La 57	8.3433	β⁻	4.15
138 Cs 55	8.3602	β⁻	3.30	141 Ce 58	8.3555	β⁻	6.45
138 Ba 56	8.3935		71.70%	141 Pr 59	8.3541		100.00%
138 La 57	8.3752	β⁺	0.09%	141 Nd 60	8.3356	β⁺	3.95
138 Ce 58	8.3771	ββ	0.25%	141 Pm 61	8.3037	β⁺	3.10
138 Pr 59	8.3393	β⁺	1.94	141 Sm 62	8.2659	β⁺	2.79
138 Nd 60	8.3256	β⁺	4.26	141 Eu 63	8.2210	β⁺	1.61
138 Pm 61	8.2700	β⁺	1.00	141 Gd 64	8.1641	β⁺	1.15
138 Sm 62	8.2359	β⁺	2.27	141 Tb 65	8.0994	β⁺	0.54
138 Eu 63	8.1634	β⁺	1.08	141 Dy 66	8.0276	β⁺	-0.05
138 Gd 64	8.1137	?					
				142 Xe 54	8.2349	β⁻	0.09
139 I 53	8.2683	β⁻	0.36	142 Cs 55	8.2649	β⁻	0.23
139 Xe 54	8.3116	β⁻	1.60	142 Ba 56	8.3109	β⁻	2.80
139 Cs 55	8.3424	β⁻	2.75	142 La 57	8.3209	β⁻	3.74
139 Ba 56	8.3671	β⁻	3.70	142 Ce 58	8.3471	ββ	11.08%
139 La 57	8.3781		99.91%	142 Pr 59	8.3364	β⁻	4.84
139 Ce 58	8.3705	EC	7.08	142 Nd 60	8.3461		27.13%
139 Pr 59	8.3495	β⁺	4.20	142 Pm 61	8.3063	β⁺	1.61
139 Nd 60	8.3238	β⁺	3.25	142 Sm 62	8.2860	β⁺	3.64
139 Pm 61	8.2857	β⁺	2.40	142 Eu 63	8.2286	β⁺	0.37

A X Z	B/A (MeV)	→	log $t_{1/2}$ or %	A X Z	B/A (MeV)	→	log $t_{1/2}$ or %
142 Gd 64	8.1936	β+	1.85	145 Tb 65	8.1788	?	
142 Tb 65	8.1148	β+	-0.22	145 Dy 66	8.1231	β+	1.00
142 Dy 66	8.0593	β+	0.36	145 Ho 67	8.0520	β+	0.38
				145 Er 68	7.9752	β+	-0.05
143 Xe 54	8.1983	β−	-0.52				
143 Cs 55	8.2439	β−	0.25	146 Cs 55	8.1579	β−	-0.49
143 Ba 56	8.2821	β−	1.16	146 Ba 56	8.2167	β−	0.35
143 La 57	8.3063	β−	2.93	146 La 57	8.2396	β−	0.80
143 Ce 58	8.3247	β−	5.08	146 Ce 58	8.2790	β−	2.91
143 Pr 59	8.3295	β−	6.07	146 Pr 59	8.2808	β−	3.16
143 Nd 60	8.3306		12.18%	146 Nd 60	8.3042	ββ	17.19%
143 Pm 61	8.3178	β+	7.36	146 Pm 61	8.2887	β+	8.24
143 Sm 62	8.2883	β+	2.72	146 Sm 62	8.2939	α	15.51
143 Eu 63	8.2466	β+	2.20	146 Eu 63	8.2620	β+	5.60
143 Gd 64	8.1992	β+	1.59	146 Gd 64	8.2496	β+	6.62
143 Tb 65	8.1420	β+	1.08	146 Tb 65	8.1889	β+	0.90
143 Dy 66	8.0752	β+	0.61	146 Dy 66	8.1482	β+	1.46
143 Ho 67	7.9996	?		146 Ho 67	8.0697	β+	0.56
				146 Er 68	8.0135	β+	0.23
144 Xe 54	8.1755	β−	0.06	146 Tm 69	7.9128	β+	-0.63
144 Cs 55	8.2122	β−	0.00				
144 Ba 56	8.2655	β−	1.06	147 Cs 55	8.1339	β−	-0.65
144 La 57	8.2818	β−	1.61	147 Ba 56	8.1915	β−	-0.05
144 Ce 58	8.3148	β−	7.39	147 La 57	8.2253	β−	0.60
144 Pr 59	8.3116	β−	3.02	147 Ce 58	8.2537	β−	1.75
144 Nd 60	8.3270		23.80%	147 Pr 59	8.2707	β−	2.91
144 Pm 61	8.3054	β+	7.50	147 Nd 60	8.2837	β−	5.98
144 Sm 62	8.3037	ββ	3.10%	147 Pm 61	8.2844	β−	7.92
144 Eu 63	8.2544	β+	1.01	147 Sm 62	8.2806		15.00%
144 Gd 64	8.2191	β+	2.43	147 Eu 63	8.2636	β+	6.32
144 Tb 65	8.1556	β+	0.00	147 Gd 64	8.2434	β+	5.14
144 Dy 66	8.1069	β+	0.96	147 Tb 65	8.2067	β+	3.79
144 Ho 67	8.0198	β+	-0.15	147 Dy 66	8.1580	β+	1.60
				147 Ho 67	8.0973	β+	0.76
145 Cs 55	8.1895	β−	-0.23	147 Er 68	8.0301	β+	0.40
145 Ba 56	8.2385	β−	0.63	147 Tm 69	7.9518	β+	-0.25
145 La 57	8.2671	β−	1.39				
145 Ce 58	8.2901	β−	2.26	148 Cs 55	8.1017	β−	-0.80
145 Pr 59	8.3022	β−	4.33	148 Ba 56	8.1675	β−	-0.22
145 Nd 60	8.3093		8.30%	148 La 57	8.1968	β−	0.02
145 Pm 61	8.3027	EC	8.75	148 Ce 58	8.2406	β−	1.75
145 Sm 62	8.2931	EC	7.47	148 Pr 59	8.2492	β−	2.13
145 Eu 63	8.2693	β+	5.71	148 Nd 60	8.2772	ββ	5.76%
145 Gd 64	8.2291	β+	3.14	148 Pm 61	8.2683	β−	5.67

A X Z	B/A (MeV)	$\rightarrow$	$\log t_{1/2}$ or %	A X Z	B/A (MeV)	$\rightarrow$	$\log t_{1/2}$ or %
148 Sm 62	8.2797		11.30%	151 La 57	8.1379	?	
148 Eu 63	8.2534	β^+	6.67	151 Ce 58	8.1778	β^-	0.01
148 Gd 64	8.2484	α	9.37	151 Pr 59	8.2079	β^-	1.28
148 Tb 65	8.2047	β^+	3.56	151 Nd 60	8.2304	β^-	2.87
148 Dy 66	8.1813	β^+	2.27	151 Pm 61	8.2414	β^-	5.01
148 Ho 67	8.1125	β^+	0.34	151 Sm 62	8.2440	β^-	9.45
148 Er 68	8.0616	β^+	0.66	151 Eu 63	8.2394		47.80%
148 Tm 69	7.9752	β^+	-0.15	151 Gd 64	8.2311	EC	7.03
148 Yb 70	7.9074	?		151 Tb 65	8.2089	β^+	4.80
				151 Dy 66	8.1847	β^+	3.03
149 Cs 55	8.0792	?		151 Ho 67	8.1456	β^+	1.55
149 Ba 56	8.1394	β^-	-0.46	151 Er 68	8.1059	β^+	1.37
149 La 57	8.1834	β^-	0.02	151 Tm 69	8.0508	β^+	0.62
149 Ce 58	8.2151	β^-	0.72	151 Yb 70	7.9892	β^+	0.20
149 Pr 59	8.2380	β^-	2.13	151 Lu 71	7.9067	p	-1.06
149 Nd 60	8.2555	β^-	3.79				
149 Pm 61	8.2616	β^-	5.28	152 Ce 58	8.1613	β^-	0.15
149 Sm 62	8.2635		13.80%	152 Pr 59	8.1852	β^-	0.56
149 Eu 63	8.2536	EC	6.91	152 Nd 60	8.2241	β^-	2.84
149 Gd 64	8.2395	β^+	5.90	152 Pm 61	8.2262	β^-	2.39
149 Tb 65	8.2099	β^+	4.17	152 Sm 62	8.2441		26.70%
149 Dy 66	8.1791	β^+	2.40	152 Eu 63	8.2267	β^+	8.63
149 Ho 67	8.1334	β^+	1.32	152 Gd 64	8.2335	$\beta\beta$	0.20%
149 Er 68	8.0736	β^+	0.60	152 Tb 65	8.2021	β^+	4.80
149 Tm 69	8.0068	β^+	-0.05	152 Dy 66	8.1930	EC	3.93
149 Yb 70	7.9298	?		152 Ho 67	8.1452	β^+	2.21
				152 Er 68	8.1197	α	1.01
150 Ba 56	8.1173	β^-	-0.52	152 Tm 69	8.0575	β^+	0.90
150 La 57	8.1551	β^-	-0.07	152 Yb 70	8.0164	β^+	0.48
150 Ce 58	8.2021	β^-	0.60	152 Lu 71	7.9301	β^+	-0.15
150 Pr 59	8.2170	β^-	0.79				
150 Nd 60	8.2497	$\beta\beta$	5.64%	153 Ce 58	8.1342	?	
150 Pm 61	8.2439	β^-	3.98	153 Pr 59	8.1719	β^-	0.63
150 Sm 62	8.2617		7.40%	153 Nd 60	8.2029	β^-	1.50
150 Eu 63	8.2414	β^+	9.06	153 Pm 61	8.2213	β^-	2.50
150 Gd 64	8.2427	α	13.75	153 Sm 62	8.2286	β^-	5.22
150 Tb 65	8.2064	β^+	4.10	153 Eu 63	8.2288		52.20%
150 Dy 66	8.1892	β^+	2.63	153 Gd 64	8.2205	EC	7.32
150 Ho 67	8.1403	β^+	1.86	153 Tb 65	8.2051	β^+	5.31
150 Er 68	8.1077	β^+	1.27	153 Dy 66	8.1858	β^+	4.36
150 Tm 69	8.0257	β^+	0.34	153 Ho 67	8.1537	β^+	2.08
150 Yb 70	7.9663	?		153 Er 68	8.1188	α	1.57
150 Lu 71	7.8685	p	-1.46	153 Tm 69	8.0714	α	0.17
				153 Yb 70	8.0226	α	0.62

A X Z	B/A (MeV)	$\rightarrow$	$\log t_{1/2}$ or %	A X Z	B/A (MeV)	$\rightarrow$	$\log t_{1/2}$ or %
153 Lu 71	7.9598	α	-0.05	156 Hf 72	7.9536	α	-1.60
				156 Ta 73	7.8743	p	-0.84
154 Pr 59	8.1462	β⁻	0.36				
154 Nd 60	8.1926	β⁻	1.41	157 Nd 60	8.1294	?	
154 Pm 61	8.2057	β⁻	2.02	157 Pm 61	8.1637	β⁻	1.03
154 Sm 62	8.2269	ββ	22.70%	157 Sm 62	8.1877	β⁻	2.68
154 Eu 63	8.2172	β⁻	8.43	157 Eu 63	8.1999	β⁻	4.74
154 Gd 64	8.2249		2.18%	157 Gd 64	8.2036		15.65%
154 Tb 65	8.1967	β⁺	4.89	157 Tb 65	8.1982	EC	9.35
154 Dy 66	8.1932	α	13.98	157 Dy 66	8.1847	β⁺	4.47
154 Ho 67	8.1507	β⁺	2.85	157 Ho 67	8.1635	β⁺	2.88
154 Er 68	8.1325	β⁺	2.35	157 Er 68	8.1364	β⁺	3.05
154 Tm 69	8.0795	β⁺	0.91	157 Tm 69	8.1029	β⁺	2.34
154 Yb 70	8.0453	α	-0.39	157 Yb 70	8.0627	β⁺	1.59
154 Lu 71	7.9701	?		157 Lu 71	8.0136	α	0.83
154 Hf 72	7.9218	β⁺	0.30	157 Hf 72	7.9610	α	-0.96
				157 Ta 73	7.8966	α	-2.00
155 Pr 59	8.1306	?					
155 Nd 60	8.1685	β⁻	0.95	158 Pm 61	8.1425	β⁻	0.68
155 Pm 61	8.1959	β⁻	1.62	158 Sm 62	8.1774	β⁻	2.50
155 Sm 62	8.2113	β⁻	3.13	158 Eu 63	8.1848	β⁻	3.44
155 Eu 63	8.2167	β⁻	8.18	158 Gd 64	8.2019		24.84%
155 Gd 64	8.2133		14.80%	158 Tb 65	8.1892	β⁺	9.75
155 Tb 65	8.2030	EC	5.66	158 Dy 66	8.1902	ββ	0.10%
155 Dy 66	8.1844	β⁺	4.55	158 Ho 67	8.1584	β⁺	2.83
155 Ho 67	8.1594	β⁺	3.46	158 Er 68	8.1422	β⁺	3.92
155 Er 68	8.1295	β⁺	2.50	158 Tm 69	8.0959	β⁺	2.38
155 Tm 69	8.0885	β⁺	1.33	158 Yb 70	8.0793	β⁺	1.95
155 Yb 70	8.0448	α	0.26	158 Lu 71	8.0237	β⁺	1.03
155 Lu 71	7.9884	α	-0.85	158 Hf 72	7.9865	β⁺	0.45
155 Hf 72	7.9317	β⁺	-0.05	158 Ta 73	7.9081	α	-1.44
				158 W 74	7.8586	α	-3.05
156 Nd 60	8.1557	β⁻	0.74				
156 Pm 61	8.1770	β⁻	1.43	159 Pm 61	8.1268	?	
156 Sm 62	8.2051	β⁻	4.53	159 Sm 62	8.1576	β⁻	1.06
156 Eu 63	8.2047	β⁻	6.12	159 Eu 63	8.1768	β⁻	3.04
156 Gd 64	8.2154		20.47%	159 Gd 64	8.1877	β⁻	4.82
156 Tb 65	8.1947	β⁺	5.66	159 Tb 65	8.1889		100.00%
156 Dy 66	8.1925	ββ	0.06%	159 Dy 66	8.1816	EC	7.10
156 Ho 67	8.1593	β⁺	3.53	159 Ho 67	8.1652	β⁺	3.30
156 Er 68	8.1435	β⁺	3.07	159 Er 68	8.1428	β⁺	3.33
156 Tm 69	8.0899	β⁺	1.92	159 Tm 69	8.1137	β⁺	2.74
156 Yb 70	8.0620	β⁺	1.42	159 Yb 70	8.0770	β⁺	1.98
156 Lu 71	7.9964	α	-0.70	159 Lu 71	8.0344	β⁺	1.08

A X Z	B/A (MeV)	→	$\log t_{1/2}$ or %	A X Z	B/A (MeV)	→	$\log t_{1/2}$ or %
159 Hf 72	7.9875	β^+	0.75	162 Ta 73	7.9694	β^+	0.55
159 Ta 73	7.9292	α	-0.24	162 W 74	7.9289	β^+	0.14
159 W 74	7.8696	α	-2.14	162 Re 75	7.8488	α	-0.97
				162 Os 76	7.7973	α	-2.72
160 Sm 62	8.1449	β^-	0.98				
160 Eu 63	8.1623	β^-	1.58	163 Eu 63	8.1158	?	
160 Gd 64	8.1831	$\beta\beta$	21.86%	163 Gd 64	8.1414	β^-	1.83
160 Tb 65	8.1775	β^-	6.80	163 Tb 65	8.1557	β^-	3.07
160 Dy 66	8.1841		2.34%	163 Dy 66	8.1618		24.90%
160 Ho 67	8.1587	β^+	3.19	163 Ho 67	8.1570	EC	11.16
160 Er 68	8.1517	EC	5.01	163 Er 68	8.1448	β^+	3.65
160 Tm 69	8.1100	β^+	2.75	163 Tm 69	8.1250	β^+	3.81
160 Yb 70	8.0926	β^+	2.46	163 Yb 70	8.0996	β^+	2.82
160 Lu 71	8.0421	β^+	1.56	163 Lu 71	8.0665	β^+	2.38
160 Hf 72	8.0067	β^+	1.13	163 Hf 72	8.0283	β^+	1.60
160 Ta 73	7.9387	β^+	0.19	163 Ta 73	7.9817	β^+	1.03
160 W 74	7.8936	α	-1.04	163 W 74	7.9312	β^+	0.44
160 Re 75	7.8124	p	-3.10	163 Re 75	7.8710	α	-0.59
				163 Os 76	7.8091	α	
161 Sm 62	8.1229	?					
161 Eu 63	8.1489	β^-	1.41	164 Gd 64	8.1303	β^-	1.65
161 Gd 64	8.1673	β^-	2.34	164 Tb 65	8.1398	β^-	2.26
161 Tb 65	8.1745	β^-	5.77	164 Dy 66	8.1588		28.20%
161 Dy 66	8.1734		18.90%	164 Ho 67	8.1480	EC	3.24
161 Ho 67	8.1632	EC	3.95	164 Er 68	8.1491	$\beta\beta$	1.61%
161 Er 68	8.1459	β^+	4.06	164 Tm 69	8.1202	β^+	2.08
161 Tm 69	8.1214	β^+	3.30	164 Yb 70	8.1093	EC	3.66
161 Yb 70	8.0926	β^+	2.40	164 Lu 71	8.0664	β^+	2.27
161 Lu 71	8.0548	β^+	1.89	164 Hf 72	8.0435	β^+	2.05
161 Hf 72	8.0088	β^+	1.23	164 Ta 73	7.9868	β^+	1.15
161 Ta 73	7.9574	β^+	0.43	164 W 74	7.9517	β^+	0.78
161 W 74	7.9021	α	-0.39	164 Re 75	7.8815	β^+	-0.42
161 Re 75	7.8361	p	-3.43	164 Os 76	7.8341	α	-1.68
162 Eu 63	8.1291	β^-	1.03	165 Gd 64	8.1101	?	
162 Gd 64	8.1591	β^-	2.70	165 Tb 65	8.1308	β^-	2.10
162 Tb 65	8.1629	β^-	2.66	165 Dy 66	8.1440	β^-	3.92
162 Dy 66	8.1735		25.50%	165 Ho 67	8.1470		100.00%
162 Ho 67	8.1555	β^+	2.95	165 Er 68	8.1400	EC	4.57
162 Er 68	8.1525	$\beta\beta$	0.14%	165 Tm 69	8.1256	β^+	5.03
162 Tm 69	8.1180	β^+	3.11	165 Yb 70	8.1041	β^+	2.77
162 Yb 70	8.1027	β^+	3.05	165 Lu 71	8.0756	β^+	2.81
162 Lu 71	8.0551	β^+	1.91	165 Hf 72	8.0430	β^+	1.88
162 Hf 72	8.0272	β^+	1.58	165 Ta 73	8.0028	β^+	1.49

A X Z	B/A (MeV)	$\rightarrow$	$\log t_{1/2}$ or %	A X Z	B/A (MeV)	$\rightarrow$	$\log t_{1/2}$ or %
165 W 74	7.9557	β^+	0.71	168 Pt 78	7.7744	α	-2.70
165 Re 75	7.9017	β^+	0.38				
165 Os 76	7.8438	α	-1.15	169 Dy 66	8.0948	β^-	1.59
				169 Ho 67	8.1091	β^-	2.45
166 Tb 65	8.1126	?		169 Er 68	8.1171	β^-	5.91
166 Dy 66	8.1373	β^-	5.47	169 Tm 69	8.1145		100.00%
166 Ho 67	8.1356	β^-	4.98	169 Yb 70	8.1045	EC	6.44
166 Er 68	8.1420		33.60%	169 Lu 71	8.0863	β^+	5.09
166 Tm 69	8.1190	β^+	4.44	169 Hf 72	8.0623	β^+	2.29
166 Yb 70	8.1124	EC	5.31	169 Ta 73	8.0315	β^+	2.47
166 Lu 71	8.0747	β^+	2.20	169 W 74	7.9947	β^+	1.88
166 Hf 72	8.0560	β^+	2.61	169 Re 75	7.9510	?	
166 Ta 73	8.0052	β^+	1.54	169 Os 76	7.9010	β^+	0.53
166 W 74	7.9750	β^+	1.27	169 Ir 77	7.8450	α	-0.40
166 Re 75	7.9138	α	0.45	169 Pt 78	7.7851	α	-2.30
166 Os 76	7.8714	α	-0.74				
166 Ir 77	7.7898	α	-1.98	170 Ho 67	8.0939	β^-	2.22
				170 Er 68	8.1120	$\beta\beta$	14.90%
167 Tb 65	8.1012	?		170 Tm 69	8.1056	EC	7.05
167 Dy 66	8.1211	β^-	2.57	170 Yb 70	8.1067		3.05%
167 Ho 67	8.1305	β^-	4.05	170 Lu 71	8.0817	β^+	5.24
167 Er 68	8.1318		22.95%	170 Hf 72	8.0707	β^+	4.76
167 Tm 69	8.1226	EC	5.90	170 Ta 73	8.0308	β^+	2.61
167 Yb 70	8.1062	β^+	3.02	170 W 74	8.0131	β^+	2.16
167 Lu 71	8.0828	β^+	3.49	170 Re 75	7.9554	β^+	0.96
167 Hf 72	8.0542	β^+	2.09	170 Os 76	7.9212	β^+	0.86
167 Ta 73	8.0158	β^+	1.92	170 Ir 77	7.8578	α	0.02
167 W 74	7.9775	β^+	1.30	170 Pt 78	7.8132	α	-2.22
167 Re 75	7.9286	β^+	0.79				
167 Os 76	7.8749	α	-0.08	171 Ho 67	8.0837	β^-	1.72
167 Ir 77	7.8127	α	-2.30	171 Er 68	8.0978	β^-	4.43
				171 Tm 69	8.1019	β^-	7.78
168 Dy 66	8.1120	β^-	2.72	171 Yb 70	8.0979		14.30%
168 Ho 67	8.1170	β^-	2.25	171 Lu 71	8.0847	β^+	5.85
168 Er 68	8.1296		26.80%	171 Hf 72	8.0661	β^+	4.64
168 Tm 69	8.1150	β^+	6.91	171 Ta 73	8.0399	β^+	3.15
168 Yb 70	8.1119	$\beta\beta$	0.13%	171 W 74	8.0086	β^+	2.16
168 Lu 71	8.0806	β^+	2.52	171 Re 75	7.9708	β^+	1.18
168 Hf 72	8.0652	β^+	3.19	171 Os 76	7.9249	β^+	0.90
168 Ta 73	8.0209	β^+	2.08	171 Ir 77	7.8726	α	0.18
168 W 74	7.9936	β^+	1.72	171 Pt 78	7.8175	α	-1.60
168 Re 75	7.9349	β^+	0.64				
168 Os 76	7.8962	β^+	0.32	172 Er 68	8.0905	β^-	5.25
168 Ir 77	7.8241	α	-0.79	172 Tm 69	8.0911	β^-	5.36

A X Z	B/A (MeV)	$\rightarrow$	$\log t_{1/2}$ or %	A X Z	B/A (MeV)	$\rightarrow$	$\log t_{1/2}$ or %
172 Yb 70	8.0975		21.90%	175 Re 75	7.9948	β^+	2.55
172 Lu 71	8.0783	β^+	5.76	175 Os 76	7.9603	β^+	1.92
172 Hf 72	8.0717	EC	7.77	175 Ir 77	7.9182	β^+	0.95
172 Ta 73	8.0385	β^+	3.34	175 Pt 78	7.8702	α	0.40
172 W 74	8.0195	β^+	2.60	175 Au 79	7.8156	α	-0.70
172 Re 75	7.9723	β^+	1.18	175 Hg 80	7.7603	α	-1.70
172 Os 76	7.9418	β^+	1.28				
172 Ir 77	7.8801	β^+	0.64	176 Tm 69	8.0447	β^-	2.06
172 Pt 78	7.8395	α	-1.02	176 Yb 70	8.0641	$\beta\beta$	12.70%
172 Au 79	7.7654	α	-2.20	176 Lu 71	8.0591		2.59%
				176 Hf 72	8.0614		5.21%
173 Er 68	8.0740	β^-	1.92	176 Ta 73	8.0393	β^+	4.46
173 Tm 69	8.0845	β^-	4.47	176 W 74	8.0303	EC	3.95
173 Yb 70	8.0875		16.12%	176 Re 75	7.9943	β^+	2.50
173 Lu 71	8.0791	EC	7.64	176 Os 76	7.9718	β^+	2.33
173 Hf 72	8.0653	β^+	4.93	176 Ir 77	7.9221	β^+	0.90
173 Ta 73	8.0395	β^+	4.05	176 Pt 78	7.8887	β^+	0.80
173 W 74	8.0119	β^+	2.66	176 Au 79	7.8246	α	0.03
173 Re 75	7.9849	β^+	2.08	176 Hg 80	7.7827	α	-1.74
173 Os 76	7.9441	β^+	1.20	176 Tl 81	7.7076	?	
173 Ir 77	7.8970	β^+	0.95				
173 Pt 78	7.8451	α	-0.47	177 Tm 69	8.0364	β^-	1.93
173 Au 79	7.7873	α	-1.23	177 Yb 70	8.0500	β^-	3.84
				177 Lu 71	8.0535	β^-	5.76
174 Er 68	8.0651	β^-	2.30	177 Hf 72	8.0519		18.61%
174 Tm 69	8.0707	β^-	2.51	177 Ta 73	8.0409	β^+	5.31
174 Yb 70	8.0839		31.80%	177 W 74	8.0252	β^+	3.91
174 Lu 71	8.0715	β^+	8.02	177 Re 75	8.0015	β^+	2.92
174 Hf 72	8.0686	$\beta\beta$	0.16%	177 Os 76	7.9718	β^+	2.23
174 Ta 73	8.0420	β^+	3.58	177 Ir 77	7.9353	β^+	1.48
174 W 74	8.0268	β^+	3.27	177 Pt 78	7.8926	β^+	1.04
174 Re 75	7.9904	β^+	2.16	177 Au 79	7.8421	α	0.07
174 Os 76	7.9635	β^+	1.64	177 Hg 80	7.7896	α	-0.89
174 Ir 77	7.9028	β^+	0.95	177 Tl 81	7.7297	?	
174 Pt 78	7.8662	α	-0.05				
174 Au 79	7.8008	α	-0.92	178 Yb 70	8.0429	β^-	3.65
174 Hg 80	7.7547	?		178 Lu 71	8.0421	β^-	3.23
				178 Hf 72	8.0495		27.30%
175 Tm 69	8.0618	β^-	2.96	178 Ta 73	8.0344	β^+	2.75
175 Yb 70	8.0710	β^-	5.56	178 W 74	8.0295	EC	6.27
175 Lu 71	8.0692		97.41%	178 Re 75	7.9989	β^+	2.90
175 Hf 72	8.0608	EC	6.78	178 Os 76	7.9814	β^+	2.48
175 Ta 73	8.0449	β^+	4.58	178 Ir 77	7.9418	β^+	1.08
175 W 74	8.0238	β^+	3.32	178 Pt 78	7.9122	β^+	1.32

A X Z	B/A (MeV)	→	log t₁/₂ or %	A X Z	B/A (MeV)	→	log t₁/₂ or %
178 Au 79	7.8498	β^+	0.41				
178 Hg 80	7.8114	α	-0.58	182 Hf 72	8.0149	β^-	14.45
178 Tl 81	7.7441	?		182 Ta 73	8.0126	β^-	7.00
178 Pb 82	7.6954	?		182 W 74	8.0183		26.30%
				182 Re 75	7.9986	β^+	5.36
179 Yb 70	8.0263	β^-	2.68	182 Os 76	7.9893	EC	4.90
179 Lu 71	8.0351	β^-	4.22	182 Ir 77	7.9542	β^+	2.95
179 Hf 72	8.0386		13.63%	182 Pt 78	7.9343	β^+	2.26
179 Ta 73	8.0336	EC	7.76	182 Au 79	7.8923	β^+	1.19
179 W 74	8.0233	β^+	3.35	182 Hg 80	7.8608	β^+	1.03
179 Re 75	8.0038	β^+	3.07	182 Tl 81	7.7968	β^+	0.49
179 Os 76	7.9789	β^+	2.59	182 Pb 82	7.7563	α	-1.26
179 Ir 77	7.9474	β^+	1.90				
179 Pt 78	7.9110	β^+	1.33	183 Hf 72	8.0000	β^-	3.58
179 Au 79	7.8654	β^+	0.85	183 Ta 73	8.0068	β^-	5.64
179 Hg 80	7.8165	α	0.04	183 W 74	8.0083		14.30%
179 Tl 81	7.7607	α	-0.80	183 Re 75	8.0010	EC	6.78
179 Pb 82	7.7016	?		183 Os 76	7.9851	β^+	4.67
				183 Ir 77	7.9620	β^+	3.54
180 Lu 71	8.0221	β^-	2.53	183 Pt 78	7.9327	β^+	2.59
180 Hf 72	8.0350		35.10%	183 Au 79	7.8984	β^+	1.62
180 Ta 73	8.0259		0.01%	183 Hg 80	7.8597	β^+	0.97
180 W 74	8.0255	$\beta\beta$	0.13%	183 Tl 81	7.8136	?	
180 Re 75	8.0000	β^+	2.16	183 Pb 82	7.7618	α	-0.52
180 Os 76	7.9875	β^+	3.11				
180 Ir 77	7.9475	β^+	1.95	184 Hf 72	7.9907	β^-	4.17
180 Pt 78	7.9227	β^+	1.72	184 Ta 73	7.9938	β^-	4.50
180 Au 79	7.8708	β^+	0.91	184 W 74	8.0051		30.67%
180 Hg 80	7.8358	β^+	0.45	184 Re 75	7.9928	β^+	6.52
180 Tl 81	7.7700	β^+	-0.15	184 Os 76	7.9887	$\beta\beta$	0.02%
180 Pb 82	7.7256	?		184 Ir 77	7.9596	β^+	4.05
				184 Pt 78	7.9427	β^+	3.02
181 Lu 71	8.0126	β^-	2.32	184 Au 79	7.8997	β^+	1.72
181 Hf 72	8.0221	β^-	6.56	184 Hg 80	7.8734	β^+	1.49
181 Ta 73	8.0234		99.99%	184 Tl 81	7.8193	β^+	1.04
181 W 74	8.0181	EC	7.02	184 Pb 82	7.7824	α	-0.26
181 Re 75	8.0041	β^+	4.85				
181 Os 76	7.9836	β^+	3.80	185 Ta 73	7.9864	β^-	3.47
181 Ir 77	7.9568	β^+	2.47	185 W 74	7.9929	β^-	6.81
181 Pt 78	7.9236	β^+	1.71	185 Re 75	7.9910		37.40%
181 Au 79	7.8845	β^+	1.06	185 Os 76	7.9813	β^+	6.91
181 Hg 80	7.8398	β^+	0.56	185 Ir 77	7.9643	β^+	4.71
181 Tl 81	7.7886	?		185 Pt 78	7.9394	β^+	3.63
181 Pb 82	7.7331	α	-1.35	185 Au 79	7.9097	β^+	2.41

A X Z	B/A (MeV)	→	$\log t_{1/2}$ or %	A X Z	B/A (MeV)	→	$\log t_{1/2}$ or %
185 Hg 80	7.8740	β^+	1.69	189 Pt 78	7.9415	β^+	4.59
185 Tl 81	7.8340	β^+	1.29	189 Au 79	7.9206	β^+	3.24
185 Pb 82	7.7871	α	0.61	189 Hg 80	7.8943	β^+	2.66
185 Bi 83	7.7299	p	-4.36	189 Tl 81	7.8627	β^+	2.14
				189 Pb 82	7.8261	β^+	1.71
186 Ta 73	7.9719	β^-	2.80	189 Bi 83	7.7795	α	-0.17
186 W 74	7.9886	$\beta\beta$	28.60%				
186 Re 75	7.9813	EC	5.51	190 W 74	7.9471	β^-	3.26
186 Os 76	7.9828		1.58%	190 Re 75	7.9496	β^-	2.27
186 Ir 77	7.9580	β^+	4.78	190 Os 76	7.9621		26.40%
186 Pt 78	7.9464	β^+	3.90	190 Ir 77	7.9475	β^+	6.01
186 Au 79	7.9097	β^+	2.81	190 Pt 78	7.9466	$\beta\beta$	0.01%
186 Hg 80	7.8878	β^+	1.92	190 Au 79	7.9191	β^+	3.41
186 Tl 81	7.8431	β^+	1.44	190 Hg 80	7.9072	β^+	3.08
186 Pb 82	7.8091	α	0.68	190 Tl 81	7.8663	β^+	2.19
186 Bi 83	7.7398	α	-1.82	190 Pb 82	7.8407	β^+	1.86
				190 Bi 83	7.7907	α	0.80
187 Ta 73	7.9631	?		190 Po 84	7.7534	α	-2.70
187 W 74	7.9751	β^-	4.93				
187 Re 75	7.9780		62.60%	191 Re 75	7.9440	β^-	2.77
187 Os 76	7.9738		1.60%	191 Os 76	7.9506	β^-	6.12
187 Ir 77	7.9616	β^+	4.58	191 Ir 77	7.9481		37.30%
187 Pt 78	7.9408	β^+	3.93	191 Pt 78	7.9387	EC	5.38
187 Au 79	7.9173	β^+	2.70	191 Au 79	7.9250	β^+	4.06
187 Hg 80	7.8871	β^+	2.16	191 Hg 80	7.9043	β^+	3.47
187 Tl 81	7.8511	β^+	1.71	191 Tl 81	7.8748	?	
187 Pb 82	7.8087	β^+	1.26	191 Pb 82	7.8418	β^+	1.90
187 Bi 83	7.7567	α	-1.46	191 Bi 83	7.7994	α	1.08
				191 Po 84	7.7542	α	-1.81
188 W 74	7.9691	β^-	6.78				
188 Re 75	7.9668	β^-	4.79	192 Re 75	7.9309	β^-	1.20
188 Os 76	7.9739		13.30%	192 Os 76	7.9485	$\beta\beta$	41.00%
188 Ir 77	7.9548	β^+	5.17	192 Ir 77	7.9390	β^-	6.80
188 Pt 78	7.9479	EC	5.94	192 Pt 78	7.9425		0.79%
188 Au 79	7.9156	β^+	2.72	192 Au 79	7.9201	β^+	4.25
188 Hg 80	7.8992	β^+	2.29	192 Hg 80	7.9137	EC	4.24
188 Tl 81	7.8536	β^+	1.85	192 Tl 81	7.8764	β^+	2.76
188 Pb 82	7.8239	β^+	1.38	192 Pb 82	7.8548	β^+	2.32
188 Bi 83	7.7647	α	-0.68	192 Bi 83	7.8041	β^+	1.57
				192 Po 84	7.7702	α	-1.48
189 W 74	7.9527	β^-	2.84				
189 Re 75	7.9618	β^-	4.94	193 Re 75	7.9243	?	
189 Os 76	7.9630		16.10%	193 Os 76	7.9363	β^-	5.03
189 Ir 77	7.9561	EC	6.06	193 Ir 77	7.9381		62.70%

A X Z	B/A (MeV)	$\rightarrow$	$\log t_{1/2}$ or %	A X Z	B/A (MeV)	$\rightarrow$	$\log t_{1/2}$ or %
193 Pt 78	7.9338	EC	9.20	197 Hg 80	7.9087	EC	5.36
193 Au 79	7.9242	β^+	4.80	197 Tl 81	7.8937	β^+	4.01
193 Hg 80	7.9080	β^+	4.14	197 Pb 82	7.8715	β^+	2.68
193 Tl 81	7.8851	β^+	3.11	197 Bi 83	7.8413	β^+	2.75
193 Pb 82	7.8544	β^+	2.08	197 Po 84	7.8060	β^+	1.73
193 Bi 83	7.8137	β^+	1.83	197 At 85	7.7626	α	-0.46
193 Po 84	7.7738	α	-0.38				
				198 Ir 77	7.8975	β^-	0.90
194 Os 76	7.9320	β^-	8.28	198 Pt 78	7.9143	$\beta\beta$	7.20%
194 Ir 77	7.9285	β^-	4.84	198 Au 79	7.9087	β^-	5.37
194 Pt 78	7.9360		32.90%	198 Hg 80	7.9116		9.97%
194 Au 79	7.9192	β^+	5.14	198 Tl 81	7.8902	β^+	4.28
194 Hg 80	7.9149	EC	10.15	198 Pb 82	7.8791	β^+	3.94
194 Tl 81	7.8837	β^+	3.30	198 Bi 83	7.8421	β^+	2.79
194 Pb 82	7.8656	β^+	2.86	198 Po 84	7.8178	α	2.03
194 Bi 83	7.8194	β^+	1.98	198 At 85	7.7695	α	0.62
194 Po 84	7.7888	α	-0.41	198 Rn 86	7.7373	α	-1.19
194 At 85	7.7373	α	-1.40				
				199 Pt 78	7.9024	β^-	3.27
195 Os 76	7.9187	β^-	2.59	199 Au 79	7.9070	β^-	5.43
195 Ir 77	7.9249	β^-	3.95	199 Hg 80	7.9054		16.87%
195 Pt 78	7.9267		33.80%	199 Tl 81	7.8942	β^+	4.43
195 Au 79	7.9215	EC	7.21	199 Pb 82	7.8758	β^+	3.73
195 Hg 80	7.9097	β^+	4.55	199 Bi 83	7.8500	β^+	3.21
195 Tl 81	7.8913	β^+	3.62	199 Po 84	7.8111	β^+	2.52
195 Pb 82	7.8574	β^+	2.95	199 At 85	7.7792	α	0.86
195 Bi 83	7.8285	β^+	2.26	199 Rn 86	7.7411	α	-0.21
195 Po 84	7.7914	α	0.67				
195 At 85	7.7465	α	-0.20	200 Pt 78	7.8993	β^-	4.65
				200 Au 79	7.8987	β^-	3.46
196 Os 76	7.9123	β^-	3.32	200 Hg 80	7.9060		23.10%
196 Ir 77	7.9142	β^-	1.72	200 Tl 81	7.8898	β^+	4.97
196 Pt 78	7.9266		25.30%	200 Pb 82	7.8818	EC	4.89
196 Au 79	7.9150	β^+	5.73	200 Bi 83	7.8485	β^+	3.34
196 Hg 80	7.9145	$\beta\beta$	0.15%	200 Po 84	7.8278	β^+	2.84
196 Tl 81	7.8881	β^+	3.82	200 At 85	7.7840	α	1.63
196 Pb 82	7.8737	β^+	3.35	200 Rn 86	7.7551	α	-0.02
196 Bi 83	7.8322	β^+	2.49				
196 Po 84	7.8049	α	0.76	201 Pt 78	7.8859	β^-	2.18
196 At 85	7.7525	α	-0.60	201 Au 79	7.8952	β^-	3.19
				201 Hg 80	7.8976		13.18%
197 Ir 77	7.9091	β^-	2.54	201 Tl 81	7.8914	EC	5.42
197 Pt 78	7.9161	β^-	4.85	201 Pb 82	7.8780	β^+	4.53
197 Au 79	7.9157		100.00%	201 Bi 83	7.8550	β^+	3.81

A X Z	B/A (MeV)	$\rightarrow$	$\log t_{1/2}$ or %	A X Z	B/A (MeV)	$\rightarrow$	$\log t_{1/2}$ or %
201 Po 84	7.8268	β^+	2.96	205 Ra 88	7.7073	α	-0.68
201 At 85	7.7938	α	1.95				
201 Rn 86	7.7573	α	0.85	206 Hg 80	7.8692	β^-	2.69
201 Fr 87	7.7114	α	-1.32	206 Tl 81	7.8718	β^-	2.40
				206 Pb 82	7.8754		24.10%
202 Au 79	7.8862	β^-	1.46	206 Bi 83	7.8534	β^+	5.73
202 Hg 80	7.8969		29.86%	206 Po 84	7.8406	β^+	5.88
202 Tl 81	7.8863	β^+	6.03	206 At 85	7.8091	β^+	3.26
202 Pb 82	7.8822	EC	12.22	206 Rn 86	7.7892	α	2.53
202 Bi 83	7.8528	β^+	3.79	206 Fr 87	7.7478	α	1.20
202 Po 84	7.8350	β^+	3.43	206 Ra 88	7.7200	α	-0.62
202 At 85	7.7954	β^+	2.26				
202 Rn 86	7.7695	α	1.00	207 Hg 80	7.8476	β^-	2.24
202 Fr 87	7.7192	α	-0.47	207 Tl 81	7.8668	β^-	2.46
				207 Pb 82	7.8699		22.10%
203 Au 79	7.8809	β^-	1.72	207 Bi 83	7.8546	β^+	9.00
203 Hg 80	7.8876	β^-	6.61	207 Po 84	7.8367	β^+	4.32
203 Tl 81	7.8861		29.52%	207 At 85	7.8141	β^+	3.81
203 Pb 82	7.8775	EC	5.27	207 Rn 86	7.7880	β^+	2.74
203 Bi 83	7.8576	β^+	4.63	207 Fr 87	7.7567	α	1.17
203 Po 84	7.8329	β^+	3.34	207 Ra 88	7.7154	α	0.11
203 At 85	7.8041	β^+	2.65				
203 Rn 86	7.7639	α	1.65	208 Tl 81	7.8472	β^-	2.26
203 Fr 87	7.7295	α	-0.26	208 Pb 82	7.8675		52.40%
				208 Bi 83	7.8499	β^+	13.06
204 Au 79	7.8674	β^-	1.60	208 Po 84	7.8394	α	7.96
204 Hg 80	7.8856	$\beta\beta$	6.87%	208 At 85	7.8118	β^+	3.77
204 Tl 81	7.8801	β^-	8.08	208 Rn 86	7.7943	α	3.16
204 Pb 82	7.8800		1.40%	208 Fr 87	7.7569	α	1.77
204 Bi 83	7.8544	β^+	4.61	208 Ra 88	7.7324	α	0.11
204 Po 84	7.8391	β^+	4.10				
204 At 85	7.8035	β^+	2.74	209 Tl 81	7.8334	β^-	2.12
204 Rn 86	7.7809	α	1.87	209 Pb 82	7.8487	β^-	4.07
204 Fr 87	7.7350	α	0.23	209 Bi 83	7.8481		100.00%
204 Ra 88	7.7042	α	-1.23	209 Po 84	7.8353	α	9.51
				209 At 85	7.8148	β^+	4.29
205 Hg 80	7.8748	β^-	2.49	209 Rn 86	7.7923	β^+	3.23
205 Tl 81	7.8785		70.48%	209 Fr 87	7.7639	α	1.70
205 Pb 82	7.8744	EC	14.68	209 Ra 88	7.7332	α	0.66
205 Bi 83	7.8574	β^+	6.12	209 Ac 89	7.6955	α	-1.00
205 Po 84	7.8363	β^+	3.78				
205 At 85	7.8104	β^+	3.20	210 Tl 81	7.8136	β^-	1.89
205 Rn 86	7.7810	β^+	2.23	210 Pb 82	7.8360	β^-	8.85
205 Fr 87	7.7454	α	0.59	210 Bi 83	7.8326	β^-	5.64

A X Z	B/A (MeV)	$\rightarrow$	$\log t_{1/2}$ or %	A X Z	B/A (MeV)	$\rightarrow$	$\log t_{1/2}$ or %
210 Po 84	7.8344	α	7.08	214 Th 90	7.6925	α	-1.00
210 At 85	7.8117	β^+	4.47				
210 Rn 86	7.7967	α	3.94	215 Bi 83	7.7614	β^-	2.66
210 Fr 87	7.7632	α	2.28	215 Po 84	7.7682	α	-2.75
210 Ra 88	7.7415	α	0.57	215 At 85	7.7679	α	-4.00
210 Ac 89	7.6987	α	-0.46	215 Rn 86	7.7639	α	-5.64
				215 Fr 87	7.7533	α	-7.07
211 Pb 82	7.8170	β^-	3.34	215 Ra 88	7.7394	α	-2.80
211 Bi 83	7.8198	α	2.11	215 Ac 89	7.7195	α	-0.77
211 Po 84	7.8189	α	-0.29	215 Th 90	7.6930	α	0.08
211 At 85	7.8114	EC	4.41	215 Pa 91	7.6578	α	-1.85
211 Rn 86	7.7940	β^+	4.72				
211 Fr 87	7.7685	α	2.27	216 Bi 83	7.7440	β^-	2.33
211 Ra 88	7.7411	α	1.11	216 Po 84	7.7589	α	-0.84
211 Ac 89	7.7076	α	-0.60	216 At 85	7.7531	α	-3.52
				216 Rn 86	7.7587	α	-4.35
212 Pb 82	7.8044	β^-	4.58	216 Fr 87	7.7425	α	-6.15
212 Bi 83	7.8034	β^-	3.56	216 Ra 88	7.7374	α	-6.74
212 Po 84	7.8103	α	-6.52	216 Ac 89	7.7114	α	-3.48
212 At 85	7.7984	α	-0.50	216 Th 90	7.6977	α	-1.55
212 Rn 86	7.7949	α	3.16	216 Pa 91	7.6597	α	-0.70
212 Fr 87	7.7670	β^+	3.08				
212 Ra 88	7.7475	α	1.11	217 Po 84	7.7412	α	1.00
212 Ac 89	7.7086	α	-0.03	217 At 85	7.7447	α	-1.49
212 Th 90	7.6824	α	-1.52	217 Rn 86	7.7445	α	-3.27
				217 Fr 87	7.7378	α	-4.66
213 Pb 82	7.7850	β^-	2.79	217 Ra 88	7.7270	α	-5.80
213 Bi 83	7.7911	β^-	3.44	217 Ac 89	7.7104	α	-7.16
213 Po 84	7.7941	α	-5.38	217 Th 90	7.6908	α	-3.60
213 At 85	7.7901	α	-6.90	217 Pa 91	7.6647	α	-2.31
213 Rn 86	7.7823	α	-1.60				
213 Fr 87	7.7685	α	1.54	218 Po 84	7.7316	α	2.27
213 Ra 88	7.7466	α	2.21	218 At 85	7.7292	α	0.18
213 Ac 89	7.7157	α	-0.10	218 Rn 86	7.7388	α	-1.46
213 Th 90	7.6841	α	-0.85	218 Fr 87	7.7268	α	-3.00
				218 Ra 88	7.7251	α	-4.59
214 Pb 82	7.7724	β^-	3.21	218 Ac 89	7.7023	α	-5.97
214 Bi 83	7.7736	β^-	3.08	218 Th 90	7.6916	α	-6.96
214 Po 84	7.7852	α	-3.79	218 Pa 91	7.6592	α	-3.92
214 At 85	7.7764	α	-6.25	218 U 92	7.6408	α	-2.82
214 Rn 86	7.7772	α	-6.57				
214 Fr 87	7.7578	α	-2.30	219 At 85	7.7196	α	1.75
214 Ra 88	7.7492	α	0.39	219 Rn 86	7.7238	α	0.60
214 Ac 89	7.7159	α	0.91	219 Fr 87	7.7212	α	-1.70

A X Z	B/A (MeV)	→	$\log t_{1/2}$ or %	A X Z	B/A (MeV)	→	$\log t_{1/2}$ or %
219 Ra 88	7.7142	α	-2.00				
219 Ac 89	7.7006	α	-4.93	225 Fr 87	7.6628	β⁻	2.38
219 Th 90	7.6838	α	-5.98	225 Ra 88	7.6676	β⁻	6.11
219 Pa 91	7.6617	α	-7.28	225 Ac 89	7.6657	α	5.94
219 U 92	7.6365	α	-4.38	225 Th 90	7.6593	α	2.72
				225 Pa 91	7.6468	α	0.23
220 At 85	7.7043	β⁻	2.35	225 U 92	7.6298	α	-1.02
220 Rn 86	7.7173	α	1.75	225 Np 93	7.6076	α	-2.22
220 Fr 87	7.7098	α	1.44				
220 Ra 88	7.7117	α	-1.74	226 Fr 87	7.6494	β⁻	1.69
220 Ac 89	7.6924	α	-1.58	226 Ra 88	7.6620	α	10.70
220 Th 90	7.6847	α	-5.01	226 Ac 89	7.6557	β⁻	5.03
220 Pa 91	7.6551	α	-6.11	226 Th 90	7.6572	α	3.26
220 U 92	7.6395	?		226 Pa 91	7.6412	α	2.03
				226 U 92	7.6320	α	-0.46
221 Rn 86	7.7013	β⁻	3.18	226 Np 93	7.6048	α	-1.46
221 Fr 87	7.7033	α	2.47				
221 Ra 88	7.7012	α	1.45	227 Fr 87	7.6408	β⁻	2.17
221 Ac 89	7.6906	α	-1.28	227 Ra 88	7.6483	β⁻	3.40
221 Th 90	7.6761	α	-2.77	227 Ac 89	7.6507	β⁻	8.84
221 Pa 91	7.6570	α	-5.23	227 Th 90	7.6475	α	6.21
221 U 92	7.6346	?		227 Pa 91	7.6395	α	3.36
				227 U 92	7.6265	α	1.82
222 Rn 86	7.6945	α	5.52	227 Np 93	7.6073	α	-0.29
222 Fr 87	7.6911	β⁻	2.93				
222 Ra 88	7.6967	α	1.58	228 Fr 87	7.6307	β⁻	1.58
222 Ac 89	7.6829	α	0.70	228 Ra 88	7.6425	β⁻	8.26
222 Th 90	7.6767	α	-2.55	228 Ac 89	7.6392	β⁻	4.34
222 Pa 91	7.6513	α	-2.54	228 Th 90	7.6451	α	7.78
222 U 92	7.6377	α	-6.00	228 Pa 91	7.6324	β⁺	4.90
				228 U 92	7.6275	α	2.74
223 Fr 87	7.6837	β⁻	3.12	228 Np 93	7.6044	β⁺	1.79
223 Ra 88	7.6853	α	5.99				
223 Ac 89	7.6792	α	2.10	229 Ra 88	7.6290	β⁻	2.38
223 Th 90	7.6687	α	-0.22	229 Ac 89	7.6333	β⁻	3.58
223 Pa 91	7.6520	α	-2.19	229 Th 90	7.6347	α	11.37
223 U 92	7.6328	α	-4.74	229 Pa 91	7.6299	EC	5.11
				229 U 92	7.6208	β⁺	3.54
224 Fr 87	7.6709	β⁻	2.30	229 Np 93	7.6062	α	2.38
224 Ra 88	7.6800	α	5.50				
224 Ac 89	7.6702	β⁺	4.00	230 Ra 88	7.6218	β⁻	3.75
224 Th 90	7.6677	α	0.02	230 Ac 89	7.6227	β⁻	2.09
224 Pa 91	7.6470	α	-0.10	230 Th 90	7.6310	α	12.38
224 U 92	7.6353	α	-3.05	230 Pa 91	7.6219	β⁺	6.18

A X Z	B/A (MeV)	$\rightarrow$	$\log t_{1/2}$ or %	A X Z	B/A (MeV)	$\rightarrow$	$\log t_{1/2}$ or %
230 U 92	7.6210	α	6.26	236 Am 95	7.5607	β^+	
230 Np 93	7.6019	α	2.44	236 Cm 96	7.5502	β^+	
230 Pu 94	7.5911	α					
				237 Pa 91	7.5699	β^-	2.72
231 Ac 89	7.6144	β^-	2.65	237 U 92	7.5761	β^-	5.77
231 Th 90	7.6201	β^-	4.96	237 Np 93	7.5750	α	13.83
231 Pa 91	7.6184	α	12.01	237 Pu 94	7.5708	EC	6.59
231 U 92	7.6135	EC	5.56	237 Am 95	7.5602	β^+	3.64
231 Np 93	7.6022	β^+	3.47	237 Cm 96	7.5465	?	
231 Pu 94	7.5866	?		237 Bk 97	7.5266	?	
232 Ac 89	7.6025	β^-	2.08	238 Pa 91	7.5589	β^-	2.14
232 Th 90	7.6151	ββ	100.00%	238 U 92	7.5701	α	99.27%
232 Pa 91	7.6095	β^-	5.05	238 Np 93	7.5662	β^-	5.26
232 U 92	7.6119	α	9.34	238 Pu 94	7.5684	α	9.44
232 Np 93	7.5969	β^+	2.95	238 Am 95	7.5556	β^+	3.77
232 Pu 94	7.5890	β^+	3.31	238 Cm 96	7.5483	EC	3.94
				238 Bk 97	7.5242	β^+	2.16
233 Th 90	7.6029	β^-	3.13				
233 Pa 91	7.6049	β^-	6.37	239 U 92	7.5586	β^-	3.15
233 U 92	7.6040	α	12.70	239 Np 93	7.5606	β^-	5.31
233 Np 93	7.5953	β^+	3.34	239 Pu 94	7.5603	α	11.88
233 Pu 94	7.5838	β^+	3.10	239 Am 95	7.5537	EC	4.63
233 Am 95	7.5665	?		239 Cm 96	7.5433	β^+	4.02
				239 Bk 97	7.5263	?	
234 Th 90	7.5969	β^-	6.32	239 Cf 98	7.5067	α	1.59
234 Pa 91	7.5947	β^-	4.38				
234 U 92	7.6007	α	0.01%	240 U 92	7.5518	β^-	4.71
234 Np 93	7.5897	β^+	5.58	240 Np 93	7.5502	β^-	3.57
234 Pu 94	7.5847	EC	4.50	240 Pu 94	7.5561	α	11.32
234 Am 95	7.5635	β^+	2.14	240 Am 95	7.5471	β^+	5.26
				240 Cm 96	7.5429	α	6.37
235 Th 90	7.5834	β^-	2.63	240 Bk 97	7.5232	β^+	2.46
235 Pa 91	7.5883	β^-	3.17	240 Cf 98	7.5101	α	1.80
235 U 92	7.5909	α	0.72%				
235 Np 93	7.5871	EC	7.53	241 Np 93	7.5443	β^-	2.92
235 Pu 94	7.5788	β^+	3.18	241 Pu 94	7.5465	β^-	8.66
235 Am 95	7.5647	β^+	2.95	241 Am 95	7.5433	α	10.13
235 Cm 96	7.5473	?		241 Cm 96	7.5369	EC	6.45
				241 Bk 97	7.5237	?	
236 Pa 91	7.5775	β^-	2.74	241 Cf 98	7.5069	β^+	2.36
236 U 92	7.5865	α	14.87	241 Es 99	7.4848	α	0.95
236 Np 93	7.5792	EC	12.69				
236 Pu 94	7.5780	α	7.96	242 Np 93	7.5334	β^-	2.52

A X Z	B/A (MeV)	$\rightarrow$	$\log t_{1/2}$ or %	A X Z	B/A (MeV)	$\rightarrow$	$\log t_{1/2}$ or %
242 Pu 94	7.5414	α	13.07	247 Es 99	7.4800	β^+	2.44
242 Am 95	7.5350	β^-	4.76	247 Fm 100	7.4650	α	1.54
242 Cm 96	7.5345	α	7.15	247 Md 101	7.4433	α	0.05
242 Bk 97	7.5189	β^+	2.62				
242 Cf 98	7.5094	α	2.32	248 Am 95	7.4874	?	
242 Es 99	7.4829	α	1.60	248 Cm 96	7.4968	α	13.03
				248 Bk 97	7.4907	α	8.45
243 Np 93	7.5253	β^-	2.03	248 Cf 98	7.4911	α	7.46
243 Pu 94	7.5310	β^-	4.25	248 Es 99	7.4756	β^+	3.21
243 Am 95	7.5302	α	11.37	248 Fm 100	7.4660	α	1.56
243 Cm 96	7.5270	α	8.96	248 Md 101	7.4416	β^+	0.85
243 Bk 97	7.5175	β^+	4.21				
243 Cf 98	7.5052	β^+	2.81	249 Cm 96	7.4856	β^-	3.59
243 Es 99	7.4857	β^+	1.32	249 Bk 97	7.4861	β^-	7.44
243 Fm 100	7.4638	α	-0.74	249 Cf 98	7.4834	α	10.05
				249 Es 99	7.4744	β^+	3.79
244 Pu 94	7.5248	α	15.41	249 Fm 100	7.4615	β^+	2.19
244 Am 95	7.5213	β^-	4.56	249 Md 101	7.4435	β^+	1.38
244 Cm 96	7.5240	α	8.76				
244 Bk 97	7.5115	β^+	4.20	250 Cm 96	7.4790	SF	11.45
244 Cf 98	7.5052	α	3.06	250 Bk 97	7.4760	β^-	4.06
244 Es 99	7.4833	β^+	1.57	250 Cf 98	7.4800	α	8.62
244 Fm 100	7.4677	SF	-2.48	250 Es 99	7.4685	β^+	4.49
				250 Fm 100	7.4621	α	3.26
245 Pu 94	7.5136	β^-	4.58	250 Md 101	7.4405	β^+	1.72
245 Am 95	7.5153	β^-	3.87				
245 Cm 96	7.5158	α	11.43	251 Cm 96	7.4668	β^-	3.00
245 Bk 97	7.5093	EC	5.63	251 Bk 97	7.4693	β^-	3.52
245 Cf 98	7.4997	β^+	3.43	251 Cf 98	7.4705	α	10.45
245 Es 99	7.4840	β^+	1.82	251 Es 99	7.4659	EC	5.08
245 Fm 100	7.4654	α	0.62	251 Fm 100	7.4569	β^+	4.28
				251 Md 101	7.4416	β^+	2.38
246 Pu 94	7.5066	β^-	5.97	251 No 102	7.4234	α	-0.10
246 Am 95	7.5050	β^-	3.37				
246 Cm 96	7.5115	α	11.17	252 Bk 97	7.4586	?	
246 Bk 97	7.5028	β^+	5.19	252 Cf 98	7.4654	α	7.92
246 Cf 98	7.4991	α	5.11	252 Es 99	7.4573	α	7.61
246 Es 99	7.4802	β^+	2.66	252 Fm 100	7.4561	α	4.96
246 Fm 100	7.4682	α	0.04	252 Md 101	7.4375	β^+	2.14
				252 No 102	7.4258	α	0.36
247 Am 95	7.4982	β^-	3.14				
247 Cm 96	7.5020	α	14.69	253 Bk 97	7.4520	?	
247 Bk 97	7.4990	α	10.64	253 Cf 98	7.4549	β^-	6.19
247 Cf 98	7.4932	EC	4.05	253 Es 99	7.4529	α	6.25

A X Z	B/A (MeV)	$\rightarrow$	$\log t_{1/2}$ or %	A X Z	B/A (MeV)	$\rightarrow$	$\log t_{1/2}$ or %
253 Fm 100	7.4485	EC	5.41	259 No 102	7.3998	α	3.54
253 Md 101	7.4377	β⁺	2.56	259 Lr 103	7.3898	α	0.80
253 No 102	7.4220	α	2.01	259 Rf 104	7.3773	α	0.49
253 Lr 103	7.4021	α	0.11	259 Db 105	7.3595	?	
				259 Sg 106	7.3386	α	-0.32
254 Cf 98	7.4493	SF	6.72				
254 Es 99	7.4436	α	7.38	260 Md 101	7.3959	SF	6.44
254 Fm 100	7.4448	α	4.07	260 No 102	7.3967	SF	-0.97
254 Md 101	7.4312	β⁺	2.78	260 Lr 103	7.3832	α	2.26
254 No 102	7.4236	α	1.74	260 Rf 104	7.3767	SF	-1.70
254 Lr 103	7.4003	α	1.11	260 Db 105	7.3561	α	0.18
				260 Sg 106	7.3424	α	-2.44
255 Cf 98	7.4382	β⁻	3.71				
255 Es 99	7.4379	β⁻	6.54	261 Md 101	7.3916	?	
255 Fm 100	7.4359	α	4.86	261 No 102	7.3882	?	
255 Md 101	7.4288	β⁺	3.21	261 Lr 103	7.3809	SF	3.37
255 No 102	7.4178	α	2.27	261 Rf 104	7.3709	α	1.81
255 Lr 103	7.4020	α	1.34	261 Db 105	7.3565	α	0.26
255 Rf 104	7.3815	SF	0.18	261 Sg 106	7.3383	α	-0.64
				261 Bh 107	7.3159	α	-1.93
256 Es 99	7.4283	β⁻	3.18				
256 Fm 100	7.4318	SF	3.98	262 No 102	7.3843	SF	-2.30
256 Md 101	7.4204	β⁺	3.67	262 Lr 103	7.3733	SF	4.11
256 No 102	7.4166	α	0.46	262 Rf 104	7.3694	SF	0.32
256 Lr 103	7.3972	α	1.45	262 Db 105	7.3512	α	1.53
256 Rf 104	7.3853	SF	-2.17	262 Sg 106	7.3403	?	
				262 Bh 107	7.3141	α	-0.99
257 Es 99	7.4221	?					
257 Fm 100	7.4222	α	6.94	263 No 102	7.3755	?	
257 Md 101	7.4176	EC	4.30	263 Lr 103	7.3704	?	
257 No 102	7.4098	α	1.40	263 Rf 104	7.3627	?	
257 Lr 103	7.3969	α	-0.19	263 Db 105	7.3506	SF	1.43
257 Rf 104	7.3806	α	0.67	263 Sg 106	7.3359	SF	-0.10
257 Db 105	7.3608	α	0.11	263 Bh 107	7.3163	?	
258 Fm 100	7.4175	SF	-3.43	264 Lr 103	7.3627	?	
258 Md 101	7.4097	α	6.65	264 Rf 104	7.3605	?	
258 No 102	7.4073	SF	-2.92	264 Db 105	7.3449	?	
258 Lr 103	7.3911	α	0.59	264 Sg 106	7.3364	?	
258 Rf 104	7.3823	SF	-1.92	264 Bh 107	7.3135	α	-0.36
258 Db 105	7.3582	α	0.64	264 Hs 108	7.2974	α	-3.07
259 Fm 100	7.4075	SF	0.18	265 Lr 103	7.3589	?	
259 Md 101	7.4048	SF	3.76	265 Rf 104	7.3537	?	

A X Z	B/A (MeV)	→	log $t_{1/2}$ or %	A X Z	B/A (MeV)	→	log $t_{1/2}$ or %
265 Db 105	7.3436	?					
265 Sg 106	7.3315	α	1.00				
265 Bh 107	7.3146	?					
265 Hs 108	7.2935	α	-3.05				
266 Rf 104	7.3504	?					
266 Db 105	7.3377	?					
266 Sg 106	7.3309	α	1.32				
266 Bh 107	7.3103	?					
266 Hs 108	7.2962	?					
266 Mt 109	7.2681	α	-3.10				

References

1. Particle Data Group: Eur. Phys. J. C **15**, 1 (2000)
2. G. Audi and A.H. Wapstra, Nucl. Phys. **A565**, 1 (1993); http://ie.lbl.gov/toimass.html.
3. R. B. Firestone, V. S. Shirley, C. M. Baglin, S.Y. F. Chu, and J. Zipkin: *Table of Isotopes* (Wiley, 1996); http://ie.lbl.gov/education/isotopes.htm.
4. D.H. Clark and F.R. Stephenson, *The Historical Supernovae*, Pergamon Press, Oxford, 1977.
5. A. Pais *Inward Bound*, Oxford University Press, Oxford, 1986.
6. E. Segré *From X rays to Quarks*, Freeman, San Francisco, 1980.
7. P.G. Hansen, A.S. Jensen and B. Jonson, Annu. Rev. Nucl. Part. Sci. **45** (1995) 591-634.
8. R. Hofstadter, Annu. Rev. Nucl. Sci. **7** (1957) 231.
9. K.S. Quisenberry, T.T. Scolam and A.O. Nier, Phys. Rev. **102** (1956) 1071-1075.
10. H. Hintenberger, W. Herr, and H. Voshage, Phys. Rev. **95** (1954) 1690-1691.
11. J. Stadlmann et al., Phys. Lett. B **586** (2004) 27-33.
12. H. Geissel et al., Nuc. Inst. Meth. Phys. Res. **B70** (1992) 286-297.
13. B. Franzke, Nuc. Inst. Meth. Phys. Res. **B24/25** (1987) 18-25.
14. G. Savard and G. Werth, Annu Rev. Nucl. Part. Sci **50** (2000) 119-52.
15. J. Van Roosbroeck et al., Phys. Rev. Lett. **92** (2004) 112501.
16. Th. Udem et al., Phys.Rev.Lett **79** (1997) 2646.
17. P.J. Mohr and B.N. Taylor, Rev. Mod. Phys. **72** (2000) 351-495.
18. E. G. Kessler et al., Phys.Lett A **255** (1999) 221.
19. See e.g. P. De Bièvre. et al., IEEE Trans. Instrum Meas. **46** (1997) 592-595; and [17].
20. J.K. Dickens, F.G. Percy and R.J. Silva, Phys. Rev. **132** (1963) 1190.
21. J. Sharpey-Schafer, Physics World **3** (September,1990) 31-34.
22. P.J. Twin et al., Phys. Rev. Lett **57** (1986) 811-814.
23. M. Lacombe, B. Loiseau, J.M. Richard, R. Vinh Mau, J. Côté, P. Pirès and R. de Tourreil, Phys. Rev. **C21**, (1980) 861.
24. G.M. Temmer, Rev. Mod. Phys. **30** (1958) 498-506.
25. I. Ahmad and P.A. Butler, Annu. Rev. Nucl Part. Sci **43** (1993) 71-116.
26. R.A. Carrigan et al., Phys. Rev. Lett. **20** (1968) 874-876.
27. J. Giovinacco et al., Eur. Phys. J. **10** (2001) 3-84.
28. C. E. Bemis et al., Phys. Rev. C **16** (1977) 1146-1158.
29. Yu. Ts. Oganessian et al., Phys. Rev. C **63** (2000) 011301.
30. Data from ENDF data base http://t2.lanl.gov/data/ndviewer.html.
31. H.Geissel, G. Münzenberg and K. Riisager, Annu Rev. Nucl. Part. Sci **45** (1995) 163-203.
32. R.F.Frosch et al., Phys. Rev. **174** (1968) 1380-1399.
33. R.M. Littauer, H.F.Schopper and R.R.Wilson, Phys.Rev. Lett. **7** (1961) 144-147.

34. Samuel S.M. Wong, *Introductory Nuclear Physics* Prentice Hall,Englewood Cliffs 1990.
35. J.M. Pendlebury, Annu. Rev. Nucl. Part. Sci **43** (1993) 687-727.
36. Dassié et al. Phys. Rev. D **51**(1995) 2090.
37. R.L. Graham et al., Can. Jour. Phys. **30** (1952) 459.
38. Loritz et al. Eur.Phys. J. **A6** (1999) 257-268.
39. R. Mössbauer, Z. Physik **151** (1959) 126.
40. C.Y. Fan, Phys.Rev. **87** (1952) 258.
41. H. Abele et al. Phys. Lett. B **407** (1997) 212-218.
42. S. Arzumanov et al. Phys. Lett. B **483** (2000) 15-22.
43. H. Behrens, J. Janecke: *Numerical Tables for Beta Decay and Electron Capture*, (Landolt-Bernstein, new Series, vol I/4, (Springer, Berlin, 1969).
44. J.R Reitz, Phys. Rev. **77** (1950) 50.
45. Weinheimer et al Phys. Lett. B **460** (1999) 219.
46. M. Goldhaber, L. Grodzins and A.W. Sunyar, Phys. Rev. **109** (1958) 1015.
47. M. Apollonio et al., Eur.Phys.J. C **27** (2003) 331-374.
48. K. Eguchi et al., Phys.Rev.Lett. **90** (2003) 021802 (KamLand Collaboration).
49. Y. Fukuda et al.: Phys. Rev. Lett. **81**, 1562 (1998)
50. Physics Today **56** (2003) 19.
51. M. Gell-Mann, Phys. Lett., **8** (1964) 214.
52. P. de Marcillac et al. Nature **422** (2003) 876-878.
53. G. Heusser, Annu. Rev. Nucl. Part. Sci **45** (1995) 543-590.
54. J.A. Simpson , Annu. Rev. Nucl. and Part. Sci **33** (1983), 323.
55. Energy-loss tables can be found in L.C. Northcliffe and R.F. Schilling, Nuclear Data Tables **A7** (1977) 233. See also F.S. Goulding and B.G. Harvey, Ann. Rev. Nucl. Sci. **25** (1975) 167.
56. R. Schneider et al. Zeit. Phys. A **348** (1994) 241-242.
57. www.epa.gov/radiation/students/calculate.html
58. W.F. Libby, in *Les Prix Nobel en 1960* Stockholm, 1961.
59. J. S. Lilley: *Nuclear Physics* (Wiley, Chichester, 2001).
60. G.W. Wetherill Annu. Rev. Nuc. Sci. **25**(1975) 283-328.
61. R. Cayrel et al. Nature **409** (2001) 691-692.
62. M. Smoliar et al., Science **271** (1996) 1099-1102.
63. F. Dyson in *Aspects of Quantum Thoery* edited by A. Salam and E.P. Wigner (Cambridge U. Press, Cambridge, 1972)
64. K. Olive et al., Phys.Rev. D **69** (2004) 027701.
65. B.L.Cohen, Rev. Mod. Phys **49** (1977) 1.
66. C.D. Bowman, Annu. Rev. Nucl. Part. Sci. **48** (1998) 505-556.
67. M. Maurette, "Fossil Nuclear Reactors", Annu. Rev. Nucl. Sci **26**, 319-350.
68. A. I. Shlyakhter, Nature **264**, 340 (1976). T. Damour and F. Dyson, Nucl. Phys. **B480** (1996) 37.
69. M. Junker et al., Phys. Rev. C **57** (1998) 2700.
70. A. Krauss et al., Nucl. Phys. **A467** (1987) 273.
71. D. Zahnow et al., Z. Phys. A **351** (1995) 229-236.
72. D. D. Clayton *Principles of Stellar Evolution and Nucleosynthesis* (University of Chicago Press, Chicago, 1983).
73. J. N. Bahcall, M.H. Pinsonneault and S. Basu, Ap. J. **555** (2001) 990-1012.
74. S. Turck-Chieze and I. Lopez, Ap. J. **408** (1993) 347.
75. F. Hoyle Ap. J.Supp. **1** (1954) 121-146.
76. M. Livio et al., Nat. **340** (1989) 281-284.
77. H. Oberhummer, A. Csótó,.and H. Schlattl, Science **289** (2000) 88-90.
78. R. Diehl and F. S. Timmes, Publ. Astron. Soc. Pac. **110** (1998) 637-659.

79. E. Anders and N. Grevess Geochimica et Cosmochimica Acta **53** (1989) 197-214.
80. E. M. Burbidge, G.R. Burbidge, W.A. Fowler and F.Hoyle, Rev. Mod. Phys. **29** (1957) 547.
81. Q.R. Ahmad et al., Phys. Rev. Lett. **89** (2002) 011301; Q.R. Ahmad et al., Phys. Rev. Lett. **89** (2002) 011302.
82. S. Fukuda et al., Phys. Rev. Lett. **86** (2001) 5651.
83. B.T. Cleveland et al., Ap. J. **496** (1998) 505-526.
84. W. Hampel et al., Phys. Lett. B **447** (1999) 127.
85. J. N. Abdurashitov et al., Phys. Rev. C **60** (1999) 055801.
86. J.N. Bahcall and M.H. Pinsonneault, Phys. Rev. Lett. **92** (2004) 121301.
87. K. Hirata et al. Phys. Rev. Lett. **58** (1987) 1490-1493.
88. R.M. Bionta et al. Phys. Rev. Lett. **58** (1987) 1494-1497.
89. J.F. Beacom, W.M. Farr and P.Vogel, Phys.Rev. **D66** (2002) 033001.
90. B.J.Teegarden, Astr. J. Supp. **92** (1994) 363-368.
91. A.F.Iyudin et al., Astron. Astrophys. **284** (1994) L1-L4.
92. J. Mather et al.: Astrophy. J. **512** (1999) 511.
93. D. Tytler et al.: Physica Scripta **T85**, 12 (2000).
94. C.L. Bennett et al., Astrophys.J.Suppl. **148** (2003) 1.
95. S. Perlmutter et al.: Astrophy. J. **517**, 565 (1999)
96. B. Schmidt et al.: Astrophy. J. **507**, 46 (1998)
97. B.F. Madore et al.: Astrophy. J. **515**, 29-41 (1999)
98. R. Cen and J. P. Ostriker: Astrophy. J. **514**, 1 (1999)
99. Combes, F. and D. Pfenniger: Astron. Astrophys. **V**, 453 (1997)
100. A.G. Cohen, A. De Rujula and S.L. Glashow: Astrophy. J. **495**, 539 (1998)
101. P. Smith and J. Lewin: Phys. Rep. **187**, 203 (1990)
102. J. Ellis et al.: Phys.Rev. D **62**, 075010 (2000)
103. D.S. Akerib et al., Phys. Rev. **D68** (2003) 82002; A. Benoit et al., Phys. Lett. B **545** (2002) 43.
104. E. Vangioni-Flam, A. Coc and M. Casse: Astron. Astrophys. **360**, 15 (2000); H. Reeves, J. Audouze, W. Fowler, and D. Schramm, Ap. J. **179** (1973) 909.
105. S. Burles and D. Tytler: *Stellar Evolution, Stellar Explosions and Galactic Chemical Evolution, Second Oak Ridge Symposium on Atomic and Nuclear Astrophysics* (IOP,Bristol, 1998).
106. K. A. Olive, G. Steigman and T. P. Walker: Phys. Rep. **333**, 389 (2000)
107. M. Rauch et al.: Astrophy. J. **489**, 7 (1997)

Index